D0023404

MODELING, SYNTHESIS, AND RAPID PROTOTYPING WITH THE VERILOG™ HDL

Michael D. Ciletti

University of Colorado, Colorado Springs

Prentice Hall
Upper Saddle River, New Jersey 07458

Library of Congress Cataloging-in-Publication Data

Ciletti, Michael D.
Modeling, synthesis, and rapid prototyping with the Verilog HDL
/ Michael D. Ciletti.
p. cm.
Includes bibliographical references and index.
ISBN 0-13-977398-3
1. Verilog (Computer hardware description language) I. Title.
TK7885.7 .C55 1999
621.39'2--dc21 99-10494
CIP

Publisher: TOM ROBBINS
Editor-in-chief: MARCIA HORTON
Production editors: IRWIN ZUCKER and EDWARD DEFILIPPIS
Managing editor: EILEEN CLARK
Manufacturing buyer: PAT BROWN
Copy editors: STEPHEN LEE AND ABIGAIL BAKER
Director of production and manufacturing: DAVID W. RICCARDI
Cover director: JAYNE CONTE
Editorial assistant: DAN DEPASQUALE

We wish to thank Aldec, Inc. and Synopsys® for allowing the use of their software in the Xilinx Student Edition.

Printed in the United States of America

10 9 8 7 6 5 4 3 2 1

ISBN 0-13-977398-3

Prentice-Hall International (UK) Limited, London
Prentice-Hall of Australia Pty. Limited, Sydney
Prentice-Hall Canada Inc., Toronto
Prentice-Hall Hispanoamericana, S.A., Mexico
Prentice-Hall of India Private Limited, New Delhi
Prentice-Hall of Japan, Inc., Tokyo
Simon & Schuster Asia Pte. Ltd., Singapore
Editora Prentice-Hall do Brasil, Ltda., Rio de Janeiro

I dedicated this book to Jerilynn, my wife, whose encouragement enabled me to undertake this journey with the confidence that she would be my companion. I also dedicate it to students everywhere, who thirst for knowledge, ask questions, and speak Verilog.

CONTENTS

Chapter 4
LOGIC SYSTEM, DATA TYPES, AND OPERATORS FOR MODELING IN VERILOG HDL, 81

PREFACE

Good things are seldom kept secret. So it is with the Verilog hardware description language. Verilog did not enjoy an official mandate at its birth, but quickly earned a widespread following of users who found they could radically improve their competitive advantage by combining the language with powerful synthesis tools. Verilog originated as a proprietary language, but was later placed in the public domain and officially approved by ballot as IEEE Standard 1364. Today, widespread foundry support, a plethora of ancillary tools, and a growing user base speak for the importance of the language in the commercial sector. The widespread use and popularity of Verilog-based design and synthesis in industry signals the need for universities to embrace this technology and introduce it to students in computer science, computer engineering, and electrical engineering.

This book introduces new users to the language and shows them how to write hardware descriptions in Verilog, and how to write them in a synthesis-friendly style. Our goal is to expose the language clearly and thoroughly. Each chapter has several examples. Some compare and contrast features of equivalent descriptions. Others take the reader down the path of poor design to create points of reference for good designs. Our presentation exposes the language gradually, and prepares the reader to enjoy using Verilog in today's design environment, where synthesis tools translate, optimize, and map Verilog descriptions to alternative physical realizations as cell-based ASICs, or mask- and field-programmable gate arrays (FPGAs). Rapid prototyping of Verilog-based designs into FPGAs can be accomplished in the desktop environment bundled with the text.

This book is written for the student and the instructor. It has a robust set of exercises at the end of each chapter. It includes student editions of the popular

PC-based Silos-III™ Verilog development and simulation environment of Simucad, Inc., and the Xilinx Foundation Series Express tools, which include Synopsys' FPGA Express™. Together, the package of tools creates an exciting opportunity for students to learn the language by writing, verifying, and synthesizing their results in to an FPGA.

This book is intended to serve the advanced/elective and graduate level courses in computer science, computer engineering, and electrical engineering, and working professionals who wish to learn more about the language through self-study. The book is written in a style that, hopefully, allows the user who has no previous background with HDLs to become skilled in the language.

The text describes both the language and its use with modern circuit synthesis tools. It is important that students of the language understand how to create hardware descriptions that can be synthesized into physical, rather than hypothetical, circuits. The text illustrates descriptive styles that synthesize, and identifies pitfalls that either prevent synthesis or lead to unexpected and undesired results (Chapters 8, 9, 10). The text is about the language, but oriented towards its practical use. The text treats the language, its use in synthesis, and its use in rapid prototyping with FPGAs. It includes a discussion of the Xilinx FPGA technology and its use as the target technology for rapid prototyping of Verilog-based designs.

Practical examples (e.g. FIFO-based data acquisition system design, microcontroller design, CPU design, electronic game design, UART design, and client-server polling circuit design) expose the utility and simplicity of the language. Examples are encapsulated and complete. Important elements of the language syntax are boxed for emphasis. An appendix includes the complete formal syntax definition of the language. Examples (1) contrast poor coding style with synthesis-friendly coding style, (2) illustrate how poorly-written hardware descriptions lead to unexpected results in synthesis (3) demonstrate how the language is used in a modern design environment using synthesis tools, (4) illustrate practical tradeoffs between alternative descriptions of the same hardware functionality, and (5) demonstrate how the Verilog HDL coupled with modern synthesis tools leverage the time and talent of the designer. The text has a generous number of figures and diagrams to supplement and enhance the presentation. Lab-oriented examples guide instructors and students in creating working Xilinx FPGA hardware solutions to design problems. The treatment of synthesis includes a discussion of the main algorithms that are used within a synthesis engine to optimize the description of a circuit.

There are several paths through the text, depending on the audience. Chapters 1 through 7 can be covered in sequence to gain a familiarity of the language, but Chapters 5 and 6 can be omitted. Chapters 8 and 9 treat synthesis of combinational and sequential logic, respectively. Chapter 10 treats the language from the perspective of how its constructs synthesize, with the aim of helping the user anticipate the results of synthesis. Chapter 11 covers the language's support for switch-level modeling. These models do not synthesize, so this chapter, too, can be skipped on the path to a deeper understanding of synthesis. Chapter 12 presents three examples: A FIFO-based data temperature monitoring system, a UART, and a bit-slice microcontroller. Elaborate models are presented, verified, and synthesized. Chapter 13 presents a discussion of the Xilinx FPGA technology, and its use in rapid prototyping of Verilog-based descriptions. It also presents an electronic roulette wheel than can serve as a basis for a hardware prototype.

Acknowledgments

This book has been shaped by several friends and colleagues, and I am grateful to all of them. My students, asked many probing questions and discovered nuances and novel solutions to exercises; readers have helped tighten-up the text and suggest clarification (Mike Baird, Charles T. Johnson-Bey, Hans van den Biggelaar, Brian Boorman, Brian Daku, K. Gopalan, J. Robert Heath, Wally Leigh, David Rensahw, Greg Sajdak, and Charlie Wang). Special thanks also to Jason Feinsmith (Xilinx), Bill Fuchs (Simucad), Jackie Patterson (Synopsys), Kevin Rust (Visteon)), and Lauren Wenzl (Xilinx) for providing in the tools and libraries used in developing the examples in the text. I am also thankful to Venk Shukla and Georgia Marszalek, who influenced my early interest in Verilog. The IEEE graciously allowed the reprinting of the formal language syntax, presented in Appendix F. Thanks, also to Tom Robbins and Irwin Zucher, my editors at Prentice Hall, and their staffs, who committed to the concept and process that led to this book.

Additional Resources

There are several sources of additional information about the Verilog language. Here are some pointers:

- Open Verilog International (OVI), comprised of representatives of EDA vendors and users, oversees the evolution of the language. They can be reached at:
- Open Verilog International (OVI), 15466 Los Gatos Blvd., Los Gatos, CA 95032

For news about Cadence Verilog-XL, see comp.cad.cadence. talkverilog@cadence.com is a email newsletter for Cadence Verilog-XL

Usenet users group: comp.lang.verilog
URL locators:
http://cadence.com
http://www.simucad.com

http://www.synopsys.com
http://www.cadmazing.com/cadmazing/pages.da.html
http://www.xilinx.com
http://www.techweb.cmp.com/eet/
ftp site: ftp.cray.com:/pub/comp.lang.verilog

The web site for the International HDL Conference, co-sponsored by Open Verilog International (OVI) and VHDL International (VI) is located at:

http://www.hdlcon.org

A web site will be hosted by Prentice Hall containing solutions to a cell library, exercises, testbenches, errata, and additional exercises.

CHAPTER 1

INTRODUCTION TO ELECTRONIC DESIGN AUTOMATION

1.1 ELECTRONIC DESIGN AUTOMATION

Electronic design automation (EDA) is the practice of using computer-based software systems to design very large-scale integrated (VLSI) circuits. EDA is relatively new, but it has already had a pervasive impact on the integrated circuit (IC) industry. EDA covers virtually all phases of the design flow of modern ICs because the size, complexity, and market window of opportunity for these circuits preclude manual design. The design challenge presented by today's ICs, often containing several hundred thousand gates, would overwhelm any attempt to manage the database manually. Even those designers attempting smaller designs (10-20,000 gates) can be at a competitive disadvantage when using manual/schematic entries, because modern EDA tools might enable other designers to produce correct designs in less time. Manual designs might even be discouraged when the design is relatively small (<10,000 gates), if the design is intended to be migrated from a programmable device to a semi-custom IC.

The productivity gains that are possible with EDA tools have dramatically shrunk the product window of opportunity in system design cycles by eliminating the need to build and test series of hardware prototypes as intermediate steps in the design process. Extensive simulation of a design, rather than prototype testing, is now the typical method for verifying its correctness. The underlying business philosophy of "getting it right the first time" is realistic in the IC industry because the infrastructure of high-quality design tools and the adoption of a design discipline together assure a high likelihood of producing a working silicon on the first pass through a fabrication process. (Formal verification tools using formal mathematical methods also exist; these do not rely on simulation to establish the correctness of a design.)

Many tools support the IC design process from the initial conceptualization of a design to the testing of parts that are produced in a fabrication facility. Some of these tools, like the SPICE simulation program,[1] enjoy widespread acceptance and have brought about changes in the overall design methodology used by the IC industry. This book, however, focuses attention on an area where a paradigm shift is still underway—the adoption of hardware description languages. Hardware description languages enable designers to write computer-based descriptions of the properties, signals, and functionality of a circuit. Traditionally, a designer would directly implement a physical realization of a design from a functional specification by working at the transistor or gate level of abstraction. Hardware description languages are having a significant impact on this design paradigm, and, in many cases, are replacing it with a new one in which designers "capture" the functionality of the design in software using higher levels of abstraction that do not prematurely occupy the designer with gate and transistor level details. This mode of capturing the design is often coupled with synthesis tools to translate and optimize the description of the design, and then map its generic form into physical parts, such as a standard cell library or a field-programmable gate array (FPGA).

Despite the powerful features of this approach, hardware description languages have not been widely accepted, even though they have enjoyed a high level of visibility since the competition between Verilog and VHDL became newsworthy. There are two possible reasons: first, many designs are of a size and complexity that allow schematic entry to be used successfully; and second, many engineers still lack working familiarity with hardware description languages. The population of designers who have not encountered either language is still significant. **The goal of this book is to introduce engineers and students to the HDL-based design and synthesis and enable them to become productive designers with the Verilog HDL.**

1.1.1 Design Flow

The design flow for creating VLSI circuits consists of a sequence of steps, beginning with design entry and culminating with the generation of a database containing geometric detail of the photomasks that will be used to fabricate the design in silicon. The important steps in typical schematic- and HDL-based design flows are summarized in Figure 1.1. In practice, the design flow is not sequential, but incremental and cyclic; as errors are discovered and features are expanded, refinements are made, and alternatives are explored.

Each step in the design flow either creates a database supporting the design flow, or verifies that the design meets specific criteria. The design flow for a cell-based target technology begins with the creation of a database describing the functionality of the design, either in schematic form or in an HDL-based text file. Step 2 addresses the issue of testability for the overall design. Can the fabricated parts be tested to detect the presence of process-induced failure modes? If the circuit is sequential, special modifications to the design might be required to ensure that it could be tested. A circuit that cannot be tested must be re-designed. Step 3 verifies that the functionality described by the database is correct, either by simulation or by formal methods. In Step 4L, test patterns are generated and applied to the gate-level circuit to conduct fault analysis, a step that assures the design team that the set of patterns will detect a specified high percentage of faults in a fabricated part. In the HDL-based flow this step is visited after the gate-level netlist is generated. Step 5L provides an initial verification that the design can operate at the speeds satisfying the timing constraints imposed by clocking strategies and interface signals. Step 6 creates a technology-dependent mask set for fabricating the design in Silicon. (This will be discussed in more detail later.) Step 7 verifies that the mask set conforms to a set of geometric and electrical rules ensuring the integrity of the design. Step 5L provided a pre-layout estimate of the timing performance of the circuit. Loading effects were ignored. Step 8 extracts information describing capacitive and resistive loads in the post-layout circuit, and then Step 5L is re-visited. If the post-layout timing margins are unsatisfactory the designer must create a new gate level design, perhaps with a different architecture, or re-run the place-and-route tool with a higher priority placed on critical paths. In the synthesis flow, the optimization engine can be run with tighter constraints on performance. Later chapters will discuss synthesis in more detail. Step 5R is a sanity check; it provides assurance that the synthesized design has the same functionality as the original behavioral description.

Successful design might require multiple, iterative passes through all or part of this flow. Other design flows are possible, and any of them can be modified to accommodate the particular technology and tools that are being used. For example, in a synthesis environment, the design flow is altered by beginning with the creation of an HDL behavioral model of the design, and then using a synthesis tool to create the gate-level design. Functional verification is performed on the behavioral model, but timing verification is postponed until the gate-level implementation is produced by a synthesis tool.

Design specifications summarize the functional behavior, timing requirements, operational considerations, and other relevant attributes of a design. These may include speed, silicon area, power and other constraints. Design entry is the step of encapsulating a representation of the design. This can be shown in a variety of forms, such as a schematic, or a state transition diagram. The testability of the design is generally addressed early in the design process to establish an overall approach to the testability issue. The addition of hardware to the circuit might be required, such as the insertion of a scan path into a sequential circuit. If the design is already in a form that is bound to a particular hardware realization, such as a library of standard cells, the timing specifications of the design can be verified by a static analysis of the paths in the circuit, or by simulation.

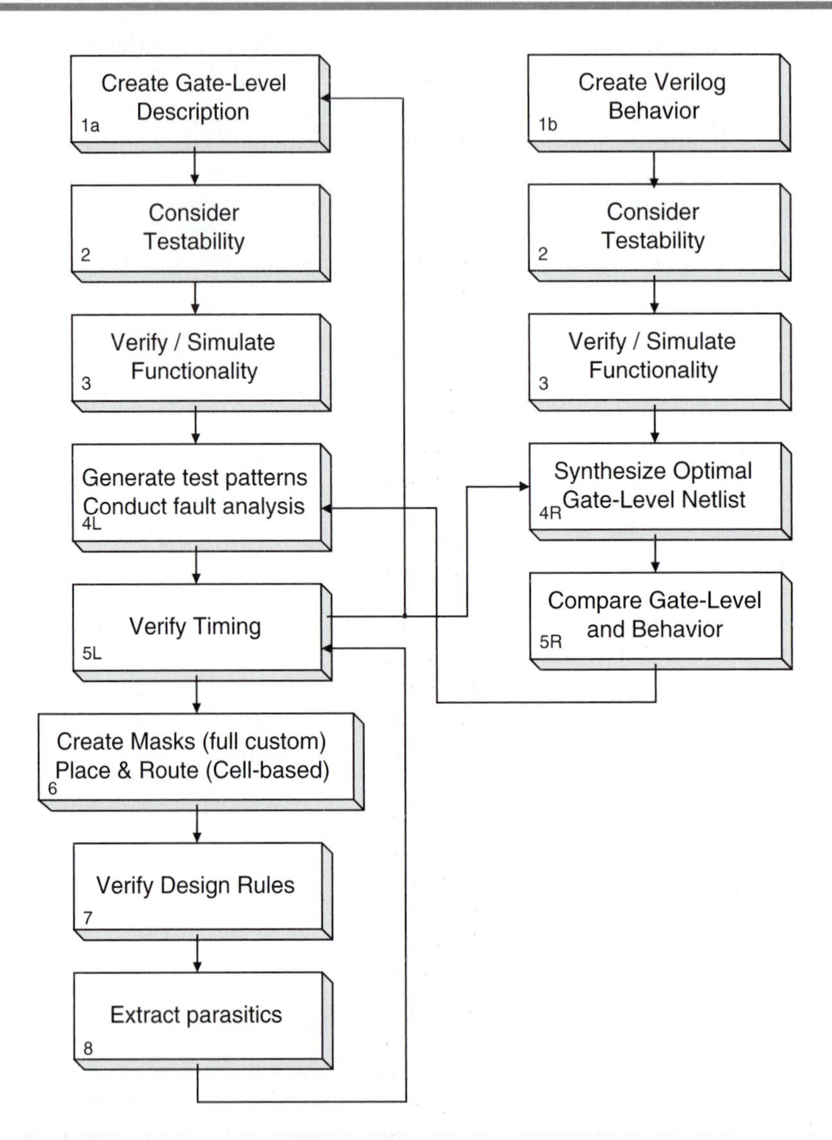

Figure 1.1 Schematic (a) and HDL-Based VLSI circuit design flow (b).

Figure 1.2 shows the various options that are available for creating the physical realization of a digital circuit in silicon, ranging from programmable logic devices (PLDs) to full-custom integrated circuits. Fixed-architecture programmable logic devices serve the low end of the market (i.e., low volume and low performance requirements). They are relatively cheap commodity parts targeted for low-volume designs. Full-custom ICs occupy the high end, where sufficient volume warrants the development time and investment required to produce fully custom designs having minimal area and maximum speed. The synthesis paradigm that will be discussed later in this text targets the broad market that spans programmable devices, field programmable and mask programmable gate arrays, and standard cells.

The physical database of a design might be implemented as a full-custom layout of high performance circuitry, or semi-custom placing and routing of standard cells or gate arrays (field or mask programmable), depending on whether the anticipated market for the application-specific integrated circuit (ASIC) offsets the cost of designing it, and the required profit. Field programmable gate arrays (FPGAs) have a fixed but electrically programmable architecture for implementing modest-sized designs. The tools supporting this technology allow a designer to write and synthesize a Verilog description into a working physical part on a prototype board in a matter of minutes. Working prototypes can be developed quickly, and design revisions can be made at very low cost. Board layout can proceed concurrently with the development of the part, because the footprint of an FPGA is known. Low-volume prototyping sets the stage for migration of a design to mask programmable and standard cell-based parts.

In mask-programmable gate array technology, a wafer is populated with an array of individual transistors which can be interconnected to create logic gates implementing a desired functionality. The wafers are pre-fabricated and later personalized with metal interconnects for a customer. All but the metallization masks are common to all wafers, so the time and cost required to complete masks is greatly reduced, and the other non-recurring engineering costs (NRE) are amortized over the entire customer base of a silicon foundry. The NRE is typically $20,000–$30,000 for programmable gate arrays and standard cell based parts. In standard cell technology, individual logic gates are predesigned to the mask level. This, too, shortens the design cycle. A place-and-route tool places the cells in channels on the wafer and interconnects them to create the necessary functionality. The mask set for a customer is specific to the logic being implemented, but the NRE associated with designing and characterizing the cell library is amortized over the entire customer base. The unit cost of the parts is relatively cheap compared to the unit cost of PLDs and FPGAs. Ultimately, the physical layout of the customized ASIC must be re-verified to confirm that the implementation not only realizes the desired functional behavior but also meets the externally-imposed timing constraints of the design and storage elements (flip-flops) used within the design itself. The mask generation step in the design flow also verifies that both electrical rules (e.g., fanout loading) and process design rules that govern geometric spacing are met, and that the physically-implemented circuit corresponds to the gate-level description from which it originated. Full-custom integrated circuits provide maximum performance at maximum cost. The volume of sales, or the peculiar needs of the customer, must be sufficient to offset the NRE costs of developing a complete mask set for the design.

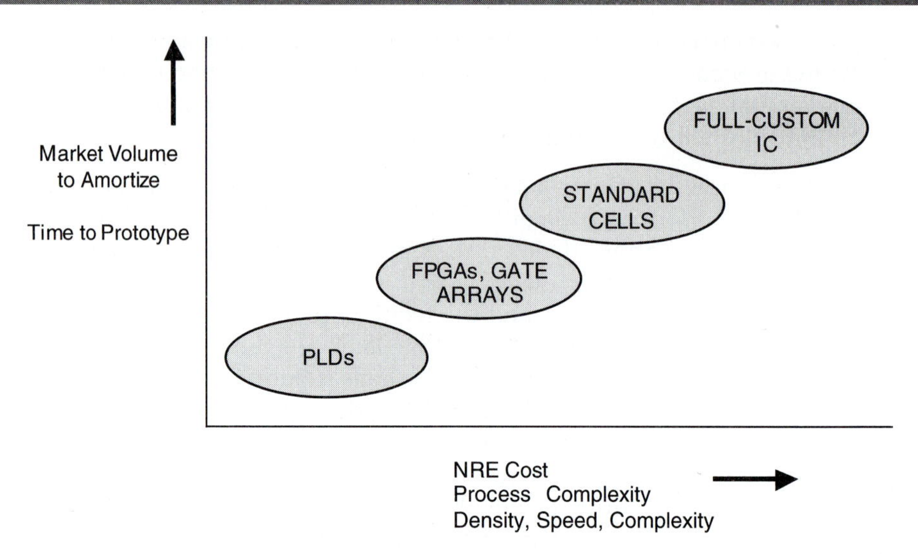

Figure 1.2 Alternative technologies for IC implementation.

1.1.2 Design Entry

Design entry encapsulates a description of the design in a database that serves subsequent steps in the design flow. A designer can perform this step by drawing paper-and-pencil schematics, using a computer-based schematic entry tool, or using a hardware description language and a variety of associated tools that capture information graphically (e.g., waveform timing diagrams and state transition graphs). Schematics and hardware description languages provide two commonly-used modes of entry.

1.1.3 Schematic-Based Design Entry

Schematic entry focuses attention on the structural detail of the design. Schematic entry is popular because designers are familiar with this traditional, visual format, and because schematic entry can be efficient for small designs. Schematics can be helpful in debugging the interconnect of a design. This can be more difficult in an HDL-based description unless the designer is careful to use descriptive names in a systematic manner, or a tool that analyzes and reveals the connectivity of the design. Schematic entry tools have achieved a high level of popularity during the past ten years as the computational power of workstations, personal computers and interactive computer graphics tools have become capable of supporting complex design tasks at reasonable cost and speeds.

Schematics display the structure of a design. Structural modeling consists of interconnecting design objects to create a structure that has a desired behavior, and a schematic "view" of a design consists of hierarchically-organized and interconnected objects. The topology of the design and the functionality of the low-level blocks and/or gate-level circuits in the schematic view implicitly define the behavior of the overall design.

Designers use a schematic entry tool to select and interconnect schematic symbols (icons) representing hardware components. Connections are made symbolically by icons of wires and buses representing physical signals in a circuit. The schematic entry software creates and manages a database containing the topological and incidental information created with the schematic, as well as interfaces that will allow other tools in the design flow to access the database. For example, the information in the topological description might be used by a simulation tool to establish a database for simulating the design represented by the schematic.

Example 1.1 A Boolean half-adder asserts a sum bit when only one of its two inputs is a logical "1," and asserts a carry when both inputs are "1." Figure 1.3 shows a block diagram symbol and schematic of a half-adder logic circuit generated by a schematic entry tool. The lines connecting the schematic symbols of gates (exclusive-or, nand and inverter) represent electrical/logical signals in the physical circuit. Other schematics could use the Add_half symbol to instantiate, or place, this circuitry in a more complex configuration of components, such as a full adder.

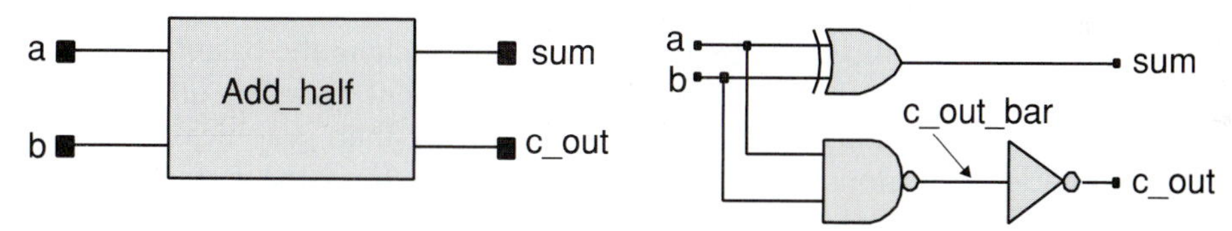

Figure 1.3 Block diagram symbol and schematic of a half-adder circuit.

A design can be viewed from several perspectives. At the schematic view of a design, the process of placing and connecting symbolic representations of components onto a drawing page is called "instantiation." The symbol of a component is called a "symbol view" of the object. In Figure 1.3, the block diagram symbol of a half adder is shown. Each object also has an associated "functional view" that specifies a relationship between the input and output signals of the component. The functional view is used by a simulator to create output waveforms from input waveforms. The simulator, using the database created by the schematic editor, determines the overall functional behavior of the aggregate (i.e., the structure formed by connecting design objects at the schematic). For example, the functionality of a simple two-input nand gate can be described by a truth table that indicates the output logic value for any pair of input logic values, or it can be described implicitly by a transistor-level schematic. A simulator typically has access to one or more libraries of components. The overall functionality of a combinational circuit would be derived from the functional information of each component, and the topological information would be contained in a schematic.

1.1.4 Design Entry Based on a Hardware Description Language

A hardware description language (HDL) is a computer-based programming language having special constructs and semantics to *model*, *represent*, and *simulate* the functional behavior and timing of digital hardware. Electrical signals are represented by variables within the language, but have semantics of both value and time, so that the temporal history of signals can be described, generated and manipulated in the language.

Recently-developed hardware description languages, such as Verilog[2–10] and VHDL[11,13] provide an alternative approach to design entry by letting the designer create a text description of the circuit without relying on a schematic. Tools exist that automatically produce a schematic drawing from the Verilog text. The text itself can be generated on an ordinary terminal (i.e., a graphics workstation is not required) and is very portable.

Schematic-based design entry is limited to describing a design in terms of a structure created by interconnected primitive design objects, such as gates or transistors. Hardware description languages can describe a design by structural models too, but they can also describe a design in terms of a software-based functional model of its behavior. Unlike schematic-based design, HDL modeling describes the functionality of a component or design using language-based constructs and/or procedural code, rather than graphical symbols and their interconnections. HDLs do not necessarily bind the model to physical hardware. A given HDL model can be instantiated into the descriptions of other designs, just as a symbol for a component can be instantiated in a schematic. Instantiation in a HDL model is accomplished by using, or referencing, the name of the model with its interface to the environment.

Example 1.2 Figure 1.4 shows the Verilog HDL counterpart of the schematic for the half adder that was presented in Figure 1.3. We will use it to illustrate terminology and basic concepts in the language.

A Verilog **module** is a software encapsulation of information representing the structure, behavior and other important properties of a hardware unit. The declaration of a **module** is always terminated by the keyword **endmodule**. A module can have ports, e.g., (*sum, c_out, a, b*), which provide an interface to the environment, and internal wires (*c_out_bar*), which are used to establish con-

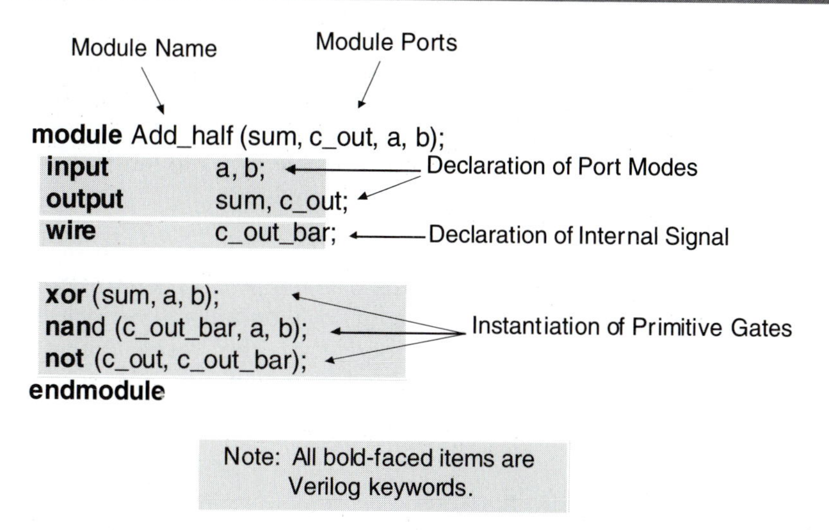

Figure 1.4 Verilog code: Structural description of a half adder.

nectivity with other instantiated objects in the model. The names (identifiers) in a port correspond to electrical signals in the physical hardware being described. The *mode* of a port specifies whether the value of the port (signal) is determined by the external environment (an **input** port) or within the module itself (an **output** port). Verilog also supports bi-directional (**inout**) ports.

Verilog can be called a "declarative language" because a Verilog **module** declares the information that describes the relationship between the inputs and outputs of a circuit. A module that contains structural information will have one or more instantiations of primitive objects (gates) and/or other modules. **A listing of primitives or modules with their ports is called a "structural declaration," and the individual items in the list are said to be instantiated, just as gates are instantiated on a schematic or placed on a circuit board.** (Such listings are terminated by a semicolon within the body of a **module**). The *Add_half* module in Example 1.1 contains instantiations of primitives (***xor, nand, not***) which are connected by their ports. For example, the declared wire, *c_out_bar,* is an output of the **nand** primitive and an input of the **not** primitive (the inverter). Electrical connectivity is represented by the signals in the ports of instantiated primitives and modules. Primitives are structural objects having pre-defined, built-in functionality. Appendix A describes all of Verilog's pre-defined primitives.

The module *Add_half* in Figure 1.4 is not a computational model in the usual software sense, in which a set of statements are executed sequentially to create and manipulate the values of variables in the host machine's memory. In fact, the order in which the primitives are instantiated, or listed, within a module is inconsequential. **It is helpful to think of a Verilog model as a representation**

of relationships between signals. When the description is purely structural, the relationship between the input and output signals (ports) is not explicit, but could be determined by simulating the behavior represented by the model, or by deriving it from low-level models and topology. Chapter 2 discusses structural models in more detail. It is also possible to write (abstract) behavioral descriptions in Verilog, as illustrated in Example 1.3.

Example 1.3 The Verilog model of a D-type flip-flop with synchronous reset in Figure 1.5. consists of a behavioral description that updates the value of the output, *q*, whenever the clock has a positive (rising) edge, provided that the reset line is not asserted. (For simplicity, the model given here omits details about signal propagation delays.)

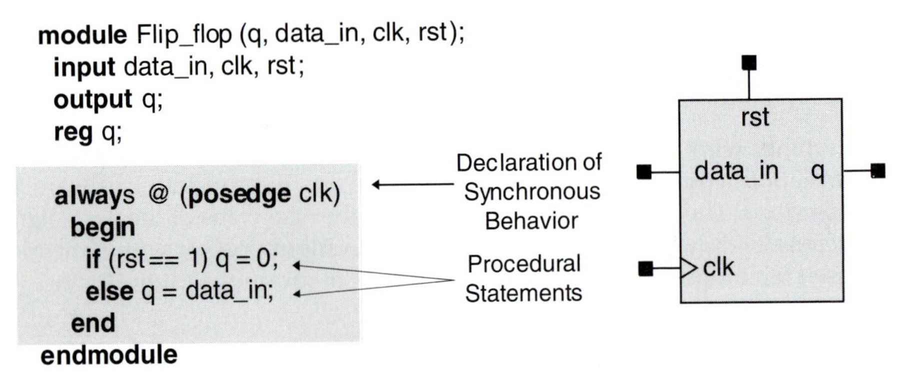

Figure 1.5 A Verilog description of a D flip-flop.

Note that the Verilog model in Example 1.3 has no internal structural detail, i.e., no objects connected to form a structure. The entire model is implemented abstractly by procedural statements. It is called a "behavioral" model. In this example, the behavior is declared by the keyword "**always**," and is accompanied by a statement which executes only when *clk* has a rising edge. The execution of the statements associated with a behavior is referred to as a "process," as is the behavior. Signal changes are called "events." The behavior is activated when simulation begins, but the activity flow, or execution, is suspended by the event control expression "@ (**posedge** clk)." When *clk* has a rising edge, the behavior is re-activated at the point where it had been suspended, and the statement executes. Upon completion of the statement, the activity flow associated with the **always** construct returns to the beginning of the behavior and is again suspended until the next active edge of the clock. In other words, the behavior declares that *q* immediately gets the value of *data_in* when the *clk* has a rising edge, provided that *rst* is not asserted. If *rst* is asserted *q* gets 0. The behavior in

Example 1.3 is ***synchronous,*** because the activity flow is synchronized with the rising edge of *clk*. It is ***cyclic*** because the activity flow is repetitive.

The behavior in Example 1.3 is said to be "event-controlled," because the evolution of the activity flow and the execution of the statement assigning value to *q* is conditioned on the occurrence of an event (i.e., the rising edge of *clk*). Thus, the event control expression defines the sensitivity of the behavioral model to external signals. In general, **the statements associated with a behavior execute sequentially, subject to any timing control expressions that might be included in the description.**

In Verilog, there are three types of timing controls: event control, delay control, and a **wait** statement. **Delay control suspends execution for a specified time; event control suspends execution until an event occurs.** In Example 1.3, the value of *q* is updated whenever the *clk* has a rising edge. The event control expression @ (**posedge** clk) sets up a mechanism to monitor *clk* during simulation. When the monitoring mechanism detects a positive-edge event for *clk*, the suspended activity flow resumes execution. Given that the behavior is suspended until it is re-activated by a rising edge of the clock, there must be a mechanism for sustaining the value of the output signal over the time interval between the clock edges, just as physical hardware maintains signals continuously. The Verilog declaration ***reg*** *q* declares that *q* is a "register" variable that retains its value once it is assigned. The value is preserved in memory until it is changed by a later assignment or until the simulation ends. In Example 1.3, *q* is updated at each clock edge. The third type of timing control, **the wait statement, suspends execution until a condition is satisfied. It is typically used to model level-sensitive behavior**. It will be discussed in later examples.

Behavioral models in HDLs describe a design's functionality abstractly, or behaviorally, without any binding to particular hardware elements. In contrast, traditional approaches to design (i.e., gate-level schematic design entry) focus attention on individual hardware elements and their interconnection. Behavioral descriptions can provide a clear, explicit description of the relationship between inputs and output signals of a design module, independently of a structural configuration that implements the behavior. In general, a Verilog model can combine structural detail with behaviors. The task of a synthesis tool is to infer gate-level detail from an abstract behavior.

Example 1.4 The Verilog text in Figure 1.6 describes a four-bit slice adder using built-in language operators for addition and concatenation. (Here { } denotes a concatenation operator, which in this example creates a five-bit wide data path from operations on the four-bit (vector) data paths and the scalar *c_in*.)

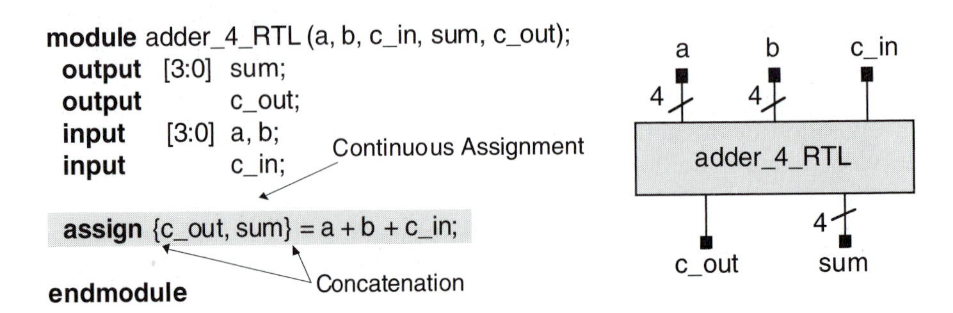

Figure 1.6 Verilog behavioral model of a four-bit slice adder.

The single statement beginning with the keyword **assign** in *adder_4_RTL*, shown in Figure 1.6, describes a behavior (addition) that requires several gates in actual hardware. This illustrates how a Verilog "continuous assignment" assigns value to the data path to create the outputs of the adder, based on the inputs to the adder. The keyword **assign,** together with the language operators (+), provides a shortcut to creating a description of behavior by decoupling a circuit's functionality from its explicit implementation by hardware components. For this reason **we will refer to the continuous assignment statement as "implicit combinational logic."** (Other features of **assign** will be discussed in Chapter 2.) More generally, the style of *adder_4_RTL* is referred to as a "data flow" description because operators act on data path variables to form results. It is also called an "RTL" style (register transfer level) because it conveniently represents operations on a data path in a synchronous circuit. We will see in later examples that Verilog readily supports the RTL operations commonly used to describe processors.[8, 17] Appendix C presents all of the Verilog operators.

Verilog's several built-in data types and operators make implicit combinational logic very easy to implement, because the designer need focus attention only on the operations being performed on the data path, not the circuit detail supporting the implementation. In general, behavioral descriptions are even more abstract than RTL descriptions, but can present a clear, readable statement of the relationship (e.g., datapath operations) between the input and output signals of a module. RTL descriptions are also referred to as behavioral descriptions.

1.1.5 Hints for Model Development

A systematic approach to developing a Verilog model of a digital circuit will reduce the amount of time spent by the developer in detecting, locating and correcting errors in the description. Syntax errors are inevitable, and syntax checkers do not necessarily pinpoint the location of the error, especially in a large fragment of code. The effort to locate a syntax error can be reduced, however, by a simple "divide and conquer" strategy. The steps listed in Figure 1.7 develop a

Verilog description systematically and incrementally, and detect and isolate errors as the description evolves. Note: in a large, complex design, a team of designers typically adopt a carefully chosen discipline of assigning meaningful names, use of capital letters, and abbreviations. For example, the standard cells of a library will have recognizable and/or encoded names, and a convention will determine how active-low or complemented signals are to be named.

1. Write the "*module... endmodule;*" design encapsulation keywords, name of the module, and the module ports.
2. Declare the module port modes. **Check/debug the syntax.**
3. Declare internal nets (e.g., **wire**) needed to support the structure of the design. **Check/debug the syntax.**
4. Declare register variables (e.g., **reg**) needed to support any behaviors within the design. **Check/debug the syntax.**
5. Write any continuous assignments to nets. **Check/debug the syntax.**
6. Instantiate and connect gates and modules nested in the design. **Check/debug the syntax.**
7. Write behavioral statements. **Check/debug the syntax.**

Figure 1.7 Steps in Systematic Model Development

1.2 A BRIEF HISTORY OF HDLs

Hardware description languages became important during the decade of the '80s, when languages like Hilo were used to describe large digital designs in a textual, rather than schematic, format for gate-level simulators used in design verification and fault simulation. A variety of languages have been developed. For example, Hardware C is a variant of the C language with additional semantics that support modeling of combinational and sequential circuits. AHPL (A Hardware Programming Language) was developed in the '70s at the University of Arizona to model synchronous data paths in computer architectures. During the same period, ISPS (Instruction Set Processor Specification) was used at Carnegie Mellon University to support software development matched to a particular computer architecture. None of these early languages ever gained widespread acceptance in the commercial ASIC design industry. This failure was likely due to their limited features, and in some cases because they were proprietary languages. Perhaps, too, they were ahead of their time, in that the current demand for HDLs has been fueled by the emergence of the synthesis paradigm of the '90s.

During the '80s, the VHSIC hardware description language (VHDL) was developed under the sponsorship of the United States Department of Defense. Jointly created by IBM, Texas Instruments and Intermetrics, this language was intended initially to serve as a vehicle for describing and documenting the design of a digital system. Although the language became an IEEE standard (IEEE 1076), it has not yet achieved widespread acceptance in industry. ASIC foundries have resisted changing to a different language after having already made substantial investments in supporting the Verilog HDL.

The Verilog HDL was developed in 1984 by Gateway Design Automation under the leadership of Phillip Moorby. It quickly became a de facto standard in industry, but was limited by its status as a proprietary language. Gateway was later acquired by Cadence Design Systems, Inc., which placed the language in the public domain under the auspices of Open Verilog International in 1990. In 1995, a ballot was passed to make Verilog a standard hardware description language (IEEE Standard 1364-1995.[2]) Since then, the EDA industry has developed a wide variety of tools supporting Verilog-based design.

1.3 THE ROLE AND REQUIREMENTS OF HARDWARE DESCRIPTION LANGUAGES IN EDA

The complexity of tasks that must be addressed in ASIC design imposes constraints on an HDL. An HDL must model hardware at different levels of abstraction and allow the designer to mix different levels of abstraction in descriptions and simulations. For ease of understanding and documentation, a language must maintain a clear relationship between language constructs and physical hardware, and must provide a convenient medium for communication of the design between various design tools, with no translation required. At an aesthetic level, a language should be simple, without needless features.

A hardware description language must be able to describe combinational logic, level-sensitive and edge-triggered storage devices, and simulate the concurrent behavior of hardware components. Support for hardware concurrency is a major feature that distinguishes a hardware description language from an ordinary procedural programming language. (More on this later.)

An HDL must also support common design practices, such as top-down design and hierarchical decomposition. The language provides a medium for describing individual components and interconnections of components forming circuits. HDLs play an important role in ASIC design, because the information produced by simulating a circuit using an HDL is cheaper to compute than the data obtained from analog simulation using languages like SPICE.[1] This increase in speed of simulation is obtained at the expense of the resolution of the waveforms being simulated. In Chapter 3, we will see that a digital simulator will report that a signal has a logical value of one or zero, but it will not describe the transients that occur when a signal changes from one value to another. Even with a more robust set of symbols describing the signals of circuit, such as the symbol "z" for a high impedance condition, logic simulation is not intended to provide a detailed simulation of analog behavior.

An HDL must easily enable a designer to work at different levels of abstraction of the design. One taxonomy of the levels of abstraction in the design of VLSI circuits is shown in Figure 1.8, which is patterned after a design hierarchy that was developed by Armstrong[11] to illustrate applications of VHDL. Levels of abstraction range from silicon detail up to processor memory switches and chips, and over that range a variety of options are available for creating structural and behavioral models of functionality. At the highest level of abstraction, designers envision an overall abstract architecture of interacting functional objects. Hardware description languages bridge the gap between the abstraction and the physical silicon implementing the functionality. Popular hardware description languages do not completely cover the range of abstractions for VLSI circuit design. Verilog and VHDL treat behavior modeling from algorithms down to Boolean equation models of gates. VHDL does not have built-in structural/functional primitives, while Verilog has both gate-level and switch-level primitives. Neither language is intended to compete with SPICE in modeling analog behavior at the level of differential equations, and neither language deals with layout detail.

LEVEL	STRUCTURAL	BEHAVIORAL
PMS CHIP	CPUs, MEMORIES MICROPROCESSORS RAM, ROM, UART PARALLEL PORT	PERFORMANCE I/O RESPONSE ALGORITHMS OPERATIONS
REGISTER	REGISTERS, ALUs COUNTERS, MUXES	TRUTH TABLES STATE TABLES OPERATIONS
GATE	GATES, FLIP-FLOPS	BOOLEAN EQUATIONS
CIRCUIT	TRANSISTORS, RLC	DIFFERENTIAL EQUATIONS
SILICON	GEOMETRICAL OBJECTS	–

INCREASING ABSTRACTION ↑ — INCREASING DETAIL ↓

Figure 1.8 Taxonomy of design abstractions.

An HDL and its simulation environment must accommodate a mixture of design descriptions at different levels of abstraction. Figure 1.9 illustrates the options that Verilog offers the designer for representing a design.

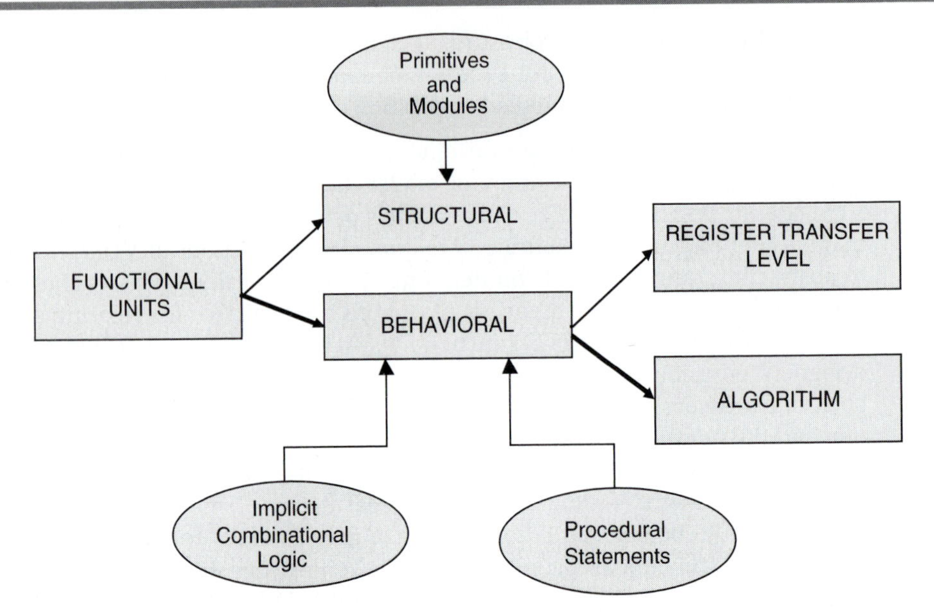

Figure 1.9 Styles of design supported by Verilog HDL.

Structural modeling consists of interconnecting design objects to create a structure having a desired behavior. For example, the half-adder presented in Figure 1.3 consists of an interconnection of primitive objects. The behavior of this half adder is defined implicitly by the structure of the design and the built-in behaviors of the primitive objects. Register Transfer Level/Data Flow modeling uses language operators to implement logic and data flow operations (register transfers) with implicit binding to hardware resources. The four-bit slice adder shown in Figure 1.6 is an example of RTL description. At a higher level of abstraction, behavioral modeling uses HDL-based procedural code to describe desired input/output behavior in the form of an algorithm, with no reference to hardware and no explicit data path. The flip-flop in Example 1.5 illustrates this style of design. In behavioral modeling, a synthesis tool must determine, allocate and schedule resources to implement the functionality.

The overarching expectation of an HDL is that it adds value to the design process. This means that the language must allow the designer to efficiently encapsulate an accurate description of the design, quickly create its physical implementation, and verify that it will meet performance specifications.

1.4 BENEFITS OF USING HDLs IN EDA

The motivation for using HDLs in electronic design automation is broad. HDLs support a compact description that is relatively easy to edit. This description is highly portable and readable by a third party without the need for a special tool. Additionally, an array of supporting tools and extensive vendor libraries exist for implementing Verilog descriptions in silicon.

Hardware description languages provide an alternative to traditional schematic-based design styles. From the standpoint of hardware support, a schematic-driven paradigm requires a color-graphic workstation (or a suitably enhanced PC); language-based entry is easily done at a terminal, and an engineer can work at a remote site without requiring local support.

Editing an HDL description can be a shorter and simpler task than editing a design that is described by schematics. The tasks of removing, relocating and rebinding schematic objects can be time-consuming if the schematic itself is very dense and/or complex.

HDLs support a higher level of abstraction than can be described by schematics. Certainly a schematic can contain a symbol of any functional unit, but it must eventually be represented in terms of lower-level detail that is ultimately expressed as a structural description. In contrast, an HDL can embody a behavior with no reference whatsoever to structural detail. Schematic entry focuses the designer's attention on structural detail; a designer using an HDL can focus on structural detail, functional behavior, or a mixture of the two. This is the HDL's real strength.

Example 1.5 Figure 1.10 shows a behavioral model for a half adder. Its functionality is described by an event-activated behavior that is sensitive to the data paths. An event on an input causes the *sum* and *c_out* bits to be recomputed. The storage variables *sum* and *c_out* are updated by the behavior, and retain their values when the behavior is inactive. (// denotes a comment.)

```
module Add_half (sum, c_out, a, b);
 input        a, b;
 output       sum, c_out;
 reg          sum, c_out;      // Declares abstract memory variables
 always @ (a or b)             // Watch for event on a or b
   begin
    sum = a ^ b;               // Exclusive or
    c_out = a & b;             // And
   end
endmodule
```

Figure 1.10 Verilog behavioral description of a half adder.

Examples in this text will demonstrate that the Verilog HDL supports a structured design methodology (SDM) that uses a "top-down," hierarchical decomposition to partition a complex design into simpler, hierarchically-organized, functional units. In SDM, the design decomposition proceeds top-down, from high to low levels of functionality and complexity; the design verification process proceeds "bottom-up". Low levels of functionality are verified before higher levels. Otherwise, the design verification task could become insurmountable. In the HDL paradigm, the design begins with behavioral descriptions that are technology-free, and proceeds to gate-level description after the behavioral description of the design has been verified. The net result of this methodology is reported to be a shortening of the design cycle for ASICs.[14,15]

HDLs support rapid prototyping of a design by focusing a designer's attention on the functionality of a design, rather than its physical/structural implementation. This dramatically shortens the time required to create and verify a design. Shortening the design cycle allows more changes to be made in less time, thereby increasing the likelihood that a design error will be found before its effects become widespread. The ease of considering design alternatives in an HDL context can stimulate and encourage consideration of design alternatives. Alternative architectures and technologies can be examined with relative ease.

Synthesis tools[16] automatically create a schematic from an HDL behavioral description. Thus, a schematic is actually a by-product of the HDL-based design flow, rather than a main focus of the design effort. This attractive feature greatly shortens the design cycle.

HDL-based models are a repository of intellectual property (IP). Once a design has been described by a behavioral model, the description is preserved. When the target technology shifts, the original IP is the basis for the re-targeted design.

HDLs are a stepping stone to productivity and quality gains when the design flow uses modern synthesis tools to transform HDL descriptions into optimized physical hardware. Verilog HDL is a key element in modern design flows that incorporate logic synthesis tools. The designer must, however, conform to a style that ensures synthesizable results (see Chapter 8).

An HDL-synthesis approach to design begins with a behavioral description of the functionality of the design, and then creates an optimal logic-level description. This description can then be mapped into a particular hardware technology (e.g. an FPGA) to meet timing and area constraints. Hardware description languages are the foundation on which this methodology rests.

1.5 SUMMARY

Hardware description languages are a relatively recent influence in EDA. They allow designers to focus their initial attention on the design's functionality, and later on the specific technology used for gate-level implementation. HDLs can be used to create a compact, portable, text-based description of a design, from which a schematic can be derived automatically. The Verilog hardware description language supports both structural and behavioral styles of design.

REFERENCES

1. Nagel, L.W., *SPICE2: A Computer Program to Simulate Semiconductor Circuits,* Memo ERL-M520, Dept. of Electrical Engineering and Computer Science, University of California at Berkeley, May 9, 1975.
2. *IEEE Standard Hardware Description Language Based on the Verilog Hardware Description,* Language Reference Manual (LRM), IEEE 1364-1995, 1996, The Institute of Electrical and Electronic Engineers, Piscataway, New Jersey.
3. Bhasker, J., *A Verilog HDL Primer,* Star Galaxy Press, Allentown, Pennsylvania, 1997.
4. Kurup, P, and Abbasi, T., *Logic Synthesis Using Synopsys,* Kluwer Academic Publishers, Boston, 1995.
5. Lee, J.M., *Verilog Quickstart,* Kluwer Academic Publishers, Boston, 1997.
6. Palnitkar S., *Verilog HDL, A Guide to Design and Synthesis,* SunSoft Press, Sun Microsystems, Inc., Mountain View, California, 1996.
7. Smith, D.J., HDL *Chip Design,* Doone Publications, Madison, Alabama, 1996
8. Smith, M. J., *Application-Specific Integrated Circuits,* Addison-Wesley Longman, Inc., Reading, Massachusetts, 1997.
9. Sternheim, E. et al., *Digital Design and Synthesis with Verilog HDL,* Automata Publishing Co., San Jose, California, 1993.
10. Thomas, D.E. and Moorby, P., *The Verilog Hardware Description Language,* Third Edition, Kluwer Academic Publishers, Boston, 1996.
11. Armstrong, J., *Chip Level Modeling with VHDL,* Prentice-Hall Inc., Englewood Cliffs, New Jersey, 1988.
12. Navabi, Z., *VHDL Analysis and Modeling of Digital Systems,* McGraw-Hill, New York, 1993.
13. Dewey, A., *Analysis and Design of Digital Systems with VHDL,* PWS Publishing Company, Boston, 1996.
14. Weber, ASIC & EDA, 1992.
15. McDougal, J.D. and Young, W.E., *Shortening the Time to Volume Production of High-Performance Standard Cell ASICs,* Hewlett-Packard Journal, February 1995, pp. 91-96.
16. Synopsys, Inc. vendor supplied documentation.
17. Mano, M. M. and Kim, C. R., *Logic and Computer Design Fundamentals,* Prentice-Hall, Inc., Upper Saddle River, N.J., 1997.

PROBLEMS

1. Discuss the tradeoffs between the various technologies available for implementing an integrated circuit.

2. Discuss the tradeoffs between schematic-based design and HDL-based design.

3. Discuss the styles of design supported by Verilog.

4. In the Verilog module shown below:

```
module machine_input_logic (y2, y1, x3, x2, x1, x0);
   input      x0, x1, x2, x3;
   output     y1, y2;

endmodule
```

   a. What is the module name?

   b. What are the module ports?

5. Answer True or False.

   a. ___ HDL-based design is most powerful at the gate level of abstraction.

   b. ___ Verilog HDL does not support hierarchical decomposition of a design.

   c. ___ Verilog HDL supports structured design methodology.

   d. ___ Verilog HDL is useful only for the initial task of design entry.

6. Answer True or False.

   a. ___ Structured design methodology requires that each important high-level unit of the design be verified before the lower-level units are verified.

   b. ___ Gate and behavioral descriptions are both supported by Verilog, but may not be mixed within the same module.

   c. ___ Verilog has constructs that model hardware concurrency.

   d. ___ Verilog is a public domain language.

7. Answer True or False.

- **a.** ___ The basic unit of design in Verilog is called a design entity.
- **b.** ___ The emerging paradigm in VLSI circuit design encourages designers to work at the gate level of design.
- **c.** ___ Verilog can model state machines.
- **d.** ___ Verilog does not support hierarchical decomposition.
- **e.** ___ Verilog models are used primarily for functional verification and cannot include timing information.

8. List three levels of abstraction in digital design.

9. Write a Verilog structural model of the circuit described by the schematic below. (See Appendix A for a listing of Verilog primitives).

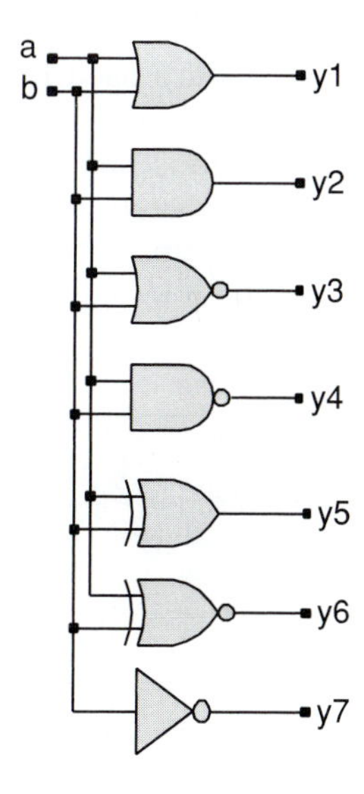

CHAPTER

2

HARDWARE MODELING WITH THE VERILOG HDL

Hardware description languages are used to create a compact, portable, and documented description of a design. The HDL model of a design is referred to as a design unit, and a design unit is said to encapsulate the design. A design unit, depicted in Figure 2.1, is ultimately an abstraction of a physical hardware unit that exhibits certain relationships between the external signals applied to the inputs of the system and the internal signals that cause manipulations of data to create its output signals.

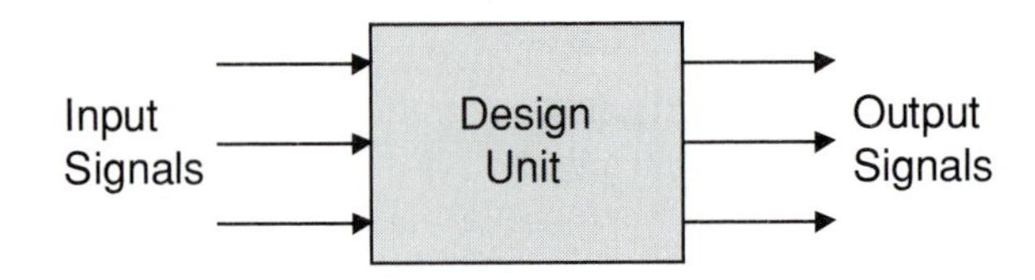

Figure 2.1 Design unit encapsulation of a digital circuit.

HDLs, like schematic entry tools, are usually bundled into larger software environments that support simulation of the behavior of the design represented in the language. By behavior we mean the relationship between the signals of the design. In this context, signals are waveforms of logic values evolving in time. The input signals create the response of the output signals through the action and interaction of the models of physical components. A design unit interacts with its environment through its input and output signals. A Verilog model of a design unit is a software representation of the unit's input-output relationship. This chapter will provide additional details about descriptive styles and modeling with Verilog.

2.1 HARDWARE ENCAPSULATION: THE VERILOG MODULE

The basic unit of design in the Verilog HDL is called a **module**. A module consists of a declarative text description using the constructs of the language. A Verilog module may contain a variety of descriptive elements that encapsulate information representing the functionality of the design, including a description of its timing characteristics. We will present a variety of examples to introduce the notion of a module and illustrate various ways of using Verilog to write the internal description of a module.

Verilog has predefined structural/functional elements called **primitives**. Primitives are said to be structural elements because they can be instantiated and connected together to create a larger design object having a more complex behavior than its constituent elements. Primitives are also said to be functional elements because they implement a (predefined) relationship between input and output signals. Many of the primitives correspond to simple combinational logic gates, such as "and" and "or" gates. Others, however, correspond to transistor switches, which are at a lower level of abstraction.

A module may contain instantiations of primitives and other modules. A primitive or a module is said to be instantiated within a module when it is listed in the module according to the syntax of the language. The contained, or instantiated, module is referred to as a "child" module; the containing module is said to be its parent.

Example 2.1 The Verilog module declaration for the half adder shown in Figure 2.2 is the parent module containing instantiations of primitives **xor, nand** and **not** (bold-faced text denotes Verilog language keywords). The signal interface between a module and its environment is through its ports; *Add_half_1* has input ports *a* and *b*, and output ports *sum* and *c_out*. The Verilog text in Figure 2.2 is called a ***module declaration***. In this example, the internal implementation of the module consists of an interconnection of primitive gates (**xor, nand, not**) which have been instantiated in the module. ***The order in which primitives are instantiated within a module has no significance—they can be instantiated in any order.*** Note: the keyword **endmodule** is required to complete the encapulation.

It is important to realize that HDL text is not a computational program in the ordinary sense. **Verilog is a descriptive language that describes a relationship between signals in a circuit.** A Verilog model *describes* a unit of digital hardware in terms of (1) interconnections of other hardware units whose models prescribe their behavior in a simulation, and (2) behavioral/procedural algorithms that abstractly describe input/output behavior that could be embodied

```
module Add_half_1 (sum, c_out, a, b);
  input a, b;
  output sum, c_out;
  wire c_out_bar;

  xor (sum, a, b);
  nand (c_out_bar, a, b);
  not (c_out, c_out_bar);
endmodule
```

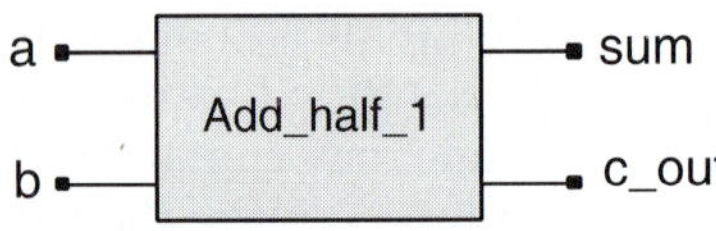

Figure 2.2 Verilog module for a structural description of a half adder.

in a hardware unit. The execution of an HDL description does not compute value in the sense that execution of a C program computes value. Instead, it uses the descriptive relationships prescribed by the text to determine an evolution of waveforms, just as hardware would produce waveforms of signals under the influence of an applied stimulus. One important distinction between an HDL and a programming language is that an HDL implicitly has a semantic of time associated with its data objects (signals), because it models their temporal existence and evolution.

In this example, the bold-faced items are keywords of the language—their meaning is predefined, and their use must conform to rules of syntax for Verilog. Each line of the description terminates with a semicolon delimiter, and the white space is cosmetic and incidental. Appendix B lists the keywords.

The contents of a Verilog **module** determine a relationship between the signals at its ports. In Figure 2.2, the signal *sum* is the output of the **xor** primitive, and the signals *a* and *b* are inputs to the **xor** primitive. *Add_half_1* in Figure 2.2 is composed of "instantiations" of the **nand**, **not** and **xor** primitives, with specific interconnections made by the presence of signal identifiers (names) in the terminal (port) list of the primitives. **The terminal list of a primitive is ordered with the outputs being the left-most entries, followed by the inputs.** Some primitives have a single output and multiple inputs; others have a single input and multiple outputs. The predefined behavior associated with these Verilog primitives, and the rules for placement of inputs and outputs within their terminal list, are given in Appendix A.

The syntax description in Figure 2.3 shows that a module declaration consists of the language key words **module** and **endmodule**, together with a module_identifier and other information (*module_item*) that specifies the ports of the module and other optional descriptions of its internal implementation. (See Appendix D for a description of Backus-Naur Form (BNF) for describing language syntax and Appendix F for the syntax of Verilog.) Note that the *list_of_ports* is an optional item in the declaration. Modules that do not need to interact with their environments do not have ports. For example, a testbench for simulating a description is usually a closed system having no ports.The *module_item_declaration* includes declarations of ports, parameters, variables, events, tasks, and functions.

Verilog Syntax: Module Declaration

```
module_declaration ::=
        module_keyword module_identifier [list_of_ports];
        {module_item}
        endmodule

module_keyword ::= module | macromodule
module_item ::=
        module_item_declaration
        | parameter_override
        | continuous_assign
        | gate_instantiation
        | udp_instantiation
        | module_instantiation
        | specify_block
        | initial_construct
        | always_contruct
```

Figure 2.3 Syntax for declaring a Verilog module.

Just as digital hardware must interface to a surrounding environment through the medium of signals, a Verilog module specifies how a design unit interfaces to its environment. With the exception of abstract events, which will be discussed in Chapter 7, all interaction between a module and its environment takes place through the ports of the module. The data objects at the ports of a module are referred to as signals.

The internal details of a module are hidden from the external world. Thus, a variety of implementations can have the same port definition and exhibit the same input/output behavior.

Example 2.2 Figure 2.4 shows *Add_half_2*, an alternative (RTL-behavioral) implementation of the half-adder that was previously implemented with primitive gates in Figure 2.2. Both models have the same ports, but their internal descriptions are different.

As shown in the previous examples, Verilog modules can be composed of declarations of ports and data objects, continuous assignment statements, instantiations of primitives (predefined and user-defined), instantiations of modules, behavioral statements, tasks and functions that are called by proce-

```
module Add_half_2 (sum, c_out, a, b);
  input          a, b;
  output         c_out, sum;

  assign {c_out, sum} = a + b + c_in;
endmodule
```

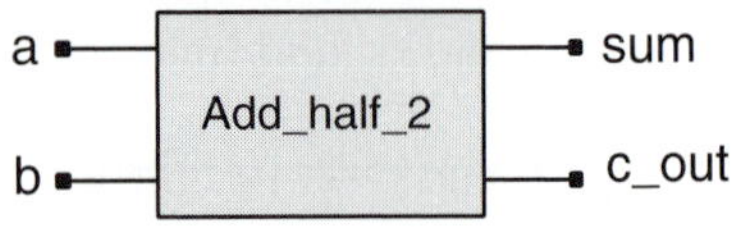

Figure 2.4 Alternative implementation of Half-Adder.

dural statements, and instantiations of other modules. **All forms of description can be mixed together in a single Verilog module.**[1] In the material that follows, we will use several examples to introduce basic elements of the Verilog language as a prelude to a fuller description of its features in Chapters 4-7.

2.1.1 Module Ports

A module interacts with its environment through the signals at its ports. Module ports may be scalar (i.e., single-valued) or vector objects (i.e., consisting of a one-dimensional array of values). For example, a reset line to a flip-flop is a scalar, but a 32-bit-wide data bus is a vector. Ports are classified by keyword as being either **input**, **output**, or **inout**, depending on whether the information exchanged at the port of the module is generated in the external environment and imported by the module, is generated within the module and exported to the environment, or both (i.e., a bi-directional device). The signals *a* and *b* in Figure 2.4 are **input** signals, denoting that the value of these data objects is made available to *Add_half_2* by the external environment; the signals *sum* and *c_out* are **output** signals because their values are determined within *Add_half_2* and made available to the external environment.

2.1.2 Module Implementation

The details of implementation of a module are hidden from the environment of the module. This means that the outside world cannot have direct access to information within a module except through its ports; all internal variables and data objects are local to the module. (Verilog also supports hierarchical dereferencing to bypass a module's ports and gain access to hidden internal variables.) Although the internal details of a module's implementation are hidden from the outside world, a module that has been declared can be instantiated, or referenced, within the implementation of another module. Module instantiation is accomplished by entering the name of the module as a module_item within a parent module and providing signal identifiers at the appropriate ports. The signals listed in a port are available for interaction with the implementation within the containing module—independent of how the module is actually implemented. **Instantiation of modules is analogous to placing and connecting circuit components on a circuit board.**

Example 2.3 In this example, we wish to implement a full adder[2] by using instantiations of half adder modules and some glue logic. First we note that the input-output relationship of a half adder is such that the *sum* and *c_out* bits are formed as the *exclusive-or* and the *and* of the data bits, respectively, as shown in the Boolean equations:

$$sum_{HA} = a \oplus b$$
$$c\_out_{HA} = ab$$

For a full adder, with "+" denoting "or," we have:

$$sum_{FA} = (a + b) \oplus c\_in$$
$$c\_out_{FA} = (a + b)\ c\_in + abc\_in$$

Note that a + b = ab' + a'b + ab, so that $c\_out_{FA} = (a \oplus b)c\_in + ab$. Part of this expression can be formed by "anding" the *sum* output of a half adder with the *c_in* bit of the full adder. Likewise, part of the logic required to form sum_{FA} is also formed by the same exclusive-or operation. In fact, using a second exclusive-or to combine the exclusive-or of the data bits with the *c_in* bit would produce the *sum* bit of the full adder. Figure 2.5 shows the schematic view and Verilog code implementing the result. The Verilog code contains a structural connection of some glue logic (Verilog primitives) and two instantiations of *Add_half*, each of which can be either *Add_half_1* or *Add_half_2* from the previous examples. We also note that the order in which signals are listed in the port of these modules determines which signal they are associated with inside the **module**.

Notice in Figure 2.5 that each instantiation of a module within another module is accompanied by a *module_identifier* (module instance names *M1* and *M2*). This feature of Verilog allows us to distinguish multiple instantiations of the same module, and is the vehicle by which the variables within a module can be referenced indirectly from anywhere in the hierarchy. (This feature supports what is called "hierarchical dereferencing," which will be discussed in Chapter 4.) The important thing to remember now is that a **module instantiation must include a module identifier.** Also note that the declaration of *Add_half* is not part of the declaration of *Add_full*. **A module is never declared within another module.** All modules are declared at the same lexicographic level in the source code file. When a module is instantiated, the order in which the ports are listed usually matches the order in which the corresponding signals were listed in the declaration of the module—unless an explicit association is made by name rather than position (see Section 2.4).

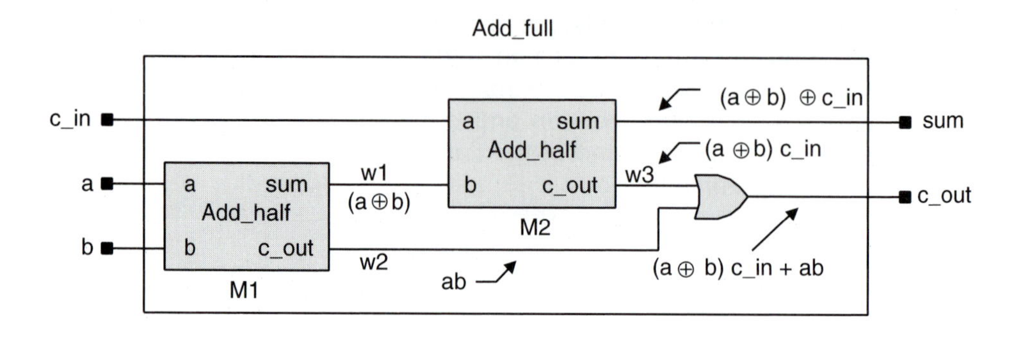


```
module Add_full (sum, c_out, a, b, c_in);        // parent module
  input a, b, c_in;
  output c_out, sum;
  wire w1, w2, w3;

  Add_half M1 (w1, w2, a, b);                    // child module
  Add_half M2 (sum, w3, w1, c_in);               // child module
  or (c_out, w2, w3);                            // primitive instantiation
endmodule
```

module instance name

Figure 2.5 A full adder module built from instantiations of half adder modules and glue logic.

2.2 HARDWARE MODELING: VERILOG PRIMITIVES

Verilog supports a robust set of structural/functional primitives that model the behavior of common digital hardware, which can simplify the task of digital design. Each of these predefined primitives is identified by a keyword, and each implements the behavior of a combinational logic function or transistor level switch. Verilog also allows the user to create primitives describing sequential behavior (like flip-flops), as well as additional combinational primitives. User-defined primitives are discussed in Chapter 5.

Primitives serve as a resource for creating more complex models of combinational hardware behavior, including sequential logic. **Primitives can be instantiated only within modules;** they are never used as stand-alone objects in a design. Instantiated primitives define concurrent hardware objects that, in a simulation, are simultaneously operative and mutually interacting through interconnecting signals at any instant of time.

An important **rule for primitives** is that the first port element of an instantiated primitive is always an output. Some primitives, like buffers (**buf**) and inverters (**not**), may have multiple outputs. In that case, the first port element must be followed by a list of the remaining outputs, then by the inputs.

Example 2.4 The Verilog text in Figure 2.6 is a declaration of module *AOI_4*, which encapsulates the functionality of a four-input and-or-invert logic gate. The module consists of a structural description in which Verilog primitives are instantiated and connected by declared signals (wires) *y1* and *y2*.

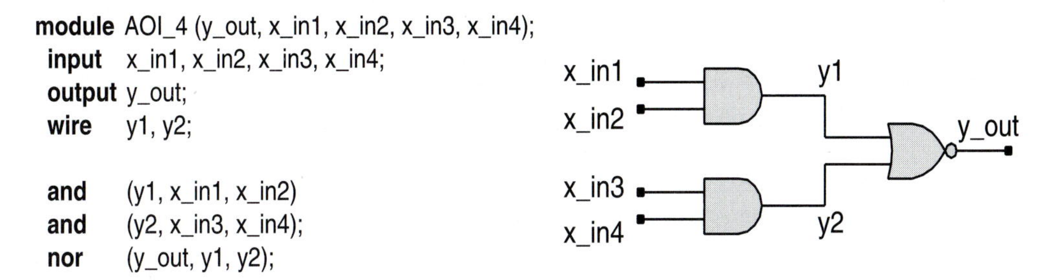

```
module AOI_4 (y_out, x_in1, x_in2, x_in3, x_in4);
 input  x_in1, x_in2, x_in3, x_in4;
 output y_out;
 wire   y1, y2;

 and    (y1, x_in1, x_in2)
 and    (y2, x_in3, x_in4);
 nor    (y_out, y1, y2);

endmodule
```


Figure 2.6 Verilog description of an and-or-invert logic gate.

Verilog includes the predefined primitives listed and categorized in Table 2.1. Their names correspond closely to physical hardware logic gates. (See Appendix A for a functional description of each primitive.) The combinational, three-state and "pull" primitives are used to create structural models at the gate level of abstraction. The other primitives are used to model functionality at the switch level of abstraction. (See Chapter 11.) Verilog primitives are not physical gates, but their functionality is obviously related to the behavior of physical gates. Under simulation, the values of the signals at the inputs of a primitive at any time during simulation determine the value of the primitive's output(s). Primitives are idealized models of behavior because they ignore the time delays exhibited in real gates, i.e., their default propagation delay is zero. However, a delay can be assigned to a primitive when it is instantiated in a module. Mechanisms for modeling propagation delay are discussed in Chapter 6.

Table 2.1 *Built-in Verilog primitives.*

Combinational Logic	Three State	MOS Gates	CMOS Gates	Bi-Directional Gates	Pull Gates
and	bufif0	nmos	cmos	tran	pullup
nand	bufif1	pmos	rcmos	tranif0	pulldown
or	notif0	rnmos		tranif1	
nor	notif1	rpmos		rtran	
xor				rtranif0	
xnor				rtranif1	
buf					
not					

The inputs and outputs of a primitive are called terminals, and are also referred to as ports. Verilog primitives are devoid of timing information, but timing information can be added to a given instance of a primitive. When delay is associated with the instantiation of a primitive, all input/output port pairs have the same delay. Functional models sometimes ignore or simplify the timing characteristics (propagation delays) of the actual hardware being modeled. This is done primarily to simplify the task of verifying digital logic.

Example 2.5 Figure 2.7 shows the module *AOI_4_unit*, obtained by adding unit delay timing to the *AOI_4* module in Figure 2.6. These time delays have effect in the simulation of the model, where they cause a unit delay between a change at the input of a primitive and the appearance of a resulting change at the output of the primitive.

```
module AOI_4_unit (y_out, x_in1, x_in2, x_in3, x_in4);
  input   x_in1, x_in2, x_in3, x_in4;
  output  y_out;
  wire    y1, y2;

  and    #1 (y1, x_in1, x_in2)      // Alternative:    and #1   (y1, x_in1, x_in2),
  and    #1 (y2, x_in3, x_in4);     //                          (y2, x_in3, x_in4);
  nor    #1 (y_out, y1, y2);
endmodule
```

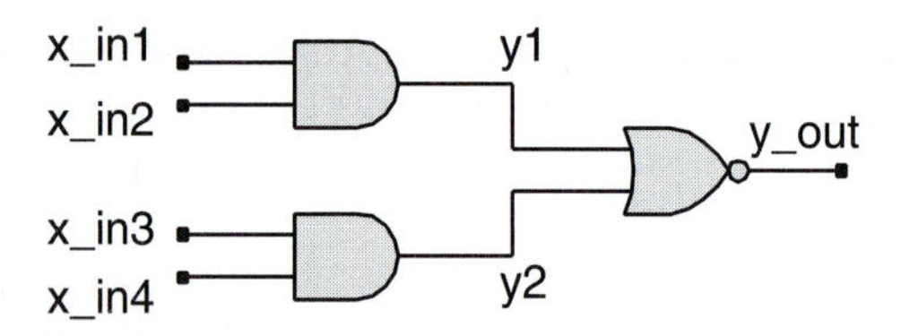

Figure 2.7 Verilog description of an and_or_invert logic with unit delays assigned to primitives.

Notice that the instantiated gates in Figure 2.7 do not have an identifier. **The use of an identifier (instance name) with an instantiated gate (primitive) is optional.**

In ASIC design, the timing information associated with a given gate model might incorporate values that have been obtained from accurate characterization of the physical gates. This information could be used by a simulator to portray a more accurate representation of the logical waveforms of the physical circuit. In the physical circuit, the propagation delay that would be measured from two distinct input pins to an output pin need not be identical. Nor is the propagation delay necessarily symmetrical for rising and falling waveforms of the signal. A model written in the Verilog language can account for these properties of physical circuits (see Chapter 6). The timing delays of fabricated parts can be represented by a triplet of values corresponding to minimum, typical and maximum propagation delays that result from variations in the fabrication process. Physical propagation delays of the individual gates may also be modified to account for the capacitive loading present in the actual physical layout of the IC. Otherwise, estimates of timing values are commonly used. Models written with min:typ:max timing values provide simulators and timing analyzers with useful information that can be exploited by the designer.

Example 2.6 Figure 2.8 shows a Verilog module for *nanf201*, a two-input nand gate implemented in 1.2 micron CMOS n-well technology, with delays given implicitly in units of nanoseconds. This gate is part of a standard cell library used for ASIC design. The functionality of the gate is provided by the instantiation of the **nand** primitive, and the timing relationship is given by the "specify block" of text. Specify blocks are enclosed by the keywords **specify** and **endspecify**. Within the specify block, the timing parameters *Tpd_0_1* and *Tpd_1_0* define min:typ:max values of the rising and falling delays of the output signal in response to an event at an input to the module. The notation (A1 => O) represents the input-output port pair with which the delay is associated. **When propagation delays are modeled with specify blocks, distinct asymmetric rising and falling delays can be assigned to input/output port pairs,** which allows the simulation of the circuit to be more accurate. We will consider specify blocks in more detail in Chapter 6.

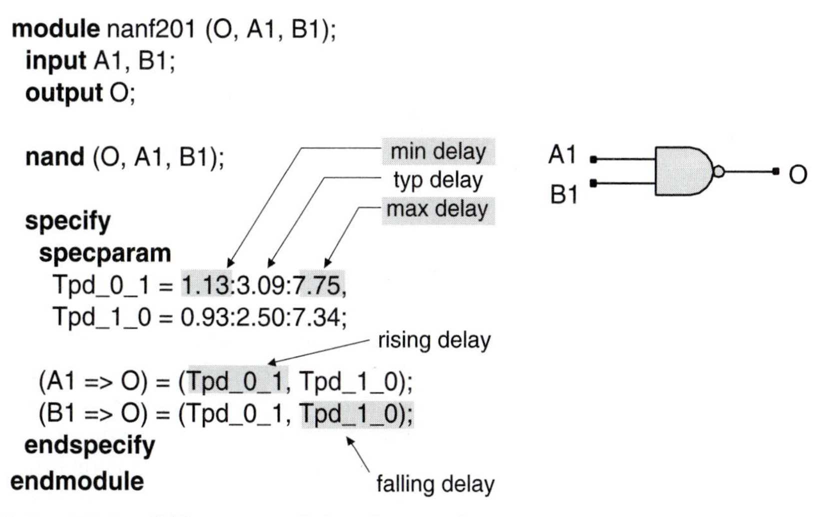

Figure 2.8 ASIC cell library module of a nand gate.

Verilog primitives are sometimes said to be "smart primitives," because the same primitive can be used to describe the behavior for any number of inputs, not just two. This simplifies the construction of Verilog models and reduces the effort to construct cells for an ASIC library. Smart primitives allow a designer to implement generic functionality without having to determine whether a particular logic gate is available in a technology library. This detail is left to a synthesis tool.

Example 2.7 In this example, we construct *nand3gate*, a module whose behavior is that of a three-input nand gate using the same "smart" **nand** primitive used in Example 2.6. The three-input nand module is shown in Figure 2.9 with its schematic symbol. The port list of the module (O, A1, A2, A3) specifies the signals that interact with the outside world. In this example, they also connect internally to the terminals of the **nand** primitive. The same primitive that was used to create *nanf201* with two inputs has been used here to create *nand3gate* with three-inputs. As in the previous example, timing values could be incorporated within the model to develop an ASIC library cell.

```
module nand3gate (O, A1, A2, A3);
  input     A1, A2, A3;
  output    O;

  nand (O, A1, A2, A3);
endmodule
```

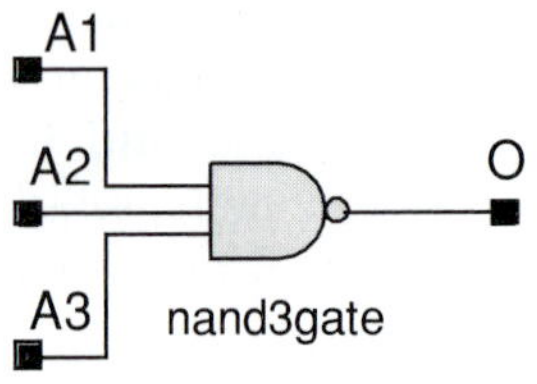

A1
A2
A3
O
nand3gate

Figure 2.9 Three-input nand module built from a "smart" primitive.

2.3 DESCRIPTIVE STYLES

Digital systems can be described in a variety of styles. The first style that we will consider is the structural style. A structural style of description consists of a systematic interconnection of design objects by signals. Strictly speaking, a structural description consists of interconnected primitives and modules, with the interconnection being made by signals that connect primitive terminals and/or module ports. Structural descriptions can be explicit or implicit. The structural style of description is analogous to placing components on a schematic drawing and interconnecting them with wires.

2.3.1 Explicit Structural Description

Explicit structural descriptions of behavior are formed by instantiating and interconnecting primitives/gates and other modules within a module. The connectivity of the structural detail is specified by the signal identifiers that appear in the ports of the instantiated primitive gates.

Example 2.8 The description of a half adder in Figure 2.10 consists of an interconnection of primitive gates. Since the description has explicit instantiations of primitive gates, the module is implemented as an explicit structural description.

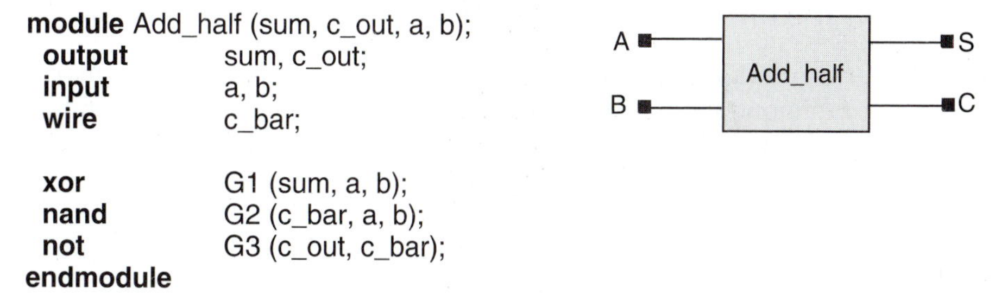


```
module Add_half (sum, c_out, a, b);
  output     sum, c_out;
  input      a, b;
  wire       c_bar;

  xor        G1 (sum, a, b);
  nand       G2 (c_bar, a, b);
  not        G3 (c_out, c_bar);
endmodule
```

Figure 2.10 Explicit structural description of half adder.

Notice that the Verilog description in Figure 2.10 contains a declaration of a **wire**, *c_bar*. This declaration makes available a wire that connects the output of the **nand** primitive and the input to the inverter. An identifier that is not declared is assumed to be a **wire**. In Verilog, objects of type **wire** are used solely to effect structural connectivity within the module.

Each primitive instance in Figure 2.10 has associated with it an identifier, or instance name (e.g., G2). Remember, **the use of an instance name with a primitive is optional.** However, **a module instance must always have a name.**

The module in Figure 2.10 is generic—it implements only the functionality of a behavior, and not the timing relationships between the signals. An alternative is to create the half adder by instantiating other modules whose descriptions include timing characteristics of the corresponding physical parts.

Example 2.9 The module *Add_half_structural* in Figure 2.11 has instantiations of modules *xorf201*, *nanf201*, and *invf101*, which are standard cells in a CMOS library. Examination of their descriptions would reveal that each includes a timing model.

```
module Add_half_structural (sum, c_out, a, b);
  output      sum, c_out;
  input       a, b;
  wire        c_bar;

  xorf201  G1 (sum, a, b);
  nanf201  G2 (c_bar, a, b);
  invf101  G3 (c_out, c_bar);
endmodule
```

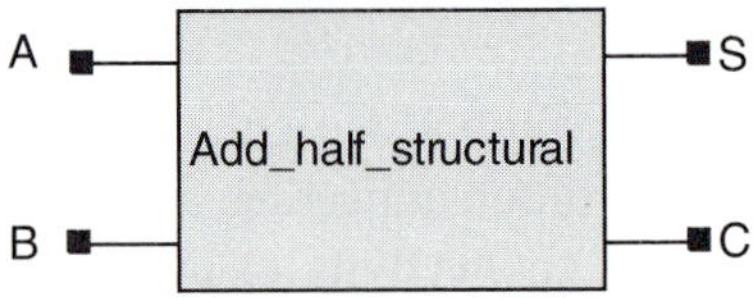

A, B → Add_half_structural → S, C

Figure 2.11 Cell-based implementation of a half adder.

2.3.2 Implicit Structural Description—Continuous Assignments

The Verilog HDL contains a rich selection of built-in operators that can be used in a "continuous assignment" statement to create an implicit structural model of combinational logic. In Chapter 7, we will see how the same operators can also be used within behavioral statements.

Example 2.10 The module in Figure 2.12 implements a two-input nand gate using a continuous assignment and the Verilog bit-wise-nand (~&) operator.

```
module nand2_RTL (y, x1, x2);
  input    x1, x2;
  output   y;

  assign y = x1 ~& x2;        // Bitwise-nand
endmodule
```

Figure 2.12 An implicit structural description of a two-input nand gate.

The model in Figure 2.12 is said to be an implicit structural model because the language operators themselves are not bound directly to physical gates or language primitives. Recall that the **assign** keyword in this example declares a Verilog continuous assignment. Continuous assignment statements correspond to combinational logic, but without requiring explicit instantiation of gates. This simplifies a description.

Implicit combinational logic (i.e., continuous assignments) can also be thought of as "event scheduling rules" that determine how an event for the identified variable (signal) on the left-hand side of the statement depends on events of the identified variables on the right-hand side of the statement. In this example, the continuous assignment determines how the output, *y*, depends on events for the variables *x1* and *x2*, just as though the relationship was implemented by a gate. That is, if *x1* or *x2* undergoes change (an event), then the value of *y* is re-computed according to the relationship specified in the continuous assignment statement. As we saw in Chapter 1, this style of description is also referred to as a "data flow" or an "RTL" description.

Continuous assignments can be made in two ways. The first method, which has already been illustrated in the previous examples, associates the keyword **assign** with a statement whose expression creates a value assigned to the target variable. The syntax is shown in Figure 2.13.

Verilog Syntax: Continuous Assignment

continuous_assign ::= **assign** [drive_strength][delay3] list_of_net_assignments;

Figure 2.13 Syntax of a continuous assignment statement.

The second method creates a continuous assignment implicitly by associating the RHS expression with the declaration of the target variable.

Example 2.11 In Figure 2.14, the output of *bit_or8_gate1* and *bit_or8_gate2*, *y*, is formed as the bitwise-or of the data words *a* and *b*. Note that, in this example, the signals are eight-bit vectors. The left-most entry of a vector identifier is the most significant bit of the word. If the data words have the values a = 1100_1100 and b = 1111_0000, then the value of *y* is 1111_1100. (Underscores may be inserted to improve the readability of numbers.) In the first version, *y* gets its value through a stand-alone continuous assignment. In the second version, the declaration of *y* and the continuous assignment to *y* are combined.

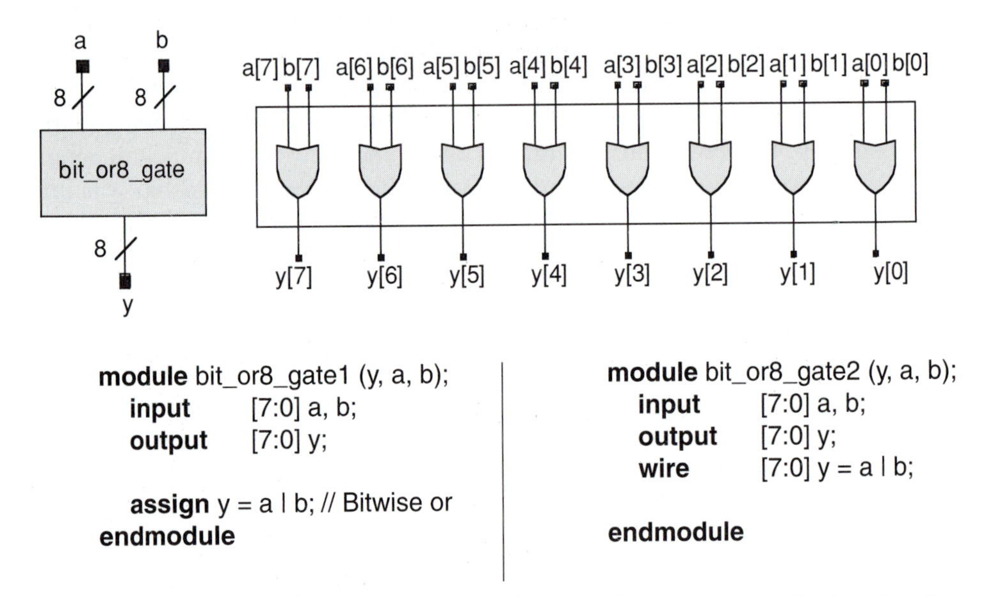
a
b
8
8
bit_or8_gate
8
y
a[7] b[7] a[6] b[6] a[5] b[5] a[4] b[4] a[3] b[3] a[2] b[2] a[1] b[1] a[0] b[0]
y[7] y[6] y[5] y[4] y[3] y[2] y[1] y[0]
module bit_or8_gate1 (y, a, b);
input [7:0] a, b;
output [7:0] y;
assign y = a | b; // Bitwise or
endmodule
module bit_or8_gate2 (y, a, b);
input [7:0] a, b;
output [7:0] y;
wire [7:0] y = a | b;
endmodule

Figure 2.14 Implicit structural description of an eight-bit or "gate", showing two alternatives for continuous assignment.

A continuous assignment statement creates an implicit combinational logic structure for the target net variable. The corresponding event scheduling mechanism is a "static binding" between the variables expression on the right-hand side (RHS) and the target net variable on the left-hand side (LHS) of the statement. The binding is "static," or continuous, because the continuous assignment construct in Verilog does not include a mechanism for eliminating or altering the binding. On the other hand, procedural continuous assignments, which will be discussed in Chapter 7, create dynamic bindings, one form of which (i.e., **force**... **release**) can, during simulation, dynamically override the binding of a continuous assignment to a net. **In the absence of an overriding procedural continuous assignment, a continuous assignment to a net is in effect for the duration of a simulation.**

2.3.3 Multiple Instantiations and Assignments

The syntax for instantiating primitives and declaring continuous assignments allows one statement to instantiate multiple primitives of the same type. It also allows multiple continuous assignments to be made with one **assign** keyword. Multiple instantiations and assignments use comma-separated lists.

Example 2.12 The Verilog description *Multiple_Gates* instantiates multiple gates with unit propagation delay specified by #1;

Mulitple_Assigns has multiple continuous assignments with unit propagation delay. Only the first gate has an instance-name, *G1*.

```
module Multiple_Gates (y1, y2, y3, a1, a2, a3, a4);
  input      a1, a2, a3, a4;
  output     y1, y2, y3;

  nand #1 G1 (y1, a1, a2, a3), (y2, a2, a3, a4), (y3, a1, a4);

endmodule

module Multiple_Assigns (y1, y2, y3, a1, a2, a3, a4);
  input      a1, a2, a3, a4;
  output     y1, y2, y3;

  assign #1 y1 = a1 ^ a2, y2 = a2 | a3, y3 = a1 + a2;

endmodule
```

2.4 STRUCTURAL CONNECTIONS

Structural descriptions are made when modules and primitives are instantiated within other modules. Instantiations of modules and primitives are connected by inserting signal identifiers in the list of the ports of a module and the terminal list of a primitive.

2.4.1 Module Port Connections

A structural connection to a port of a Verilog module can be made in two ways. The first is to include the name of the connecting signal at the appropriate location in the module port list.

Example 2.13 The module *parent_mod* shown in Figure 2.15 has within it an instantiation of *child_mod*. The structural connections to *child_mod* are made by placing the bit lines of the identifier *g* at the desired location in the port of *child_mod*. The order in which the ports are placed determines the binding between the elements of *g* and the internal implementation of *child*_mod. In this case, *g[3]* and *g[1]* are associated with the inputs *sig_a* and *sig_b*; *g[0]* and *g[2]* are associated with (bound to) the outputs *sig_c* and *sig_d*.

```
module parent_mod;
  wire [3:0] g;

  child_mod G1 (g[3], g[1], g[0], g[2]); // Listed order is significant

endmodule

module child_mod (sig_a, sig_b, sig_c, sig_d);
  input sig_a, sig_b;
  output sig_c, sig_d;

  // module description goes here.
endmodule
```

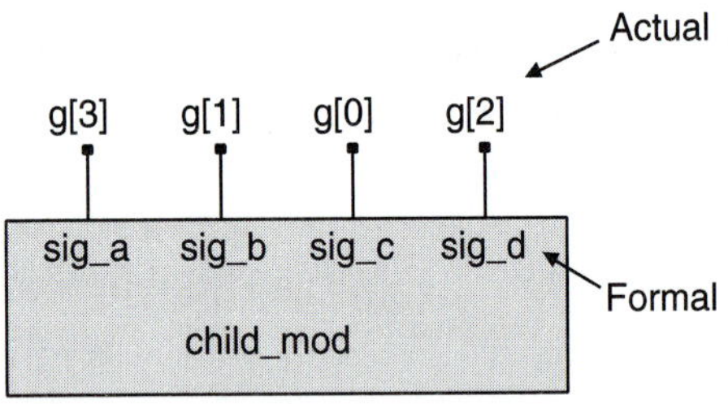

Figure 2.15 Port connection by position association.

Connection by position is the usual method of establishing connectivity in small circuits. However, it is also possible to establish module connectivity by name, without regard to position. This is done by explicitly substituting signal name pairs in the module port. Connection by name is convenient when the size of the port list is large and its order is hard to remember.

Example 2.14 In Figure 2.16 the syntax *.sig_c(g[3])* matches the actual signal *g[3]* to the formal signal *sig_c* in *child_mod*, rather than to *sig_a*. The order of the elements in the port list of instance *G1* of *child_mod* is insignificant, because the binding of a signal is accomplished by name association.

```
module parent_mod;
  wire [3:0] g;

  // The order in the module port list below has no significance.
```

```
        child_mod G1  (
                      .sig_c(g[3]),
                      .sig_d(g[2]),
                      .sig_b(g[0]),
                      .sig_a(g[1])
                      );
endmodule

module child_mod (sig_a, sig_b, sig_c, sig_d);
   input          sig_a, sig_b;
   output         sig_c, sig_d;

   // The module description goes here.

endmodule
```

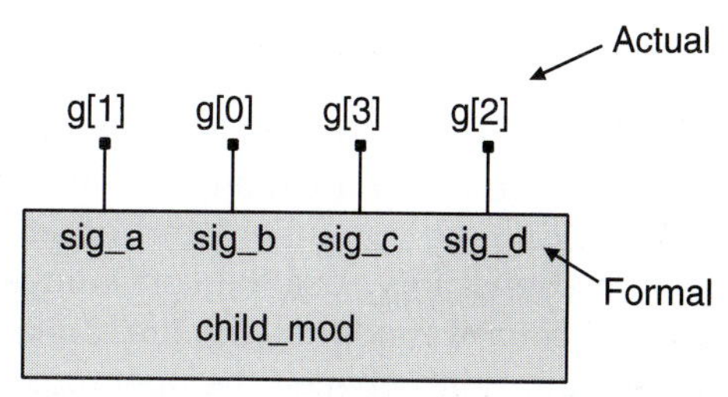

Figure 2.16 Port connection by name association.

2.4.2 Primitive Terminal Connections

A structural connection to a terminal of a Verilog primitive is made by including the name of the connecting signal at the appropriate position in the terminal list of the gate.

2.4.3 Empty Port Connections

If an instance of a module is to have an empty port, the position of the port in the instantiation is left empty. For example, a module representing the behavior of a four-input nand gate might have the declaration header *module nand4 (y, x1, x2, x3, x4)*. If the input at the position of *x3* is to be left unattached, an instance of *nand4* would be *nand4 M1 (y, x1, x2, , x4)*. An unconnected module input is automatically driven to the "z" state (high impedance), and unconnected outputs are not used.

2.5 BEHAVIORAL DESCRIPTIONS IN VERILOG

A behavioral description consists of procedural statements that determine a relationship between the input and output signals of the module, without reference to hardware or structure. As mentioned previously, there are two basic styles of behavioral description. The first, called RTL (register transfer level) style, describes an input/output relationship in terms of dataflow operations on signals and registers values. If registers are involved, the transfers may occur synchronously. The second style consists of an abstract, algorithmic description of operations that need not conform to a dataflow, and which might not be synthesizable.

Modeling with behavioral constructs is gaining greater importance because it allows the designer to quickly create a design that can be synthesized automatically by modern synthesis tools, thereby eliminating the need to deal immediately with gate-level detail. In this sense, Verilog and other HDLs allow the design to be addressed at a higher level of abstraction than traditional schematic-based methods that rely on gate-level detail.

2.5.1 RTL/DATA FLOW DESCRIPTIONS

A register transfer level (RTL) description uses Verilog language operators to determine the values of signals in a circuit. It characterizes the flow of data through computational units in computers and other machines. Verilog RTL descriptions of combinational logic use the continuous assignment statement to create an implicit structural model of combinational logic. The assignment declares a Boolean relationship between the operands of the statement's RHS and the LHS' target variable. Under simulation, the simulator monitors the activity of the operands and determines whether and when to schedule activity for the target variable.

Example 2.15 The module in Figure 2.17, which implements the behavior of a four-input "and" gate, can be described in a dataflow style using Verilog's built-in bitwise AND operator (&) to "and" the bits of the data word.

```
module and4_rtl (y_out, x_in1, x_in2, x_in3, x_in4);
  input           x_in1, x_in2, x_in3, x_in4;
  output          y_out;

  assign          y_out = x_in1 & x_in2 & x_in3 & x_in4;
endmodule
```

An alternative implementation in Figure 2.18 uses the unary reduction "AND" operator to accomplish the same result.

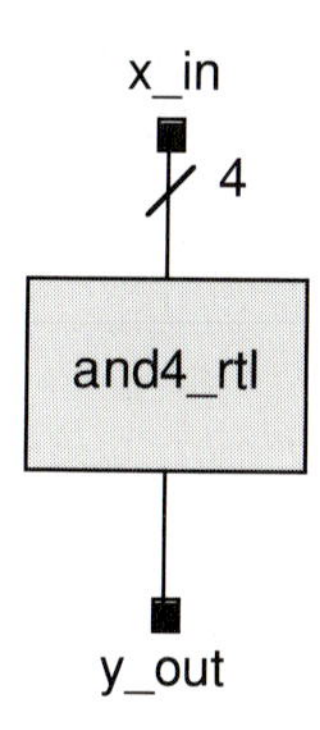
x_in
4
and4_rtl
y_out

Figure 2.17 An RTL description of a four-input and gate.

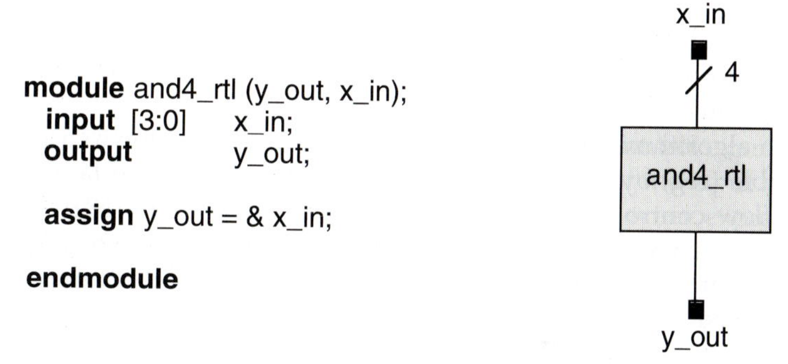
module and4_rtl (y_out, x_in);
input [3:0] x_in;
output y_out;
assign y_out = & x_in;
endmodule
x_in
4
and4_rtl
y_out

Figure 2.18 An alternative RTL description of a four-input and gate.

Example 2.16 The next example of an RTL description is the flip-flop shown in Figure 2.19. At every rising edge of the clock, the value of the output is taken from *data_in* if the active-low set and reset are not asserted. In general, data flow implementations form one or more outputs based on logical operations involving the inputs. Here, the flip-flop output, *q*, is updated synchronously, and retains its value between the clock edges. It is necessary that *q* be declared to be a ***reg*** variable.

2.5.2 Algorithm-Based Descriptions

In Verilog, the register types of data objects retain an assigned value until another assignment is made to them. They are like variables in a procedural programming language. In fact, they can be assigned value only by a procedural statement within a behavior. Data types are discussed in more detail in

2.10 LANGUAGE CONVENTIONS

Our discussion of the Verilog HDL will use conventions that govern the language. Verilog is case-sensitive, so distinct identifiers can be written using the same characters in upper and lower case text (some simulators have a case-insensitive option). Do not use the same name for different items within the same scope (e.g., within the same module). Each instance of a module must be named, but the same name cannot be used for both a module and an instance of the module. All Verilog keywords are lower case. Identifiers (e.g., the name of a wire) may use upper and lower case alphabetical characters, the decimal digits (0,1,...,9) and the underscore (_). An identifier may contain up to 1024 characters, but its first character must not be a digit. A Verilog "program" consists of text statements formed by a stream of one or more characters (tokens) and terminated by a semicolon. The text is written in a free format, with white space (blanks and tabs) and newline characters having no syntactical significance. Blanks and white space can be used to enhance the readability of code. Blanks and tabs do have significance when they appear within a quoted character string. Similarly, newline and formfeed characters have significance within quoted strings, where these characters are used to control the format of text output. Character strings in Verilog are enclosed by double quotes and must be confined to a single line of text.

Single-line comments are made using a double slash: "//". All characters that follow "//" are ignored. A block (multi-line) comment is made using "/*" to begin the block, and "*/" to terminate the block. Block comments may not be nested.

Compiler directives begin with the grave accent: "`". One use of a compiler directive is to define text string substitutions (e.g., "`***define*** *state0 3'b001*"). The compiler will make a literal substitution of the three-bit binary string "*001*" wherever "*state0*" appears in the source code. Such substitutions make the source code more readable. Compiler directives are active from the point at which they appear in the source code, until they are overridden by another directive.

All names that begin with the character "$" denote a built-in system task or function (e.g., ***$monitor***). Strings are enclosed in double quotes and must be written on a single line (some systems allow strings to span more than one line if the backslant "\" is used at the break point).

2.11 REPRESENTATION OF NUMBERS

Numbers may be represented in decimal (d or D), hex (h or H), octal (o or O) and binary (b or B) formats. All number formats can be represented internally in a sized format, <size><base_format><number>, where *size* is a decimal value that determines how many bits will represent the stored value in the host machine, *base_format* determines the arithmetic base of *number*, and *number* is the value expressed in the indicated base. For example, 8'b1010 is stored internally as an eight-bit value, equivalent to decimal value 10 ($(0000\_1010)_2$). The sized number 4'hA is interpreted as a hexadecimal value and saved as four bits $(1010)_2$. The default base is decimal; numbers whose *base_format* is not given must be decimal values.

Numbers can be represented in an unsized format (*size* is optional), in which the user does not specify the number of bits required to store the value. The manner in which an unsized number is stored is machine-dependent, but guaranteed to be no less than a 32-bit value.

The representation of numbers may include the use of underscore characters to improve the readability of the text. Signed numbers (+ or -) are allowed. Real numbers can be represented in scientific notation (using e or E). The characters of a real number are restricted to the digits (0 ... 9) and the underscore. The exponent of a number in scientific notation must be an integer. The exponent and mantissa may contain only numeric values. If the mantissa contains a decimal point, there must be at least one number on either side of it. Some examples are shown in Table 2.2.

Number	#Bits	Base	Dec. Equiv.	Stored
2'b10	2	Binary	2	10
3'd5	3	Decimal	5	101
3'o5	3	Octal	5	101
8'o5	8	Octal	5	00000101
8'ha	8	Hex	10	00001010
3'b5	Not Valid!			
3'b01x	3	Binary	-	01x
12'hx	12	Hex	-	xxxxxxxxxxxx
8'hz	8	Hex	-	zzzzzzzz
8'b0000_0001	8	Binary	1	00000001
8'b001	8	Binary	1	00000001
8'bx01	8	Binary	-	xxxxxx01
'bz	unsized	Binary	-	z ... z (32 bits)
8'HAD	8	Hex	173	10101101

Table 2.2 *Representation of numbers in Verilog*

If the most significant bit of a sized number is "x" or "z", the representation automatically extends x or z values to complete the representation of the value. For example, 12'bx0001 has value xxxx_xxxx_0001.

2.12 SUMMARY

Hardware description languages resemble procedural programming languages, but they have additional constructs that allow them to describe digital systems and simulate their behavior in a simulation environment. Special language constructs allow a design to be decomposed systematically into a simplified hierarchical family of interacting design units, each of which has less complexity than the original design. The descriptions of the design units can be formed from predefined Verilog "gate" primitives, predefined operators, or behavioral statements. All of these various styles of implementation can be incorporated into a single design.

REFERENCES

1. Sternheim, E. at al., *Digital Design and Synthesis with Verilog HDL,* Automata Publishing Co., San Jose, California, 1993.
2. Comer, David J., *Digital Logic and State Machine Design*, Third Edition, Saunders College Publishing, New York, 1995.
3. Cadence Design Systems, Inc.—Vendor supplied documentation.
4. Synopsys, Inc.—Vendor supplied documentation.

PROBLEMS

1. Write the binary equivalent of the following Verilog numbers.
   - a. 8'h7
   - b. 8'O36
   - c. 8'b1010_0011
   - d. 12
   - e. 8'bx
   - f. 6'h2E
2. What is the syntax error in the following real numbers?
   - a. .4e3
   - b. 4.61E6.5
   - c. 34.2x3E3
   - d. 2.e2
3. Using Verilog primitives, write a structural description (with primitive gates) of the circuit having inputs *A0, A1, A2, B0, B1,* and *B2* and outputs *y1* and *y2* described by the Boolean equations below (overscore denotes logical complement of a literal).

$y1 = \overline{A0}\, B2 + \overline{A2}\, A0\, B2 + A0\, \overline{B1}\, B0$

$y2 = A2\, \overline{B1} + A1\, \overline{B2}\, \overline{B0} + \overline{A1}\, A0\, B1$

4. Repeat Problem 3 using continuous assignments and Verilog language operators.

5. Find the syntax errors in the following declarations:

a.
```
nand (y_out1, x_in1, x_in2); #1 (y_out2, x_in3, x_in4);
```

b.
```
module nand_gate3 (y_out, x_in1, x_in2, x_in3);
    input x_in1, x_in2, x_in3;
    output y_out;
    nand (x_in1, x_in2, x_in3, y_out);
endmodule
```

6. Using Verilog language operators, write a Verilog description of a module that produces signals *y1* and *y2* according to the Boolean equations below, where and, or, not, xor denote Boolean operations.

y1 = x1 or x2 and (not x3)

y2 = x4 or (x2 xor x3)

7. Write an instantiation of a three-input "or gate having the following input-output propagation delays: tpd_0_1 = 2, tpd_1_0 = 1.

8. Using Verilog primitives, write a Verilog structural model for a 4:1 mux.

9. Use the results of Problem 8 to create a hierarchical structure that implements a 16:1 mux.

10. Discover the syntax errors in the Verilog description given below.

```
module Has_bugs (sig_3, sig_2, sig_1, sig_0);
   input sig_3, sig_2, sig_1, sig_0;
   output sig_3;
endmodule;
```

11. Using Verilog language primitives, write a description of the circuit shown below, where *rst[0]* and *set[0] form q[0]* and *qb[0]*, etc.

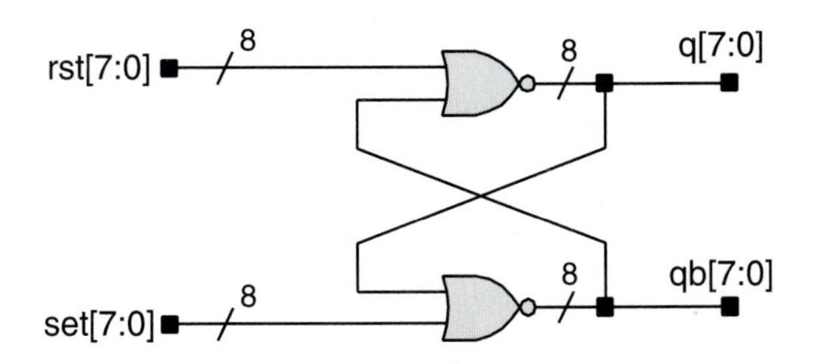

12. Repeat Problem 11 using Verilog operators and continuous assignment statements.

13. Write the following alternative descriptions of a combinational circuit that will produce an output representing the number of 1s in a four-bit word. The input four-bit word is available in parallel, and the result is to be encoded as a three-bit word.

    a. A gate level description

    b. Continuous assignments

14. List the allowed modes of ports in Verilog.

15. Answer True or False.

    a. ___ Verilog is case sensitive.

    b. ___ Verilog modules may not be nested.

    c. ___ The IEEE 1164 standard establishes a logic system for the Verilog HDL.

    d. ___ Verilog includes primitives for transistor/switch level design.

    e. ___ Verilog supports module pin-pin timing delay.

    f. ___ A Verilog module may not be declared within another module.

    g. ___ Verilog supports port connection by position and name.

    h. ___ The **endmodule** keyword is followed by a semicolon.

    i. ___ A primitive must have an instance-name.

    j. ___ A module may have an optional instance-name.

16. Write a Verilog description for a module that uses language primitives to implement a four-input nand gate.

17. Repeat Problem 16 using continuous assignments.

18. Draw the waveforms produced by the following behavior.

```
always begin
   #0 clock = 0; phase_1 = 1;
   #75 clock = ~ clock;
   #20 phase_1 = ~phase_1;
   #5;
end
```

(Note that the "#" operator delays the execution of a statement. Also, remember that the **always** behavior is cyclic.)

19. Using Verilog language operators, write an implicit structural description of the 4:2 priority encoder named *encode4_2* depicted below. Line *y0* is to have the highest priority. The output lines are to be encoded in sequence, with "0" associated with the assertion of *y0*.

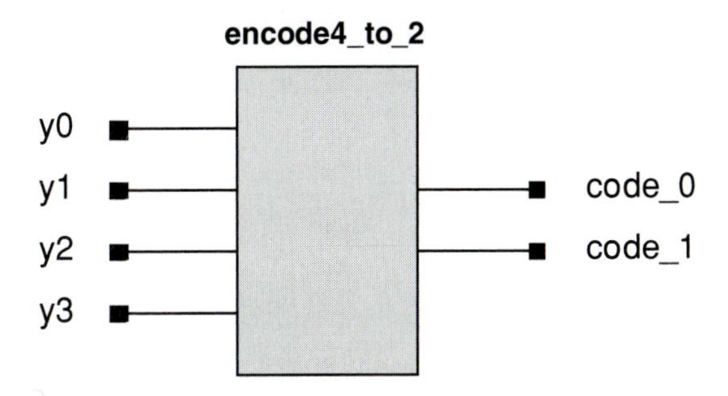

20. Using Verilog language operators, write an implicit structural description of the 8:3 decoder depicted below. The output lines are to be encoded in sequence, with *code[0]* asserted when *y[7:0] = 0*.

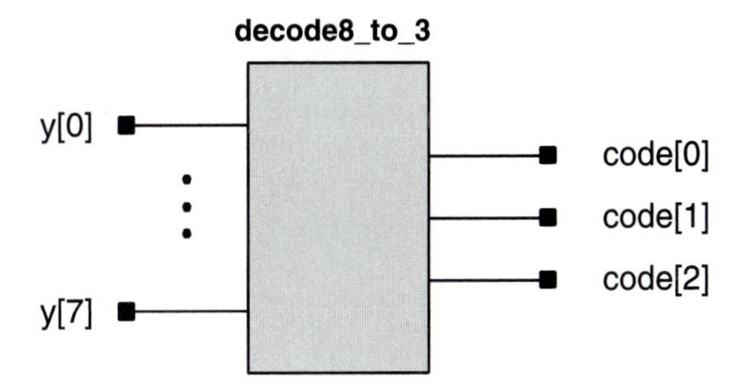

21. Which of the follow assignments have correct syntax? Of those that do, what will be stored in memory?

a. A = 8'b101

b. A = 5'o9

c. C = 12'HAA

d. D = 4'BBB

e. E = 4d'x3

f. F = 4'hz

g. G = 4'O____8

h. H = 8h'z9

22. What are the values of A, B, and C after the following code has executed?

```
 begin
    A = 4'd2;
    B = 3;
    C = A + B;
end
```

CHAPTER 3

EVENT-DRIVEN SIMULATION AND TESTBENCHES

Verifying the functionality of a design is a critical step in the EDA design flow. Once a model has been created, it is essential that it be known to function correctly for all cases of the signal patterns that could be applied to its inputs. A small-scale analog circuit can be verified by simulating and examining the circuit's signals to confirm that the circuit behaves correctly. Simulating the analog waveforms in a circuit requires intensive computations in order to determine waveform values accurately at small time steps. Analog circuit simulators automatically adjust the time step increment to bound the numerical convergence error between the simulated and actual behavior of the circuit. However, the accuracy provided by analog simulators might be expensive and unnecessary.

Modern EDA/CAD systems include logic simulators for verifying the correctness of an HDL-based design. Logic simulation provides an alternative to the intense computation required to simulate the analog waveforms of a circuit. A logic simulator uses models of the (Boolean) logical behavior of a circuit. In the simplest case, the waveforms in these models can assume only the values of 0 and 1; any intermediate values of the waveforms are ignored. This simplification has the effect of greatly reducing the computational effort and time required to verify the behavior of a circuit.

Logic simulation relies on a user-defined testbench that includes a specification of the signals that must stimulate a circuit, and a specification of how the circuits response to the stimuli will be observed.

3.1 SIMULATION WITH VERILOG

The voltages at the nodes of an integrated circuit may attain a continuum of values. A digital model of the signals in a circuit restricts the values that a signal may attain to a finite set of values. In the simplest case, a signal may attain a value of logical 0 or 1, which might correspond to 0 and 5 volts, respectively. It is convenient in simulation to expand the set of signal values to provide additional information. In Verilog, the set of signal values consists of 0, 1, x and z. The additional values, x and z, model situations where a signal's value is ambiguous (it could be 0 or 1 as a result of signal contention) and effectively an open circuit, respectively. Given a digital model, a simulator must determine the logic waveforms that result from the application of input signals to the circuit.

3.1.1 Event-Driven Simulation

Logic simulators provide a fast, efficient, visual representation of the behavior of a digital circuit by computing and displaying logic values corresponding to electrical waveforms in physical hardware. Logic simulation is usually done on event-driven simulators.

At any instant of time, most signals (gate outputs) in digital hardware are quiescent; i.e., they do not change value. Since only a few gates change at any time, logic simulators exploit this topological latency by using an "event-driven" scheme in which computational effort is expended only at those times at which one or more signals change their value. The local effect and finite duration of event-driven activity is the main reason why it is feasible to simulate the logical behavior of circuits containing several thousand logic gates. On the other hand, analog simulation, e.g., SPICE, is inherently memory-latent, because the arrays required to support simultaneously-interacting, differential-equation models of the interaction and evolution of voltages and currents in a circuit are usually sparse. But analog simulation is intensive in CPU demands, because these simulators generate accurate approximations to physical (i.e., analog) waveforms.

In logic simulation, an event is said to occur when a signal undergoes a change in value. A simulation of an HDL model of a digital circuit is said to be "event-driven" when the activity of the simulator is initiated only at those times when the signals in the model experience a change. Rather than re-compute the values of all signals at prescribed time steps, as in analog simulation, event-driven digital simulation computes new values of only those signals affected by the events that have already occurred, and only at those times when change can occur. For example, a change on one or both of the inputs to the **and** gate in Figure 3.1 might cause its output to change value (according to the input/output truth table for the **and** gate in Verilog's four-value logic system). Subsequently, this change causes the output of the **not** gate to change. The simulator monitors signals *A* and *B*, and when they change, it determines whether to schedule a change for signal *C*. When the scheduled change in signal *C* occurs, the simulator schedules an event for signal *D*, and so on. It is characteristic of event-driven simulation that events on the circuit's input signals propagate through the circuit, and possibly to its outputs.

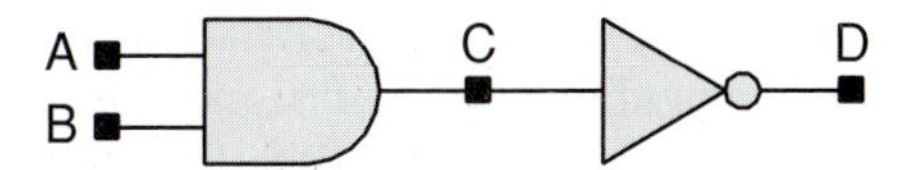

Figure 3.1 Circuit for event-driven simulation.

3.1.2 Simulation Data Structures

The organization of an event-driven logic simulator is shown in Figure 3.2. Simulators have internal data structures that represent the causal relationships between events on signals in the hardware model, so that a minimal amount of computation is performed to determine the value at the output of a logic gate. When a signal in the circuit changes value, an elaborate set of data structures enables the simulator to consider updating only those signals that could be affected by the event. The remaining signals are ignored because there is no need to re-compute their values. A logic simulator creates and manages an ordered list of "event-times", i.e., those times at which events have been scheduled to occur. An "event queue" (i.e., "signal-change" list, *sig_ch(t)*) is associated with each event time. It consists of the names and new values of those signals that are to change at that time.

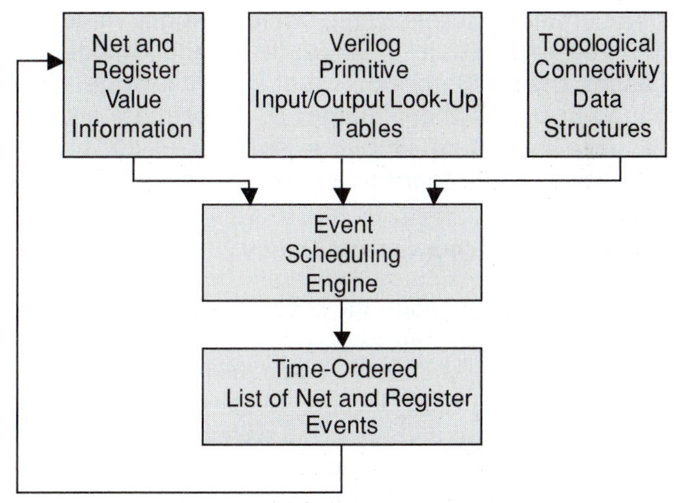

Figure 3.2 Organization of an event-driven simulator.

At the beginning of a simulation, a simulator automatically creates an initial event-time list at time $t_{sim}= 0$; the associated signal-change list assigns "x" to all variables (nets and registers) in the circuit. This logic value indicates that the value of the signal is initially unknown. When simulation begins, the simulation engine expands the event-list to include entries for value changes of the circuit's input signals (e.g., *A, B*) at appropriate times. It then considers the next event-time and updates the values of signals in the corresponding signal-change list. Then it updates the event-time list to include new entries for signals whose values were affected by the changes that were just made (e.g., *sig_ch(10)* is augmented by the event *C = 0*). Data structures are removed from a signal-change list as the associated variables are evaluated and possibly assigned their values. When *sig_ch(t)* becomes empty, the engine proceeds to the next event-time and repeats the process. When the event-time list is empty, the simulation ends.

Example 3.1 Figure 3.3 shows the output waveforms produced by an *and-invert* circuit having zero propagation delays when its input waveforms are as shown (a shaded area denotes the "x" value of a signal). The "event-time" list and its associated signal-change lists are depicted. We'll show how to describe the stimulus of *A* and *B* later. Note that the signal-change list contains entries for only those signals that actually change value. At a given event-time, the signal-change list has been partitioned to illustrate the causal relationships between the scheduled changes. For example, at time $t_{sim} = 20$, the change of signal *B* causes the change in signal *C*, which causes the change in signal *D*. The simulator suspends t_{sim} while it assigns value to *B*, detects the need to schedule *C*, schedules *C*, and changes *C*. When *C* changes, the simulator notes that *D* must change, schedules the change in *D*, and then changes *D*. All of these actions occur at the same instant of simulator time $t_{sim} = 20$, but they occur sequentially with regard to a single processor's CPU time.

3.1.3 Effect of Propagation Delay

A logic simulator can simulate the behavior of a circuit under the condition that the propagation delays are zero. This is done, for example, when verifying only the functionality of the circuit. When the propagation delays are zero, the propagation of events through the logic can be obscured. So "unit delay" simulations are used to reveal in more detail the evolution of events. They display the functionality correctly, but do not display the true timing of the circuit.

A realistic simulation takes into account the actual propagation delays of the gates. A logic simulator must manage the scheduling of events for all the signals in the circuit being simulated. Each primitive and continuous assignment statement may have a propagation delay associated with their action. The simulator uses these delay times to schedule events. Thus, Verilog primitives and continuous assignment statements can be thought of as "event scheduling rules" that specify when the value change of a signal should occur.

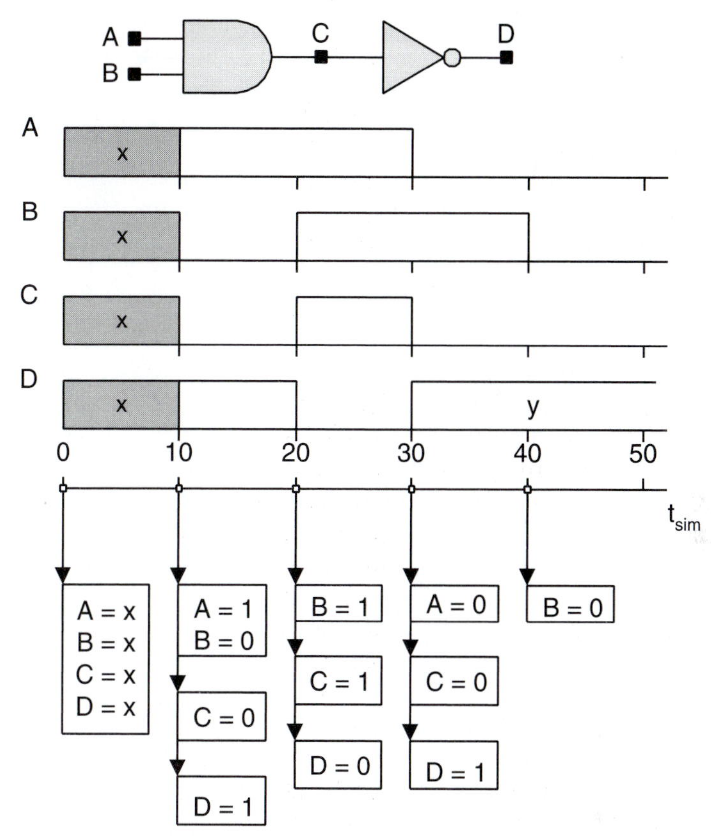

Figure 3.3 Representation of data structures for event-driven simulation.

Example 3.2 The primitive gates in the circuit in Figure 3.4 have the indicated propagation delays between the times when their input signals change and their output is affected by the change. The logic waveforms and data structures are also depicted. Notice that changes to *C* occur three time units after changes to *A* and/or *B*, and signal *D* changes two time units after *C*. **When propagation delays are modeled, signal changes do not propagate instantaneously through the circuit.**

3.1.4 INERTIAL DELAY AND EVENT DE-SCHEDULING

Signal transitions in a digital circuit correspond to the accumulation or dissipation of charge at a node in the circuit. Since charge cannot accumulate or dissipate instantly, the physical behavior of a signal transition is said to have "inertia." The propagation delay of primitive gates in Verilog obeys an "inertial"

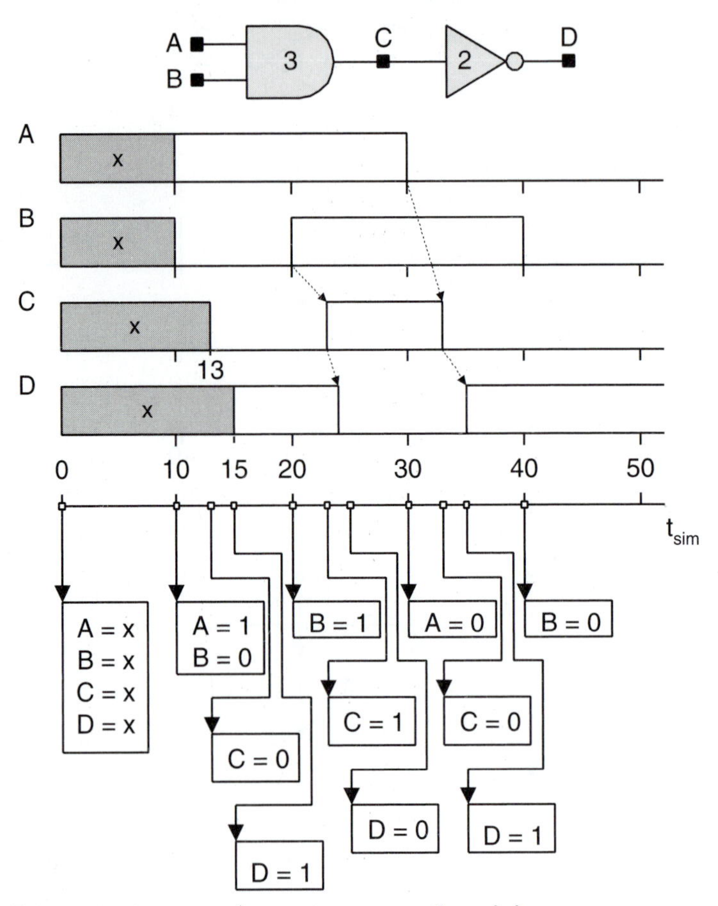

Figure 3.4 Data structures and event propagation delay.

delay model. This model attempts to account for the fact that charge must accumulate in the physical circuit before a voltage level can be established corresponding to a "0" or "1". If an input signal is applied and then removed before sufficient charge has accumulated, the output signal will not achieve a voltage level corresponding to a transition. For example, if all the inputs to an "and" gate are at value "1" for a long time before one of them is changed momentarily to "0", the output will not change to "0" unless the input is held to "0" for a long enough time. The amount of time that the input pulse must endure is the inertial delay of the gate. The width of a pulse must be at least as long as the inertial delay of the gate. The Verilog simulation engine must detect whether the duration of an input has been too short, and then de-schedule previously scheduled output. Verilog uses the propagation delay of a gate as the minimum width of an input pulse that could affect the output; i.e., the propagation delay is also used as the inertial delay. Inertial delay has the effect of suppressing input pulses whose duration is shorter than the inertial delay of the gate.

Example 3.3 The input to the inverter in Figure 3.5 changes at t_{sim} = 3. Because the inverter has a propagation delay of 2, the effect of this change is to cause the output to be *scheduled* to change at t_{sim} = 5. However, for pulsewidth Δ = 1, the input changes again at t_{sim} = 4. The simulator cannot anticipate this activity. The effect of the two successive changes is to create a narrow pulse at the input to the inverter. Because the pulse width is less than the propagation delay of the inverter, the simulator must de-schedule the previously scheduled output events corresponding to the leading edge of the narrow input pulse. De-scheduling is required because the simulator cannot anticipate the falling edge and must wait until it occurs. As a result, *y_out1* stays at "1." On the other hand, the pulse having Δ = 6 persists sufficiently long for the output to be affected.

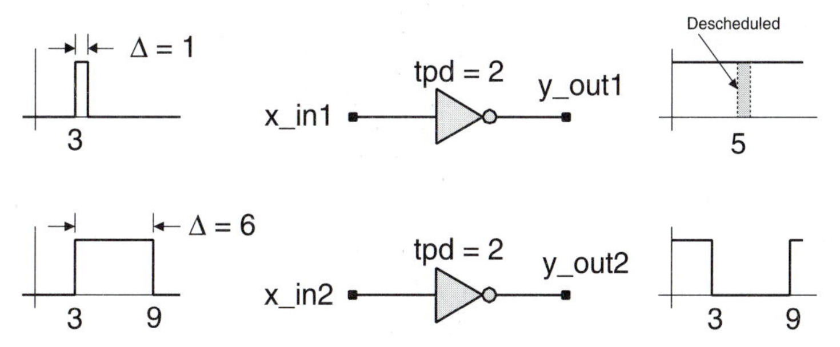

Figure 3.5 Event de-scheduling.

Figure 3.6 shows the data structures for a nand-inverter combination, and illustrates the scheduled activity that must be de-scheduled. The event changing *B*→*1(30)* schedules the events *C* → *0(33)* (i.e., *C* goes to 0 at t_{sim} = 33). The annotated data structure of each event shows the time of its causal antecedent (i.e., the time of the prior event that caused it). The causal antecedent of the event *C* → *0(33)* is the event *B* → *1(30)*. Note that in the absence of inertial delay effects, the event *C* → *0(33)* would cause the event *D* →*1(35)*, and the event *A* → *0 (33)* would cause the event *C* → *1(35)*. However, the inputs to the nand gate are applied over a two-unit interval (from t_{sim} = 30 to t_{sim} = 32), which is shorter than the inertial delay of the part. Consequently, the event *C* → *0(33)* must be de-scheduled. The events *D* → *1(35)* and *C* → *1(35)* are unscheduled and never enter the event queue.

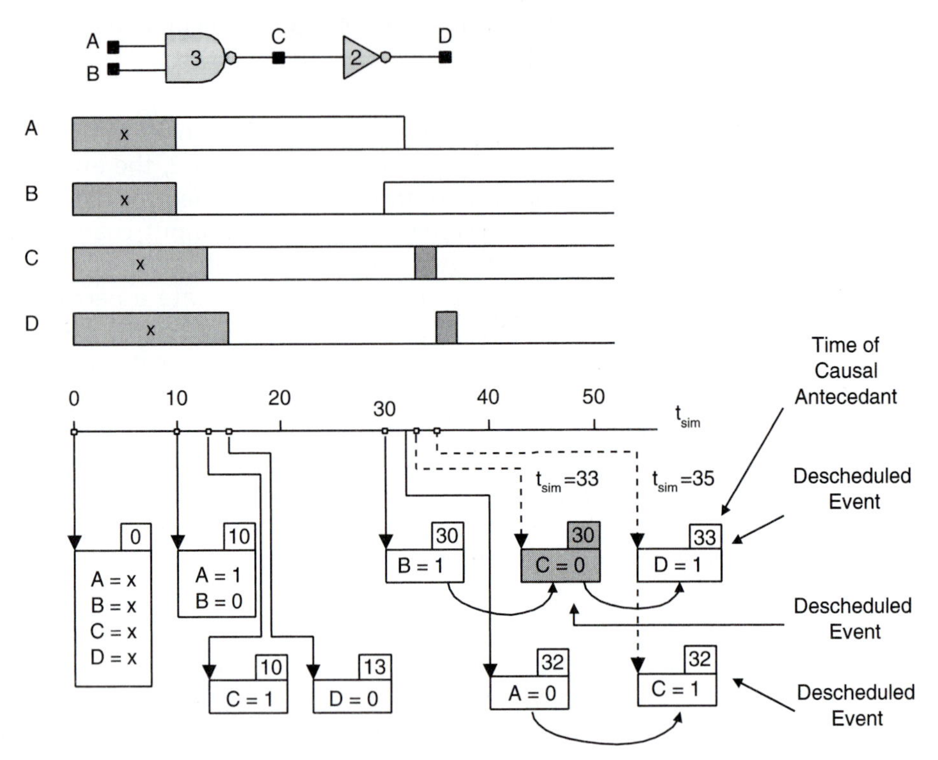

Figure 3.6 Example of event de-scheduling.

3.2 DESIGN UNIT TESTBENCH

The Verilog description of a circuit can be verified by simulating its behavior in a specially-written module called a "design unit testbench" (DUTB). The testbench contains an instantiation of the unit under test (UUT) (i.e., the model to be verified) and additional Verilog behaviors (implemented with procedural code) that (1) generate the input waveforms that are applied to the UUT, (2) monitor the response of the UUT under the influence of the stimulus, and (3) issue reports. This methodology separates the description of the design from the description of the tester of the design. (It is poor practice to mix the two together in the same module.) Figure 3.7 depicts the overall test methodology implemented with HDLs.

At a minimum, the DUTB should contain the definition of the test waveforms and a specification of which signals in the design are to be observed during the test. The stimulus generator is typically implemented as one or more Verilog behaviors, in which procedural statements assign value to register variables to create waveforms at the input ports of the UUT. Input waveforms,

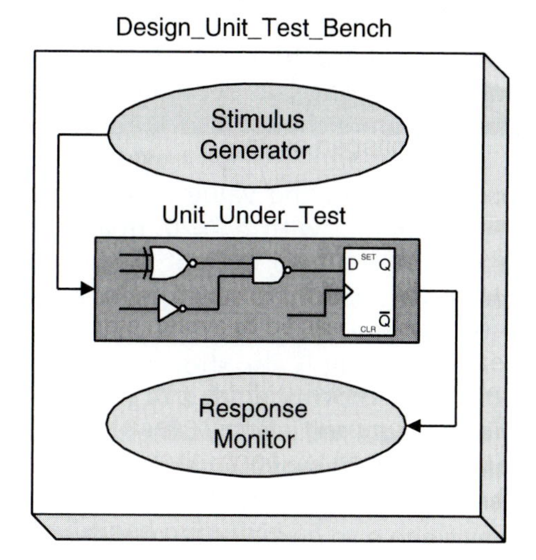

Figure 3.7 HDL-based simulation methodology.

internal waveforms and waveforms at the output ports are observed by a response monitor. Testbenches may also contain automated features that compare actual circuit responses to a set of expected waveforms and issue diagnostic messages. For example, the waveforms generated from a gate-level synthesis product can be compared to waveforms produced by its seed behavioral description. Testbenches for large circuits or complex behaviors should be carefully documented to indicate the steps of the testing regime. (Some simulation environments include interactive features and graphical input tools that enhance the flexibility and simplicity of the test scheme.)

Example 3.4 The simple, unit-delay, cross-coupled nand gates in Figure 3.8 form a hardware latch circuit with active-low *preset* and *clear,* and having the logic truth table shown below.

Preset	Clear	q_n	q_{n+1}	$qbar_n$	$qbar_{n+1}$
0	0	-	1	-	1
0	1	-	1	-	0
1	0	-	0	-	1
1	1	-	q_n	-	$qbar_n$

The first three lines of the table show that the *preset* and *clear* signals cause the outputs to enter a particular state, independently of the previous state. When the *preset* is 0, *q* must be 1; when the *clear* input is 0, *qbar* must be 1. When both inputs are 1, the value of the outputs depends on the value they had before the inputs were changed to 1. Thus, the fourth line indicates that, if both inputs assume a value of 1, then the state is retained, or latched, to the value it had before the transition occurred. For example, with $q_n = 0$, $qbar_n = 1$, a transition of (*preset, clear*) from (1,0) to (1,1) at the inputs would result in $q_{n+1} = 0$, $qbar_{n+1} = 1$; similarly, with $q_n = 1$, $qbar_n = 0$, a transition from (0,1) to (1,1) would result in $q_{n+1} = 1$, $qbar_{n+1} = 0$. The Verilog model of this circuit is also shown in Figure 3.8.

```
module Nand_Latch_1 (q, qbar, preset, clear);
  output q, qbar;
  input  preset, clear;

  nand   G1 (q, preset, bar),
         G2 (qbar, clear, q);
endmodule
```

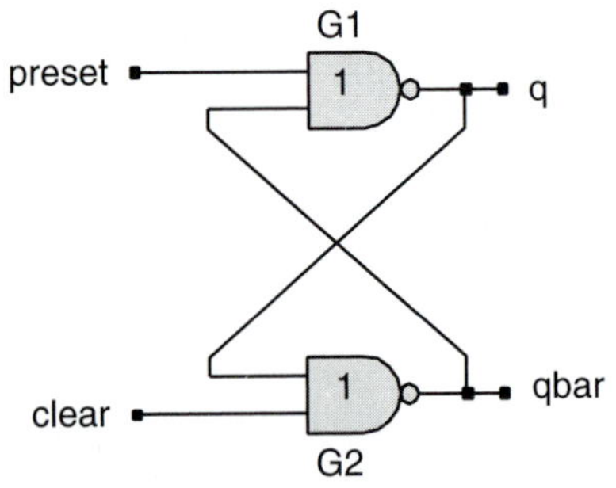

G1
preset
1
q
1
qbar
clear
G2

Figure 3.8 Cross-coupled nand latch and Verilog description.

Notice that a single statement in *Nand_Latch_1* instantiates a comma-separated list of two nand primitives. **Be aware that each primitive in the list has the same propagation delay.**

A scheme for simulating the *Nand_Latch_1* module is depicted in Figure 3.9. Simulating the latch involves (1) generating signals that can be attached to its input ports, and (2) observing the signals at its output ports under the influence of the inputs. The signals *preset, clear, q* and *qbar* are all local to *test_Nand_Latch_1*. We have chosen the names of these DUTB signals to be identical to those of their counterparts within the UUT ports (i.e., those of the *Nand_Latch_1* module), but it is not necessary to do so.

Example 3.5 The module shown below, *test_Nand_Latch_1*, is a simple DUTB for the *nand latch*. Notice that **reg** variables are declared in the DUTB to hold the values of the *preset* and *clear* signals that are the inputs to the latch. Recall that a variable of type **reg** can be assigned value only by a procedural statement. A **reg** variable retains its value until it is assigned a new value. A **wire** is defined for *q* and *qbar* local to the *test_Nand_Latch_1* module. These nets are driven by the output ports of the latch

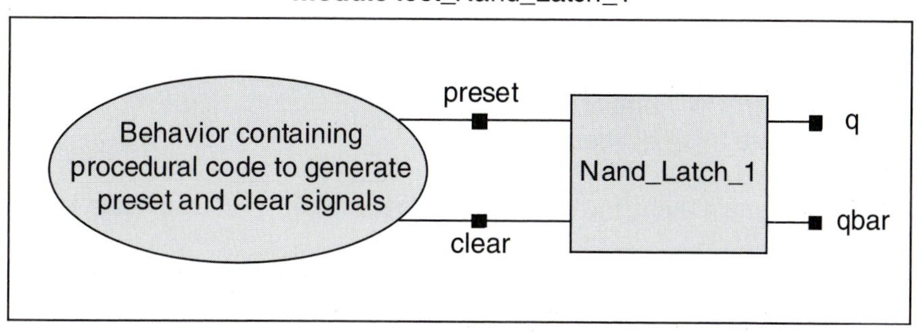

Figure 3.9 Design Unit Testbench scheme for testing the latch in Figure 3.8.

module. The UUT, *Nand_Latch_1*, is instantiated within the DUTB. The DUTB also contains two **initial** behaviors. The first **initial** behavior has an associated block of procedural code containing the Verilog built-in task **$monitor**(). The **$monitor** task is used to observe events on *preset*, *clear*, *q* and *qbar* and print a formatted line of output when any argument changes value. The quoted string within the task specifies that the output is to be presented in a binary format (%b); Verilog allows the output to be in binary (b or B), octal (o or O), decimal (d or D) or hexadecimal (h or H) format. Only one **$monitor** task is in effect at a time; subsequent calls to **$monitor** overwrite the format string. **$time** is a built-in Verilog function that returns the simulator time. The second **initial** behavior has an associated block of procedural code assigning value to the *preset* and *clear* signals. These statements are executed sequentially. The values "#10" have the effect of delaying the activity flow for 10 time steps of the simulator; this results in the waveforms shown in Figure 3.10. The **$stop** task stops the execution and puts the simulator into an interactive mode, allowing the user to control its activity. The simulator resumes activity where the operator enters "." at the keyboard, or invokes an appropriate command at the user interface of the tool. The testbench includes a third behavior that acts as a stopwatch to ensure termination of the simulation. The system task **$finish** returns control to the operating system when it executes in the sequential activity flow. Here, this behavior is actually redundant, because the **initial** behavior that creates the waveforms for *preset* and *clear* will automatically expire when it executes the last procedural statement. But in general, it is a recommended practice to include a "stopwatch." (Note: that the system function **$time** returns the time base variable of the simulation engine. It is automatically available, but cannot be manipulated.)

```
module test_Nand_Latch_1;                              // Design Unit Testbench
  reg       preset, clear;
  wire      q, qbar;

  Nand_Latch_1 M1 (q, qbar, preset, clear);    // Instantiate UUT

  initial                          // Create DUTB response monitor
    begin
      $monitor ($time, "preset = %b clear = %b q = %b
                            qbar = %b", preset, clear, q, qbar);
    end

  initial
    begin                          // Create DUTB stimulus generator
      #10   preset   = 0;    clear = 1;
      #10   preset   = 1;    $stop;     // Enter. to proceed
      #10   clear    = 0;
      #10   clear    = 1;
      #10   preset   = 0;
    end

  initial
    #60   $finish ;                // Stop watch
endmodule
```

When a Verilog simulator is invoked, it will parse the source code file(s), make substitution of text, and check for syntax errors. Source files may contain multiple modules, and the simulator must link hierarchically-organized modules together with unambiguous and consistent references to all identifiers.

Example 3.6 The results of executing the testbench source file of Example 3.5 on a commercial simulator are shown in Figure 3.10 for the case where the gates have zero delay. Observe that the signals have the value "x" initially until they are explicitly changed during the simulation. **The simulator automatically initializes all variables to have the value "x".** The testbench determines when the input signals change; the output signals change in response to the inputs driving the gates of the circuit. A signal in a physical circuit will have an initial value of "0" or "1."

Zero-delay simulation provides a quick check on the functionality of a circuit. However, it also masks the temporal propagation of events through the

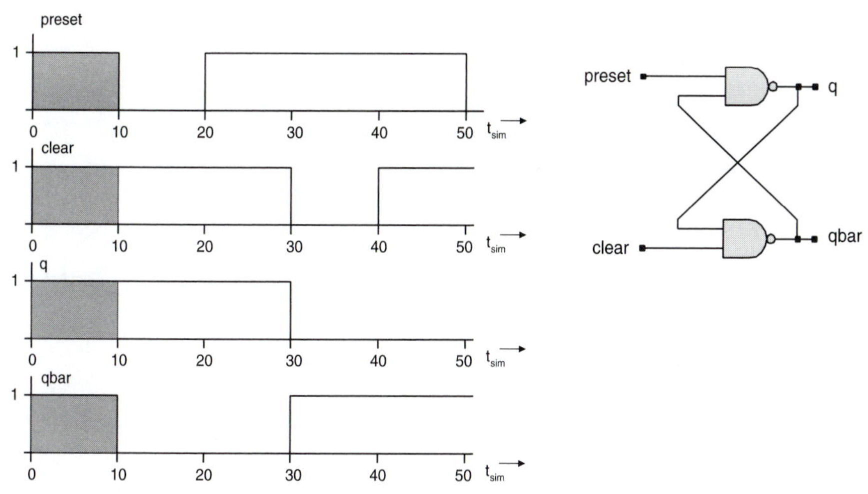

Figure 3.10 Nand latch simulation with zero delay.

circuit. Unit delay simulation reveals this additional detail. The results of simulating with unit delay on the gates are shown in Figure 3.11. **Notice that the effect of the unit delay at each primitive is apparent in the results of the simulation.** Changing *preset* and *clear* at time t = 10 causes *q* to change to *q* = 1 at time t = 11. This value is fed back though the companion nand gate, which changes value at t = 12. (This example also illustrates the ability of event-driven simulators to handle circuits that contain feedback. However, caution must be taken to avoid feedback loops that are interminable.)

A more elaborate verification strategy is required when a design is large. In this case, a formal design verification document should be written to identify the functional features that are to be tested. It should also provide a statement of how the test will be conducted, a general description of the test stimulus that will be applied by the DUTB (e.g., exhaustive vs. selective pattern generation), a discussion of any known limitations to the chosen test patterns (e.g., incomplete) and a description of the test patterns and expected results, with particular attention to a correlation between subsets of the patterns and functional features they test. In the case of the *Nand_Latch_1*, Figure 3.12 shows a simplified state transition graph that could be used as a guide in developing a testbench that fully exercises the model. The notation 0/0 at the vertex of the graph denotes the value of the pair *q/qbar*, and 1/1 at an edge of the graph indicates a value for the pair *preset/clear*. Transitions involving "x" values are not shown in this example.

Figure 3.13 depicts the general scheme for comparing the results of simulating the behavioral description with a gate-level description. First, the designer creates a behavioral description of the circuit and a companion testbench, including a stimulus generator and a response monitor. The behavioral description is debugged, then verified to confirm that its functionality is cor-

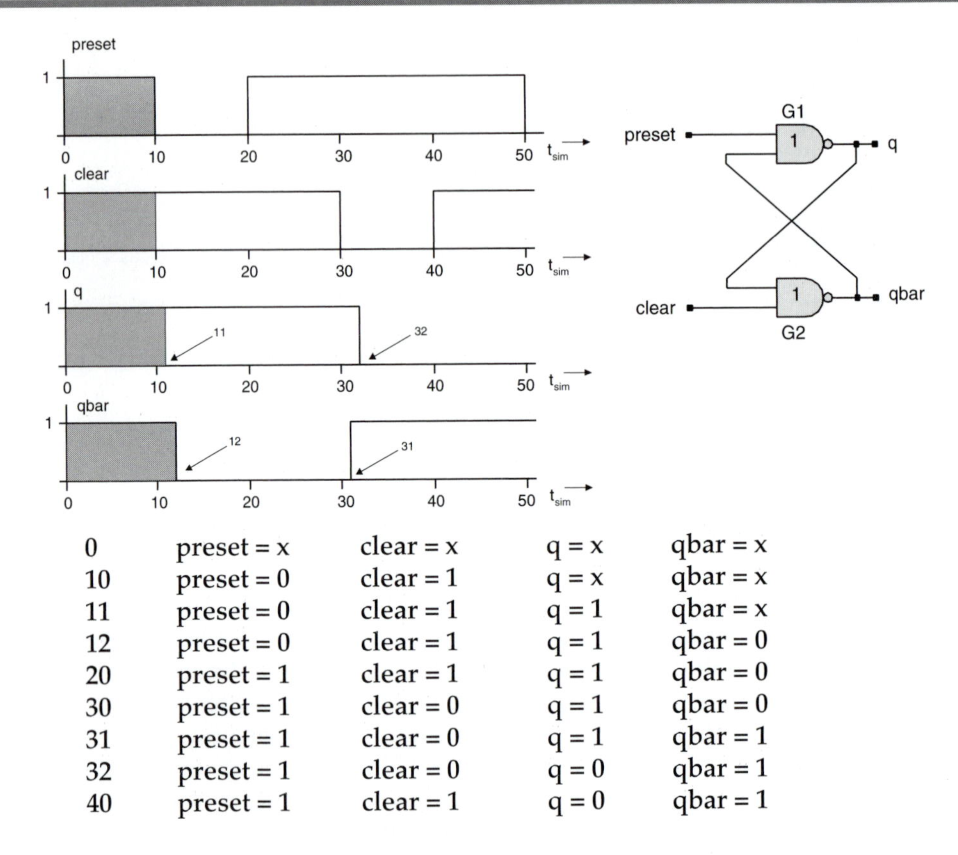

0	preset = x	clear = x	q = x	qbar = x
10	preset = 0	clear = 1	q = x	qbar = x
11	preset = 0	clear = 1	q = 1	qbar = x
12	preset = 0	clear = 1	q = 1	qbar = 0
20	preset = 1	clear = 1	q = 1	qbar = 0
30	preset = 1	clear = 0	q = 1	qbar = 0
31	preset = 1	clear = 0	q = 1	qbar = 1
32	preset = 1	clear = 0	q = 0	qbar = 1
40	preset = 1	clear = 1	q = 0	qbar = 1

Figure 3.11 Signal data produced by simulation of *test_Nand_Latch_1*.

rect. After the behavioral description has been synthesized, the gate-level description is instantiated in the testbench as a second UUT. The DUTB is modified to stimulate both circuits with identical patterns and compare their responses. Discrepancies between the two responses must be resolved. (Note: simulators automatically create an output text file named verilog.log.)

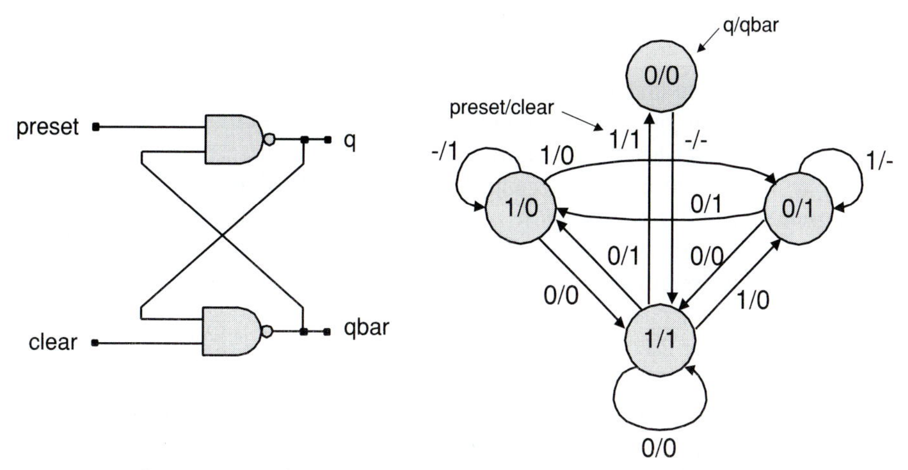

Figure 3.12 Schematic and state transition graph for a nand latch.

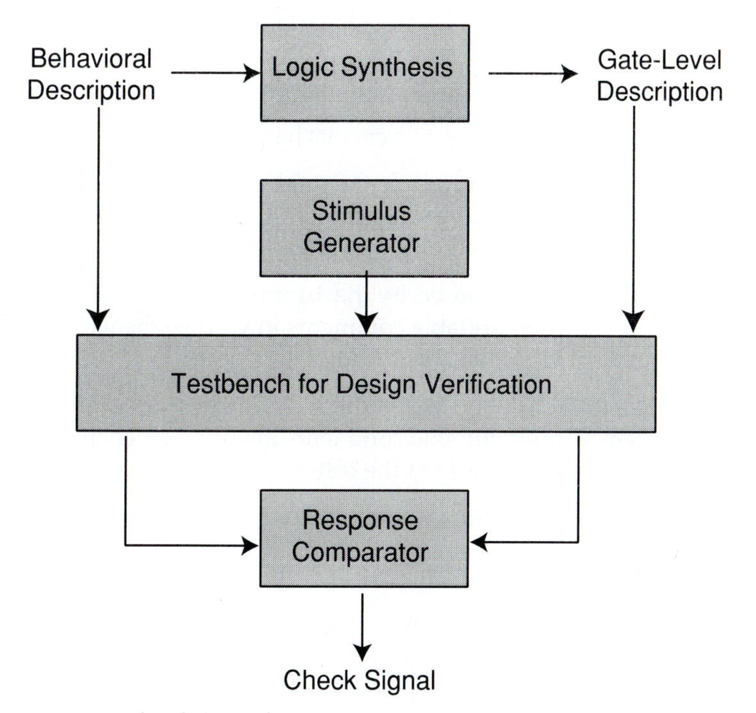

Figure 3.13 Test methodology for comparing gate-level and behavioral descriptions of a circuit.

3.3 SUMMARY

Verilog modules encapsulate the behavior of digital circuits. Verification of the model is typically accomplished by embedding the unit to be tested in a testbench that includes Verilog behaviors to create signals to stimulate the circuit, and Verilog tasks that monitor the signals during an event-driven simulation. Systematic verification gives a high level of confidence in the model. Verilog's built-in system tasks enable testbenches to easily display simulation results in text and graphical formats. Results can also be captured to files and displayed in hardcopy.

PROBLEMS

Note: Where appropriate in the following exercises requiring testbenches, consider inputs having all possible values, i.e., "0", "1", "x", and "z".

1. Answer True or False.
   - **a.** ___ The logic system in Verilog includes only the values 0, 1, and x.
   - **b.** ___ A net of type **wirereg** can store information.
   - **c.** ___ Verilog simulators initialize all variables to the value "x".
   - **d.** ___ Verilog can describe circuits having combinational feedback loops.
2. Write a testbench and verify the modules in Problems 3 and 4 in Chapter 2. Document your testbench with comments that create a truth table showing the expected outputs for given inputs.
3. Write a testbench and verify the 4:1 mux described in Problem 8 of Chapter 2. Insert suitable comments in your testbench to identify the objective of each phase of the test.
4. Write a testbench and verify the 16:1 mux described in Problem 9 of Chapter 2. Insert suitable comments in your testbench to identify the objective of each phase of the test.
5. Find the syntax error(s) in the code below:

```
assign y1 = a ^ b; y2 = a + b;
```

6. Write a testbench to verify the circuit in Problem 9 of Chapter 2. Insert suitable comments in your testbench to identify the objective and results of each phase of the test.
7. Write a testbench that will generate a clock signal having a period of 200 ns.
8. Write a testbench for the half adder given in Figure 2.2 and simulate its behavior for all possible combinations of the inputs.

9. Write a testbench and verify the *Add_half_2* module in Figure 2.4.
10. Add unit delays to the primitive gates in Figure 2.2. and write a testbench to verify the *Add_half_1* module.
11. What are the elements of a testbench?
12. Write a well-documented testbench and verify the *Add_full* module in Figure 2.5 for all possible combinations of input bits.
13. Write testbenches to hierarchically verify the functionality of *Add_rca_16* in Figure 2.23.
14. Write a testbench and verify the *AOI_4* module in Figure 2.6.
15. Write a testbench and verify the *AOI_4_unit* module in Figure 2.7.
16. Write a testbench and verify the *nand3gate* module in Figure 2.9.
17. Write a testbench and verify the *bit_or8_gate2* module in Figure 2.14
18. Write a testbench and verify the *and4_rtl* module in Figure 2.16.
19. Write a testbench and verify the *and4_rtl* modules in Figure 2.17 and Figure 2.18.
20. Write a testbench and verify the *Flip_flop* module in Figure 2.18. Be sure to consider how the model behaves when its inputs have "x" and "z" values.
21. What is the difference between an "initial" behavior and an "always" behavior?
22. Write a testbench that exhaustively verifies the behavior of the *Add_rca_4* module in Figure 2.23. Use this result to develop and execute a verification strategy for the *Add_rca_16* module in Figure 2.23 and the *Add_mix_16* module in Figure 2.24. Your testbench must implement an "exclusive" or of the gate level and behavioral results to create a message indicating the result of the simulation.
23. Write a testbench and verify the *compare_2_str, compare_2a, compare_2b* and *compare_2_algo* modules in Example 2.19.
24. Using primitive gates, write a Verilog structural description of the master-slave flip-flop shown in the figure below. Use the following module header:

```
module flop (data, clock, clear, q, qb);
...
endmodule
```

Assign the following delays to the indicated gates:

```
10 units:  nd1, nd2, nd4, nd5, nd6, nd8, iv1, iv2
9 units:   nd3, nd7
```

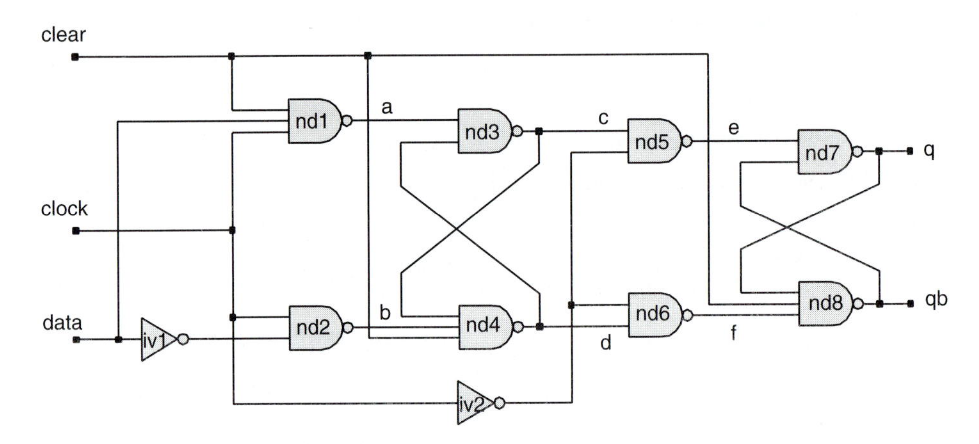

Write a testbench to verify the behavior of the flip-flop. Discuss what happens when the inputs don't have values assigned to them until sometime after the simulation begins. Also consider all possible values for each input. Using simulation, determine the rising and falling delays from the *clock* to *q*, and again from *clock* to *qbar*. Generate graphical and text output.

25. A transparent latch[2] has the property that its output (*q*) has the same value as the input while the gating signal (*G*) is asserted, and holds its state when the gating signal is de-asserted. Using Verilog primitives, write a Verilog structural description of the transparent latch shown below. Draw the state transition diagram for the latch and develop a well-documented testbench having stimulus that covers all transitions in the graph. Examine the response of the circuit when the inputs are not applied until sometime after the simulation begins.

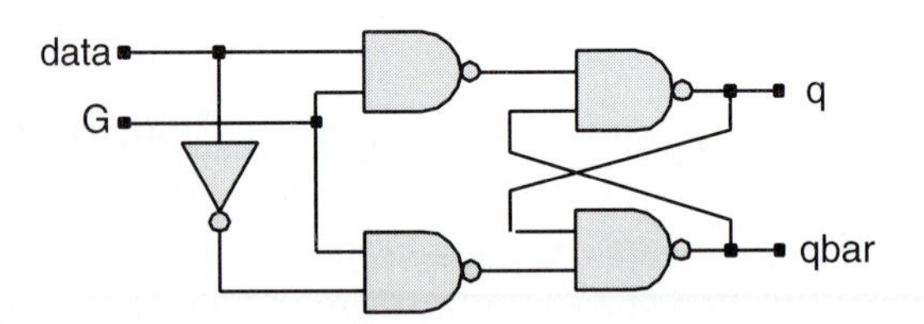

26. Write a testbench and verify the *encode4_to_2* module in Problem 19 of Chapter 2.

27. Write a testbench and verify the *decode8_to_3* module in Problem 20 of Chapter 2.

28. Write a description of a module that has two four-bit input words, *A* and *B*, and produces an eight-bit output word having the value *B[3], B[2], B[1,], B[0], A[3], A[2], A[1], A[0].*

CHAPTER 4

LOGIC SYSTEM, DATA TYPES, AND OPERATORS FOR MODELING IN VERILOG HDL

Verilog models determine the evolution of waveforms representing the logical signals in a circuit under simulation. The values of these signals are variables which change during the simulation, as inputs are applied to the model. The Verilog primitives used to compose structural models have a built-in description of the models' logical behavior. Verilog continuous assignments determine signal values according to the rules of the language's built-in operators. Verilog behaviors determine the values of storage variables according to the computational activity of the simulation. During simulation, logical values are created and scheduled for assignment to the variables within a description, and their evolution under event-driven simulation corresponds to the waveforms that would be observed under appropriate instrumentation, such as a logic analyzer.

4.1 VARIABLES

Verilog models use variables to represent the values of signals in a physical circuit in a digital format. There are only two kinds of variables: nets and registers. Nets are used solely to establish and represent structural connectivity in a circuit; register variables are abstractions of storage elements. The name (identifier) of a variable distinguishes it from other variables. The value of a variable can change during simulation. Nets and registers are both referred to loosely as "signals," even though a register variable in a given model may have no counterpart to a physical signal. Nets are always used to represent structural connectivity; registers are always used to store encoded information during simulation of a circuit model. Nets and registers may be scalar or vector quantities.

A storage variable (register) can be one of the following data types: **reg, integer, real, realtime** and **time**. A register variable is an abstraction of a hardware storage element,[1] but it need not correspond directly to physical storage elements in a circuit. For example, the physical realization of an HDL algorithm that represents combinational logic will not require memory in its hardware implementation, even though its description uses a data storage variable (see Example 2.17).

4.2 LOGIC VALUE SET

The format used to store the value of a variable in the host computer depends upon the variable's type. All values in Verilog are expressed in terms of a predefined four-valued logic set, using the symbols shown in Table 4.1.

Table 4.1 *Verilog's predefined four-value logic system.*

LOGIC VALUE	INTERPRETATION
0	Logical 0, or FALSE condition
1	Logical 1, or TRUE condition
X	Represents an unknown logic value
Z	Represents a high impedance condition

Logic simulators use the "x" value in the logic set to denote ambiguity when they cannot determine whether the value of a signal is "0" or "1". The "z" value represents a condition in which the driver of a net is disabled or disconnected. Figure 4.1 illustrates the response of the **and** primitive for input signals ranging over the four-valued logic system.

The logic values in Verilog may also have a "strength" associated with them. This optional feature models transistor switch-level descriptions of hardware

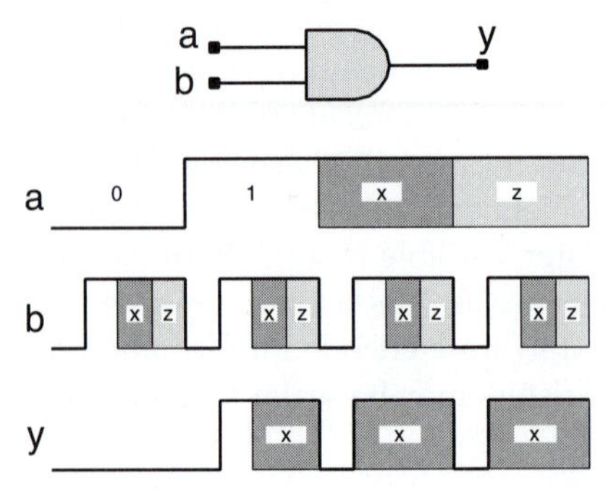

Figure 4.1 Four-valued waveforms for the and primitive.

implemented in pMOS and nMOS technology, where pulldown and pullup resistors are implemented as "always-on" transistors. Net strengths can be used in Verilog models to resolve contention when signals are simultaneously trying to pull a node up and down under simulation. For a discussion of net strengths, see Chapter 11.

The logic value of a net variable is associated with its name, or identifier, in the model of the circuit. Likewise, the value of a register variable is associated with an identifier to represent the data that would be stored in a physical storage element, independently of whether the implemented circuit actually requires memory.

How variables are assigned value depends upon their type. **During simulation a net variable may be driven by a primitive, module, continuous assignment, or force ... release procedural continuous assignment.** Our introductory examples have illustrated all but the procedural continuous assignment to a net. This advanced topic will be considered in Chapter 7.

4.3 DATA TYPES

Table 4.2 lists the built-in Verilog data types for connectivity and storage variables.

Table 4.2 *Predefined Verilog data types for nets and registers.*

Nets (Connectivity)		Registers (Storage)
wire	supply0	reg
tri	supply1	integer
wand	tri0	real
wor	tri1	time
triand	trireg	realtime
trior		

These predefined data types implement structural connectivity (nets) and data storage (**reg**), and support procedural computation (**integer**, **time**, **real** and **realtime**). All register variables in Verilog are static variables—their values exist throughout simulation—subject to assignments made under program flow. (Note that "net" is not a keyword of the Verilog language. It refers to a family of data types, all the members of which establish connectivity within the structure of a design.) Objects of a given type are assigned value explicitly (e.g., by behavioral statements) or implicitly (e.g., driven by a gate) within a Verilog module. A net may be assigned value explicitly only by a continuous assignment statement or a **force ... release** procedural continuous assignment (see

Chapter 7). It may also be assigned value implicitly as a consequence of being the output terminal of a primitive within a module, or by being connected to an output port of a module. **A register variable may be assigned value only within a behavior.**

4.3.1 Net Data Types

A net variable may be used only to create connectivity within a model, such as establishing a connection between the output of one primitive gate and the input of another. In this case, the value of the variable is determined by the gate and its inputs. The simulation environment automatically enforces this relationship so that, when the inputs to a primitive gate undergo change, the output of the gate may also change, depending upon the nature of the input change.

Example 4.1 In Figure 4.2, signal (net variable) *y1* connects the output of the **not** gate to the input of the **nand** gate. Under simulation, the value of *y1* is determined by the inverter and its input, *A*.

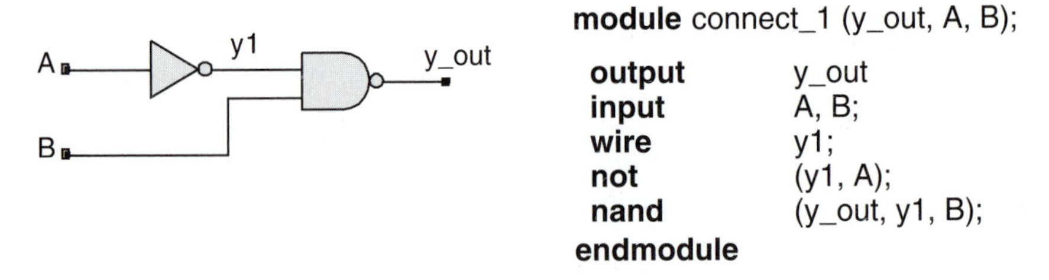
A
B
y1
y_out
module connect_1 (y_out, A, B);
output y_out
input A, B;
wire y1;
not (y1, A);
nand (y_out, y1, B);
endmodule

Figure 4.2 Connectivity established by nets.

All structural connections are made with nets. Verilog provides a variety of predefined nets to allow the semantics of a model to more closely conform to the hardware reality being represented. The semantics of the various types of nets are listed in Table 4.3 below.

With the exception of a **trireg** net, the logic value associated with a net is determined by the driver of the net. A net declared to be **supply0** or **supply1** has its logic value fixed for the duration of simulation. Nets declared to be **supply0** or **supply1** correspond to power and ground in the physical implementation. The **tri0** and **tri1** nets model resistive connections to ground and power, respectively. These nets have a strength of "pull". When no stronger driver is applied to the **tri0** net, its value is 0; when no stronger driver is applied to the **tri1** net, its value is 1. (For details, see Chapter 11.) Resistive terminations are used to simplify devices like programmable logic arrays (PLAs), where only the pull-down logic might be implemented, and a pull-up resistor precharges the output to 1.

Table 4.3 Semantics of Verilog nets.

wire	Establishes connectivity, with no logical behavior or functionality implied.
tri	Establishes connectivity, with no logical behavior or functionality implied. This type of net has the same functionality as wire, but is identified distinctively to indicate that it will be tri-stated in hardware.
wand	A net that has multiple drivers. It models the hardware implementation of a "wired and," e.g., open collector technology.
wor	A net that has multiple drivers. It models the hardware implementation of a "wired or," e.g., emitter coupled technology.
triand	A net that has multiple drivers. It models the hardware implementation of a "wired and." The net is tri-stated.
trior	A net that has multiple drivers. It models the hardware implementation of a "wired or." The net is tri-stated.
supply0	A global net that is connected to the circuit ground.
supply1	A global net that is connected to the power supply.
tri0.	A net that is connected to ground by a resistive pulldown connection.
tri1	A net that is connected to the power supply by a resistive pullup connection.
trireg	A net that models the charge stored on a physical net.

The **trireg** net models capacitive storage on a node when its drivers are three-stated (i.e., disconnected). The amount of charge stored on the node determines the strength of the net. Three charge strengths are possible: **small**, **medium** and **large**. The default strength is **medium**. When the driver(s) of the net turn off, the net retains the logic value it had before they turned off (i.e., 0, 1, or x). The only way for a **trireg** net to have a "z" value is for it to have been initialized to "z" or forced to "z" by a "**force**" procedural continuous assignment (see Chapter 7). See Chapter 11 for information about the charge decay characteristics of a **trireg** net.

The general syntax of a net declaration is given in Figure 4.3, with *net_type* being one of the types listed in Table 4.3, except for **trireg**. The default size of a net is scalar. (The default net type can be overridden by a compiler directive.) Note that only a **trireg** net may have charge strength in its declaration. A delay (i.e., delay3) may be associated with a net to model its *transport* delay, the time taken for a signal to propagate over the physical length of the net from its source to its destination. This will be discussed in Chapter 6. The *list_of_net_decl_assignments* is a list of one or more expressions (possibly empty) that associate an expression with a net variable. This implicitly associates a continuous assignment with the declared net, but does not use the keyword **assign** (see Example 2.12).

Verilog Syntax: Net Declaration

net_declaration ::=
 net_type [**vectored** | **scalared**] [range] [delay3] list_of_net_identifiers;
 | **trireg** [**vectored** | **scalared**][charge_strength]
  [range] [delay3] list_of_net_identifiers;
 | net_type [**vectored** | **scalared**] [drive_strength][range][delay3]
 list_of_net_decl_assignments;

Figure 4.3 Syntax for declaring a net.

Verilog nets are declared within a module, and the scope of an identifier that is the name of a net is local to the module in which the net is declared. Vector nets are declared using a range specification notation [msb_expr : lsb_expr], where *msb_expr* and *lsb_expr* are non-negative constant expressions determining the integer indices of the most and least significant bits of the word represented by the binary value of the net. The default size of a net is a scalar (i.e., a single-bit). Multiple objects of the same type and size can be declared simultaneously, as a comma-separated list.

The declaration of a net may include the optional keyword **vectored** to indicate that a net is to be treated as a single object. The individual bits of such a net may not be referenced; nor may they be the output of a primitive or target of a continuous assignment. The optional keyword **scalared** explicitly denotes a net whose bits can be referenced individually or as a part-select. By default, a net is **scalared** unless it is declared to be **vectored**.

A net is accessed by reference to its identifier. A reference to a vector net can include a bit-select (i.e., a single bit, or element) or part-select consisting of a range of contiguous bits enclosed by square brackets (e.g., [7:0]). An expression can be the index of a part-select. Note that, if a declared vector identifier has an ascending (descending) order from its LSB to its MSB, a referenced part-select of that identifier must have the same ascending (descending) order from its LSB to its MSB.

Example 4.2 Some examples of net declarations and references are given below:

```
wire [7:0] data_bus;          // A 8-bit wide vector wire.
                              // data_bus[7] is the MSB.
```

```
wire [0:3] control_bus;            // A 4-bit wide vector wire.
                                   // control_bus[0] is the MSB.

data_bus[5]                        // Access to bit 5 of data_bus.
data_bus[3:5]                      // Access to part-select of bits
                                   // 3 to 5 of data_bus.
data_bus[k+2]                      // Accesses an element of
                                   // data_bus,is dependent on
                                   // evaluation of the expression k+2.
wire scalared [31:0] Bus_A;        // A bus that will be expanded.
wire vectored [31:0] Bus_B;        // A bus that will not be expanded.

wire y1, z_5;                      // Declares y1 and z_5 as scalar
                                   // wires.
wand A, B, C;                      // Declares three "wired and" nets.
trireg [7:0] A;                    // Declares 8-bit charge storage
                                   // net.

wire A = B + C, D = E + F;         // Multiple Declarations and
                                   // continuous assignments.
```

Logic values are associated with nets. In simulation, a net automatically assumes the logic value of the output of the primitive or continuous assignment that drives it, or the logic value of the **input** or **inout** port to which it is connected. If a **wire** or **tri** type net is driven by multiple drivers (having the same strength), the resolved value of the net is determined by pairwise application of the rules in Table 4.4. For example, if both drivers have a value of "0" the value on the wire is "0". However, if one driver has a value of "0" and another driver has a value of "1" or a value of "x" the value on the wire is "x". If the other driver has a value of "z", i.e., disconnected, the value on the wire is "0".

Table 4.4 *Resolution of values on a wire with multiple drivers.*

wire / tri	0	1	x	z
0	0	x	x	0
1	x	1	x	1
x	x	x	x	x
z	0	1	x	z

Simulators issue a warning if a net has multiple drivers. This alerts the designer to what might not be the intended structure. If multiple drivers are intended, signal protocols may be required to preclude contention. Figure 4.4 illustrates how the contention between the values from multiple drivers on nets is resolved by a simulator. The simulator produces a value of "x" when a "0" and "1" are in contention and have the same strength. A physical circuit would likely be damaged by the short circuit.

Also, signals *out3, out4, out5,* and *out6* have been added to the schematic and to the display of waveforms to show the outputs of the individual drivers of *out1* and *out2*.

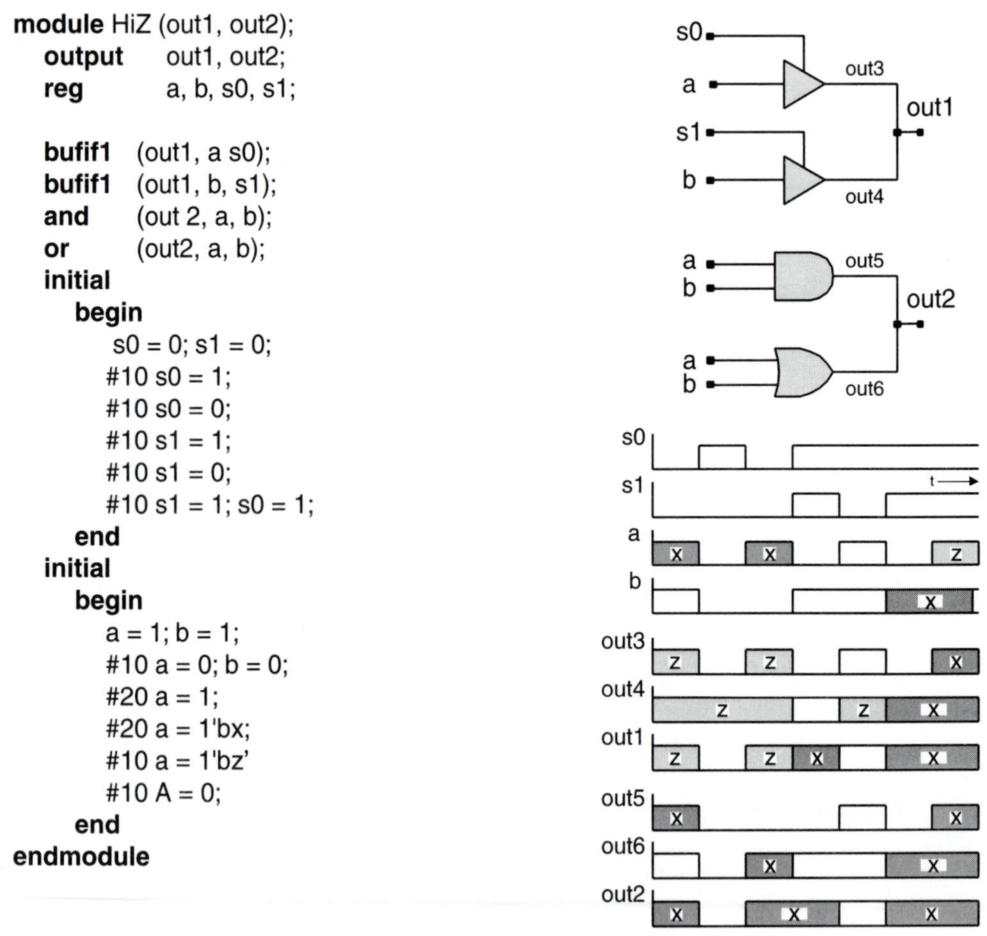

```
module HiZ (out1, out2);
  output    out1, out2;
  reg       a, b, s0, s1;

  bufif1  (out1, a s0);
  bufif1  (out1, b, s1);
  and     (out 2, a, b);
  or      (out2, a, b);
  initial
    begin
      s0 = 0; s1 = 0;
      #10 s0 = 1;
      #10 s0 = 0;
      #10 s1 = 1;
      #10 s1 = 0;
      #10 s1 = 1; s0 = 1;
    end
  initial
    begin
      a = 1; b = 1;
      #10 a = 0; b = 0;
      #20 a = 1;
      #20 a = 1'bx;
      #10 a = 1'bz'
      #10 A = 0;
    end
endmodule
```


Figure 4.4 Resolution of multiple drivers on a net.

Also note that *HiZ* has two "**initial**" behaviors specifying stimulus waveforms. A module may contain multiple behaviors.

4.3.2 Initial Value of a Net

At time $t_{sim} = 0$, nets driven by a primitive, module, or continuous assignment have a value determined by their drivers, which defaults to the "x" (ambiguous) logic value. The simulator assigns the default logic value "z" (high impedance) to all nets that are not driven. (The initial value for a register variable is also "x"). These initial assigned values remain until they are changed by subsequent events during simulation.

4.3.3 Wired Logic

Verilog provides built-in support for designs that must have multiple drivers on nets incorporating either *wired-or* logic or *wired-and* logic. In physical hardware, a *wired-and* construction is implemented in open-collector logic; a *wired-or* is implemented with emitter-coupled logic. The behavior of wired nets having multiple drivers of the same strength is automatically resolved by the simulator. If a *wired-and* net has at least one driver that is "0" the net is driven to "0". If a *wired-or* net has a driver that is "1", the net is driven to 1, independently of the other drivers. The Verilog model of a net having these characteristics must be implemented with either a declared **wand** or **wor** net. (Otherwise, the Verilog compiler will issue a error message or warning to the effect that a wire has multiple drivers.) A net of type **wire** cannot resolve the contention between multiple drivers having different values. If the multiple drivers of a **wand**, **wor**, **triand**, or **trior** net have different strengths, the rules in Chapter 11 must be used. Table 4.5 summarizes the logic for wired nets having multiple drivers with the same strength.

Table 4.5 *Truth tables for resolving multiple drivers on (a) wired-and and (b) wired-or nets.*

triand / wand	0	1	x	z
0	0	0	0	0
1	0	1	x	1
x	0	x	x	x
z	0	1	x	z

(a)

trior / wor	0	1	x	z
0	0	1	x	0
1	1	1	1	1
x	x	1	x	x
z	0	1	x	z

(b)

Example 4.3 The model in Figure 4.5 illustrates the distinction between a **wire** and **wand**. Note that the types of *w_mult* and *w_wand* are **wire** and **wand**, respectively. When the output of the nor gate and buffer conflict, *w_mult* carries a value of "x," but the net *w_wand* resolves the conflict and carries a value of "0."

```
module Good_Wand;
  reg     a, b, c;
  wire    w_nor, w_buf, w_mult;
  wand    w_wand;

  nor  (w_nor, a, b);
  buf  (w_buf, c);

  nor  (w_mult, a, b);
  buf  (w_mult, c);

  nor  (w_wand, a, b);
  buf  (w_wand, c);
endmodule
```

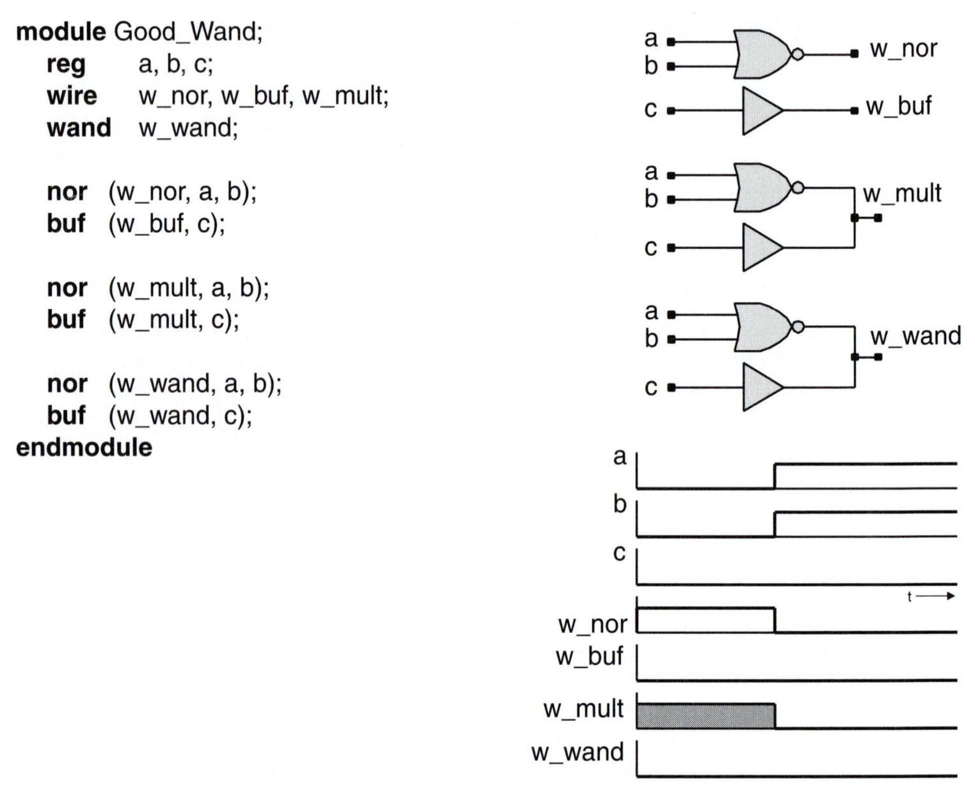

Figure 4.5 Examples of correct and incorrect wired-and logic.

4.3.4 UNDECLARED NETS—DEFAULT NET TYPE

Verilog does not require that all nets in a description be declared. Nets that are not explicitly declared will default implicitly to type **wire**. The *default_nettype* compiler directive may set the type of an undeclared net to be any one of the other values in Table 4.3, with the exception of **supply0** and **supply1**.

Example 4.4 The cross-coupled configuration of primitive nand gates in Figure 4.6 implements a latch. The declarations of wires *preset, clear, q* and *q_bar,* and the use of identifiers (instance names) with the primitive instantiations, are optional.

```
module   Nand_Latch (q, q_bar, preset, clear);
  output q, q_bar;
  input  preset, clear;
  wire   preset, clear, q, q_bar;

  nand #1
    G1 (q, preset, q_bar);
    G1 (q_bar, clear, q);
endmodule
```

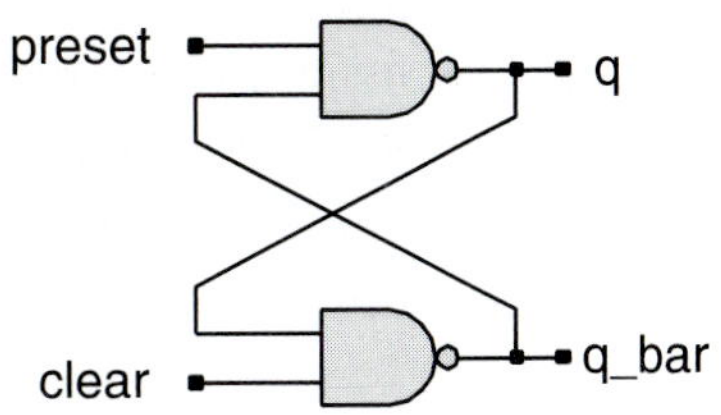

preset, q, clear, q_bar

Figure 4.6 Latch formed from cross-coupled nand gates.

4.3.5 Register Data Types

The register family of variable types has five members: **reg, integer, time, real** and **realtime**. The semantics of register data types are summarized in Table 4.6. A **reg** data type models the feature of hardware that allows a logic value to be stored in a flip-flop or latch. A **reg** is an abstraction of a hardware storage element, but it does not correspond directly to physical memory. The default size of a register variable is a single bit.

***Table 4.6** Semantics of register variable data types.*

Register type	Usage
reg	Stores a logic value
integer	Supports computation
time	Stores time as a 64-bit unsigned quantity
real	Stores values (e.g. delay) as real numbers
realtime	Stores time values as real numbers

Two examples of declarations of **reg** objects are given below:

```
reg A_Bit_Register;
reg [31:0] A_Word;
```

A register object may be assigned value only within a procedural statement, a user sequential primitive, task, or function. A **reg** object may never be the output of a primitive gate or the target of a continuous assignment. Example 4.5 illustrates the use of **reg** variables.

Example 4.5 A Verilog behavior consists of a sequence of procedural statements which execute sequentially. The code fragment below declares a behavior that is activated at $t_{sim} = 0$. It assigns value to *A* and *B*, which must be of type **reg**. The assignment to *A* is made before the assignment to *B*.

The indicated values of *A* and *B* are stored in memory immediately when the statements assigning value to *A* and *B* execute.

```
reg A, B;
initial
 begin
   A = 0;
   B = 1;
 end
```

4.3.6 Initial Value of a Register Variable

In simulation, all register variables have an initial value of "x". The value may be changed by subsequent execution of a procedural statement within a behavior.

4.3.7 Undeclared Register Variables

Verilog has no mechanism for undeclared register variables. Keep in mind that all references to an identifier that has not been declared are assumed to be references to a net of the default type. If you make such a reference within a behavior, the compiler will report an error saying, in effect, that you have illegally attempted to assign value to a net as though it were a register variable. To eliminate the error, you must declare the identifier to be a register variable or eliminate the illegal assignment.

4.3.8 Addressing Net and Register Variables

The most significant bit of a part-select of a register is always the leftmost array index; the least significant bit is the rightmost array index. A constant or variable expression can be the index of a part-select. If the index of a part-select is out of bounds, the value "x" is returned by a reference to the variable.

Example 4.6 If an eight-bit word *vect_word* has a stored *bit pattern* for the binary value of decimal 4, then *vect_word[2]* has a value of 1; *vect_word[3:0]* has a value of 4; *vect_word[5:1]* has value 2, i.e., *vect_word[7:0]* = 0000_0100_2, and *vect_word[5:1]* = 0_0010_2.

4.3.9 Passing Variables Through Ports

There are important rules to remember about declarations of nets and registers and their use in ports of modules and primitives. Figure 4.7 summarizes the rules that apply to nets and registers that are port objects in a Verilog module. For example, a register variable may not be declared to be an **inout** port.

	PORT MODE		
VARIABLE TYPE	input	output	inout
NET VARIABLE	YES	YES	YES
REGISTER VARIABLE	NO	YES	NO

Figure 4.7 Rules for nets and registers.

A variable that is declared as an **input** port of a module is implicitly a net variable within the scope of the module, but a variable that is declared to be an **output** port may be a net or register variable. A variable declared to be an **input** port of a module may not be declared to be a register variable. An **inout** port of a module may not be a register type. A register variable may not be placed in an **output** port of a primitive gate, and may not be the target (LHS) of a continuous assignment statement.

These rules are a consequence of the fact that the value of a register variable may be changed only within a behavior, task, or function within a module. The external environment may not alter a register variable, either explicitly (by an attempt to assign value to it outside of a procedural statement) or implicitly (by passing it into the module through a port).

4.3.10 Two-Dimensional Arrays (Memories)

Verilog does not have a distinct data type for true two-dimensional arrays with access to each array element. Instead, it has an extension of the declaration of a register variable to provide a "memory" (i.e., multiple addressable cells of the same word size). The syntax for a memory of register variables is shown in Figure 4.8. The same form of syntax applies to the other types of register variables.

Verilog Syntax: Declaration of a Memory of reg Variables

reg_declaration ::= **reg** [range] register_name {, register_name};

register_name ::= register_identifier | memory_identifier

[upper_limit_constant_expression : lower_limit_constant_expression]

Figure 4.8 Syntax for declaring memories.

Example 4.7 The code fragment in Figure 4.9 shows how the syntax for declaring a **reg** memory variable simplifies to the form: ***reg*** *word_size array_name memory_size* for an array of 1024 32-bit words:

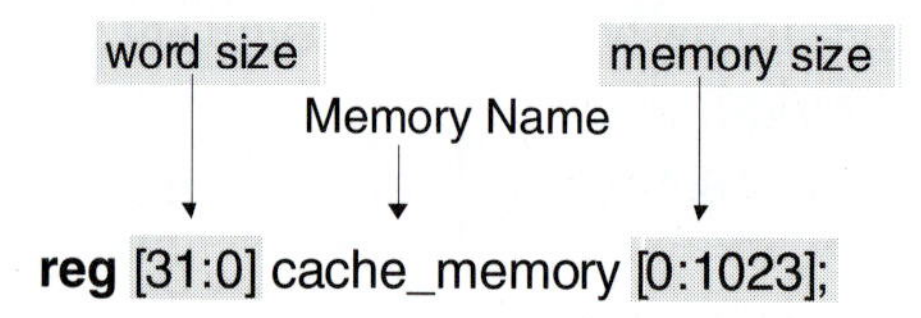

Figure 4.9 Declaration of a memory having 1024 32-bit words.

Declarations of memories must obey the rules in Figure 4.10:

RULES: Implementation of Memories
1. Bit-select and part-select are not valid with memories.
2. Reference may be made to only a word of memory.

Figure 4.10 Rules for implementing memories.

The individual bits within a memory cannot be addressed directly because a memory is not a true two-dimensional array. Instead, a word (element) of the memory must first be assigned to a one-dimensional array. Then the bit(s) of interest can be referenced from the array.

Example 4.8 Correct access to the data in a memory is illustrated below. Note that *cache_memory [17]* is first assigned to *a_word_register*, then a bit from *a_word_register* is loaded into *instr_register*.

```
...
reg [31:0] cache_memory [0:1023];
reg [31:0] a_word_register;
reg [7:0]  instr_register;
...
a_word_register = cache_memory [17];
...
// a loop
...
instr_register[k] = a_word_register [k+4];
...
```

The code fragment below illustrates an *improper* access to a memory to an array element.

```
cache_memory [0:4];
instr_register = a_word_register [4:15];
```

4.3.11 Data Type: integer

The **integer** data type supports numeric computation in procedural code. Integers are represented internally to the wordlength of the host machine (at least 32 bits). A negative integer is stored in 2's complement format. Verilog operators operate on integers with 2's complement arithmetic, with the MSB indicating the sign of the value. For example, the negative integer -4_{10} is stored as 1111_1111_1111_1111_1111_1111_1111_1100. When the size of a number is less than the length of the word used by the machine to store an integer, the number is padded with 0s to the left. Since integers have fixed wordsize, they may not be declared to have a range specification. Some examples of valid declarations of integers and arrays of integers are shown below:

a. **integer** A1, K, Size_of_Memory;

b. **integer** Array_of_Ints [1:100];

An integer will be interpreted as a signed value in two's complement form if it is assigned a value without a base specifier. If the assigned value has a specified base, the integer is interpreted as an unsigned value. For example,[2] if A is an integer, the result of A = –12/3 is –4; the result of A = 'd12/3 is 1431655761.

4.3.12 Data Type: real

Accurate modeling of delay values might require the use of **real** data objects. Real objects are stored in double precision, typically a 64-bit value. Real variables can be specified in decimal and exponential notation. An object of type **real** may not be connected to a port or terminal of a primitive. The language includes two system tasks that convert data types to permit real data transfer across a port boundary in a hierarchical structure: **$realtobits** and **$bitstoreal**. The Language Reference Manual[2] describes limitations on the use of operators with real operands. (Also see Appendix C.)

4.3.13 Data Type: time

The data type **time** supports time-related computations within procedural code in Verilog models. Time variables are stored as unsigned 64-bit quantities. A variable of type **time** may not be used in a module port; nor may it be an input or output of a primitive.

Example 4.9 Variables of type **time** are used to store simulation time values in the code below.

a. **time** T_samples [1:00];

b. **time** A_Time_Value;

4.3.14 Data Type: realtime

This data type stores time values in real number format.

4.3.15 Scope of a Variable

The scope of a variable is the module, task, function, or named procedural (begin ... end) block in which it is declared. In Figure 4.11, a net at the input port of *child_module* can be driven by a net or register in the enclosing *parent_module*, and a net or register at the output port of *child_module* can drive a net in *parent_module*.

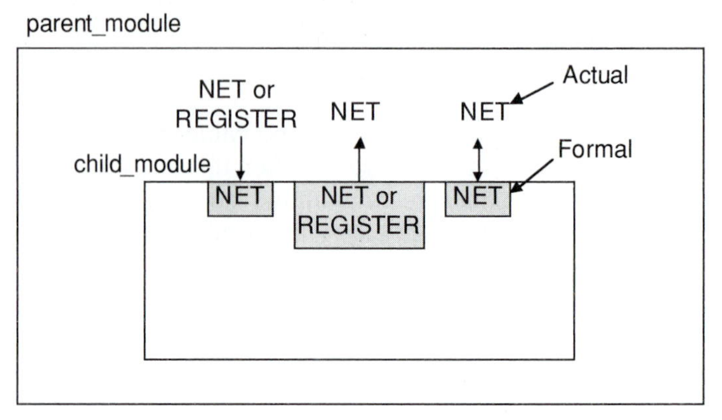

Figure 4.11 Allowed types for port objects.

4.3.16 Variable References and Hierarchical De-referencing

An identifier must be associated with only one object within a scope or domain within which the identifier has unique meaning (i.e., within a module, named procedural block, task, or function). Consequently, a variable may be referenced directly by its identifier within the scope in which it is declared. Verilog also supports hierarchical de-referencing by a variable's hierarchical path name. This feature lets testbenches monitor the activity of variables at any location within the hierarchical decomposition of the unit under test. If a variable is referenced but not declared locally, Verilog will search upwards through the boundaries of named blocks, tasks and functions to resolve the identifier, but it will not search beyond a module boundary.

Example 4.10 The testbench *test_Add_rca_4* in Figure 4.12 uses hierarchical de-referencing to monitor the internal carries of a 4-bit slice ripple carry adder and reveal the activity of the carry structure. For example, *M1.c_in3* is the carry out of the full adder named *G2* within the scope of *M1*.

```
module test_Add_rca_4();
  reg    [3:0]     a, b;
  reg              c_in;
  wire   [3:0]     sum;
  wire             c_out;

  initial
    begin
      $monitor ($time,,"c_out= %b c_in4= %b c_in3= %b
                 c_in2= %b c_in= %b",
                 c_out, M1.c_in4, M1.c_in3, M1.c_in2, c_in);
    end

  initial
    begin
    // stimulus patterns for data paths go here
    end

  Add_rca_4 M1 (sum, c_out, a, b, c_in);  // module declaration
endmodule

module Add_rca_4 (sum, c_out, a, b, c_in);
  output   [3:0]     sum;
  output             c_out;
  input    [3:0]     a, b;
  input              c_in;
  wire               c_out, c_in4, c_in3, _in2;

  Add_full G1 (sum[0], c_in2, a[0], b[0], c_in);
  Add_full G2 (sum[1], c_in3, a[1], b[1], c_in2);
  Add_full G3 (sum[2], c_in4, a[2], b[2], c_in3);
  Add_full G4 (sum[3], c_out, a[3], b[3], c_in4);
endmodule
```

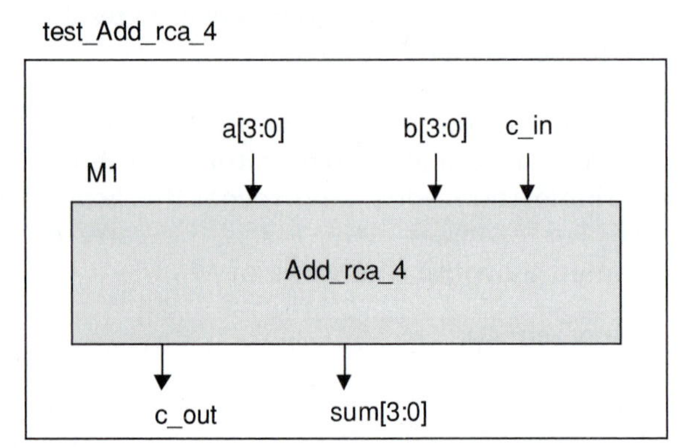

Figure 4.12 Testbench using hierarchical de-referencing to access internal variables.

4.4 STRINGS

Verilog does not have a distinct data type for strings. Instead, a string must be stored within a properly-sized register by a procedural assignment statement. A properly sized **reg** (array) has eight bits of storage for each character of the string that it is to hold.

Example 4.11 A declaration of a register variable, *string_holder*, that will accommodate a string having *num_char* characters is given below.

```
reg [8*num_char-1 : 0] string_holder;
```

The declaration implies that each of the *num_char* characters will be allocated eight-bits to encode the character. If the string "Hello World" is assigned to string_holder, it is necessary that num_char be at least 11 to ensure that a minimum of 88 bits are reserved. If an assignment to an array consists of less characters than the array will accommodate, zeros are automatically filled in the unused positions, beginning at the position of the MSB (i.e., the left-most position).

4.5 CONSTANTS

A constant in Verilog is declared with the keyword **parameter**, which declares and assigns value to the constant. The value of a constant may not be changed during simulation.

Example 4.12 The examples below illustrate declarations of parameters in Verilog.

a. **parameter** high_index = 31; // integer

b. **parameter** width = 32, depth = 1024; // integers

c. **parameter** byte_size = 8, byte_max = byte_size -1; // integer

d. **parameter** a_real_value = 6.22; // real

e. **parameter** av_delay = (min_delay + max_delay)/2; // real

f. **parameter** initial_state = 8'b1001_0110; // reg

Note that constant expressions may be used in the declaration of the value of a constant.

4.5.1 Direct Substitution of Parameters

Although the value of a constant may not be changed during simulation, Verilog provides a facility for changing the value of a constant in a module during compilation. One method for doing this overrides the value of the parameter on a module instance basis.

Example 4.13 In the descriptions below, the parameters declared within the *G2* instance of *modXnor* are overridden by including #(4,5) in the instantiation of the module. The values given in the instantiation replace the values of *size* and *delay* that were given in the declaration of *modXnor.* The replacement is made in the order which the parameters were originally declared.

```
module modXnor (y_out, a, b);
  parameter    size = 8, delay = 15;
  output       [size-1:0]     y_out;
  input        [size-1:0]     a, b;
  wire         [size-1:0]     #delay y_out = a ~^ b;   // bitwise xnor
endmodule

module Param;
  wire   [7:0] y1_out;
  wire   [3:0] y2_out;
  reg    [7:0] b1, c1;
  reg    [3:0] b2, c2;

  modXnor G1 (y1_out, b1, c1);               // Uses default parameters
  modXnor #(4, 5) G2 (y2_out, b2, c2);       // Overrides default parameters

endmodule
```

A common mistake is to confuse the syntax for overriding the parameters of a module with that for assigning delay to a primitive. The important rules to remember are that a module instantiation may not have delay associated with it, and a UDP declaration may not contain parameter declarations. Lastly, parameters may not be associated with a primitive gate.

specifies rules for determining the value that results from an assignment using the conditional operator.

RULES: CONDITIONAL OPERATOR

1. Logic value "z" is not allowed in the *conditional_expression* .
2. Zeros are automatically filled if the operands have different lengths.
3. If the *conditional_expression* is ambiguous, then both *true_expression* and *false_expression* are evaluated, and the result is calculated on a bitwise basis according to the truth table in Figure 4.14.

Figure 4.14 Rules for the conditional assignment operator.

Table 4.14 *Truth table for the conditional operator.*

? :	0	1	x
0	0	x	x
1	x	1	x
x	x	x	x

Note that the truth table in Table 4.14 assigns "0" ("1") to the expression when both *true_expr* and *false_expr* have the value "0" ("1"). In these cases, the evaluation does not depend on *conditional_expression*.

4.6.8 Concatenation Operator

The concatenation operator forms a single word from two or more operands. This operator is particularly useful in forming logical buses. The concatenation result follows the same order in which the words are given. The concatenation operator nests to any depth and repetition.

Example 4.27

a. If the operand A is the bit pattern 1011 and the operand B is the bit pattern 0001, then {A,B} is the bit pattern 1011_0001.

b. {4{a}} = {a, a, a, a)

c. {0011, {{01}, {10}}} = 0011_0110

One restriction that applies to the use of {} is that no operand may be an unsigned constant. Otherwise, the compiler would not be able to size the result.

4.7 EXPRESSIONS AND OPERANDS

Verilog expressions combine operands with the language's operators to produce a resultant value. A Verilog operand may be composed of nets, registers, constants, numbers, bit-select of a net, bit-select of a register, part-select of a net, part-select of a register, memory element, a function call, or concatenation of any of these. The result of an expression may be used to determine an assignment to a net or register variable, or to effect a choice among alternatives. The value of an expression is determined by performing the indicated operations on its operands. An expression may consist of a single identifier (operand) or some combination of operands and operators that conforms to the allowed syntax of the language. The evaluation of an expression always produces a value represented by one or more bits.

Example 4.28 Some examples of expressions are given below.

a. **assign** THIS_SIG = A_SIG ^ B_SIG;

b. **assign** y_out = (select) ? input_a : input_b;

c. @ (SET **or** RESET) **begin ... end**;

4.8 OPERATOR PRECEDENCE

Verilog evaluates expressions from left to right, and the evaluation of a Boolean expression is terminated as soon as the expression is determined to be true or false. The precedence of the binary Verilog operators determines the order of evaluation in an expression according to Table 4.15. The operators in the same row have the same precedence, and the rows are ordered from top to bottom.

Table 4.15 *Verilog operators and their precedence.*

Operator Precedence	Operator	Operator Symbol
Highest	Unary	+ - ! ~
	Multiplication, Division, Modulus	* / %
	Add, Subtract	+ -
	Shift	<< >>
	Relational	< <= > >=
		== != === !==
Lowest	Conditional	? :

The result produced by a compiler may not correspond to the intent expressed in an expression. As a precaution, parentheses should be used to eliminate ambiguity in expressions. (See Appendix C for additional information about operators.)

Example 4.29 The expressions below produce different results:

```
A < SIZE-1 && B!= C && INDEX != LASTONE
(A < SIZE-1) && (B!= C) && (INDEX != LASTONE)
```

Example 4.30 The speed at which simulators execute Verilog code depends on the style in which the code was written. A simple illustration of how two equivalent descriptions have different execution speeds is given by the following two statements:

a. **if** (inword == 0) A = B;   // simulates faster

b. **if** (!inword) A = B;

The first statement will execute faster than the second, because its logical test can terminate as soon as a non-zero bit is found. Suppose the simulator tests *inword* from its MSB to its LSB. If *inword* = 8'b0000_100x, the test will terminate after the fifth bit is tested. Not so with the second statement. For the same data word, the test of !*inword* is a test to verify that *inword* is false. That is, the test must verify that *inword* does not have the value of a nonzero integer. Finding the first 1 in the word does not reveal the presence of "x" in the last bit, which renders *inword* false.

4.9 SUMMARY

Verilog implements a fixed, four-value logic system. The Verilog HDL contains a rich assortment of language primitives and operators that represent typical combinational logic. The language's data types allow the user to model the connectivity and memory exhibited by hardware. The net family of variables are used solely to establish connectivity between the structural elements of a design. The register family of variables is used to provide memory within procedural code that implements an abstract model of behavior.

REFERENCES

1. Thomas, D. E. and Moorby, P., *The Verilog Hardware Description Language*, Third Edition, Klowerm Academic Publishers, Boston, 1996.
2. *IEEE Standard Hardware Description Language Based on the Verilog Hardware Description*, Language Reference Manual (LRM), IEEE 1364-1995, 1996, The Institute of Electrical and Electronic Engineers, Piscataway, New Jersey.

PROBLEMS

1. Answer True or False.
   - **a.** ___ A Verilog register variable can be the output of a predefined primitive.
   - **b.** ___ A Verilog net variable cannot be assigned value by a continuous assignment.
   - **c.** ___ A Verilog net variable can be assigned value by a conditional assignment
   - **d.** operator.
   - **e.** ___ A Verilog net may not be assigned value within a Verilog behavior.
   - **f.** ___ All module ports are scalars.
2. Write the following declarations:
   - **a.** A 32-bit vector net named *B* of type **wire**.
   - **b.** A 64-bit **reg** named *G*.
   - **c.** A scalar **reg** variable named *A*.
   - **d.** A cache_memory, an array consisting of 128 16-bit words.
   - **e.** A scalar **wire** named *y_out*.
   - **f.** An eight-word array of four 32-bit integers, *a, b, c, d*.
   - **g.** A 32-bit **wire** named data_bus.
   - **h.** Four 16-bit wires named *sig_a, sig_b, sig_c, sig_d*.
   - **i.** A four-bit register with a LSB index of 4.
3. Determine whether the following construction is correct:

```
wire Y;
nand (Y, A, B);
nor (Y, x1, x2);
buf (R, Y);
```

de-asserted. Sample waveforms illustrating this behavior are shown for a latch whose *enable* is asserted active high.

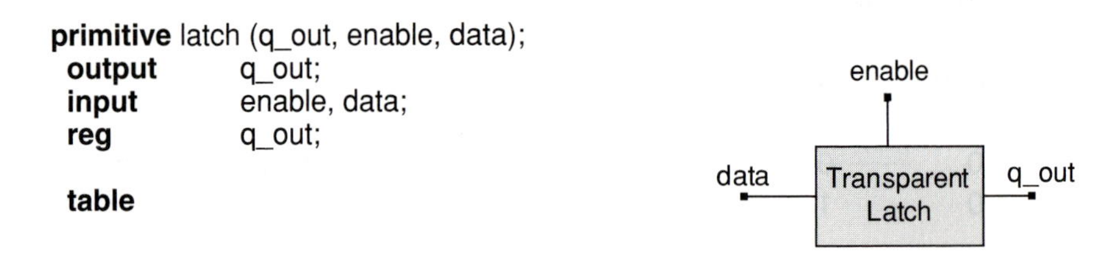


```
primitive latch (q_out, enable, data);
 output      q_out;
 input       enable, data;
 reg         q_out;

 table

//     en    data        state       q_out/next_state

       1     1     :     ?     :     1 ;
       1     0     :     ?     :     0 ;
       0     ?     :     ?     :     - ;

// Note: '-'  denotes "no change."
// The state is the residual value of q_out

 endtable
endprimitive
```

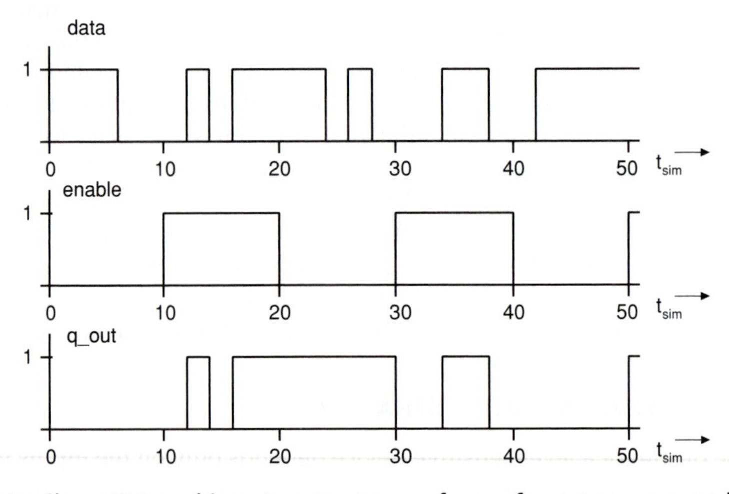

Figure 5.2 Verilog UDP and input-output waveforms for a transparent latch.

Example 5.4 An alternative realization of a transparent latch is given by *latch_rp*. This model deals with the possibility that, under simulation, the input *enable* might acquire an "x" value. The first three lines in the UDP table describe the behavior of the latch when the *enable* line is "0" or "1". The last two lines describe the behavior when *enable* is ambiguous. If *enable* changes to a value of "x", the model should not schedule an event for the output when the data and the state are matched. If the model did not explicitly account for this condition, a simulator would propagate an "x" value to the output when *enable* changes to an "x" value. Adding these two lines to the model creates a more useful and accurate model that has less pessimism under simulation.

```
primitive latch_rp (, enable, data);
   output      q_out;
   input       enable, data;
   reg         q_out;

   table
//    en data      state      q_out/next_state
      1   1    :    ?    :        1 ;
      1   0    :    ?    :        0 ;
      0   ?    :    ?    :        - ;

// Above entries do not deal with enable = x.
// Ignore event on enable when data = state:

      x   0    :    0    :        - ;
      x   1    :    1    :        - ;

// Note: The table entry "-" denotes no change.
   endtable
endprimitive
```

Figure 5.3 A UDP with reduced pessimism.

5.2.2 Edge-Sensitive Behavior

The transparent latch modeled in Example 5.3 exhibits level-sensitive behavior; i.e., the value scheduled for the output is determined only by the value, or level, of the "enable" and "data" inputs. A UDP describing edge-sensitive behavior will be activated by a particular transition of an input signal. UDPs can describe behavior that is sensitive to either the positive or negative edge (transition) of a "clock" signal, with built-in semantics for positive (**posedge**) and negative (**negedge**) signal transitions.

Example 5.5 The UDP *d_prim1* in Figure 5.4 describes the behavior of an edge-sensitive D-type flip-flop. The input signal *clock* synchronizes the transfer of *data* to *q_out*.

The table entry notation for a sequential UDP uses a pair of parentheses to enclose the defining logic values of a signal whose transition affects the

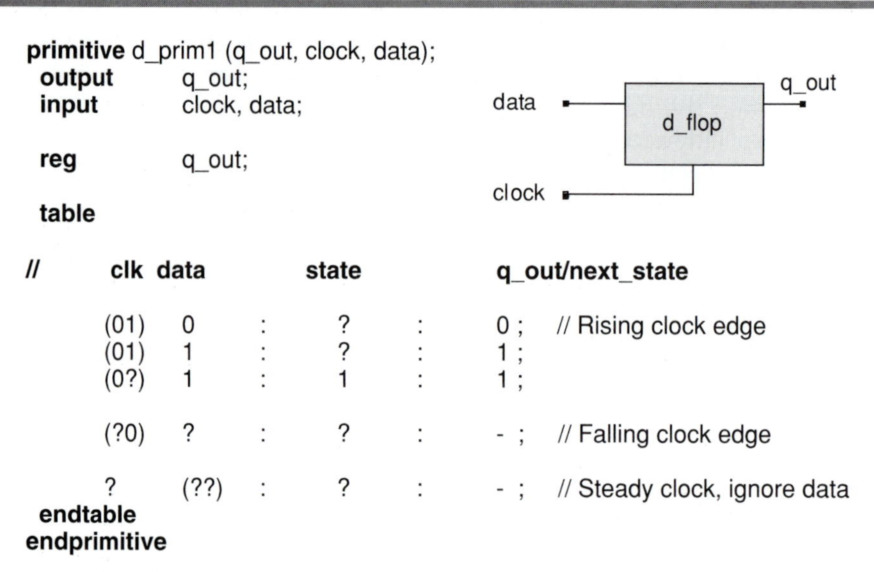

```
primitive d_prim1 (q_out, clock, data);
 output      q_out;
 input       clock, data;

 reg         q_out;

 table

//     clk  data        state         q_out/next_state

       (01)   0    :      ?     :     0 ;   // Rising clock edge
       (01)   1    :      ?     :     1 ;
       (0?)   1    :      1     :     1 ;

       (?0)   ?    :      ?     :     - ;   // Falling clock edge

        ?    (??)  :      ?     :     - ;   // Steady clock, ignore data
 endtable
endprimitive
```


Figure 5.4 UDP for a D-type flip-flop.

output (i.e., the synchronizing input signal). In the table in Figure 5.4, the (01) entry in the column for *clock* denotes a low-to-high transition of the signal *clock* (i.e., a value change). Note also that the row corresponding to the entry of (?0) for *clock* actually denotes 27 input possibilities and replaces 27 rows of entries. For example, (?0) represents (00), (10) and (x0). Each of these is combined with three possibilities for the data; each of the resulting nine possibilities is combined with three possibilities for the state. In effect, this row explicitly specifies that the output should not change in any of these situations. Since the model represents the physical behavior of a "rising edge" sensitive behavior, the output should not change on a falling edge. Were this row omitted, the model would propagate an "x" value under simulation. It is desirable that the UDP table be as complete and unambiguous as possible.

5.2.3 Mixed Behavior (Level- and Edge-Sensitive)

The model of a sequential UDP may contain a mixture of level-sensitive and edge-sensitive behavior to model synchronous behavior with asynchronous set and reset conditions. If events are scheduled by both level-sensitive and edge-sensitive table entries, the level-sensitive behavior dominates, just as in the case of a physical flip-flop having asynchronous override of the clocked behavior.

Example 5.6 A J-K flip-flop having asynchronous preset and clear with edge-sensitive sequential behavior is described in Figure 5.5. The *preset* and *clear* inputs are "active-low", and the output is sensitive to the rising edge of the clock. The signal *clock* synchronizes the changes of *q_out*, depending upon the values of *j* and *k*. Depending upon the values of the *preset* and *clear* signals when the clock edge occurs, *q_out* does not change (*j* = "0", *k* = "0"), *q_out* gets a value of "0", (*j* = "0", *k* = "1"), *q_out* gets a value of "1", (*j* = "1", *k* = "0"), or *q_out* is toggled (*j* = "1", *k* = "1"). In this table, the dash symbol, "-", denotes no change, and "*" denotes all transitions of the input.

```
primitive jk_prim (q_out, clock, j, k, preset, clear);
  output      q_out;
  input       clock, j, k, preset, clear;
  reg         q_out;

  table

  //      clk    j    k    pre  clr      state      q_out/next_state
     // Preset Logic
          ?      ?    ?    0    1    :   ?      :   1 ;
          ?      ?    ?    *    1    :   1      :   1 ;

     // Clear Logic
          ?      ?    ?    1    0    :   ?      :   0 ;
          ?      ?    ?    1    *    :   0      :   0 ;
  // Normal Clocking
  //      clk    j    k    pre  clr      state      q_out/next_state

          r      0    0    0    0    :   0      :   1 ;
          r      0    0    1    1    :   ?      :   - ;
          r      0    1    1    1    :   ?      :   0 ;
          r      1    0    1    1    :   ?      :   1 ;
          r      1    1    1    1    :   0      :   1 ;
          r      1    1    1    1    :   1      :   0 ;
          f      ?    ?    ?    ?    :   ?      :   - ;

  // j and k cases
  //      clk    j    k    pre  clr      state      q_out/next_state
          b      *    ?    ?    ?    :   ?      :   - ;
          b      ?    *    ?    ?    :   ?      :   - ;
```

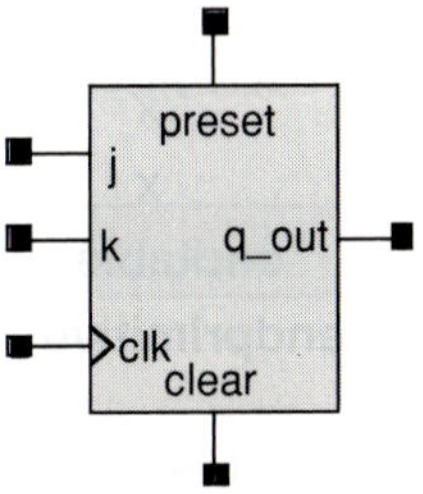

6.4 TIME SCALES FOR SIMULATION

The physical units of the time delays in a Verilog description are determined by a compiler directive, according to the format:

Verilog Syntax: Times Scale Directive

*timescale*_compiler_directive ::= `timescale <time_unit> / <time_precision>

Figure 6.3 Syntax for timescale compiler directives.

The *time_unit* and *time_precision* are integer values taken from the set {*1, 10, 100*}, followed by an abbreviation for a physical unit of measure, according to Table 6.2.

Table 6.2 *Verilog time units for timescale directives.*

Unit of Measured Time	Abbreviation
seconds	s
milliseconds	ms
microseconds	us
nanoseconds	ns
picoseconds	ps
femtoseconds	fs

The *time_unit* specifies the physical unit of measure, or time scale, of a numerical time delay value; the *time_precision* specifies the resolution of the time scale; i.e., the minimum step size of the scale during simulation. (Note: The *time_precision* must not exceed the *time_unit*.) It determines how numerical values are to be rounded ***before*** being used in a simulator. For example, a compiler directive of `*time_scale 10 ns / 1 ns* sets the resolution of the time scale to a nanosecond, and specifies that the numerical values in the source code are to be interpreted in units of tens of nanoseconds. For example, with a timescale directive of *10 ns / 1 ns*, the delay specification *#20* corresponds to 200 ns.

Example 6.2 Table 6.3 shows additional examples illustrating how rounding of value occurs before simulation. In each case, the delay value used in simulation is obtained by multiplying the delay specification by the *time_unit*, and then rounding according to the value of the *time_precision* relative to the *time_unit*. For example, for a timescale directive of 10 ns/1 ns, a delay specification of 4.629 is interpreted as 46.29 ns, and rounded to the specified precision of 1 ns, which gives a value of 46 ns (time steps) after rounding. The result after conversion is in the units specified by *time_precision*.

Table 6.3 *Examples of timescale directives.*

Timescale Directive unit / precision	Delay Specification	Simulator Time Step	Delay Value Used in Simulation	Conversion
1 ns / ns	#4	1 ns	4 ns	Multiply 4 by 1
1 ns / 100 ps	#4	100 ps	40 ps	Multiply 4 by 10
10 ns / 100ps	#4	100 ps	400 ps	Multiply 4 by 10 by 10
10 ns / ns	#4	1 ns	40 ns	Multiply 4 by 10
100 ns / ns	#4	1 ns	400 ns	Multiply 4 by 100
10 ns / 100ps	#4.629	100 ps	46.3 ns	Round 46.29 to 46.3
10 ns / 1ns	#4.629	1 ns	46 ns	Round 46.29 to 46
10 ns / 10ns	#4.629	10 ns	5 ns	Round 4.629 to 5

The timescale directive lets a simulator scale the time base of individual models to a common precision determined by the smallest value (finest resolution) specified by a compiler directive in the description being simulated. A timescale directive can be placed before any module in a source description, and applies to the modules that follow it. If multiple directives are given, the scope of a directive's effect extends to the location of a subsequent directive.

Example 6.3 The first timescale directive in the fragment of code below applies to *The_first_module*; the second applies to *The_second_module* and to any code that follows in the source file, up to the location of another directive. The numerical values in *The_first_module* are in units of ns; those in the second are in units of tens of ns. The units of the data reported by **$realtime**, with a %f format specifier, are tens of ns, because the **$monitor** task is located with a module, which has a time-unit of tens of ns. Therefore, the output is to be interpreted in units of tens of ns. The first assignment of value to *x1* occurs at t = 50 ns, and the transition of the output, *y*, occurs at 53.23 ns. The final transition of *y* occurs at t_{sim} = 203.23 ns. This time value is obtained by adding the rising delay of the output, 3.23 ns (3.225 ns, rounded to the precision of tens of ps), to the time of the input transition, 200 ns. Notice that the units of time as reported by $time have been rounded to the nearest tens of ns.

```
`timescale 1 ns / 10 ps
module The_first_module (y, x1, x2);
  input     x1, x2;
  output    y;

  nand #(3.213: 3.225: 3.643, 4.112: 4.237: 4.413) (y, x1, x2);
endmodule

`timescale 10 ns / 10 ns

module The_second_module ();
  reg     x1, x2;
  wire    y;

  The_first_module M1 (y, x1, x2);
  initial begin
    $timeformat (-10,1," x100ps",10);
    $monitor ($time,,"%f x1= %b x2= %b y= %b ",$realtime, x1, x2, y);
  end
  initial #30 $finish;
  initial begin
    #5 x1 = 0; x2 = 0;
    #5 x2 = 1;
    #5 x1 = 1;
    #5 x2 = 0;
    #5 $stop;
  end
endmodule
```

$t	$real_t	x1	x2	y
0	0.000000	x1= x	x2= x	y= x
5	5.000000	x1= 0	x2= 0	y= x
5	5.323000	x1= 0	x2= 0	y= 1
10	10.000000	x1= 0	x2= 1	y= 1
15	15.000000	x1= 1	x2= 1	y= 1
15	15.424000	x1= 1	x2= 1	y= 0
20	20.000000	x1= 1	x2= 0	y= 0
20	20.323000	x1= 1	x2= 0	y= 1

Notice that the units of time as reported by **$time** in Example 6.3 have been rounded to the nearest ns. The resolution of *The_first_module*, being the smallest, determines the precision of the simulation. The delay values of *The_first_module* will be rounded to (3213:3225:3643, 4112:4237:4413); the assignment statements in the second behavior execute at intervals of 5000.

Replacing the **`timescale** directive in *The_first_module* by **`timescale** 1 ns / 1 ps gives the following output:

```
$t      $real_t         x1      x2      y
---------------------------------------
0       0.000000        x1= x   x2= x   y= x
5       5.000000        x1= 0   x2= 0   y= x
5       5.322500        x1= 0   x2= 0   y= 1
10      10.000000       x1= 0   x2= 1   y= 1
15      15.000000       x1= 1   x2= 1   y= 1
15      15.423700       x1= 1   x2= 1   y= 0
20      20.000000       x1= 1   x2= 0   y= 0
20      20.322500       x1= 1   x2= 0   y= 1
```

The precision of 1 ps is apparent in the listed data. The first assignment to *x1* occurs at 50 ns, and the transition of y occurs at 53.225 ns, accurate to 1 ps.

The **$timeformat** system task (see Appendix E) provides an alternative format of the output generated by **$time**, which has a default format of the time units of the module in which it is invoked.

Example 6.4 In both of the preceding examples, the format of the output generated by **$time** is its default (%t), with rounding to tens of ns. The code below includes a **$timeformat** task with arguments that specify that the output is to be listed in units of ps, although *The_second_module* has a time_unit of 10 ns.

```
`timescale 1 ns / 1 ps
module The_first_module (y, x1, x2);
  input        x1, x2;
  output       y;

  nand #(3.213: 3.225: 3.643, 4.112: 4.237: 4.413) (y, x1, x2);
endmodule

`timescale 10 ns / 10 ns

module The_second_module ();
  reg  x1, x2;
  wire y;
```

limits on its outputs. Here, the optional error limit specifies that pulses having a duration between the error limit and the rejection limit will propagate as "x" values, instead of being suppressed.

```
specify
   (clk => q)                    = 10;
   (data => q)                   = 7;
   (clr, preset *> q)            = 3;

   specparam
      PATHPULSE$clk$q            = (3, 8);
      PATHPULSE$clr$q            = (0, 5);
      PATHPULSE$                 = 4;

endspecify
```

As a result of this specification, the path (*clk* => *q*) has a rejection limit of three and an error limit of eight. The path (*clr* => *q*) has a rejection limit of 0 and an error limit of five. That is, the explicit specifications have precedence for these paths. Paths not having an explicit declaration of a rejection limit and an error limit will have a rejection limit defined by the last non-specific **PATHPULSE$.** The path from *data* to *q* will therefore have a rejection and error limit of four.

6.9 SUMMARY

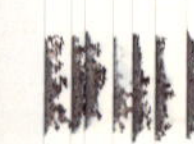

Verilog has built-in constructs for modeling the propagation delay of gates and the transport delay of nets. Module path delays extend the notion of propagation delay by associating a delay between the input and output ports of a module. Module paths can be simple, edge-sensitive, or state-dependent. The inertial model commonly used for propagation delay of gates can be extended to module paths, where the **PATHPULSE$** construct determines whether narrow pulses will propagate on a module path.

REFERENCE

IEEE Standard Hardware Description Language Based on the Verilog Hardware Description, Language Reference Manual (LRM), IEEE 1364-1995, 1996, The Institute of Electrical and Electronic Engineers, Piscataway, New Jersey.

PROBLEMS

1. Develop a gate-level implementation (in a cell library) of a four-bit ripple-carry adder, and by simulation or by using a static timing analysis tool:

   a. Create a distribution of the path lengths.

   b. Find the longest path. Identify it below and indicate its delay.

   Path begins at: __________

   Path ends at: __________

   Path delay: __________ nsec

   c. Find the shortest path. Identify it below and indicate its delay.

   Path begins at: __________

   Path ends at: __________

   Path delay: __________ nsec

2. Develop a gate-level implementation (using a cell library) of a four-bit carry look-ahead adder, and by simulation or by using a static timing analysis tool:

   a. Create a distribution of the path lengths.

   b. Find the longest path. Identify it below and indicate its delay.

   Path begins at: __________

   Path ends at: __________

   Path delay: __________ nsec

   c. Find the shortest path. Identify it below and indicate its delay.

   Path begins at: __________

   Path ends at: __________

   Path delay: __________ nsec

3. What is the default propagation delay used by Verilog primitives?

4. Describe inertial delay.

5. Compare the ripple-carry and carry look-ahead adders of Problems 1 and 2. Discuss their relative silicon area and performance. Clearly state any assumptions affecting your conclusions.

6. Write a testbench to verify the *Add_rca_16* module in Example 2.18. (Use low-level modules having physical delays.)

7. Draw the waveform of *y_out* when *a* and *b* have the waveforms shown in Figure 4.4.

```
module wand_of_assigns (y_out, a, b);
   input       a, b;
   output      y_out;
   wand        #3 y;
   assign      #5 y = ~a;
   assign      #3 y = ~b;
   buf         (y_out, y);
endmodule
```

8. Answer True or False.

a. ___ Verilog supports an inertial delay model for nets.

b. ___ In the following example, if y =1 at t=0, then the value of *buf_out* changes from x to 1 at time 4.

```
wire #3 y_out;
and #4 (y_out,x_in,y_in);
buf #2 (buf_out,y_out)

initial
   begin
      x=1;y=1
end
```

9. Using the indicated propagation delays, write two Verilog modules describing the circuit shown below.

a. Use module parameters to implement the delays of the instantiated primitives.

b. Use specify blocks to declare module pin-pin delays.

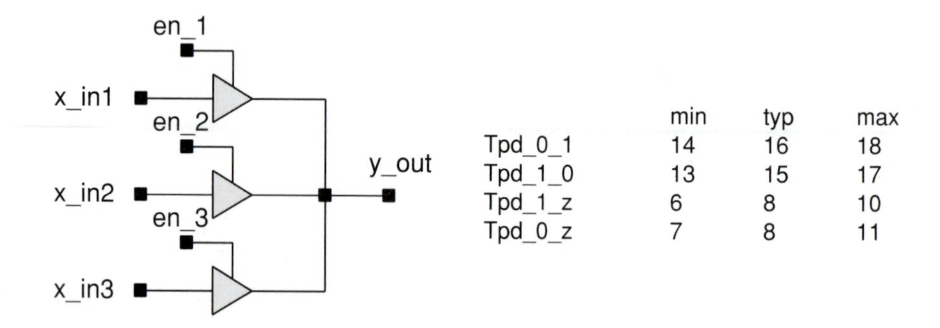

	min	typ	max
Tpd_0_1	14	16	18
Tpd_1_0	13	15	17
Tpd_1_z	6	8	10
Tpd_0_z	7	8	11

10. Write a pin-pin timing model for the input-output paths of *mod_timing* shown below

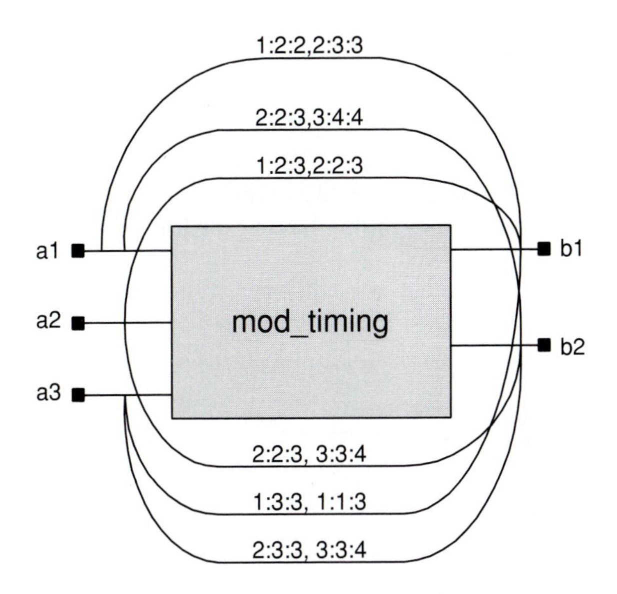

11. Find the error in the description below

```
module something_wrong (y_out, x_in1, x_in2);
   output      y_out;
   input       xin1, x_in2;

   nand #(3,4,5) (y_out, x_in1, x_in2);
endmodule
```

12. Write a declaration of a 16-bit bus named *some_bus* with transport delay of eight time units.

13. Determine the time values that a simulator will create from the following numbers and timescale directives:

`timescale	number	value
10 ns / 1 ps	2.447	
10 / 10 ns	2.447	
10 ns / 1 ps	2.334	
10 ns / 10 ps	2.334	

14. If a timescale directive has been set to 10 ns / 10 ps, what delay will be caused by the delay control #9?

15. When does *buf_out* change in the description below?

```
wire #2 y_out;
and #5 (y_out, x_in, y_in);
```

```
buf #2 (buf_out, y_out);
initial begin
   x_in = 1;
   y_in = 1;
end
```

16. Encapsulate the functionality of the circuit shown below in a module, and include pin-pin timing having the following values:

Input	Output	Rising Delay	Falling Delay
a	y	5	4
b	y	4	3
c	y	3	2

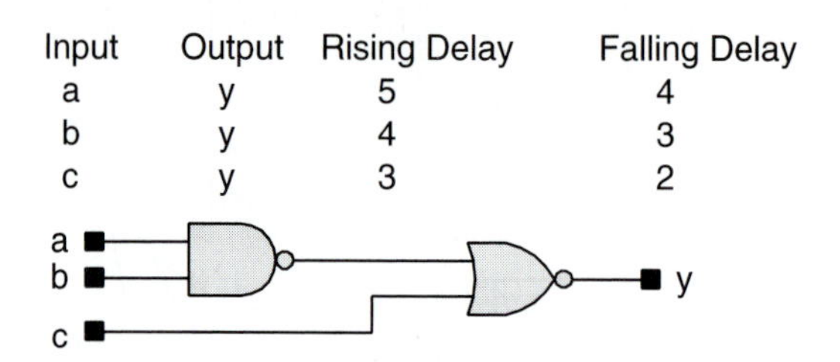

17. What is the typical falling delay of *a1* to *y2* in the specification below?

(a1, a2 *> y1, y2) = (15:16:18, 20:23,25);

18. How many paths are represented by the specification below?

(a, b, c *> y1, y2, y3)

CHAPTER 7

BEHAVIORAL DESCRIPTIONS IN VERILOG HDL

The paradigm of ASIC design that has evolved over the past ten years encourages designers to rapidly create a prototype of a design by first developing a behavioral description without binding it to a structure of physical hardware. This initial stage of the design flow focuses on functionality, not gate-level detail. Later, after functional verification, a synthesis tool optimizes and maps the design into a selected physical technology. This approach to design focuses attention on the functionality that is to be implemented, rather than individual logic gates and their interconnections, and provides the freedom to explore alternatives to a design before a commitment has been made to expend resources.

Traditionally, designers have had to design at the gate level by creating and assembling prototypes of small working units, using low-level logic gates. In the new paradigm, a complex behavior can be decomposed and represented at a meaningful level of abstraction, and then synthesized into gates. Synthesis tools map the HDL description into a physical technology, such as an ASIC standard cell library, or a library of programmable parts, such as field programmable gate arrays (FPGAs). Today's designers must understand behavioral modeling and the synthesis tools that map their Verilog descriptions into hardware. Not all descriptions synthesize, and not all synthesized descriptions are desirable. Synthesis issues will be discussed in Chapter 8 and later chapters.

Aside from its importance in synthesis, behavioral modeling provides flexibility to a design project by allowing portions of the design to be modeled at different levels of abstraction and completeness. The Verilog language accommodates mixed levels of abstraction so that portions of the design implemented at the gate level (i.e., structurally) can be integrated and simulated concurrently with other portions of the design represented by behavioral descriptions.

Previous chapters have presented three kinds of abstract (i.e., non-structural) behaviors in Verilog: continuous assignments, **initial** behaviors, and **always** behaviors. Continuous assignments (See Chapter 2) implement implicit combinational logic by establishing static bindings between expressions and target nets. The two behavioral constructs, **initial** and **always**, declare a description of functionality in terms of a set of concurrent, communicating behaviors (i.e., processes), or computational activity flows, modeling the relationship between the input and output ports (signals) of a module. The **initial** construct declares a one-shot sequential activity flow, and the **always** construct declares a cyclic sequential activity flow. (Synthesis ultimately associates this behavior with actual hardware.)

Example 7.1 An **initial** behavior is declared by associating the keyword **initial** with a procedural statement. The **initial** behavior in the Verilog module *demo_1* has a procedural assignment statement assigning a value to the **reg** *sig_a* at $t_{sim} = 0$. It then terminates. In simulation, the value of *sig_a*, "0", will be retained indefinitely, because *sig_a* is of type **reg** and no further assignments are made to it.

```
module demo_1 (sig_a);
  output      sig_a;
  reg         sig_a;              // Type declaration
                                  An "initial" Behavior
  initial
    sig_a = 0;                    // Procedural assignment to sig_a
endmodule
```

In this chapter, the term "behavior" will be used exclusively to mean an **initial** or **always** behavior. The terms *behavior* and *process* will be used interchangeably to refer to the evolution of the values of register variables within a module under the influence of the external simulation environment, and the other net and register variables in the module. Both constructs use procedural statements to describe the declared behaviors by controlling the activity flow and determining assignments to register variables. The statements implementing a declared behavior will be referred to as "procedural" or "behavioral" statements. Together with a module's external stimuli and internal structural detail, they affect assignment of value to the register variables within the module.

Earlier examples noted that procedural statements execute sequentially within a behavior, subject to time control and flow control constructs (**if**, **case**,**?** ... **:**). In Verilog, procedural assignment can be made only to register variables; i.e., objects having type **reg**, **integer**, **real**, **realtime**, and **time**.

Figure 7.1 depicts the elements of a Verilog module and shows the relationship between declarations, instantiated primitives, modules, continuous assignments, **initial** behaviors, and **always** behaviors. Instantiated primitives and modules are structural objects. Solid lines denote wires connecting primitives and modules; dashed lines denote references made to net and register variables. Modules may also contain declarations of port elements, internal nets, register and net variables, events, tasks, functions, and specify blocks.

Continuous assignments and primitives are similar—they respond to inputs and produce outputs according to a functional specification, possibly with propagation delay. When they are triggered by activity (events) at their inputs, they update the value of their output. A behavior only assigns value to a register variable when an assignment statement executes in the activity flow of the behavior. A given statement is not sensitive to the activity in the circuit. Its sole role is to execute when control is passed to it. The statements and timing controls before it in the listing of statements determine whether and when that execution ever happens.

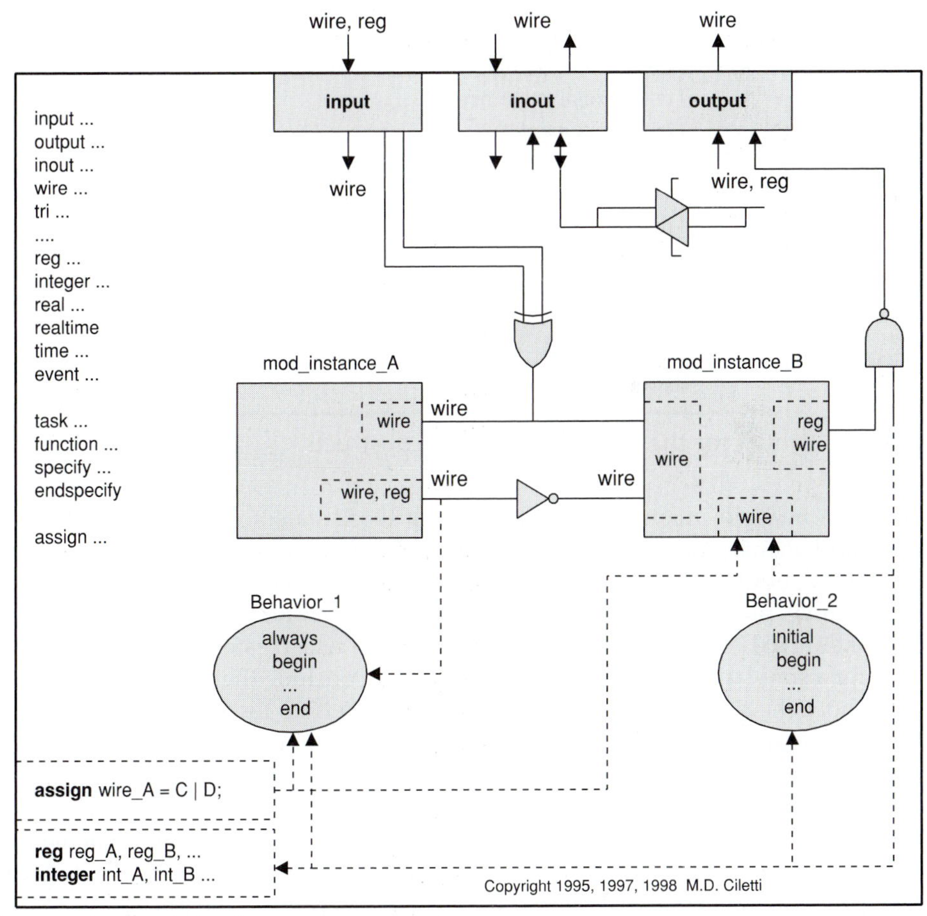

Figure 7.1 Module elements.

7.1 VERILOG BEHAVIORS

Verilog has two kinds of behaviors, denoted and declared by the keywords "**initial**" and "**always**," which model one-shot and cyclic behaviors, respectively. The syntax of the **initial** and **always** constructs is shown in Figure 7.2.

Verilog Syntax: initial and always Constructs

initial_construct ::= **initial** statement
always_construct ::= **always** statement
statement ::=
    blocking _assignment ;
    | non_blocking_assignment ;
    | procedural_assignment ;
    | procedural_timing_control_statement
    | conditional_statement
    | case_statement
    | loop_statement
    | wait_statement
    | disable_statement
    | event_trigger
    | seq_block
    | par_block
    | task_enable
    | system_task_enable

Figure 7.2 Syntax for the **initial** and **always** constructs.

A behavior may consist of a single or block statement. An **initial** behavior implements a one-shot activity flow activated at $t_{sim} = 0$ (the beginning of the simulation), and expires after all procedural statements have completed execution. As previous examples have already demonstrated, designers typically use the **initial** behavior to initialize a simulation and create stimulus waveforms for a simulation testbench. Do not, however, conclude that the activity of a one-shot behavior must cease before the simulation itself terminates.

An **always** behavior declares a cyclic activity flow activated at $t_{sim} = 0$. The procedural statements describing the cyclic activity flow re-execute after the last procedural statement has executed. Re-execution of an **always** behavior continues indefinitely until the simulation is terminated. The procedural statements of a behavior execute sequentially, in the order in which they are listed in the source code, subject to optional timing control and flow control constructs.

A behavior is declared within a module by using the keyword is **initial** or **always**, followed by a single statement or a **begin** ... **end** (compound) block statement. The statements associated with a behavior are called "behavioral" or "procedural" statements. A block statement is just a set of procedural statements enclosed by the keywords **begin** and **end**; within a block statement, the individual statements execute sequentially. Block statements may be nested within other blocks, and named blocks may contain local variables (of the register type).

It can be helpful to view the keywords **initial** and **always** and their accompanying procedural statements as encapsulating abstract activity-generators—abstract gates and/or sequential logic (see Figure 7.1). These behaviors (activity generators) exist concurrently with any gate instantiations that have been made in the containing module. There is a difference, though—**gates effect changes only in the value of net variables, and behavioral statements primarily effect changes in the value of register variables**. (The **force ... release** procedural continuous assignment, which can assign value to nets and registers, is an exception to this general rule. See Section 7.4.2.)

A module may contain any number of behaviors, but they may not be nested. Behaviors describe hardware implicitly, having no explicit association with physical gates. This feature lets designers focus attention on functionality, independently of hardware. In simulation, behaviors interact and execute concurrently with other behaviors and instantiations (e.g., primitives) within a module.

7.2 BEHAVIORAL STATEMENTS

Example 7.2 Module *demo_2* assigns initial values (i.e., at $t_{sim} = 0$) to several register variables in the same behavior. This **module** contains no other behaviors, so these assigned values will persist for the duration of a simulation.

```
module demo_2 (sig_a, sig_b, sig_c, sig_d);
 output        sig_a, sig_b, sig_c, sig_d;
 reg           sig_a, sig_b, sig_c, sig_d;       // Type declaration

                                   An "initial" Behavior
 initial                      <----
  begin
   sig_a = 0;                       // Procedural assignments
   sig_b = 1;                       // execute sequentially.
   sig_c = 1;
   sig_d = 0;
  end
endmodule
```

These same assignments can also be made in a variety of ways—for example, by using four behaviors rather than a single behavior—but a proliferation of several "small" behaviors will create needless overhead for simulation.

Example 7.3 The behavior encapsulated by the **always** keyword in *clock_gen1* implements a simple clock generator and generates the waveform shown in Figure 7.3. The **initial** behavior assigns the initial value to *clock*; the **always** behavior delays a *Half_cycle* before toggling the value of the *clock*. It then re-executes indefinitely. A second **initial** behavior assures that the simulator terminates after 1000 time steps (10 cycles of *clock*).

```
module clock_gen1 (clock);
  parameter Half_cycle = 50;
  parameter Max_time = 1000;
  output clock;
  reg clock;

  initial
    clock = 0;
  always
    begin
      #Half_cycle clock = ~ clock;
  end

  initial
    #Max_time $finish;
endmodule
```


Figure 7.3 A clock generation behavior.

The procedural statements in an **initial** or **always** behavior look like statements in high-level programming languages such as "C" and Pascal. Procedural statements support sequential computations that manipulate the values of data objects in memory. However, Verilog behaviors are not used just to compute the value of data objects; they also implicitly govern the activity flow of a simulation by influencing whether and when events will be scheduled in the event queue of the simulator.

A distinction must be made between the notions of simulator time and CPU time. Simulation time, denoted by t_{sim}, refers to the time axis managed by the simulator. It does not correspond to real time, or to the CPU time of the computer on which the simulator is executing. Simulator time is used to schedule and manage the relative time between events on data objects; i.e., when the data objects (e.g., a register variable) undergo an assignment of value.

When a procedural assignment statement executes, a value is immediately assigned to a register variable. In Example 7.2, the statements are executed in sequential order just as they are listed, and *sig_a*, *sig_b*, *sig_c*, and *sig_d* are all assigned value at simulation, time $t_{sim} = 0$. In event-driven simulation more than one assignment may be made at the same simulator time. (Multiple assignments made at the same simulator time will be ordered with respect to each other. The order of assignment cannot be controlled by the model.)

The procedural constructs in Verilog can be organized into the general categories shown in Table 7.1. Other conceptual groupings are also possible. The conditional operator "(**?** ... **:**)" is included because it can be used in both procedural and continuous assignments.

Table 7.1 *Organization of Verilog procedural constructs.*

ASSIGNMENTS	CODE MANAGEMENT
Conditional (? ... :)	Function Calls
Procedural Assignment (=)	Task Calls
Procedural-continuous **assign** ... **deassign** **force** ... **release**	Prog. lang. Interface (PLI)
Non-Blocking Assignment (<=)	
TIMING & SYNCHRONIZATION	**FLOW CONTROL**
Assignment Delay Control	Conditional (? ... :)
Intra-Assignment Delay	Case
Event Control	Branching
Wait	Loops
Named Events	Parallel Activity (**fork** ... **join**)
Pin-Pin Delay	

7.3 PROCEDURAL ASSIGNMENT

A statement that assigns value to a register variable is called a "procedural assignment." There are three kinds of procedural assignments. The first, which uses the "=" operator, is referred to simply as procedural assignment. A second form, called "procedural continuous assignment," uses the keyword **assign** with the same operator. A third form uses the "<=" operator. This form is called a "non-blocking" assignment.

The procedural assignment statement can assign value only to register variables (i.e., to data objects of type **reg**, **integer**, **real**, **realtime**, and **time**). Assignments to register variables obey different rules than assignments to nets. When the input to a primitive or continuous assignment statement changes, the output is evaluated and possibly scheduled to change in the future. Register variables do not behave that way—they can only get value when a procedural statement executes. Further, a statement can only execute when control is passed to it within the activity flow of a behavior. Thus, the mere appearance of a procedural assignment statement in a declared behavior does not guarantee that the target register will ever be assigned value by the statement—it depends on whether the statement ever executes in simulation (i.e., control must be passed to the statement). Likewise, the event of a variable in the RHS expression of a procedural assignment does not cause the statement to execute. The assignment of value to a register depends upon the execution of a statement, rather than some variable changing. When the statement executes, the value is assigned to the register variable immediately. It is not scheduled by an event queue.

Recall that a net is used solely to implement structural connectivity within and between Verilog modules. *With the exception of the* ***force*** *...* ***release*** *procedural continuous assignment* (see Section 7.4.2), *a net may not be assigned value by a behavioral statement.* A common error is to make a net or undeclared variable the target of a procedural assignment. Compilers will report this error condition because an undeclared variable is, by default, of type net.

Important distinctions between net and register variables are summarized by the rules in Figure 7.4.

The righthand side of a procedural assignment can be any expression that evaluates to a value. The lefthand side target of the assignment can be the identifier of a **reg**, **integer**, **real**, **realtime**, or **time** variable. In the case of a vector, any bit can be selected from it; a part-select of contiguous bits can also be selected. The field of a part-select of a register variable must be a constant width (i.e., it cannot be determined dynamically during simulation). Even more complex righthand targets for procedural assignments can be formed by using the concatenation operator. A memory element may also be the target of a procedural assignment.

RULES: Nets and Registers

A register variable can be referenced anywhere in a module.

A register variable can be assigned value only within a procedural statement, task, or function.

A register variable cannot be an **input** or **inout** port of a module.

A net variable can be referenced anywhere in a module.

A net variable may not be assigned value within a behavior, task, or function, except by a **force** ... **release** procedural continuous assignment.

A net variable within a module must be driven by a primitive, continuous assignment, **force** ... **release** procedural continuous assignment, or module port.

Figure 7.4 Distinctions between nets and registers.

7.4 PROCEDURAL CONTINUOUS ASSIGNMENT

Physical hardware can exhibit level-sensitive and edge-sensitive behavior. For example, the output of a transparent latch will follow the data input when the latch is enabled, but when the latch is disabled, it must ignore any changes on its data input and retain its last output value until it is again enabled. Verilog can model this behavior with procedural continuous assignment (PCA) statements (i.e., continuous assignments made within a behavior).

Recall that a continuous assignment establishes a static binding of an RHS expression to a LHS net variable. The binding, or relationship, is defined for the duration of a simulation. Continuous assignments can be made only to nets, never to register variables. On the other hand, a procedural continuous assignment creates a dynamic binding to a variable. Procedural continuous assignments are created dynamically, during execution, when the assignment statement executes within the activity flow of a Verilog behavioral statement.

There are two forms of procedural continuous assignment. The first creates a dynamic binding only to a target register variable; the second creates a dynamic binding to a target register variable or target net variable.

7.4.1 assign ... deassign Procedural Continuous Assignment

An **assign ... deassign** procedural continuous assignment to a register variable resembles the continuous assignment that can be made to a net, but it is not absolutely continuous - its binding can be removed. This type of PCA use the same operator, =, as a procedural assignment, but accompanies it with the keyword **assign**. PCAs using the keyword **assign** are used to model the level-sensitive behavior of combinational logic, transparent latches, and asynchronous control of sequential parts. Note that, although the PCA construct is convenient, some synthesis tools do not support it. Also note that the **deassign** construct is optional, but it has effect in simulation.

The next example will illustrate how the level-sensitive mechanism of a PCA can be used to model combinational logic.

Example 7.4 The multiplexer in Figure 7.5, *mux4_PCA*, uses a procedural continuous assignment to bind an assignment expression (event scheduling rule) to the target register variable, just as a continuous assignment binds an expression to a net.

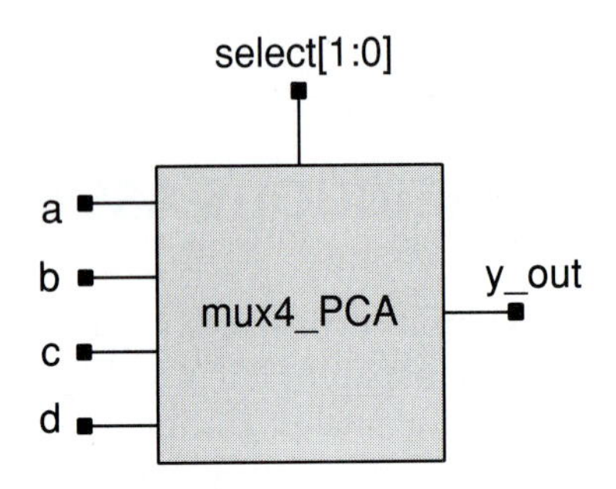


```
module mux4_PCA (a, b, c, d, select, y_out);
 input        a, b, c, d;
 input [1:0]  select;
 output       y_out;
 reg          y_out;

 always @ (select)
   if (select == 0)  assign y_out = a;  else
   if (select == 1)  assign y_out = b;  else
   if (select == 2)  assign y_out = c;  else
   if (select == 3)  assign y_out = d;  else y_out = 1'bx;
endmodule
```

Figure 7.5 Symbol and description of a four-channel multiplexer using a PCA.

The binding established by a continuous assignment is a static binding—it remains in effect for the duration of simulation. On the other hand, the binding established by a procedural continuous assignment takes effect when the assignment statement executes, and remains in effect until another procedural assignment is made to the same target register variable, or until a **deassign** statement is made to the register. In Example 7.4, the output of *mux4_PCA* will follow the activity of the input channel specified by the select signal. When select changes, a new channel binding supercedes the previous one.

The **assign** ... **deassign** procedural continuous assignment can be made only to a register variable, never to a net. When an **assign** of a procedural continuous assignment statement executes, it establishes an event scheduling rule between the target register variable and the value assignment expression on the RHS of the assignment. This rule remains in effect indefinitely (hence the term "continuous"), but it can also be de-assigned or overridden by a subsequent PCA. **While a PCA is in effect, it overrides all procedural assignments to the target variable**. The mechanism of a procedural continuous assignment models behavior in which an asynchronous control signal must override a synchronizing signal—the behavior of a flip-flop that has asynchronous control inputs as well as a synchronizing clock signal. A PCA also effectively models the behavior of a latch, which must respond to input events when it is enabled, but ignore them when it is not. Note in Example 7.4 that the **assign** keyword is used without the **deassign** keyword. Consequently, in this model, a PCA binding will always be associated with *y_out* during simulation.

In the previous example, the PCA bound an *expression* to the output of a multiplexer. In the next example, a PCA is used to model the *constant* binding established by the asynchronous control signals of a flip-flop.

Example 7.5 In synchronous operation the *data* input of a D-type flip-flop in Figure 7.6 is transferred to the output *q* at the synchronizing edge of *clock* (i.e., at the falling edge of the synchronizing signal). If either the *preset* or the *clear* signal are asserted, this (synchronous) clocking action is ignored, and the output is held at a constant value. The Verilog model of this behavior is shown for active-low *preset* and *clear.*

First, notice that the output of the flip-flop is declared to be a **reg**, a requirement in anticipation of a behavioral description of its activity. In this model, the first **always** behavior is activated at $t_{sim} = 0$ and executes cyclically, subject to the event control (synchronization) imposed by the falling edge of the clock. However, an event on the *clear* or *preset* signal (active-low) activates the second **always** behavior. The logic of the **if** statement determines which asynchronous control is asserted (with preference to the active-low *clear* signal), and creates a procedural continuous assignment to *q*. This PCA overrides the procedural assignments made in the behavior synchronized by the clock. The output

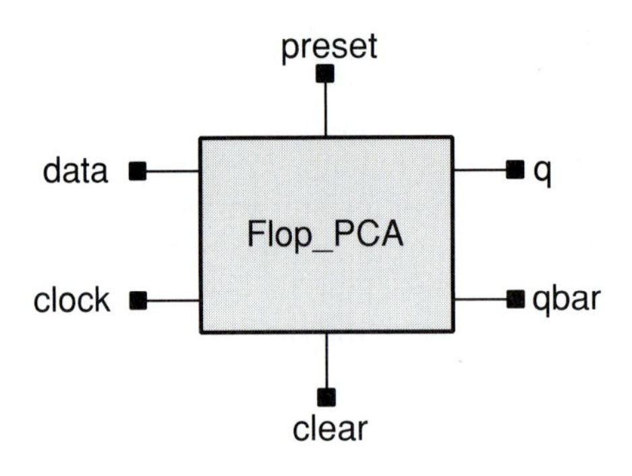


```
module Flop_PCA (preset, clear, q, qbar, clock, data);
  input     preset, clear, clock, data;
  output    q, qbar;
  reg       q;

  assign qbar = ~q;

  always @ (negedge clock)
    q = data;

  always @ (clear or preset)
    begin
     if (!clear) assign q = 0;
      else if (!preset) assign q = 1;
        else deassign q;
    end
endmodule
```

Behavior for synchronous procedural assignments.

Behavior for asynchronous PCA overriding synchronous procedural assignments.

Figure 7.6 Symbol and description of a flip-flop with asynchronous preset and clear.

remains fixed at the value specified by *preset* or *clear* until the control input is de-asserted. When that happens, the second behavior is again activated. If neither *preset* or *clear* are asserted, the PCA to *q* will be de-assigned. The keyword **deassign** causes whatever PCA is in effect on *q* to be removed. Beginning with the next falling edge of *clock*, the input *data* will be transferred synchronously to the output. Note, too, that the procedural continuous assignment to *q* is executed (i.e., the binding is established) when control is passed to the statement in the sequential activity flow, and remains in effect until a new PCA is established or the assignment is deassigned. This model is efficient in simulation because it does not check the asynchronous controls within the synchronous behavior, but it might not be accepted by a synthesis tool.

7.4.2 force ... release PROCEDURAL CONTINUOUS ASSIGNMENT

The **force** ... **release** form of procedural continuous assignment is similar to the **assign** ... **deassign** PCA, but it applies to nets as well as register variables. When the **force** assignment is made to a net, the expression assigned to the target net overrides all other drivers of the net until the **release** is executed. Thus, a **force ... release** PCA can override a primitive driver and continuous assignment driver of a net; it can override a procedural assignment and an **assign ... deassign** PCA to a register. The effect of a **force** ... **release** PCA to a net is the same as if the circuit was instantaneously modified by removing the net's driving logic and inserting new combinational logic driving the net. When the target of a **force ... release** PCA is a register variable, the residual value of the register is overridden. The "forced" event scheduling rule remains in effect, and overrides all procedural assignments and **assign ... deassign** procedural continuous assignments to the target register until the binding is released. When the binding is released, the register will remain at its (new) residual value until a different assignment is made to it by a procedural statement. The **force** ... **release** form of PCA is used primarily with hierarchical de-referencing in testbenches. Do not expect your synthesis tool to support this construct.

The next example illustrates how a **force ... release** PCA can be used to sensitize a circuit being tested.

Example 7.6 Suppose the testbench in Figure 7.7 has an objective requiring *mod_def* to be toggled by *sig_g* through the action of *sig_in1*. If *mod_abc* and *mod_xyz* are complex or sequential, it might be difficult and/or time-consuming to determine values for *in1*, ..., *in5* that will sensitize the path from *sig_in1* to *sig_g* through the gates. The ***force ... release*** procedural continuous assignment provides a simple alternative. Include the following text in the testbench:

```
...
force sig_a = 1;
force sig_b = 1;
force sig_c = 0;
sig_in1 = 0;
#5 sig_in1 = 1;
#5 sig_in1 = 0;

// Insert code to conduct tests

release sig_a;
release sig_b;
release sig_c;
...
```

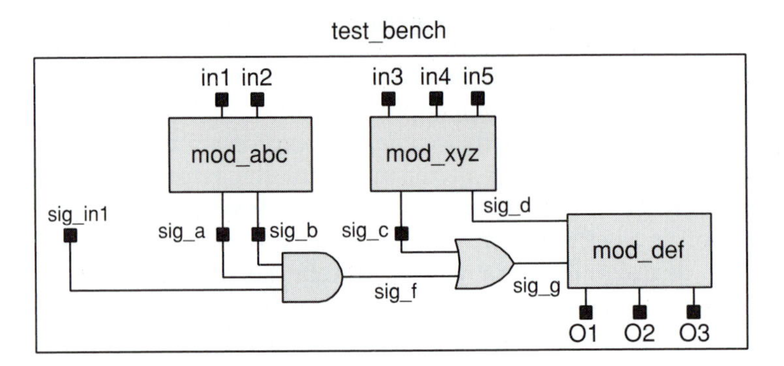

Figure 7.7 Testbench using **force ... release** to sensitize a path.

Figure 7.8 identifies the variables that may be driven by primitives or assigned value by continuous assignments, procedural assignments, and procedural continuous assignments. The entry for the output of a sequential primitive indicates that it may be a register. Within the scope of a sequential user-defined primitive, the output must be a scalar register variable. Within the scope of the module in which the UDP is instantiated, the output is a net variable.

	Mode of Assignment				
Variable Type	Output of Primitive	Continuous Assignment	Procedural Assignment	**assign** ... **deassign** PCA	**force** ... **release** PCA
Net	Yes	Yes	No	No	Yes
Register	Comb - No Seq - Yes	No	Yes	Yes	Yes

Figure 7.8 Modes of assignments to nets and variables.

7.5 PROCEDURAL TIMING CONTROLS AND SYNCHRONIZATION

Four mechanisms provide explicit control over the time of execution of a procedural statement in a behavior: delay control, event control, named events, and the **wait** construct. Previous examples have already demonstrated how behaviors use delay control of procedural assignments to suspend an activity flow within a behavior for a prescribed amount of time (e.g., to create stimulus waveforms). Event control, named events, and the **wait** construct are event-sensitive mechanisms that synchronize activity within and between behaviors in the same or different modules. When a behavior executes, it continues until it encounters a delay control operator, "#," event control operator, "@," or **wait** construct. Then it suspends and allows other processes to execute.

7.5.1 Delay Control Operator (#)

The delay control operator (mentioned in Chapter 3) temporarily suspends the activity flow within a behavior by postponing the execution of a procedural statement. There are two forms for delay control, depending upon the placement of the operator in a procedural assignment. The first form is implemented by placing the delay control operator, "#," and a *delay_value* to the left of a procedural statement. This form of the syntax for delay control is given in Figure 7.9.

Verilog Syntax: Delay Control

```
delay_control ::= # delay_value
                  | # (mintypmax_expression)
```

Figure 7.9 Syntax for delay control in a procedural assignment.

The effect of the # operator depends on where it is located in a procedural statement. When the operator precedes an assignment statement (including a null statement), it causes the simulator to postpone the execution of the statement until the specified amount of time has elapsed. The delay operator "blocks" execution of the statement that follows it, and suspends the activity flow at the location of the "#" operator. This also affects the execution time of all subsequent procedural statements in the **initial** or **always** block containing the "#." **A blocked statement must execute before the statements that follow it can execute.**

Example 7.7 In the behavior shown below, the register variable *IN3* will be assigned value "1" at time 100, and *IN5* will be assigned value "0" at time 600. Note: at time "0" IN3, IN4, and IN5 will be initialized to the value of "x".

```
initial
  begin
    #0     IN1 = 0;    IN2 = 1 ;  // Executes at t_sim = 0
    #100   IN3 = 1;               // Executes at t_sim = 100
    #100   IN4 = 1;    IN5 = 1;   // Executes at t_sim = 200
    #400   IN5 = 0;               // Executes at t_sim = 600
  end
```

Example 7.8 The module *Simple_Clock* implements a dual-output signal generator that could be used to provide clocks to other modules in a simulation. This module could be included in a testbench—not in a design that is to be synthesized. It contains a cyclic (**always**) behavior. The statements in the behavior complete execution for the first time at $t_{sim} = 200$. Then the activity flow returns to the **always** construct to repeat the cycle of execution.

```
module Simple_Clock (clock_1, clock_2);
  output      clock_1, clock_2;
  reg         clock_1, clock_2;

  always
    begin
      #0      clock_1 = 0; clock_2 = 1;
      #100    clock_1 = 1;
      #25     clock_2 = 0;
      #75;
    end
endmodule
```

Two cycles of the simulated waveforms of *clock_1* and *clock_2* are shown in Figure 7.10. The behavior generates *clock_1* and *clock_2* as long as the simulation is active because the block of statements associated with a cyclic behavior re-execute indefinitely. On the second pass, *clock_1* gets the value of "0" and *clock_2* get the value of "1" at $t_{sim} = 200$. At $t_{sim} = 300$ *clock_2* gets the value of "1". Then at $t_{sim} = 325$ *clock_2* gets the value of "0".

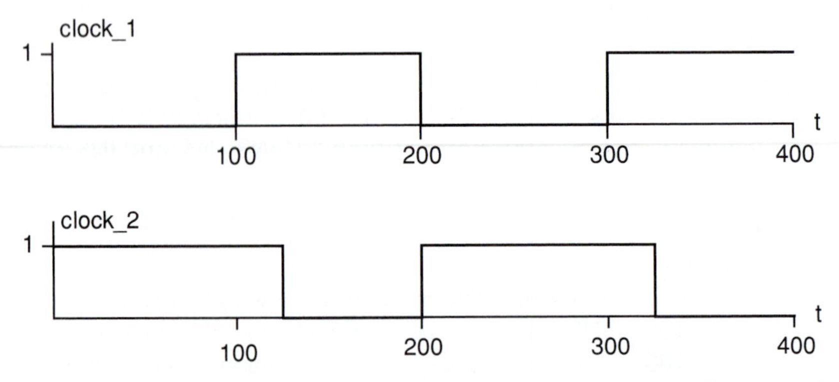

Figure 7.10 Waveforms generated by a cyclic behavior repeatedly executing procedural statements.

Example 7.9 The Verilog description in *clock_2phase* uses delay control to implement two clocks that have the same duty cycle, but opposite phase. The **initial** behavior initializes *clock_a* and *clock_b* at $t_{sim} = 0$, then expires. The **always** (cyclic) behavior also activates at $t_{sim} = 0$, but immediately suspends until a half-cycle of the clock has elapsed. Then the statements toggling *clock_a* and *clock_b* execute. The activity flow returns to the **always** construct and suspends activity for another half-cycle of the clock, and so on.

```
module clock_2phase (clock_a, clock_b)
  output            clock_a, clock_b;
  reg               clock_a, clock_b;
  parameter         clock_period = 50;

  initial
    begin
      clock_a = 0;
      clock_b = 1;
    end

  always
    begin
      #clock_period/2;
      clock_a = ~clock_a;
      clock_b = ~clock_b;
    end
endmodule
```

In the syntax for delay control, the delay can be expressed as a number, identifier (with implicit numeric value), or expression. When a *delay_value* expression ***explicitly*** creates a value of zero delay (e.g., #0) the effect is to cause the statement to execute at the end of the current simulation cycle.

7.5.2 Event Control Operator (@)

An event control operator synchronizes the execution of a procedural statement (or a block of procedural statements) to a change in the value of either an identifier or an expression. There are two forms of *event control.* The first form uses the operator "@" placed before either an *event_identifier* or *event_expression,* followed by a statement. The syntax of this form of the event control construct is shown in Figure 7.11.

Verilog Syntax: Event Control

```
event_control ::= @ event_identifier statement_or_null
          | @ (event_expression) statement_or_null

statement_or_null ::= statement | ;
```

Figure 7.11 Syntax of an event control statement.

The first form of event control synchronizes the activity flow of a behavior to an event of a net or register variable, recognized by its identifier. The second form synchronizes the activity flow to an event of an *event_expression.*

Example 7.10 In the code fragment below, when *Signal_1* changes value (an event), the statement assigning value to *register_A* executes. Until then, the activity flow of the associated behavior is suspended at the procedural statement.

```
begin
  ...
  @ Signal_1      register_A = register_B;
  ...
end
```

When the activity flow of a behavior reaches the event control operator, @, the activity flow is suspended and the *event_expression* or *event_identifier* is monitored by the simulator to detect an event. An event occurs when the *event_expression* or *event_identifier* changes value. When the event occurs, the statement that follows the *event_expression* or *event_identifier* executes. **It is important to note that the activity flow must be suspended at the event control operator in order for an event to activate execution of the statement associated with the event control expression.** While it is suspended, other processes may execute. For example, in the code fragment below, if *event_A* has occurred, but not *event_B*, the behavior will be suspended at the second event control expression. If *event_A* occurs a second time while the process is suspended, it will be ignored. Similarly, if the activity flow is suspended by *@ (event_A)* the occurrence of *event_B* will be ignored.

```
...
  @ (event_A) begin
    ...
    @ (event_B) begin
    ...
  end
end
```

Given that event control suspends the host process until an event has occurred, the anticipated event must be caused by activity in some process other than the one that is suspended. Otherwise, the simulator cannot resume activity.

The behavior of synchronous sequential circuits is synchronized to either the rising or falling edge of a "clock" signal. Verilog has predefined edge qualifiers (**posedge** and **negedge**) that condition the execution of a procedural statement on either a rising or falling waveform transition. Verilog semantics treat the following transitions as **posedge**: 0 --> 1, 0 --> x, and x --> 1; it treats the transitions 1 --> 0, 1 --> x, and x --> 0 as **negedge** transitions. This edge detection mechanism is built into the Verilog language. When coupled with the event control operator, "@," it provides synchronization to the activity flow within a behavior.

Example 7.11 In the description below, the assignments to *q_register* are synchronized to the positive-edge transitions of *clock*, and occur 10 time units after the rising edge; i.e., *q_register* (*clock* +10) gets *data_path*(*clock* + 10), not the value of *data_path*(*clock*).

```
always @ (posedge clock) #10 q_register = data_path;
```

Note: This behavior is not necessarily a good model of a D-type flip-flop, because it samples *data_path* 10 units after the edge of *clock*.

Example 7.12 A D-type flip-flop can be modeled by a cyclic behavior synchronized to an edge of a clock. In *df_behav*, the action of *reset* is synchronous. (The reset behavior of the flip-flop can also be synchronized to the same edge or a different edge of the same clock.)

```
module df_behav (data, clk, q, q_bar, set, reset);
  input          data, clk, set, reset;
  output         q, q_bar;
  reg            q;

  assign q_bar = ~ q;

  always @ (posedge clk) // Flip-flop with synchronous set/reset
    begin
      if (reset == 0) q = 0;
        else if (set ==0) q = 1;
          else q = data;
    end
endmodule
```

The *event_expression* and *event_identifier* used in event control must reference a net or register (not a parameter, which is a constant). Remember: A register variable referenced in an event control expression cannot be assigned value by the behavior that it synchronizes. In simulation, it must be assigned in some other concurrently executing behavior.

7.5.3 EVENT or

The Verilog language also allows a complex *event_expression* to be formed as the disjunction (logical "or") of other variables. This form of the *event_expression* is referred to as "event or-ing."

Example 7.13 The behavior below synchronizes a register transfer to a simulation event in *Signal_1* or *Signal_2*.

```
always @ (Signal_1 or Signal_2)  register_A = register_B;
  ...
```

Example 7.14 In the synchronous cyclic behavior in the model *asynch_df_behav1*, the rising edge of the clock synchronizes the flip-flop, subject to asynchronous *set/reset* behavior.

```
module asynch_df_behav1 (data, clk, q, q_bar, set, reset);
  input     data, clk, set, reset;
  output    q, q_bar;
  reg       q;

  assign q_bar = ~q;

  always @ (set or reset or posedge clk)
    begin
      if (reset == 0) q = 0;
        else if (set == 0) q = 1;
          else if (clk == 1) q = data;
    end
endmodule
```

Notice that ***this behavior does not model the behavior of a flip-flop correctly.*** Consider what happens when either *set* or *reset* is de-asserted during the active half-cycle of *clk*. *q* gets *data* even though there is no clock edge! The model described by *asynch_df_behav2* presents a solution to the problem.

```
module asynch_df_behav2 (data, clk, q, q_bar, set, reset);
  input     data, clk, set, reset;
  output    q, q_bar;
  reg       q;
```

```
  assign q_bar = ~q;

  always @ (negedge set or negedge reset or posedge clk)
    begin
      if (reset == 0) q = 0;
      else if (set == 0) q = 1;
        else q = data;
    end
endmodule
```

In this version, *q* retains its residual value during the interval between deassertion of a control signal and the next active edge of *clk*.

Example 7.15 The output of a transparent latch, depicted in Figure 7.12, follows the input when the latch is enabled, and holds the residual value of the latch when it is disabled. The Verilog description *t_latch* uses the "event-or" construct to describe this behavior.

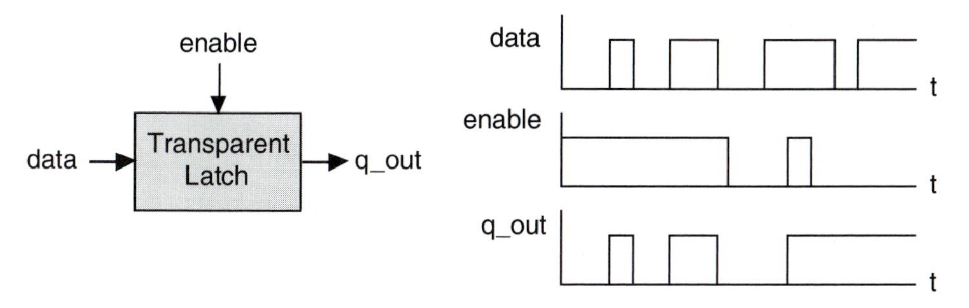

Figure 7.12 Transparent latch.

```
module t_latch (q_out, enable, data);
  output q_out;
  input enable, data;
  reg q_out;

  always (@ (enable or data))
    begin
      if (enable) q_out = data;
    end
endmodule
```

The behavior is activated at $t_{sim} = 0$ and immediately suspends until the event control expression changes. When *enable* is asserted, the process is re-activated and *q_out* immediately gets the value of *data*. Then the activity flow returns to the **always** construct and suspends to await the next change of the event control expression. If *data* changes while *enable* is asserted, the cycle of *q_out* getting *data* repeats. When *enable* is de-asserted, the control flow of the **if** construct has no branch. So *q_out* retains the value it had immediately before *enable* was de-asserted. While *enable* is de-asserted, the events of *data* re-activate the process, but no assignment is made to *q_out*.

7.5.4 NAMED EVENTS

The Verilog "named event" provides a high-level mechanism of communication and synchronization within and between modules. A named event, sometimes referred to as an "abstract event," provides inter-process communication without requiring details about physical implementation. In the early stages of design, this can free the designer from having to pass signals between modules explicitly through their ports. **A named event can be declared only in a module,** with keyword **event**; it can then be referenced within that module directly, or in other modules by hierarchically de-referencing the name of the event. The occurrence of the event itself is determined explicitly by a procedural statement using the "event-trigger" operator, "**->**".

Example 7.16 A typical implementation of a named event is shown in *Demo_mod_A*:

```
module Demo_mod_A ( ... );
...
  event something_happens;    // Declaration of an abstract event
  always
    begin
    ...
    -> something_happens      // Triggering of an abstract event
    end
endmodule

module Demo_mod_B( ... );

  always @ (Top_Module.Demo_mod_A.something_happens) // Event monitor
    ...
    begin
    ...  // do something when something_happens in Demo_mod_A
    end
  end
endmodule
```

When *something_happens* is triggered (i.e., abstractly asserted true) in *Demo_mod_A*, the event control in *Demo_mod_B* detects the event and then executes the statement that follows. Notice that no synchronizing signals were passed between the ports of the two modules.

Example 7.17 In the description in Figure 7.13, the abstract event *up_edge* is triggered when *clock* has a positive edge transition. A second behavior detects the event of *up_edge* and assigns value to the flip-flop's output, subject to an asynchronous *reset* signal.

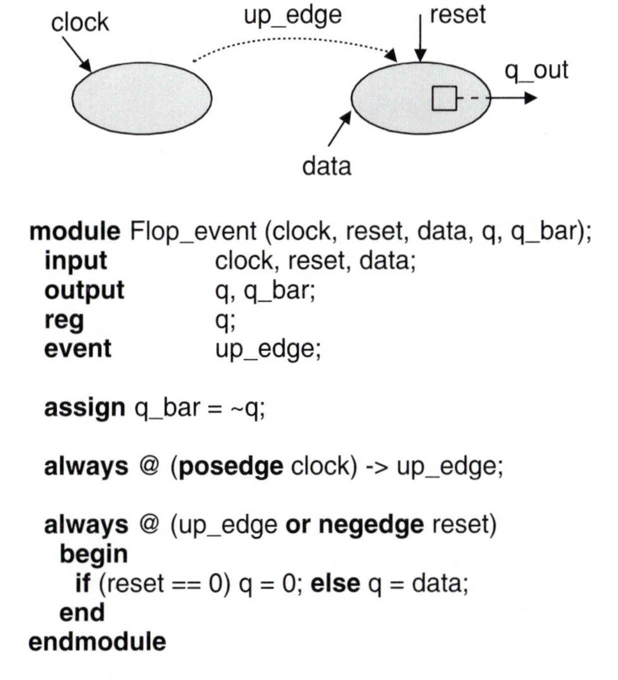


```
module Flop_event (clock, reset, data, q, q_bar);
  input        clock, reset, data;
  output       q, q_bar;
  reg          q;
  event        up_edge;

  assign q_bar = ~q;

  always @ (posedge clock) -> up_edge;

  always @ (up_edge or negedge reset)
    begin
      if (reset == 0) q = 0; else q = data;
    end
endmodule
```

Figure 7.13 Flip-flop implemented with a named event.

The mechanism of named events can be used across module boundaries anywhere in a design hierarchy. Example 7.18 demonstrates the use of named events across module boundaries.

Example 7.18 In Figure 7.14, activity in a module within *Top*, M1, triggers activity in a sibling module, M2. M2 can de-reference the event, *do_it*, in M1 by its hierarchical path, *Top.M1.do_it*.

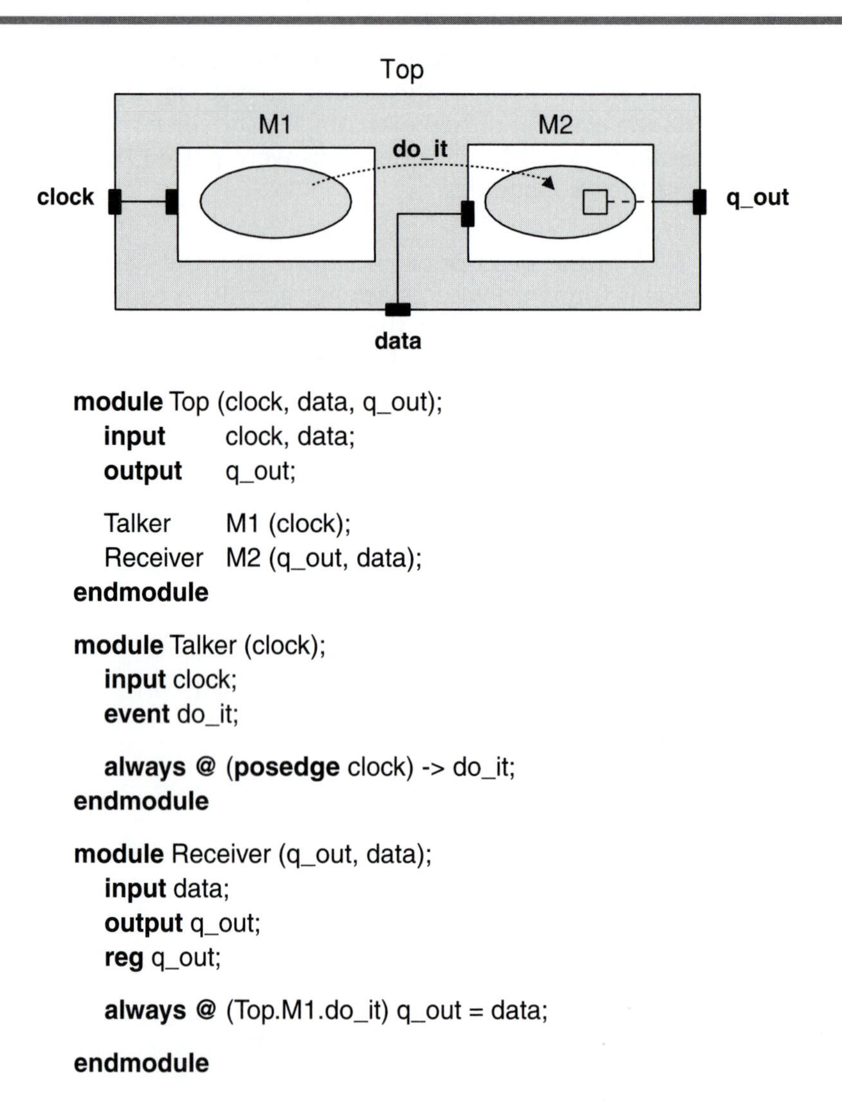


```
module Top (clock, data, q_out);
   input      clock, data;
   output     q_out;

   Talker     M1 (clock);
   Receiver   M2 (q_out, data);
endmodule

module Talker (clock);
   input clock;
   event do_it;

   always @ (posedge clock) -> do_it;
endmodule

module Receiver (q_out, data);
   input data;
   output q_out;
   reg q_out;

   always @ (Top.M1.do_it) q_out = data;

endmodule
```

Figure 7.14 Flip-flop behavior implemented with a named event across module boundaries. Hierarchical de-referencing makes *Top.M1.do_it* visible to *Receiver.*

A named event must be declared before it is used. No hardware needs to be specified to hold the value of the named event (i.e., no **reg** is needed.) or transmit the value to another module (i.e., no net is needed). Thus, a suspended behavioral (procedural) statement in one module can resume execution on the basis of behavior that occurs in another module, or in another behavioral statement within the same module, with no need for additional data objects supporting the transaction. As such, the Verilog named event is a more abstract mechanism for communicating and signaling than is event control. When coupled with hierarchical de-referencing, named events enable designers to quickly develop models of interaction between remotely located modules in a design hierarchy.

7.5.5 The wait Construct

The **wait** construct models level-sensitive behavior by suspending (but not terminating) the activity flow in a behavior until an *expression* is "TRUE". If *expression* evaluates to "TRUE" when the statement is encountered, the execution is not suspended. If *expression* is "FALSE", the simulator suspends the activity thread that contains the **wait** and establishes a monitoring mechanism that observes a condition *expression*. Other processes may execute while the process with the **wait** is suspended. When the condition *expression* becomes "TRUE," execution resumes with the *statement_or_null* associated with **wait**. The syntax of the **wait** statement is shown in Figure 7.15.

Verilog Syntax: wait Construct

wait_statement ::= wait (expression) statement_or_null;

statement_or_null ::= statement | ;

Figure 7.15 Syntax for the **wait** statement.

Example 7.19 When the activity flow of the behavior within *example_of_wait* reaches the wait statement, the enable is evaluated to determine whether it is TRUE or FALSE. If it is TRUE the activity flow continues. Otherwise, it suspends until *enable* becomes TRUE.

```
module example_of_wait ( );
...
  always
    begin
  ...
      wait (enable) register_a = register_b;
      #10 register_c = register_d;
  ...
    end
endmodule
```

7.6 INTRA-ASSIGNMENT DELAY—BLOCKED ASSIGNMENTS

When a timing control operator (# **or** @) appears in front of a procedural statement, the delay is referred to as a "blocking" delay, and the statement that follows the operator is said to be "blocked." The statement that follows a blocked statement in the sequential flow cannot execute until the preceding statement has completed execution. Verilog supports another form of delay in which a timing control is placed to the righthand side of the assignment operator within an assignment statement. This type of delay, called "intra-assignment delay," evaluates the righthand side expression of the assignment, and then schedules the assignment to occur in the future at a time determined by the timing control. Ordinary delay control postpones the execution of a statement; intra-assignment delay postpones the occurrence of the assignment that results from executing a statement. The execution of a statement in a list of blocked procedural assignments (i.e., those using the = operator) must be completed before the next statement can execute.

Example 7.20 When the first statement is encountered in the sequential activity flow below, the value of *B* is sampled and scheduled to be assigned to *A* five time units later. The statement does not complete execution until the assignment occurs. After the assignment to *A* is made, the next statement can execute (*C* = *D*).

```
...
A = #5 B;
C = D;
...
```

Thus, *C* gets *D* five time units after the first statement is encountered in simulation. (This type of delay might be used in the model of a flip-flop.)

Intra-assignment delay control (#) has the effect of causing the righthand side expression of an assignment to be evaluated immediately when the procedural statement is encountered in the activity flow. However, the assignment that results from the statement is not executed until the delay specified by "#" has elapsed. Thus, **referencing and evaluation are separated in time from the actual assignment of value to the target register variable**.

Intra-assignment delay can also be implemented with the event control operator and an event expression. In this case, the execution of the statement is scheduled subject to the occurrence of the event specified in the event expression.

Example 7.21 In the description below, *G* gets *ACCUM* when *A_BUS* changes. As a result of the intra-assignment delay, the procedural assignment to *G* cannot complete execution until *A_BUS* changes. The statement *C* = *D* is blocked until *G* gets value. The value that *G* gets is the value of *ACCUM* when the statement is encountered in the activity flow. This may differ from the actual value of *ACCUM* when *A_BUS* finally has activity.

```
...
G = @ (A_BUS) ACCUM;
C = D;
...
```

7.7 NON-BLOCKING ASSIGNMENT

Procedural assignments using the "=" operator execute sequentially and are called "blocking" assignments. The statement that follows a procedural assignment in the sequential activity flow of a behavior cannot execute until the procedural assignment's execution is complete. Verilog also includes a non-blocking procedural assignment construct; it uses the operator "<=" to distinguish this kind of assignment from blocking procedural assignments (i.e., those that use the "=" operator to denote assignment). **Non-blocking assignments behave differently than blocked assignments.** A non-blocking assignment does not block the execution of the statements that follow it in the listed source code. The syntax of the non-blocking assignment is shown in Figure 7.16.

Verilog Syntax: Non-Blocking Assignment
non_blocking_assignment ::= reg_lvalue <= [delay_or_event_control] expression;

Figure 7.16 Syntax for a non-blocking assignment.

Non-blocking assignments execute in two steps. First, the RHS expression is evaluated; then the simulator schedules an assignment to the LHS at a time determined by an optional intra-assignment delay control or event control (the default is 0). At the end of the time step in which either the indicated delay control has expired or the event specified by the intra-assignment event control has occurred, the assignment is made to the LHS variable. This means that non-blocking assignments are ordinarily the last assignments made in a given time step during simulation. However, if their execution triggers additional blocking assignments, the latter will be executed after the scheduled non-blocking assignments. These causally-related blocked assignments are scheduled for execution after their antecedent non-blocking assignments at the same time step. The two-step mechanism of non-blocking assignment ensures that there are no races between the assignments in a **begin ... end** block of non-blocking assignments in the same behavior.

Example 7.22 The blocking and non-blocking assignments in the behaviors below are executed at $t_{sim} = 0$. The blocked assignments ("=") execute first because they are listed first in the order of the statements. They must execute before the non-blocking assignments ("<=") are encountered. The RHS expressions of the non-blocking assignments are evaluated concurrently and independently of their order with respect to each other, and then the assignments are made to the RHS. The behaviors are equivalent.

```
initial
  begin
    A = 1;
    B = 0;
    ...
    A <= B;       // Uses B = 0
    B <= A;       // Uses A = 1
  end
```

```
initial
  begin
    A = 1;
    B = 0;
    ...
      B <= A;       // Uses A = 1
      A <= B;       // Uses B = 0
  end
```

The result of simulating the behaviors is that A and B swap values, leaving A = 0 and B = 1. The order of the non-blocking statements has no effect. On the other hand, if the "<=" operator is replaced by the "=" operator in the behaviors, the results depend on the order of the statements. The two behaviors below differ only in the order in which the non-blocking assignments are listed.

```
initial
  begin
    A = 1;
    B = 0;
    ...
    A = B;       // Uses B = 0
    B = A;       // Uses A = 0
  end
```

```
initial
  begin
    A = 1;
    B = 0;
    ...
    B = A;    // Uses A = 1
    A = B;    // Uses B = 1
  end
```

Non-blocking assignments are useful in modeling concurrent transfers of data in synchronous circuits. The assignments in a list of non-blocking assignments in a behavior execute concurrently—they schedule assignments to target variables without blocking the activity flow to succeeding statements in the sequential activity flow. The result of executing a list of non-blocking assignments does not depend upon the order in which the statements are listed. This form of assignment models concurrent assignments to a set of register variables (i.e., at the same time step) without regard to the order in which the assignments are written. Synthesis tools recommend the use of this assignment operator to describe current transfers of data to registers.

In general, the sampling of the righthand side of a list of non-blocking assignments occurs before any assignments are made, and guarantees that the outcome (i.e., the resulting values of register variables) does not depend on the order of the list, with one exception. The Language Reference Manual (LRM) specifies that the scheduling of non-blocking assignments *in the same behavior* is to be made in the same order in which the statements are listed. Therefore, if two or more statements in the list assign value to the ***same*** register variable, the last statement in the list will determine the final value of the register. The assignments are all scheduled in the same time-step of the simulator.

Example 7.23 When an LRM-compliant simulator encounters the fragment of code below, the RHS of each is sampled. Then the first statement assigns value to B. Next, the second statement assigns value to B. The result is that B holds the value 1.

```
...
B <= 0;
B <= 1;
...
```

Note: Although synthesis tools do not support a mixture of blocking and non-blocking assignments in the same behavior, the language supports this level of complexity

7.8 INTRA-ASSIGNMENT DELAY: NON-BLOCKING ASSIGNMENT

When a procedural assignment includes intra-assignment delay, the evolution of the activity flow in a behavior depends upon whether the affected statement involves a procedural or non-blocking assignment. A delay in non-blocking assignments affects only the scheduling of the assignments, not the completion of execution of the statements. The statements complete execution concurrently at the same time step at which they are encountered.

Example 7.24 In the non-blocking assignments below, the value of *accum* is sampled at the rising edge of *clock*; the assignment of the sampled value to *G* is scheduled to occur when *a_bus* changes. This differs from the behavior in Example 7.21, where the assignments were blocking. Here the execution of the statement *C* <= *D* is not blocked. So *D* is sampled when *accum* is sampled, and *C* gets the value of *D* immediately. *G* will get the sampled value of *accum* when *a_bus* has activity, even though the behavior has already expired.

```
initial begin
 @ (posedge clock)
   G <= @ (a_bus) accum;
   C <= D;
end
```

Now consider the behavior shown below:

```
always begin
 @ (posedge clock)
   G <= @ (a_bus) accum;
end
```

At the first active edge of the clock, *accum* is sampled. Then a monitoring mechanism is established to watch *a_bus,* and the activity flow returns to the beginning of the behavior, where it is suspended to wait for the next active edge of *clock*. What happens if *a_bus* does not have an event before the next active edge? Are assignments overwritten? That is, does *G* get assigned the value of *accum* that was sampled at the first clock edge, followed immediately by the value of *accum* at the second clock edge? Or is a queue of assignments built, with *G* getting the sampled values of *accum* at successive events of *a_bus*? Neither! The simulator issues a warning that queuing of non-blocking assignments is not supported; meaning that G gets only the current sample of *accum* (i.e., the value sampled at the last active edge of the clock before the event of *a_bus*), not some previously sampled and queued value. This means that the simulator repeatedly overwrites its memory holding the sample of *accum* as it is taken at each active edge of the clock. For the waveforms shown in Figure 7.17, the value of *accum* sampled at the edge of *clock* immediately before the event of *a_bus* is always "1". The previously sampled values of *accum* are overwritten at each active edge of *clock*.

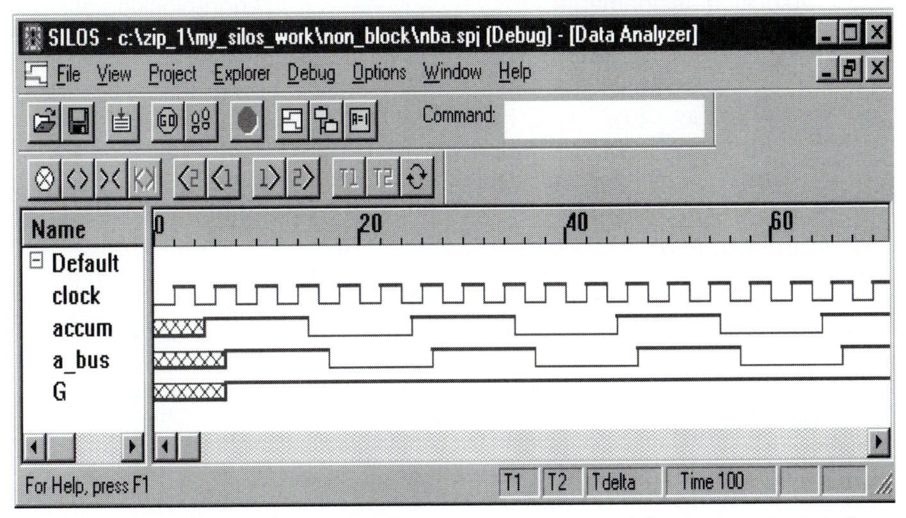

Figure 7.17 Simulation of non-blocking assignments without event queuing.

Example 7.25 Now we will compare the effect of intra-assignment delay in a sequence of blocked assignments to the effect in a sequence of non-blocked assignments. The two behaviors shown below in module *nb1* differ only in the type of assignment statements they use. The simulation results illustrate the difference between blocking and non-blocking assignments with intra-assignment delay control. The statement assigning value to *a* completes execution at t_{sim} = 10; the statement assigning value to *b* is encountered at t_{sim} = 10, after the preceding statement completes execution. The statement assigning value to *c* is encountered at t_{sim} = 12 and completes execution with an assignment to *c* at t_{sim} = 15. On the other hand, the non-blocking assignments to *d*, *e*, and *f* all execute at t_{sim} = 0. The non-blocking assignment to *d* schedules an event for *d* at t_{sim} = 10; the non-blocking assignment to *e* schedules an event for *e* at t_{sim} = 2, and the non-blocking assignment to *f* schedules an event for *f* at t_{sim} = 3.

```
module nb1;
 reg a, b, c, d, e, f;

//blocking assignments                  // non-blocking assignment
 initial                                  initial
  begin                                    begin
   a   = #10  1;                            d   <= #10  1;
   b   = #2   0;                            e   <= #2   0;
   c   = #3   1;                            f   <= #3   1;
  end                                      end;
                                          endmodule
```

t	a	b	c	d	e	f
0	x	x	x	x	x	x
2	x	x	x	x	0	x
3	x	x	x	x	0	1
10	1	x	x	1	0	1
12	1	0	x	1	0	1
15	1	0	1	1	0	1

The results of the simulation show that a non-blocking assignment does not block the procedural flow; it evaluates and schedules the assignment, but does not block the execution of the subsequent statements in the flow.

Example 7.26 Combinational logic can be modeled by both one-shot and cyclic forms of Verilog behavior. Consider *bit_or8_gate1* and *bit_or8_gate2*, which described an eight-bit wide bitwise-or datapath operation by a continuous assignment statement in Example 2.11. A one-shot behavior describes the same functionality in *bit_or8_gate3*. The behavior is activated at $t_{sim} = 0$, creates a procedural continuous assignment to y, and then expires. The assignment remains in effect after the behavior expires. **Although this is a valid description, it is not a preferred style, and will not be accepted by a synthesis tool**. It uses more code than is necessary to describe the functionality. Another alternative, *bit_or8_gate4*, is an acceptable style, but is not as simple as the descriptions using continuous assignment in Example 2.11.

```
module bit_or8_gate3 (y, a, b);
  input    [7:0]    a, b;
  output   [7:0]    y;
  reg      [7:0]    y;
```

```
   initial begin
      assign y = a | b;
   end
endmodule

module bit_or8_gate4 (y, a, b);
   input    [7:0]    a, b;
   output   [7:0]    y;
   reg      [7:0]    y;

   always @ (a or b) begin
      y = a | b;
   end
endmodule
```

When behavioral models containing delay control are used to describe combinational logic, the behavior of the model may not match that of a physical gate.

Example 7.27 The delay control in *bit_or8_gate5* is intended to represent propagation delay between an input event and the resulting output event.

```
module bit_or8_gate5 (y, a, b);
   input    [7:0]    a, b;
   output   [7:0]    y;
   reg      [7:0]    y;

   always @ (a or b) begin
      #5 y = a | b;
   end
endmodule
```

Two things are wrong. First, the assignment of value to *y* is determined by old data (i.e., the values of *a* and *b* at five time steps after the activating event). The second issue is that **while the delay control is blocking the procedural assignment to *y*, the event control expression does not respond to events of *a* or *b*.** Physical combinational parts do exhibit this behavior.

Alternatively, if intra-assignment delay is used the data that activates the behavior in *bit_or8_gate6* (below) will be sampled to form *y*. This too is a flawed model. The assignment is blocked, so the event control expression cannot respond to input events until the statement assigning value is executed.

```
module bit_or8_gate6 (y, a, b);
  input    [7:0]    a, b;
  output   [7:0]    y;
  reg      [7:0]    y;

  always @ (a or b) begin
    y = #5 a | b;
  end
endmodule
```

If intra-assignment delay is used with a non-blocking assignment, the data that activates the behavior in *bit_or8_gate7* will be sampled to form *y*. The non-blocking assignment executes and returns control to the event control operator, anticipating the next event on the datapath, even though *y* has not yet received a new value from the original event. This corresponds more closely to the reality of propagation delay in combinational logic.

```
module bit_or8_gate7 (y, a, b);
  input    [7:0]    a, b;
  output   [7:0]    y;
  reg      [7:0]    y;

  always @ (a or b) begin
    y <= #5 a | b;
  end
endmodule
```

Figure 7.18 shows the result of simulating the various implementations of the eight-bit datapath operation. Notice that *y5*, the output from *bit_or8_gate5*, shows bb_h as the output in response to *a[7:0]* changing from 00_h to aa_h to bb_h while *b[7:0]* is 00_h. That corresponds to the delayed sampling of the inputs resulting from the delay control value #5. (The first change in *a* triggers the **always** behavior, and the result of the second change is sampled.) For the same input waveforms, the value of *y6* is aa_h, which is correct. However, *y6* ignores the input transitions when *a[7:0]* changes from aa_h to bb_h, because it occurs while the activity flow is suspended by the delay control, #5, and before the activity flow returns control to the event control expression. The output *y7* responds to the input activity and effects the correct delay on the assignments to the output. The input pulse is shorter than the delay, yet affects the output. In fact, there is no inertial delay mechanism built into delay control within behaviors. (This is not a limitation, because short pulses can be suppressed on module path delays—see Chapter 6). However, *bit_or8_gate5* acts like combinational logic with inertial delay.

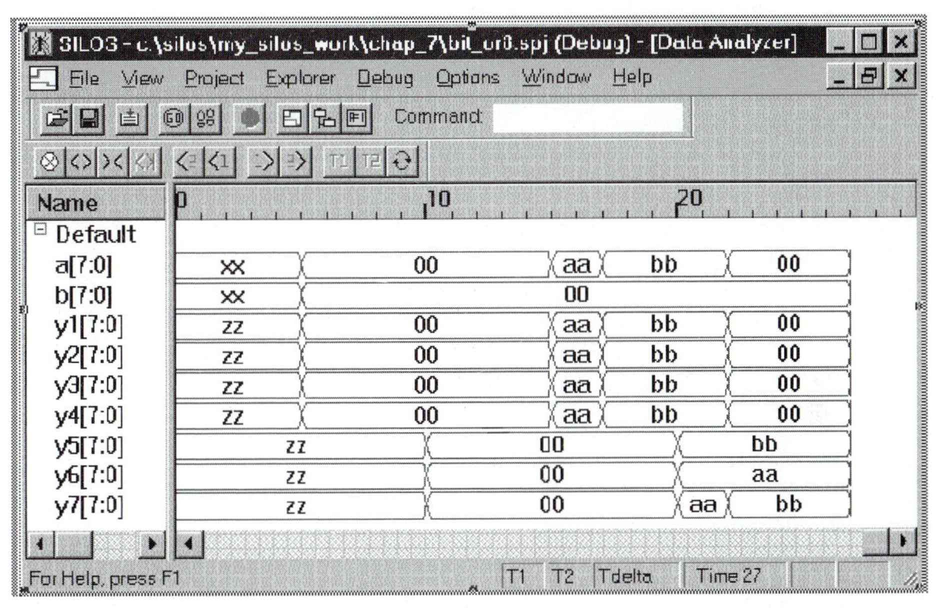

Figure 7.18 Results of simulating *bit_or8_gate1 (y1) ... bit_or8_gate7 (y7).*

7.9 SIMULATION OF SIMULTANEOUS PROCEDURAL ASSIGNMENTS

In general, multiple behaviors may assign value to the same register variable at the same time step. A simulator must determine the outcome of these multiple assignments, and distinguish between blocking and non-blocking assignments. The processing steps of the simulator are organized to establish a "stratified event queue" to order the assignments to variables in simulation. Consequently, the stratified event queue manages the assignments to registers when non-blocking and blocking assignments are being made simultaneously to the same target. At a given time step, the simulator will (1) evaluate the expressions on the RHS of all the assignments to register variables in statements that are encountered at that time step, (2) execute the blocking assignments to registers, (3) execute non-blocking assignments that do not have intra-assignment timing controls (i.e., they execute in the current time step), (4) execute past procedural assignments whose timing controls have scheduled an assignment for the current simulator time, and (5) advance the simulator time (t_{sim}). The language reference manual for Verilog refers to this organization of the simulation activity as a "stratified event queue." That is, the queue of pending simulation events is organized into five different regions, as shown in Figure 7.19.

CLASS OF EVENT	TIME OF OCCURRENCE	ORDER OF PROCESSING
ACTIVE EVENTS	CURRENT t_{sim}	ANY ORDER
INACTIVE	CURRENT t_{sim}	AFTER ALL ACTIVE EVENTS
NON-BLOCKING ASSIGN UPDATE	EVALUATED DURING A PREVIOUS SIMULATION TIME AND TO BE ASSIGNED AT THE CURRENT t_{sim}	AFTER ALL ACTIVE AND INACTIVE EVENTS
MONITOR	CURRENT t_{sim}	AFTER ALL ACTIVE, INACTIVE AND NON-BLOCKING ASSIGN UPDATE EVENTS
FUTURE 1. INACTIVE 2. NONBLOCKING ASSIGNMENT UPDATE	A FUTURE SIMULATION TIME	

Figure 7.19 Organization of the stratified event queue.

The first region, the active region, consists of events that are scheduled to occur at the current simulation time, and which have top priority for execution. These events result from (1) evaluating the LHS of a continuous assignment and updating the RHS, (2) evaluating the inputs of a primitive and changing the output, (3) executing a procedural (blocked) assignment to a register variable, and (4) evaluating and executing **$display** and **$write** system tasks. Any procedural assignments blocked by a #0 delay control are placed in the inactive queue, and execute in the next simulation cycle at the current time step of the simulator.

The order of processing active events associated with procedural assignments, continuous assignments and primitives is not specified by the LRM and is tool-dependent. For example, if an input to a module at the top level of the design hierarchy has an event at the current simulation time, as prescribed by a testbench, the event would reside in the active area of the queue. Now suppose that the input to the module is connected to a primitive having zero propagation delay and whose output is changed by the event on the input port. This event would also be scheduled to occur at the current simulation time, but would be placed in the inactive areas of the queue. If a behavior is activated by the module input and the behavior generates an event by means of a non-blocking assignment, that event would be placed in the inactive area too, but after all procedural assignments. Events that were scheduled to occur at the current simulation time, but which originated in non-blocking assignments at an earlier simulation time, would be placed in the "non-blocking assignment update" area. A fourth area, the monitor area, contains events that are to be processed after the active, inactive, and non-blocking assignment update events, such as the **$monitor** task. The last area of the stratified event queue builds a queue which consists of events to be executed in the future.

Given this organization of the event queue, the simulator executes all of the active events in a single simulation cycle. As it executes, it may add events to any of the regions of the queue, but it may only delete events from the active region. After the active region is empty, the events in the inactive region are activated (i.e., they are added to the active region) and a new simulation cycle begins. After the active and inactive regions are both empty, the non-blocking assignments that originated at previous simulation times are activated, and a new simulation cycle begins. After the monitor events have executed, the simulator advances time to the next time at which an event is scheduled to occur. Whenever an explicit #0 delay control is encountered in a behavior, the associated process is suspended and added as an inactive event for the current simulation time. The process will be resumed in the next simulation cycle at the current time.

Simulators also obey a fundamental rule: blocked assignments are placed ahead of non-blocking assignments in the active area of the queue, i.e., blocked assignments execute before non-blocked assignments that are at the same time step. There is an exception: blocking assignments triggered by blocking assignments will be scheduled after any non-blocking assignments that are already scheduled.

The distinction between the tasks **$display** and **$monitor** needs to be understood by the designer writing a testbench for simulation. The **$display** task executes immediately when it is encountered in the sequential activity flow of a behavior, but the **$monitor** task executes automatically at the end of the current time step (i.e., after the non-blocking assignments have been updated). Thus, in the code below, *execute_display* assigns value to *a* and *b*, samples the current RHS of *a* and *b*, displays the current values of *a* and *b*, then updates *a* and *b*. The values of *a* and *b* at the end of the behavior are not the values that were displayed (i.e., **$display** executes before the non-blocking assignments). On the other hand, *execute_monitor* assigns value to *c* and *d*, samples *c* and *d*, updates *c* and *d*, and then prints the values of *c* and *d*. The values of *c* and *d* when the behavior expires are the same as the values that were printed.

```
initial begin: execute_display
  a = 1;
  b = 0;
  a <= b;
  b <= a;
  $display ("display: a = %b  b= %b", a, b);
end
```

```
initial begin: execute_monitor
  c = 1;
  d= 0;
  c <= d;
  d <= c;
  $monitor ("monitor: c = %b  d= %b", c, d);
end
```

The standard output produced by the code is printed below:

```
display: a = 1   b = 0
monitor: c = 0   d = 1
```

Example 7.28 The code fragments below summarize blocking and non-blocking assignments with intra-assignment delay. It is assumed that the first statement in each set is encountered at $t_{sim} = t^*$.

Non-Blocking Assignment with Intra-Assignment Delay

```
A <= #5 B_reg;        // A(t* + 5) gets B_reg(t*)
G <= #20 D_Acc;       // G(t* + 20) gets D_Acc(t*)
```

Procedural Assignment (Blocking)

```
#5 A = B_reg;         // A(t* + 5) gets B_reg(t*+5)
#20 G = D_Acc;        // G(t* + 25) gets D_Acc(t*+25)
```

Procedural Assignment (Inter-Assignment Delay)

```
A = #5 B_reg;         // A(t* + 5) gets B_reg(t*)
G = #20 D_Acc;        // G(t* + 25) gets D_Acc(t*+5)
```

7.10 REPEATED INTRA-ASSIGNMENT DELAY

The *event_expression* in intra-assignment delay can be repeated a specified number of times, with the effect that the delay of a statement's assignment (execution) is extended.

Example 7.29 In the statement below, the **repeat** keyword specifies that the assignment to *reg_a* will be made after five falling edges of the clock. The value of *reg_b* assigned to *reg_a* is its residual value when the statement is encountered and suspended in the sequential flow, not necessarily the value it has when the first (or the last) clock edge occurs.

```
reg_a = repeat (5) @ (negedge clock) reg_b;
```

The statement is equivalent to the following statements:

```
begin
   temp = reg_b;
   @ (negedge clock);
   @ (negedge clock);
   @ (negedge clock);
   @ (negedge clock);
   @ (negedge clock);
   reg_a = temp;
end
```

Example 7.30 The effect of repeated intra-assignment delay is demonstrated in Figure 7.20 for *repeater.* Notice, that when t_{sim} = 55 *reg_b* gets the value *reg_a* had at t_{sim} = 10, not the value of *reg_a* at t_{sim} = 55.

```
module repeater;
   reg clock;
   reg reg_a, reg_b;

   initial
      clock = 0;

   initial begin
      #5   reg_a = 1;
      #10  reg_a = 0;
      #5   reg_a = 1;
      #20  reg_a = 0;
   end

   always
      #5 clock = ~ clock;

   initial
      #100 $finish;
   initial
      begin
         #10 reg_b = repeat (5) @ (posedge clock) reg_a;
      end
endmodule
```

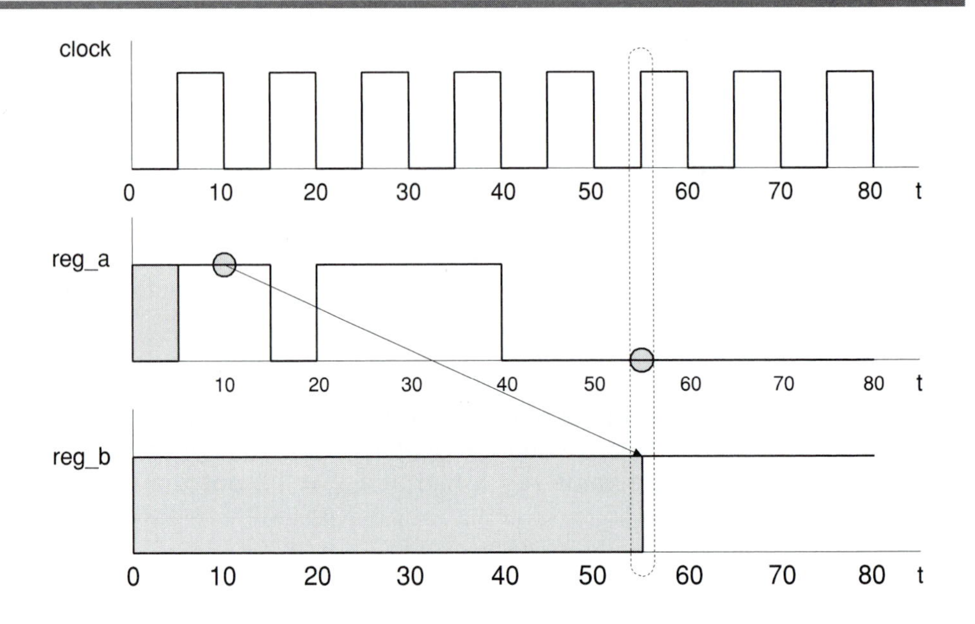

Figure 7.20 Waveforms for repeated intra-assignment delay.

7.11 INDETERMINATE ASSIGNMENTS AND AMBIGUITY

Multiple concurrent behaviors may assign value to the same register *in the same time step*. When multiple non-blocking assignments are *made by different behaviors* to the same target variable in the same time step of the simulator, the order in which the assignments are made, and therefore the outcome, is indeterminate. For a given simulator, it depends upon the order in which the code is written, and may differ between vendors for the same code. The LRM does not specify an order in which multiple behaviors are to be executed. Therefore, an order cannot be imposed on the assignments they make to a target register in the same time step. This is not a fault of the Verilog language or the simulator. It is a fact of life that race-conditions exist in hardware and in realistic software models of behavior.

Example 7.31 In *multi_assign* below, *c* and *d* differ only in the order in which their assignments are made in the source code. The outcome depends upon how a given simulator develops the queue of assignments to *c* from the source code. Changing the order in which the two non-blocking assignments to *c* are listed will change the outcome, as shown in Figure 7.21.

```
module multi_assign();
   reg a, b, c, d;
   initial #50 $finish;
   initial begin
      #5 a = 1; b = 0;
   end

   always @ (posedge a) begin
      c = a;
   end

   always @ (posedge a) begin
      c = b;
   end

   always @ (posedge a) begin
      d = b;
   end

   always @ (posedge a) begin
      d = a;
   end
endmodule
```

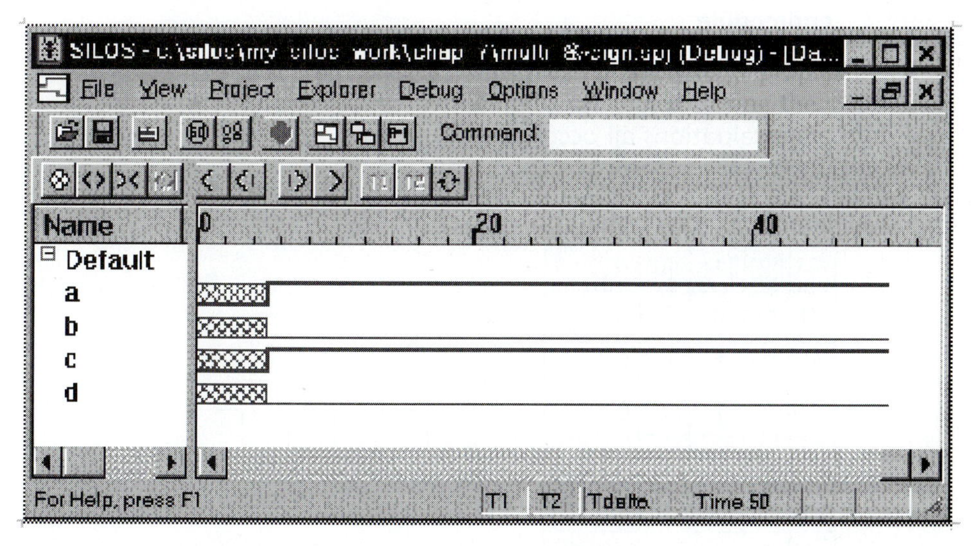

Figure 7.21 Simulation of multiple behaviors assigning value to the same register in the same time step.

Non-blocking assignments with intra-assignment delay create a schedule of assignments to the target register variable. The next example demonstrates that multiple non-blocking assignments to the same target can create non-conflicting event schedules.

Example 7.32 In *multiple_non_block_1*, the **for** loop creates multiple non-blocking assignments to *wave*, but they are stretched out in time and do not overwrite. The RHS of the multiple assignments are all sampled at $t_{sim} = 0$, but the value they sample depends upon the loop iterant, i[2:0]. The sampled values are scheduled in assignments that occur at future times depending on the loop iterant, because each assignment has a delay of # (i * 10), where * is the operator for multiplication. (Remember, the value of the delay can be determined by an expression, not necessarily a constant.)

```
module multiple_non_block_1;
  reg wave;
  reg [2:0] i;

  initial
    begin
      for (i =0; i <= 5; i=i+1)
        wave <= #(i*10) i[0];
    end
endmodule
```

Figure 7.22 shows the evolution of the successive evaluations of *i[0]* and the resulting assignments to the register holding *wave*. The evaluations all occur at $t_{sim} = 0$, and each evaluation schedules an event for *wave*.

i	i[2]	i[1]	i[0]	t_{sim}
0	0	0	0	0
1	0	1	1	0^+
2	0	1	0	0^{++}
3	0	1	1	0^{+++}
4	1	0	0	0^{++++}
5	1	0	1	0^{+++++}

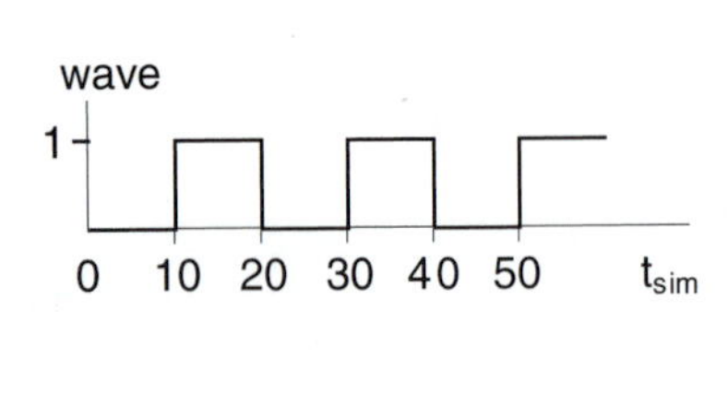

Figure 7.22 Waveform generated by repeated assignments.

Example 7.33 In *multiple_non_block_2,* multiple non-blocking assignments are made to *wave2,* but they do not overwrite because the values are distributed over the time axis. The code produces the waveforms shown in Figure 7.23 under simulation.

```
module multiple_non_block_2;
  reg wave1, wave2;
  initial
    begin
      #5   wave1 = 0;
           wave2 = 0;
           wave1 <= #5 1;
           wave2 <= #10 1;
           wave2 <= #20 0;
      #10  wave1 = 1;
           wave1 <= #5 0;
    end
endmodule
```

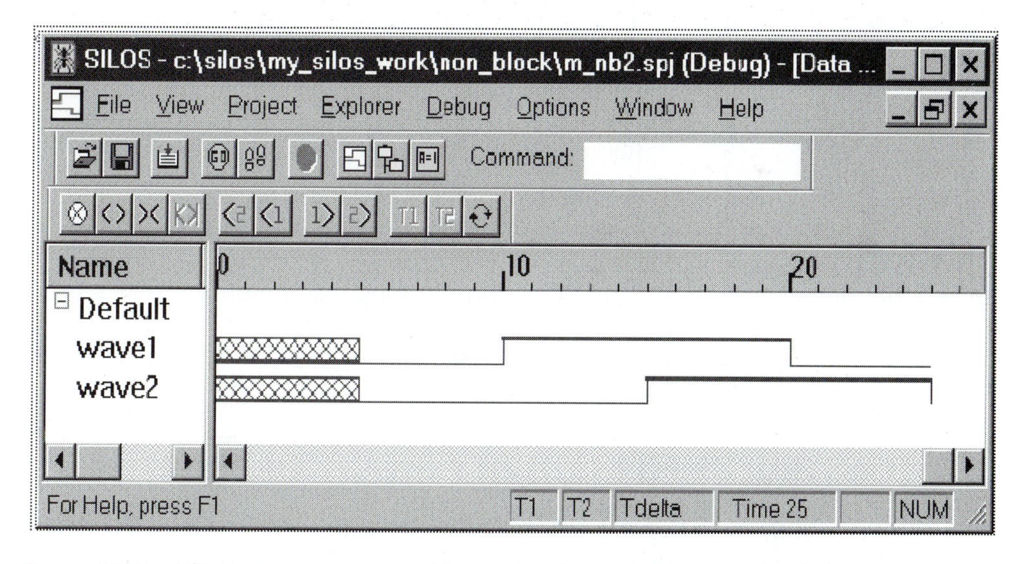

Figure 7.23 Waveform generated by multiple non-blocking assignments.

Example 7.34 In *multiple_non_block_3,* non-blocking assignments are used to derive a waveform pair from a reference signal, as shown in Figure 7.24.

```
module non_block(sig_a, sig_b, sig_c);
  reg sig_a, sig_b, sig_c;
  initial
    begin
      sig_a = 0;
      sig_b = 1;
      sig_c = 0;
    end

  always sig_c = #5 ~sig_c;

  always @ (posedge sig_c)
    begin
      sig_a <= sig_b;
      sig_b <= sig_a;
    end
endmodule
```

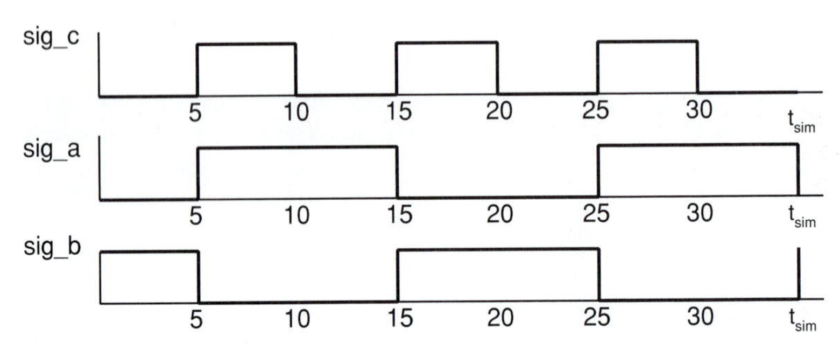

Figure 7.24 Signal generator formed by non-blocking assignments.

Figure 7.25 summarizes the styles that can be used to describe combinational and sequential logic with the Verilog HDL.

7.12 CONSTRUCTS FOR ACTIVITY FLOW CONTROL

Procedural flow control statements modify the activity flow within a behavior by determining whether to select or branch among alternative statements (**? ... :**, **case**, **if**), execute certain computations repeatedly (**repeat**, **for**, **while**, **forever**), suspend activity flow for a specified time or until a condition has been met (**wait**), branch into parallel activity flows (**fork** ... **join**), or terminate execution of a task or named block of procedural statements within a behavior (**disable**). The activity flow created by procedural statements can be sequential or parallel.

STYLE	COMBINATIONAL	SEQUENTIAL
STRUCTURAL		
Pre-defined Primitives	•	
User-Defined Primitives	•	•
Instantiated Modules	•	•
BEHAVIORAL		
Continuous Assignment (assign)	•	
Procedural assignment (=)	•	•
Non-Blocking Assignment (<=)	•	•
Procedural Continuous Assignment		
assign ... deassign	•	•
force ... release	•	•

Figure 7.25 Descriptive styles for behavior modeling.

7.12.1 Activity Flow Control: Conditional Operator (? ... :)

The conditional operator (**?** ... **:**) was discussed in Chapter 4, where it was used in continuous assignment statements. The operator can also be used in procedural statements assigning value to a register within a Verilog behavior.

Example 7.35 The conditional operator in *mux_behavior* implements a selection between two operations on datapaths *a* and *b*.

```
module mux_behavior (y_out, clock, reset, sel a, b);
  input             clock, reset, sel;
  input  [15:0]     a, b;
  output [15:0]     y_out;
  reg    [15:0]     y_out;

  always @ (posedge clock or negedge reset)
    if (reset == 0) y_out = 0; else y_out = (sel) ? a + b : a - b;
endmodule
```

7.12.2 Activity Flow Control: The case Statement (case, casex, casez)

The **case** statement controls an activity flow by executing one of several statements, depending upon the value of a match between an *expression* and a *case_item*, according to the syntax shown in Figure 7.26. The **case** statement requires an exact bitwise match in selecting the *case_item*.

Verilog Syntax: case Statement

case_statement ::= **case** (expression) case_item {case_item} **endcase**
| **casex** (expression) case_item {case_item} **endcase**
| **casez** (expression) case_item {case_item} **endcase**

case_item ::= expression {, expression} : statement_or_null
| **default** [:] statement_or_null

Figure 7.26 Syntax for the **case** statement.

Notice that the syntax of the *case_item* may include a comma-separated list of items, all of which decode to the same expression. Only the first is executed (i.e., multiple executions do not occur). The statement associated with a *case_item* may be a single statement or a block statement.

Example 7.36 The description of a four-channel multiplexer in Figure 7.27 has a single behavior using a **case** statement with a **case** *expression* consisting of the *select* word. An event on a datapath or *select* causes *y_out* to be computed. This model is correct, but inefficient in simulation—it reacts to all channel activity independently of the channel selection.

```
module mux4_case(a, b, c, d, select, y_out);
  input          a, b, c, d;
  input  [1:0]   select;
  output         y_out;
  reg            y_out;

  always @ (a or b or c or d or select)
  begin
    case (select)
      0: y_out = a;
      1: y_out = b;
      2: y_out = c;
      3: y_out = d;
      default y_out = 1'bx;
    endcase
endmodule
```

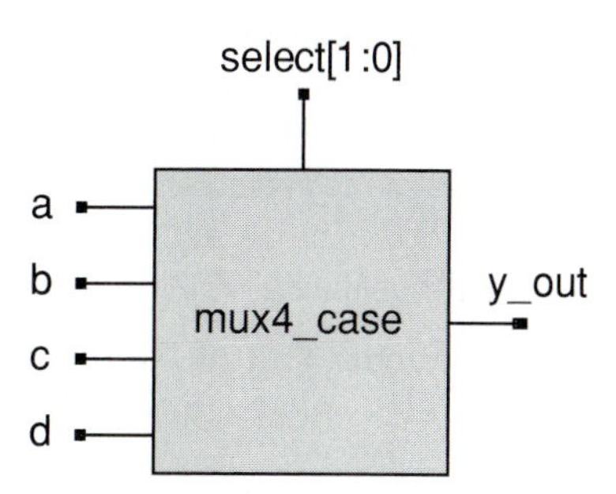

Figure 7.27 Four-channel multiplexer.

In simulation, the **case** *expression*, which may be a constant, is evaluated in Verilog's four-value logic system. The *case_item* expressions are examined in the order in which they are listed, and if a match is found between the value of the case *expression* and a *case_item*, the corresponding statement executes. The search terminates if a match is found. If no match is found and the (optional) default *case_item* is given, the default statement will execute. The **case** construct requires a complete bitwise (0, 1, x, z) match between the **case** *expression* and *case_item* in Verilog's four-valued logic system. This requires that the *expression* and each *case-item* have the same bit-length.

Two other variations of the **case** construct treat "don't care" situations in simulation. The **casex** construct ignores values in those bit positions of the case *expression* or the *case_item* that have the value "x" or "z"; this effectively treats those bit positions as "don't cares," because a match will be found if an "x" value occurs in the case_item (**case** *expression*) independently of the value in the case *expression* (*case_item*). The **casez** construct ignores any bit position of the **case** *expression* or *case_item* that have value "z". The **casez** construct also uses the literal "**?**" character as an explicit "don't care". A typical use is in a machine having an instruction decoder that must decode a large number of "don't care" conditions.

Example 7.37 Suppose a machine includes the instruction-decode behavior shown below:

```
always @ (decode_pulse)
  casez (instruction_word)
    16'b0000_????_????_???? : ;        // Null statement for no-op
    ...

endcase
```

Here, the effort of writing several additional lines of code to represent all possible bit patterns is eliminated by the **casez** statement with the "**?**" don't-care.

The **case** statement executes a statement on the basis of evaluation/match of a single expression, and is less general than the **if** statement, which can evaluate different expressions on each branch. Note: In simulation, the **default** *case_item* is optional. However, in synthesis it is essential that each hardware-compatible *case_item* (i.e., those involving patterns of 0 or 1) be considered. Otherwise, the result of synthesis might be unwanted latches. (See Chapter 8.)

Table 7.2 summarizes the treatment of "don't care"s by the case constructs. Note that the "?" option applies only to the **casez** construct.

Table 7.2 *Treatment of "don't care"s in* ***case*** *constructs.*

Expression or case_item	case	casex	casez
0	0	0	0
1	1	1	1
x	x	0 1 x z	x
z	z	0 1 x z	0 1 x z
?	*	*	0 1 x z
default	0 1 x z	0 1 x z	0 1 x z

* not applicable

7.12.3 Activity Flow Control: Conditional Statement (if ... else)

Conditional statements, or branch statements, alter the ordinary sequential flow of activity within a behavior by allowing a choice of execution among alternative statements, based upon the Boolean value of an expression. The three variants of the "**if**" statement given in Figure 7.28 provide flexibility in describing the conditions that determine the activity flow within a behavior.

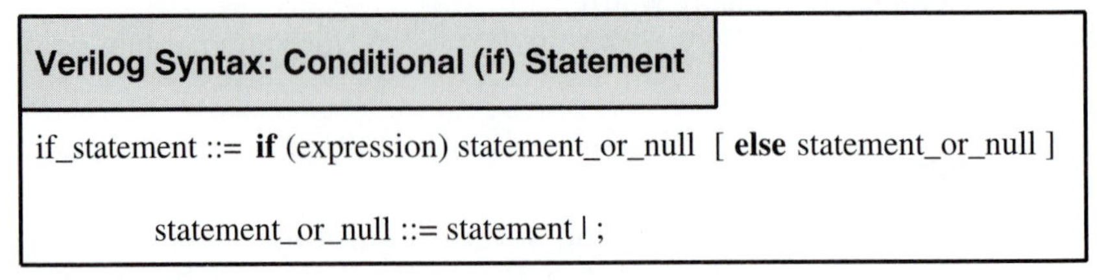

Verilog Syntax: Conditional (if) Statement
if_statement ::= if (expression) statement_or_null [else statement_or_null]
statement_or_null ::= statement | ;

Figure 7.28 Syntax of the conditional statement.

The statement executed with the "**if**" can be a single or block statement; a null statement (used for readability) must terminate with a semicolon.

Example 7.38 The simplest form of a conditional branch executes a single statement if a Boolean condition is true.

```
(a)  if (A < B) some_register = some_value + 1;
(b)  if (C < D);              // null statement
```

A block statement can also be executed:

```
(c)  if (k==1)                // Note use of the == logical operator
       begin : A_Block        // for case equality. See Chapter 4.
         sum_out = sum_reg(4);
         c_out = c_reg(2);
       end
```

The value of the Boolean expression evaluated in the **if** statement is treated as false if it has the numerical value of "0", or the values of "x", or "z". A non-zero numeric expression will evaluate to "true". If the evaluation is "false," the execution defaults to the next statement in the activity flow.

The second form of the "**if**" statement provides statements for both branches of the "**if**".

Example 7.39

```
...
   if (Value_1 < Value_2)
     sum = sum + 1;
   else
     sum = sum + 2;
...
```

Example 7.40

```
always @ (a or b)
  begin
    if (a >= b)
    begin
      neg_result = 0;
      diff = a-b;
    end
  else
    begin
      neg_result = 1;
      diff = b-a;
    end
  end
...
```

The "**if**" statement can be nested to direct the activity flow among several paths.

Example 7.41

```
if (A == 1) Sig_out = reg_a; else
   if (A == 2) Sig_out reg_b; else
      if (A ==3) Sig_out = reg_c;
```

7.12.4 Activity Flow Control: Loops

Verilog has four looping mechanisms that allow procedural statements to be executed repeatedly within an activity flow: **repeat**, **for**, **while**, and **forever**.

7.12.4.1 The **repeat** Loop

The **repeat** loop executes an associated statement or block of statements a specified number of times. The syntax of the **repeat** loop is shown in Figure 7.29.

Verilog Syntax: The repeat Loop
repeat_loop ::= **repeat** (expression) statement

Figure 7.29 Syntax of the **repeat** loop.

When the activity flow within a behavior reaches the **repeat** keyword, *expression* is evaluated once to determine the number of times that the statement is to be executed. If expression evaluates to "x" or "z", the result will be treated as 0, and no statement shall be executed. Otherwise, execution repeats for the specified number of times unless it is prematurely terminated by a "**disable**" statement within the activity flow.

Example 7.42 A **repeat** loop is used in the fragment of code below to initialize a memory array.

```
...
word_address = 0;
repeat (memory_size)
   begin
      memory [word_address] = 0;
      word_address = word_address + 1;
   end
...
```

7.12.4.2 The **for** Loop

The **for** loop executes an associated statement or block of statements repeatedly under the condition that an "end_of_loop" *expression* evaluates true. The syntax of the **for** loop is shown in Figure 7.30.

Verilog Syntax: The for Loop

for_loop ::= **for** (reg_assignment ; expression ; reg_assignment) statement

Figure 7.30 Syntax of the **for** loop.

At the beginning of execution of the **for** loop, the *reg_assignment* statement executes once, usually to initialize a register variable that controls the loop. If the *expression* is "TRUE", the *statement* will execute. After the *statement* has executed, the second ("loop_update") *reg*_assignment statement will execute. Then the activity flow will return to the beginning of the **for** statement and check the value of *expression* again. If *expression* is "FALSE", the loop is complete and the activity flow will proceed to the statement that immediately follows the **for** statement. (Note: The value of the register variable governed by the second assignment in the **for** loop may be changed in the body of the loop during execution.) The register variable must be an **integer** or a **reg**.

Example 7.43 In this example, the **for** loop is used to assign value to two fields of bits within a register after it has been initialized to "x". The results of executing the loop are shown in Figure 7.31.

```
reg [15:0] demo_register;
integer K;

...
for (K = 4; K; K = K-1)
  begin
    demo_register [K+ 10] = 0;
    demo_register [K+2] = 1;
  end
...
```

15	14	13	12	11	10	9	8	7	6	5	4	3	2	1	0
x	0	0	0	0	x	x	x	x	1	1	1	1	x	x	x

Figure 7.31 Register contents after execution of the **for** loop.

At the beginning of execution, the initial *reg_assignment* executes and assigns the value "4" to K. Thus, the end of loop *expression* is a "TRUE" value. The assignments to *demo_register* are made, and then K is decremented. This process continues until a decrementation assigns the value $K = 0$. This produces a "FALSE" value for the end_of_loop *expression* and control passes to whatever statement follows the **for** loop statement.

Example 7.44 Now consider what happens in the code fragment below:

```
reg [3:0] K
for (K=0; K<=15; K = K+1) ...
```

The size of register K is such that, at the end of the iterant with K = 15, the loop update mechanism will set K to 0, which satisfies the expression that determines whether to re-execute the statement. Consequently, the loop will execute endlessly.

Example 7.45 Adders are important elements of data paths in digital hardware. Their performance (speed) plays a critical role in determining the overall performance of the host system. Various adders can be evaluated on the basis of their computational speed and silicon area. When speed is important, carry look-ahead adders are often used in place of ripple-carry adders.

The description of a carry look-ahead adder is based upon the observation that the value of the "carry" into any stage of a multi-cell adder depends upon the data bits of the previous stages and the carry into the first stage. This relationship can be exploited to improve the speed of the adder by using additional logic to implement the carry, rather than waiting for the value to propagate through the cells of the adder.

A given cell is said to "generate" a carry if both of the cell's data bits are "1". A cell is said to "propagate" a carry if either of the cell's data bits could combine with the carry into the cell to cause a carry out to the next stage of the adder. Let a_i and b_i be the data bits at the i-th cell of the adder; let c_i be the carry *into* the i-th cell, let s_i be the sum bit *out* of the i-th cell, and let c_{i+1} be the carry out of the cell. We define "generate" and "propagate" bits g_i and p_i using the bitwise-and operator ("&") and the exclusive-or operator ("^") as follows:

```
gi = ai & bi
pi = ai ^ bI
```

The Venn diagram in Figure 7.32 shows where p_i and g_i are asserted, depending upon a_i and b_i. Note that p_i and g_i are mutually exclusive.

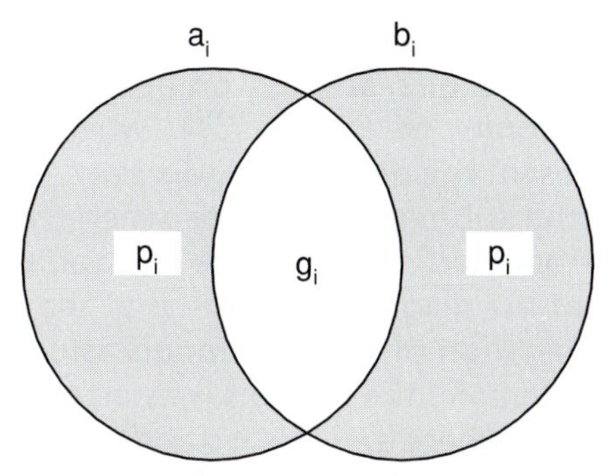

Figure 7.32 Venn diagrams for assertion of p_i (propagate) and g_i (generate) for an adder cell.

The ***logical*** expressions forming the sum and carry bits at each stage of the adder can be written in terms of bitwise operators as follows:

$s_i = (a_i \wedge b_i) \wedge c_i = p_i \wedge c_i$
$c_{i+1} = (a_i \wedge b_i) \,\&\, c_i \mid a_i \,\&\, b_i = p_i \,\&\, c_i \mid g_i$

Note that, because p_i and g_i are mutually exclusive, the algorithm can also be expressed in ***arithmetic*** terms as:

$s_i = (a_i \wedge b_i) \wedge c_i = p_i \wedge c_i$
$c_{i+1} = (a_i \wedge b_i) \,\&\, c_i + a_i \,\&\, b_i = p_i \,\&\, c_i + g_i$

The carry bit can be formed using either the bitwise "or" operator or the arithmetic sum (modulo 2). The schematic for the subcircuit implementing the algorithmic version of the carry is shown in Figure 7.33.

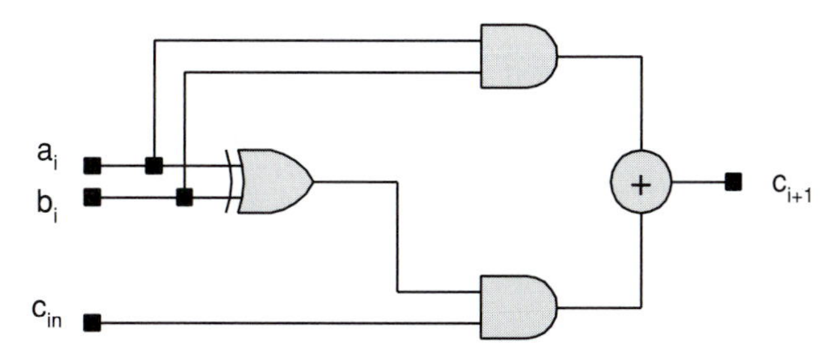

Figure 7.33 Schematic for arithmetic implementation of carry bit.

The carry out of the i-th cell is formed by ***adding*** the bit propagated by the cell with the bit generated by the cell. Only one of the terms will be "1", because p_i and g_i are mutually exclusive. This second form of the equation for c_{i+1} produces the same result as the first one.

The dependencies of s_i and c_{i+1} on a_i, b_i and c_i are depicted in the Venn diagram of Figure 7.34, where the three circular regions represent the data inputs to a cell of the adder, a_i, b_i and c_i; each of the sub-regions denote where the data outputs, s_i and c_{i+1}, are asserted. The presence of a variable's label within a sub-region of the diagram indicates that the variable is asserted for the combination of the data inputs associated with the sub-region. For example, the sum bit is asserted in four sub-regions of the diagram: where $a_i = 1$, $b_i = 0$ and $c_i = 0$; where $a_i = 0$, $b_i = 1$ and $c_i = 0$; where $a_i = 1$, $b_i = 1$ and $c_i = 1$; or where $a_i = 0$, $b_i = 0$ and $c_i = 1$.

The first three cells of the adder have:

$s_0 = p_0$ ^ c_0
$c_1 = (p_0$ & $c_0) + g_0$

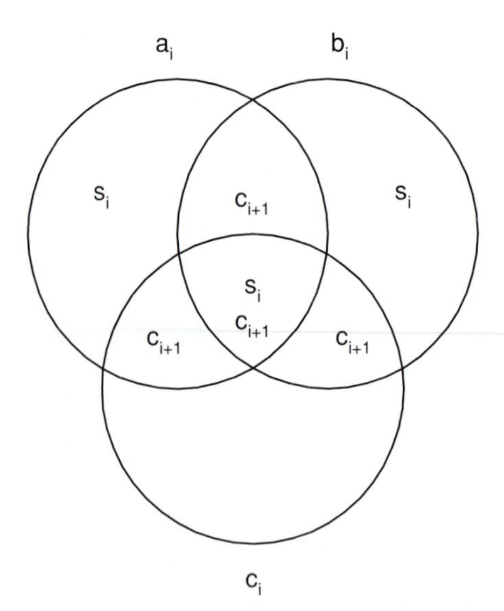

Figure 7.34 Data input/output relationships of an adder cell.

$s_1 = p_1 \wedge c_1 = p_1 \wedge [(p_0 \& c_0) + g_0] = p_1 \wedge (p_0 \& c_0) + p_1 \wedge g_0$

$c_2 = p_1 \& c_1 + g_1 = p_1 \& p_0 \& c_0 + p_1 \& g_0 + g_1$

$s_2 = p_2 \wedge c_2 = p_2 \wedge [p_1 \& p_0 \& c_0 + p_1 \& g_0 + g_1]$
$c_3 = p_2 \& c_2 + g_2 = p_2 \& [p_1 \& p_0 \& c_0 + p_1 \& g_0 + g_1] + g_2$
$= p_2 \& p_1 \& p_0 \& c_0 + p_2 \& p_1 \& g_0 + p_2 \& g_1 + g_0$

This exposes the fact that the sum and carry-out bits of each cell can be expressed in terms of the data bits of that cell and the previous cells, and the carry into only the first cell of the adder chain. All of this data is available simultaneously, so there is no need to wait for a carry bit to propagate through the adder to a particular cell. This allows the adder to operate faster, but the cost of this improvement is the extra logic needed to compute the sum and carry-out of each stage. The gate-level implementation of the look-ahead adder requires considerably more silicon area than the ripple-carry adder implemented in the same technology. (Look-ahead carry is usually implemented on a bit slice of a word.)

Another important observation is that the sum and carry bits at each cell can be computed recursively. To expose this, we write:

$s_0 = p_0 \wedge c_0$
$c_1 = p_0 \& c_0 + g_0$

$s_1 = p_1 \wedge c_1$
$c_2 = p_1 \& c_1 + g_1$

$s_2 = p_2 \wedge c_2$
$c_3 = p_2 \& c_2 + g_2$
...

The algorithm to implement the addition will take as many steps as there are cells in the adder. The computation at each step of the recursion depends upon the data bits of the corresponding cell and the carry that was calculated at the immediately previous step. If the propagate bits of an n-bit adder are arranged in a vector $p = (p_{n-1}, ..., p_2, p_1, p_0)$, the results of the recursion can be used to form an n+1 dimensional vector: $(c_n, ..., c_2, c_1, c_0)$ such that the output word and *c_out* bit are obtained as:

sum = p ^ (c_{n-1}, ..., c_2, c_1,c_0).

c_out = c_n

The algorithm[1] for a four-bit carry look-ahead adder is implemented below. Figure 7.35 depicts the overall scheme to implement the behavioral/algorithmic description described in *Add_Prop_Gen*. The placements of the wire declarations/assignments in the figure are organized to correspond to the data flow in Figure 7.35.

```
module Add_prop_gen (sum, c_out, a, b, c_in); // 4-bit look-ahead carry adder
                                              // behavioral model
  output [3:0]   sum;
  output         c_out;
  input  [3:0]   a, b;
  input          c_in;

  reg    [3:0] carrychain;
  wire   [3:0] g = a & b; // carry generate, continuous assignment, bitwise and
  wire   [3:0] p = a ^ b; // carry propagate, continuous assignment, bitwise xor

  always @ (a or b or c_in)                      // event "or"
    begin : carry_generation                     // usage: block name
      integer i;
      #0 carrychain[0] = g[0] + (p[0] & c_in);   // needed for simulation
      for(i = 1; i <= 3; i = i + 1)
        begin
          carrychain[i] = g[i] | (p[i] & carrychain[i-1]);
        end
    end

  wire [4:0] shiftedcarry = {carrychain, c_in} ; // concatenation
  wire [3:0] sum = p ^ shiftedcarry;             // summation
  wire c_out = shiftedcarry[4];                  // carry out, usage: bit select
endmodule
```

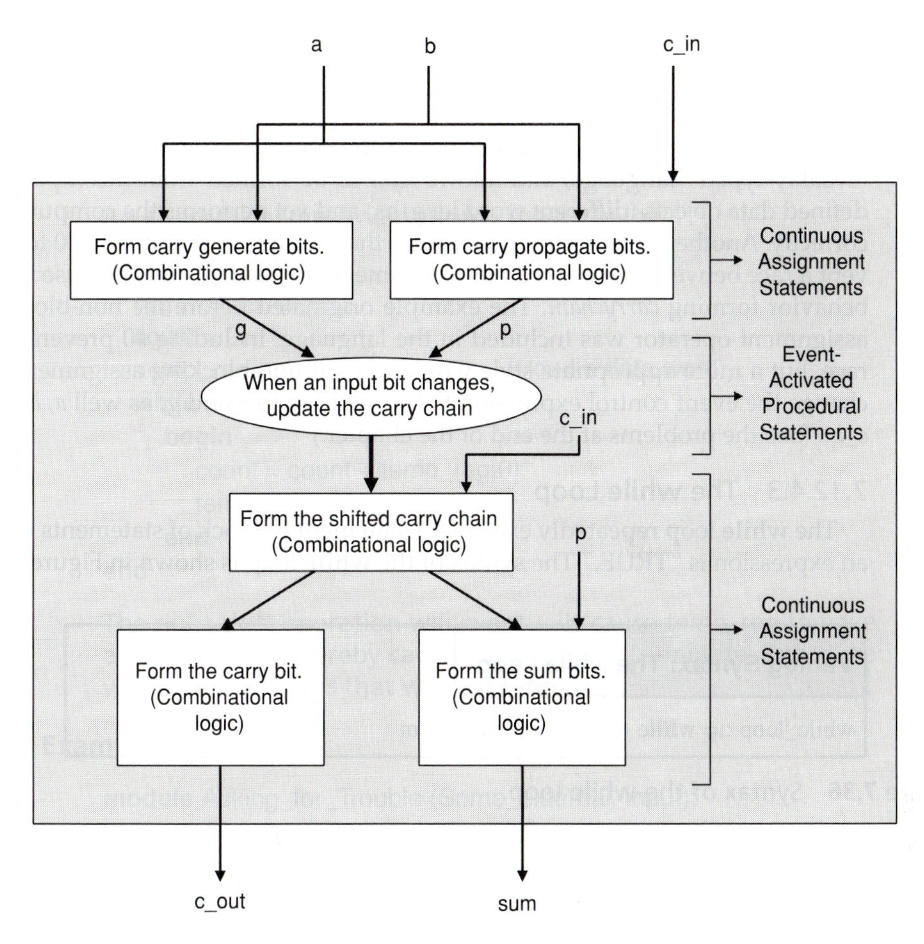

Figure 7.35 Organization of a carry look-ahead algorithm.

This description of an adder combines continuous assignment statements with procedural code. The generate vector is formed from the bitwise "and" of the data words; the propagate vector is formed as the "exclusive or" of the data words. The **always** statement is activated at the beginning of simulation and repeatedly re-executes throughout the simulation, subject to the flow control statements within it. The behavior (***always* @** *(a* ***or*** *b* ***or*** *c_in)* ***begin*** *...* ***end***) begins execution when there is a change in value in the data words or the *c_in* bit, and then computes a new value for the *carrychain* register. If the new value differs from the residual value, the event will be detected and new values of the *shifted-carry, sum* and *c_out* values will be computed by continuous assignments. Note: *{carrychain, c_in}* forms a five-bit result from

7.12.4.4 The **forever** Loop

The **forever** loop causes unconditional repetitive execution of statements, possibly subject to the **disable** statement.

Example 7.48 Clocks and pulsetrains in testbenches are easily described with a **forever** loop. The code below produces the waveforms in Figure 7.37 under simulation. This example also illustrates how the activity of an **initial** behavior may continue for the duration of a simulation.

```
parameter half_cycle = 50;

initial
  begin: clock_loop
    clock = 0;
    forever
      begin
        #half_cycle clock = 1;
        #half_cycle clock = 0;
      end
  end

initial
  #350 disable clock_loop;
```

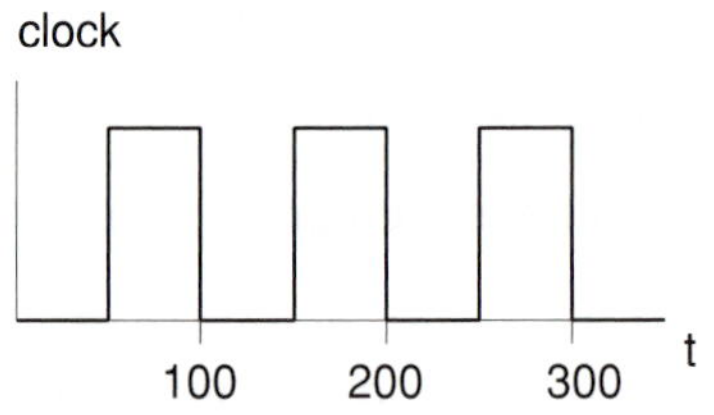

Figure 7.37 Pulse train described by a **forever** loop.

The syntax of the **forever** loop is given in Figure 7.38.

Verilog Syntax: The forever Loop
forever_loop ::= **forever** statement

Figure 7.38 Syntax of the **forever** loop.

7.12.4.5 Comparison of Loops

In many situations, loops can be constructed using any of the four basic looping mechanisms of Verilog. Table 7.3 compares the features of loops and their mechanisms for terminating during simulation.

Table 7.3 *Features and mechanisms of termination for loops.*

LOOP	FEATURE	TERMINATION
repeat	Expression determines a constant number of repetitions	"false" expression or "disable"
for	Loop variable determines number of repetitions	"end" of "disable"
while	Iterates while expression evaluates "true"	"end" or "disable"
forever	Iterates Indefinitely	"disable

7.12.4.6 Comparison of "**always**" and "**forever**"

Do not be misled into thinking that "**always**" and "**forever**" are one and the same construct just because they both are cyclic. First, the "**always**" construct declares a concurrent behavior. The **forever** loop is a computational activity flow. Its execution is not necessarily concurrent with any other activity flow. The second significant distinction is that "**forever**" loops can be nested; behaviors may not be nested. Finally, a "**forever**" loop executes only when it is reached within a sequential activity flow. An "**always**" behavior becomes active and can execute at the beginning of simulation.

7.12.5 THE disable STATEMENT

The **disable** statement is used to prematurely terminate a named block of procedural statements or a task. The effect of executing the **disable** is to transfer the activity flow to the statement that immediately follows the named block or task in which **disable** was encountered during simulation. (Do not confuse **disable** with **$finish**, which terminates simulation.)

Example 7.49 The *find_first_one* module below finds the location of the first "1" in a 16-bit word. When **disable** executes, the activity flow exits the **for** loop, proceeds to **end**, and then returns to the **always** to await the next event on *trigger.* At that time, *index_value* holds the value at which *A_word* is one.

```
module find_first_one(A_word, trigger, index_value);
  input [15:0] A_word;
  input trigger;
  output [3:0] index_value;
  reg [3:0] index_value;

  always @ trigger
  begin
    index_value = 0;
    for (index_value=0; index_value <= 15; index_value = index_value + 1);
      if (A_word [index_value] == 1) disable;
  end
endmodule
```

Example 7.50 The *stand_alone_digital_machine* module shown below executes endlessly. When a condition causes an internal **disable** to execute, *Perform_machine_tasks* aborts, returning control back to the activity of the *Perform_power_up_resets* task.

```
module stand_alone_digital_machine;
  always
    begin
      Perform_power_up_resets;       // User-defined task
      forever
        begin
          Perform_machine_tasks;   // User-defined task
        end
    end
endmodule;
```

7.12.6 PARALLEL ACTIVITY FLOW: THE fork ... join STATEMENT

Verilog behavioral statements are not limited to modeling serial/sequential activity flow. The **fork ... join** keyword pair encapsulates procedural statements and creates parallel threads of activity (see *par_block* in the formal syntax), each executing concurrently with the others. The syntax of the **fork ... join** construct is shown in Figure 7.39.

Verilog Syntax: The fork ... join Parallel Block

```
fork
  statement_1
  statement_2
  ...
join
```

Figure 7.39 Syntax of the **fork ... join** construct.

The **fork** ... **join** construct is not supported by synthesis tools. Nonetheless, it is an important feature of the language because it conveniently supports waveform generation in testbenches.

When the **fork** keyword is encountered in simulation, *statement_1* and *statement_2*, etc., begin their activity. Each statement may be a block statement, and when all of them have expired the activity flow will continue with the statement that immediately follows the **join** keyword. The statement that follows the **fork ... join** cannot execute until all the activity of the parallel threads within the **fork ... join** is complete.

Example 7.51 The **fork ... join** can be used to implement absolute timing within a signal generator. Each assignment to *sig_wave* below is scheduled relative to $t_{sim} = 0$.

```
...
fork        // t_sim = 0
   #50      sig_wave = 'b1;
   #100     sig_wave = 'b0;
   #150     sig_wave = 'b1;
   #300     sig_wave = 'b0;     //Executes at t_sim = 300
join
...
```

The simulated waveform of sig_wave is shown in Figure 7.40.

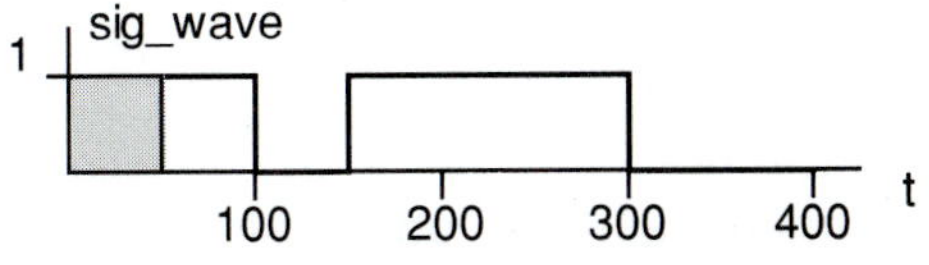

Figure 7.40 Waveform generated by a **fork** ... **join** statement.

In contrast, a sequential block is implemented with the same procedural assignments in the code below:

```
begin
   #50    sig_wave = 'b1;
   #100   sig_wave = 'b0;
   #150   sig_wave = 'b1;
   #300   sig_wave = 'b0;    // Executes at t_sim = 600
end
```

Notice the change in the waveform of sig_wave, as shown in Figure 7.41.

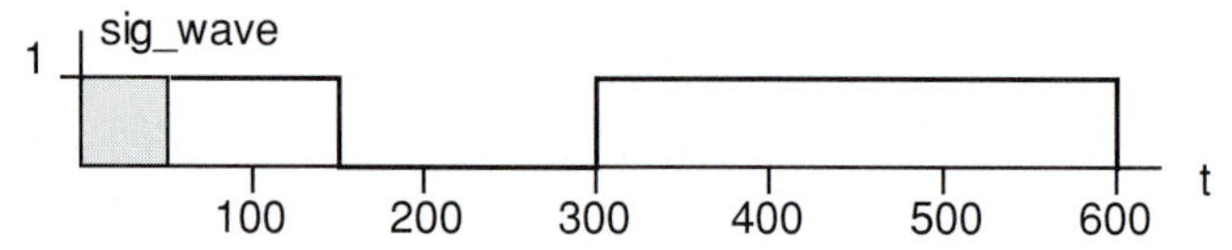

Figure 7.41 Waveform generated by procedural assignment statements.

In a sequential block, the delay value associated with a statement by the "#" delay control operator is ***relative*** to the time of execution of the previous statement. In the **fork ... join**, the delay associated with a statement is relative to the simulation time of the enclosing **fork** and the delays are local to each thread.

It is important to note that the order in which the statements within **fork ... join** are executed does not depend upon the order in which they are written. Changing the order of procedural assignments in the first code fragment in Example 7.50 will produce the same simulation results; changing the order of the statements in the sequential block changes their order of execution, and the result.

7.12.7 RACE CONDITIONS AND THE fork ... join STATEMENT

Care must be taken to avoid race conditions when the same register variable is assigned value and referenced simultaneously in multiple activity threads in a parallel block. The outcome of simulation will be indeterminate (i.e., vendor-dependent and unpredictable).

Example 7.52 Note that *register_a* in *race_condition* is simultaneously assigned value in one statement and referenced in another. Therefore, a software race exists, because the outcome depends upon the order of execution, and there is no way to determine which statement will execute first. The LRM does not specify which statement should execute first.

```
module race_condition ( ... );
  ...
  fork                                  // Parallel activity
    #150 register_a = register_b;       // Delayed execution
    #150 register_c = register_a;       // Delayed execution
  join
endmodule
```

The race condition can be eliminated by placing intra-assignment delay in the statements:

```
module fix_of_race_condition( ... );
  ...
    fork                                    // Parallel activity
      register_a = #150 register_b;         // Intra-assignment delay
      register_c = #150 register_a;         // Intra-assignment delay
    join
endmodule
```

The values of *register_a* and *register_b* are sampled before assignments are made.

7.13 TASKS AND FUNCTIONS

Verilog has two types of sub-programs that encapsulate and organize a description: tasks and functions. Tasks create a hierarchical organization of the procedural statements within a Verilog behavior; functions substitute for an expression. Tasks and functions let designers manage a smaller segment of code. Both facilitate a readable style of code, with a single identifier conveying the meaning of many lines of code. Encapsulation of Verilog code into tasks or functions hides the details of an implementation from the outside world. Overall, tasks and functions encourage the readability, portability, and maintainability of code.

7.13.1 TASKS

Tasks are declared within a module and may be referenced only from within a behavior. A task can have parameters passed to it, and the results of executing the task can be passed back to the environment. When a task is called, copies of the parameters in the environment are associated with the inputs, outputs and inouts of the task, according to the order in which they are declared. The variables in the environment are visible to the task. Additional, local variables may be declared within a task. A word of caution: A task can call itself, but the memory supporting the variables of a task is shared by all calls. The language does not support recursion, so we should anticipate side-effects. The syntax of a task declaration is shown in Figure 7.42.

Verilog Syntax: Task Declaration

task_declaration ::=
  task task_identifier;
  { task_item_declaration }
  statement_or_null
  endtask

Figure 7.42 Syntax of a task declaration.

A task must be named, and may include declarations of any number or combination of the following: **parameter**, **input**, **output**, **inout**, **reg**, **integer**, **real**, **time**, **realtime**, and **event**. All declarations of variables are local to the task. The arguments of the task retain the type they hold in the environment that invokes the task. For example, if a wire bus is passed to the task, it may not have its value altered by an assignment statement within the task. All arguments to the task are passed by value—not by a pointer to the value. When a task is invoked, its formal and actual arguments are associated in the order in which the task's ports have been declared.

Example 7.53 A task that receives a binary word and returns the number of ones in the word is declared in *Bit_Counter*. The environment in which the task is invoked passes the value of *reg_a* to *count_ones_in_word* and receives count when the task completes execution.

```
module Bit_Counter (data_word, bit_count);
  input  [7:0]     data_word;
  output [3:0]     bit_count;
  reg    [3:0]     bit_count;

  always @ (data_word)
    count_ones_in_word (data_word, bit_count);

  task count_ones_in_word;
    input  [7:0]   reg_a;
    output [3:0]   count;
    reg            count;
    reg    [7:0]   temp_reg;
```

```
        begin
          count = 0;
          temp_reg = reg_a;     // load a data word
          while (temp_reg)
            begin
              count = count + temp_reg[0];
              temp_reg = temp_reg >> 1;
            end
        end
      endtask
    endmodule
```

7.13.2 Rules for Tasks

The rules that must be observed when using Verilog tasks are summarized in Figure 7.43.

7.13.3 Functions

Functions may implement only combinational behavior; i.e., they compute a value on the basis of the present value of the parameters passed into the function. Consequently, they may not contain timing controls (no delay control, event control, or **wait** statements), and may not invoke a task. They may, however, call other functions, but not recursively. A function must be declared within a module, and may contain a declaration of inputs and local variables. Verilog functions are referenced within an expression anywhere that an expression is valid—for example, in the RHS of a continuous assignment statement. A function is also implemented by expression and returns a value at the location of the function's identifier. The value of a function is returned by its name when the expression calling the function is executed. Consequently, a function may not have any declared **output** or **inout** port (argument). It must have at least one **input** argument. The execution, or evaluation, of a function takes place in zero time; i.e., in the same time step that the calling expression is evaluated by the host simulator. The definition of a function implicitly defines an internal register variable having the same name, range, and type as the function itself; this variable must be assigned value within the function body.

The syntax for a function declaration and function call are shown in Figure 7.44. If the optional *range_or_type* is omitted, the value returned by the function is a scalar **reg** variable.

7.15 SYSTEM TASKS FOR TIMING CHECKS

The timing characteristics of the gates used by a synthesis tool are determined by a particular fabrication process technology. On the other hand, the behavioral description from which the design was synthesized is independent of technology. The post-synthesis design flow must verify that the timing of the synthesized circuit satisfies the timing constraints of the design. Synthesis tools consider the timing delays of individual gates, but ignore global timing issues caused by interconnect delays in the physical layout. The realized timing of a circuit is determined by layout (e.g., placement and routing in a cell-based design) within a particular technology. Timing analysis must be performed on the placed-and-routed circuit to ensure that the design operates at the specified clock speeds. Timing analysis increases the confidence that (1) the setup and hold conditions of flip-flops are satisfied under all operating conditions, (2) clocks are not compromised by skew, (3) flip-flops resume synchronous operation with acceptable recovery times after a reset or clear condition has been de-asserted, and (4) hierarchical interface timing specifications are met.

Today's synthesis tools are limited in their ability to take into account the capacitive loading that results from the interconnect delays and fanout loading of gates in the actual circuit (after the place-and-route step). Consequently, the delays used in synthesis are only estimates of the actual delays in the working silicon. Parasitic capacitance values must be extracted from the physical layout and used to back-annotate the gate-level model of the circuit to provide more accurate estimates of the timing characteristics of critical paths. This verification can be done at the gate level and, if necessary, the analog transistor level along critical paths in the circuit. In deep sub-micron design, parasitic interconnect capacitance plays a dominant role, yet synthesis tools lack the ability to be driven by these effects a-priori.

There are two basic forms of timing analysis. Static analysis considers the structural topology of a circuit, enumerates the paths along which signals can propagate, and determines whether certain timing constraints are met. Dynamic timing analysis verifies the timing of a circuit through simulation. Both methods have advantages and limitations.

Static timing analysis has the advantage that it considers all possible paths through the circuit. However, unless the analysis tool is carefully written, the results of static analysis can overwhelm the designer by generating timing reports on so-called "false paths" in the circuit (i.e., paths which are never exercised in actual operation). But static analysis does not miss reporting a timing violation, within the accuracy of the data base describing signal paths in the circuit.

On the other hand, dynamic timing analysis is only as good as the stimulus set used to simulate the circuit. If the stimulus set fails to exercise all functional paths, timing violations can be missed. But dynamic timing analysis does not report false alarms. As circuits have become larger, this approach has become less practical and reliable for global verification, but it is still used on the critical paths of circuits.

Verilog has several predefined system tasks for dynamic and static timing analysis. These tasks can be invoked within a behavior in a testbench, or invoked within a **specify** block of a module. In the latter case, they cause simulators to check for timing violations at any instance of the module in a design. This approach is satisfactory for checking for timing violations on the setup, hold, and pulsewidth constraints of a flip-flop. Other constraints, such as the skew between two signals, are more appropriately checked within a testbench. (Remember, hierarchical de-referencing can be used to gain access to signals deep within the hierarchy of the structure being tested.)

7.15.1 SETUP AND HOLD CONDITIONS ($setup, $hold, $setuphold)

In synchronous circuits, data signals propagate from primary inputs and registers and travel through combinational logic to destination registers and/or primary outputs. A signal must not arrive late. If it does, the destination flip-flop may enter the so-called "metastable state",[4] where it can remain for an indeterminate time. This can lead to incorrect operation of the circuit. The *setup* time constraint of a flip-flop specifies a time interval before the active edge of the clock. If the data input changes during the setup interval, a violation occurs. Figure 7.47 shows two waveforms, *sig_a* and *sig_b*, and the synchronizing signal, *sys_clk*. The setup interval is shown in front of the rising (active) edge of the clock. A timing analysis tool must (1) determine whether a signal could have a transition within the setup interval, and (2) report a timing violation. The absolute time difference between the early boundary of the setup interval and the arrival of the data is called the *slack* on that path to the input of the flip-flop. A circuit with negative slack has a timing violation, and the circuit will not operate correctly. Long paths (i.e., relatively large propagation delays) cause setup violations. The path lengths calculated during the synthesis process are lower bounds on the physical path length that results from layout-induced parasitic loading. If all the paths to flip-flops have positive slack the circuit satisfies the setup constraints. In Figure 7.47, *sig_b* has a timing violation. The following system task could be written into a specify block within a model of the flip-flop to detect a setup timing violation:

```
$setup (data, posedge clk, 5);
```

where *data* and *clk* are (formal) input ports of the model of the flip-flop. Under simulation, the actual ports, *sig_a* and *sys_clk*, will be checked for a setup timing violation.

A "hold" violation occurs at a flip-flop when the data is not stable for a sufficiently long interval after the active edge of the clock. Hold violations can be caused by short paths in the circuit. When a signal propagates from a source flip-flop to a destination flip-flop, the output of the source flip-flop must not reach the destination too soon. Otherwise, the wrong data could propagate to

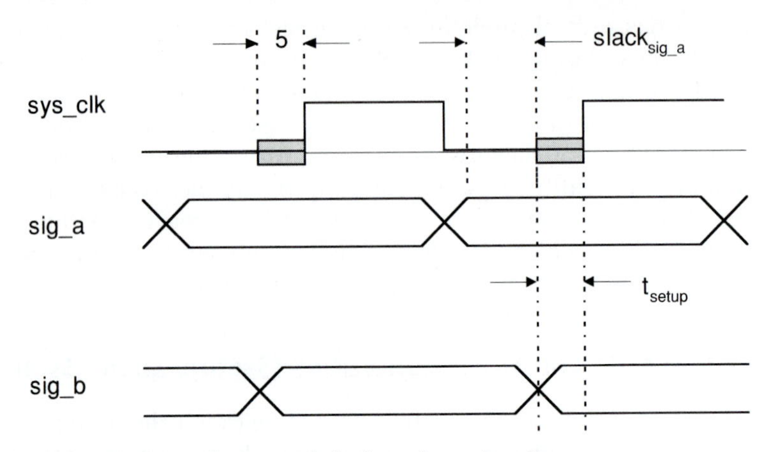

Figure 7.47 Illustration of setup violations in a circuit.

the output of destination flip-flop. In Figure 7.48, *sig_a* violates the hold timing constraint, but *sig_b* satisfies it. A separate Verilog system task can be called to incorporate a hold constraint timing check in the model of a flip-flop. The syntax is illustrated by:

$hold (data, **posedge** clk, 2);

where *data* and *clk* are (formal) input ports of the model of the flip-flop. Under simulation, the actual ports, *sig_a, sig_b* and *sys_clk,* will be checked for a hold timing violation at their respective flip-flops.

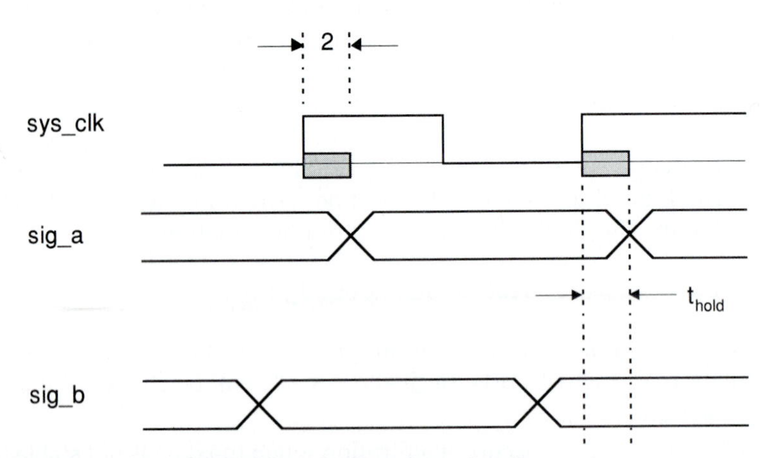

Figure 7.48 Illustration of hold violations in a circuit.

Verilog includes a system task that combines the setup and hold timing checks into a single task. For example, the previous timing checks for setup and hold conditions can be represented by the task:

$setuphold (data, **posedge** clk, 5, 2);

The combined timing constraints are illustrated in Figure 7.49.

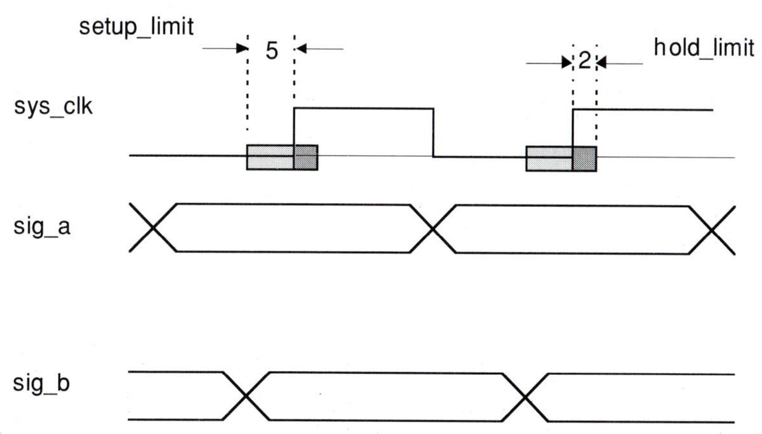

Figure 7.49 Illustration of setup and hold timing constraints.

7.15.2 Signal Period ($period)

The **$period** system task is used to check whether the period of a signal is sufficiently long. It does so by measuring the time interval between successive active edges of the signal. In Figure 7.50, $t_{period} > t_{limit}$, so the period of *clock_a* meets the timing constraint. As an example, the following behavior invokes the mechanism for monitoring the period of *clock_a*:

$period (**posedge** clock_a, t_limit).

where t_limit is a parameter specifying the minimum acceptable interval between two successive rising edges of *clock_a*. This task could be included within a behavior in a testbench that monitors the activity of the signals in the circuit.

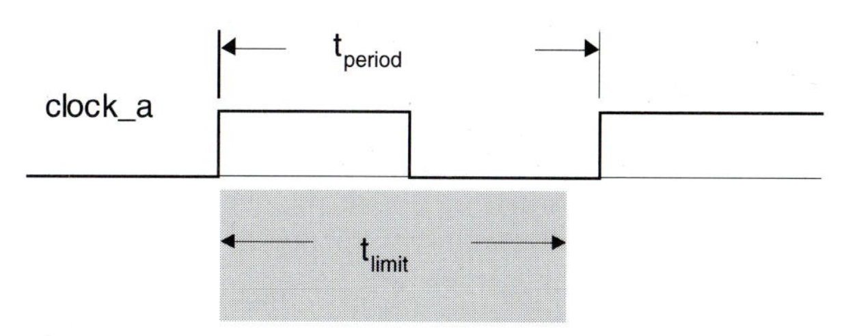

Figure 7.50 Illustration of signal period constraints.

7.15.3 Minimum Pulse Width ($width)

The width of the active-edge pulse of the clock (synchronizing) signal of a flip-flop must not be too small. If it is, the flip-flop will not operate correctly. In Figure 7.51, the system task for checking the width of *clock_a* would be invoked by the statement **$width** (**posedge** clock_a, t_mpw), where t_mpw is a parameter specifying the minimum acceptable pulse width. The signal *clock_a* satisfies the constraint, but *clock_b* does not. The task monitors the time interval between the specified edge of a signal and its next (following) opposite edge. The statement invoking the task would normally be included within a **specify** block in the description of the flip-flop.

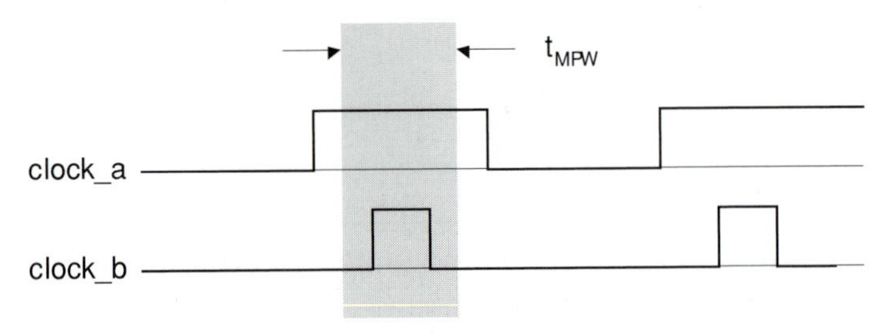

Figure 7.51 Illustration of minimum pulse width violations in a circuit.

7.15.4 Signal Skew ($skew)

Signal skew is a term used to denote the time interval between reference features of two waveforms. For example, the active edge transitions of a clock signal at different physical locations in a synchronous circuit must not have excessive skew. Otherwise, the synchronicity of the circuit might be compromised. The system task **$skew** reports a violation when the skew between two signals exceeds a specified value. For example, in Figure 7.52, the task **$skew** (**negedge** clk1, **negedge** clk2, t_skew) would report a violation when the magnitude of the time difference between rising edges of *clk1* and *clk2* exceeds the value of the parameter t_{skew}.

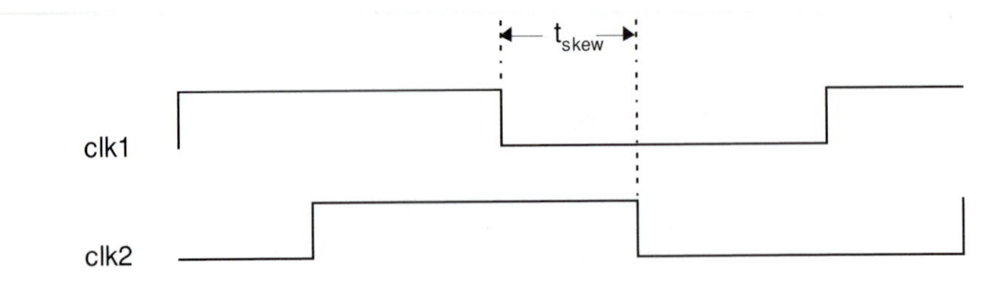

Figure 7.52 Illustration of skew violations in a circuit.

7.15.5 RECOVERY TIME ($recovery)

The **$recovery** task reports a timing violation if the time interval between an edge-triggered reference event and a data event exceeds a limit. For example, the output of a three-stated device driving a bus can be checked to verify that the time required for it to reach a known state after the disabling control is de-asserted is less than a specified limit. The task **$recovery** (**negedge** *bus_control, bus_driver, t_rec*) would check the recovery time of the waveforms in Figure 7.53, where t_{rec} is a parameter specifying the maximum recovery time.

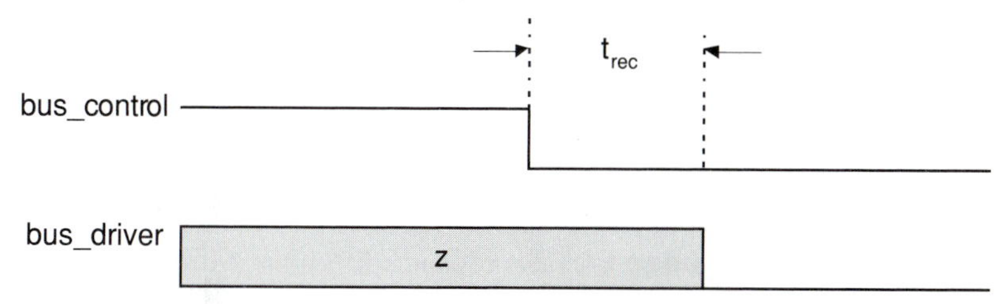

Figure 7.53 Illustration of recovery time check in a circuit.

7.15.6 NO SIGNAL CHANGE ($nochange)

The timing check task **$nochange** reports an error if a data signal changes value while a reference signal has a specified value, including an edge transition. In general, the task would be invoked, for example, by the procedural statement **$nochange** (**posedge** *reference_signal, data_signal, start_edge_offset, end_edge_offset*). The duration of the reference event specifies an interval within which *data_signal* must be fixed. Optionally, *start_edge_offset* and *end_edge_offset* modify the interval to expand or contract its size. For example, the task call **$nochange** (**posedge** *clk, data –1,* 2) checks whether *data* is stable in the interval (–1, 2) relative to the rising edge of the *clk*.

7.15.7 EDGE SEMANTICS FOR TIMING CHECKS

Recall that the **posedge** semantics in Verilog include the following transitions 01, 0x, and x1; the **negedge** semantics include the transitions: 10, x0, and 1x. Edge transitions involving "z" values are treated the same as transitions involving "x". Specific edges of a signal can be specified in the event control expression used in the timing check system tasks. The edges 01, 10, 0x, x1, 1x, and x0 can be used with the specifier **edge**. This feature allows for a finer-grain timing check. For example: **$setuphold** (sig_a, **edge** 01 clk, 5, 2).

7.15.8 Conditioned Events for Timing Checks

The false alarms reported by a timing simulation can be reduced by suspending the timing checks when they are not important. For example:

```
$setup (data, posedge clock &&& (!reset), 4)
```

would check the setup constraint between *data* and *clock* only when *reset* is not asserted. The special operator &&& is used to qualify the timing check event.

Example 7.56 The specify block from the description of an ASIC library cell for a D-type flip-flop is given below. Some information has been removed, but the system tasks for timing verification are evident.

```
specify                         // Module paths
  specparam Area_Cost = 4268.16;

  if (RST3(negedge CLK2 => (Q +: DATA1)) = (0.537:1.239:3.221,
                                             0.713:1.406:3.657);

  specparam RiseScale$CLK2$Q = 0.00082:0.00168:0.00431;
  specparam FallScale$CLK2$Q = 0.00078:0.00159:0.00406;

  if (RST3)
    (negedge RST3 => (Q +: 1'b0)) = (0.576:1.164:3.076);

  specparam RiseScale$RST3$Q = 0.00000:0.00000:0.00000;
  specparam FallScale$RST3$Q = 0.00072:0.00158:0.00145;

  specparam
    t_SETUP$DATA1 = 0.74:0.88:1.61,
    t_HOLD$DATA1 = 0.76:0.30:0.55,
    t_PW_H$CLK2 = 0.37:0.78:1.98,
    t_PW_L$CLK2 = 0.37:0.78:1.98,
    t_RELEASE$RST3$CLK2 = 1,
    t_RELEASE$SET4$CLK2 = 1,
    t_PW_L$RST3 = 0.67:1.76:3.29;

  $setuphold (negedge CLK2 &&& RST3, DATA1, t_SETUP$DATA1,
    t_HOLD$DATA1, notify);

  $width (posedge CLK2 &&& RST3, t_PW_H$CLK2, 0, notify);
  $width (negedge CLK2 &&& RST3, t_PW_L$CLK2, 0, notify);
  $width (negedge RST3, t_PW_L$RST3, 0, notify);

  specparam inputCap$RST3 = 37.50,
    inputCap$DATA1 = 30.80,
```

```
        inputCap$CLK2 = 37.50;

    specparam InputLoad$RST3 = 0.008:0.019:0.050,
        InputLoad$DATA1 = 0.007:0.016:0.041,
        InputLoad$CLK2 = 0.008:0.019:0.050;

endspecify
```

7.16 VARIABLE SCOPE REVISITED

The scope of variables that are declared within a **begin ... end** block is local to the block. However, such blocks can also be given an optional name, which makes the description more readable. If a block is named, the variables that are declared within it can be hierarchically de-referenced from any location in the design (e.g., from within a testbench).

Example 7.57 If variable *K* is declared within procedural *Block2* that is nested within module *Block1* that is nested within a design called *Top* in Figure 7.54, then *K* can be referenced anywhere within the design by the name *Top.Block1.Block2.K*.

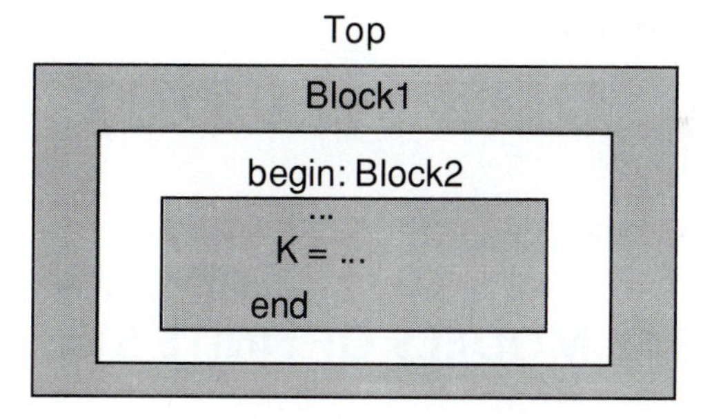

Figure 7.54 Hierarchical de-referencing of a variable within a block.

7.17 MODULE CONTENTS

A module is the basic unit of design in the Verilog HDL. At the beginning of this chapter, Figure 7.1 presented a pseudo-structural view of the elements of a module. Figure 7.55 lists the various elements that may be included within a Verilog module, and which have been discussed in this chapter. It also suggests an overall organization to be followed by the user in writing the Verilog code for a model.

- **module** THE_DESIGN
  - Port declaration
  - Parameter declaration and override
  - Type declaration
  - Event declaration
  - Continuous assignment
  - Primitive instantiation
  - Module instantiation
  - Specify Block
  - Task declaration
  - Function declaration
  - Behavioral statements
    initial
    always
      Procedural assignment
      Non-blocking assignment
      Procedural continuous assignment
      Loops (for, repeat, while, forever)
      Flow control (if, conditional, case, wait, disable)
      System tasks and function
      Event trigger
      Task calls
      Function calls
- **endmodule**

Figure 7.55 Elements of a Verilog module.

7.18 BEHAVIORAL MODELS OF FINITE STATE MACHINES

Synchronous sequential machines having a finite number of states form the core of many important applications. Finite state machines have two basic forms: Mealy machines and Moore machines. The structure of the Mealy machine (Figure 7.56a) consists of a state register and a combinational logic unit that forms the next state and outputs from the state and inputs. The Moore machine (Figure 7.56b) forms the next state from the input and the present state, but forms the output only from the state. The outputs of a Mealy machine are asynchronous, and the outputs of a Moore machine are synchronous. The outputs of Mealy machines are subject to glitches in the inputs, so in practice, many designs are implemented as Moore machines. Verilog can describe the various types of state machines in a variety of ways.

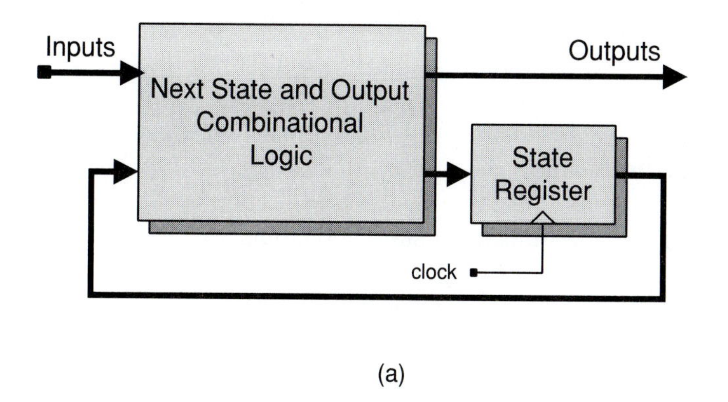

(a)

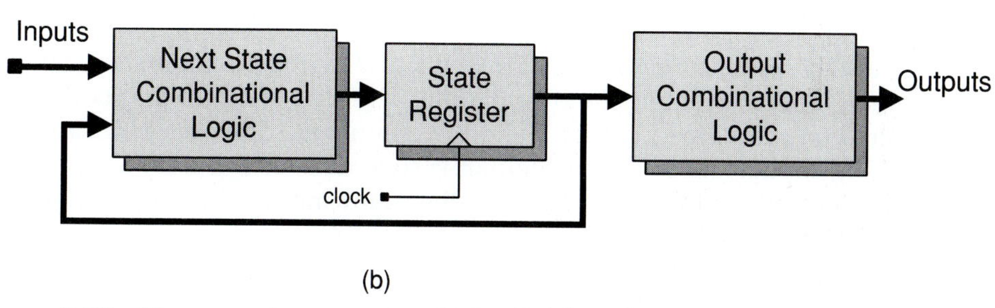

(b)

Figure 7.56 The generic structure of Mealy (a) and Moore (b) finite state machines.

There are two fundamental descriptive styles of finite state machines. The first, called the explicit style, declares a state register to encode the machine's state. A Verilog behavior explicitly assigns values to the state register to govern a synchronous sequence of state transitions. The second, or implicit, style uses multiple event controls within a cyclic behavior to implicitly describe an evolution of states. The transitions of the states occur at clock edges, which serve as implicit state boundaries.

7.18.1 Explicit Finite State Machines

A common form of a Verilog model of an explicit Mealy finite state machine consists of (1) continuous assignment statements that implement the next state logic, (2) continuous assignments that implement the output logic, and (3) a synchronous cyclic (**always**) behavior that updates the state. This style is illustrated below for a machine that has an active-low reset control input and a clock:

```
module FSM_style1 ( ... );
  input ... ;
  output ... ;
  parameter size = ... ;
  reg [size-1 ;0] state, next_state;

  assign the_outputs = ... // a function of state and inputs

  assign next_state = ... // a function of state and inputs

  always @ (negedge reset or posedge clk)
    if (reset == 1'b0) state = start_state; else
      state <= next_state;
endmodule
```

This style uses continuous assignments to describe the combinational logic forming both the outputs and next state. The *start_state* represents the state entered on assertion of the *reset,* and the outputs are derived combinationally from the inputs and state. An alternative style replaces the continuous assignment generating the next state with the following asynchronous (combinational) behavior:

```
module FSM_style2 ( ... );
  input ... ;
  output ... ;
  parameter size = ... ;
  reg [size-1 ;0] state, next_state;

  assign the_outputs = ... // a function of state and inputs

  always @ (state or the_inputs)
    begin
      // decode for next_state with case or if statement
    end


  always @ (negedge reset or posedge clk)
    if (reset == 1'b0) state = start_state; else
      state <= next_state;
endmodule
```

Both styles are acceptable, but the latter can exploit the **case** statement and other procedural constructs, and might be easier to manage when the description is complex. In both styles, the outputs are asynchronous. In some applications, it is desirable to have the outputs registered. The description of the machine can take the form shown below:

```
module FSM_style3 ( ... );
  input ... ;
  output ... ;
  parameter size = ... ;
  reg [size-1 ;0] state, next_state;


  always @ (state or the_inputs)
    begin
      // decode for next_state with case or if statement
    end


  always @ (negedge reset or posedge clk)
    if (reset == 1'b0) state = start_state; else begin
      state <= next_state;
      outputs <= some_value (inputs, next_state);
                    // synchronous assignment of outputs
    end
endmodule
```

Notice that the outputs are formed from the inputs and *next_state*. Given the non-blocking assignments, the value of *state* and *next_state* are sampled at the values held immediately before the clock transition, with the value of *next_state* being the value assumed by *state* after the clock transition.

State machines can be described in tabular format (state transition table), conventional graphical format (state transition graph) or an algorithmic state machine (ASM) chart. The tabular format lists states and identifies the next states and outputs that will be reached upon application of the inputs. The state transition graph represents the table in a format using vertices to represent states, and directed edges to denote synchronous transitions between states. Each vertex of the graph is labeled with the name and/or binary code of the state; each edge is labeled with a pair of symbols. The first in the pair indicates the input value that causes the transition from the state at the origin of the directed edge to the state at the end of the edge; the second symbol in the pair indicates the output that is asserted in the current state (before the transition) while the indicated input value is asserted. The clock signal is not shown on the graph, but it is understood that the transitions occur at the active edge of a synchronizing signal, whatever its name. In hardware, the current state is stored in a register of flip-flops driven by the synchronizing signal. The machine typically has either a synchronous or asynchronous reset signal to assure that the machine can be placed in a known state (e.g., at power up). Otherwise, the state transition graph must be decoded exhaustively to specify transitions from every state under every combination of inputs. There will be a tradeoff here between the additional logic and the simpler/smaller flip-flops.

Example 7.58 Mealy machine. The state transition graph in Figure 7.57 depicts the behavior of a finite state machine controller for a moving vehicle. The machine is modeled with only four states, and the transitions between states depend on the state and control inputs (brake and accelerator). The brake overrides the accelerator in the event that both are asserted. The model uses parameters to define the bit patterns of the states and to make the description more readable.

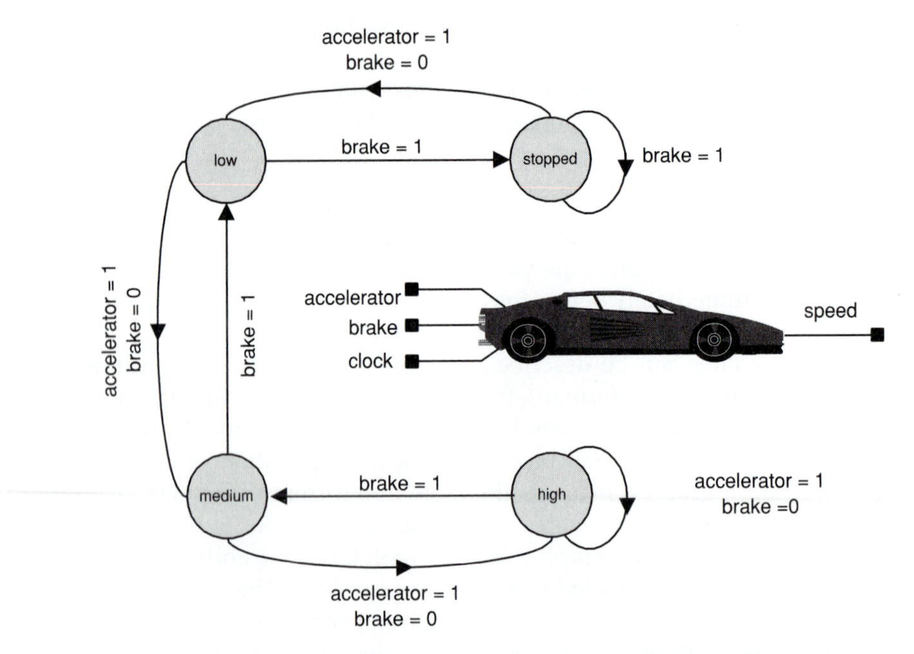

Figure 7.57 State transition graph for a speed-controlled moving vehicle.

```
module speed_machine_1 (clock, accelerator, brake, speed); // Explicit FSM
  // Style: Decode the inputs and the state

  input                 clock, accelerator, brake;
  output    [1:0]       speed;
  reg       [1:0]       state, next_state;

  parameter stopped     = 2'b00;
  parameter S_low       = 2'b01;
  parameter S_medium    = 2'b10;
  parameter S_high      = 2'b11;

  assign speed = state;

  always @ (posedge clock)
    state <= next_state;

  always @ (state or accelerator or brake)
    if (brake == 1'b1) case (state)
        stopped:    next_state <= stopped;
        S_low:      next_state <= stopped;
        S_medium:   next_state <= S_low;
        S_high:     next_state <= S_medium;
        default:    next_state <= stopped;
      endcase
    else if (accelerator == 1'b1) case (state)
        stopped:    next_state <= S_low;
        S_low:      next_state <= S_medium;
        S_medium:   next_state <= S_high;
        S_high:     next_state <= S_high;
        default:    next_state <= stopped;
      endcase
    else            next_state <= state;
endmodule
```

This style implements the speed controller with two behaviors. The first is synchronized with the clock; it governs the state transitions. The second behavior is activated by an event of a control input or by an event of the state. It updates the next state value in response to these inputs. It is an asynchronous, combinational behavior. Figure 7.58 shows the results of simulating the machine. The state is initially *S_stopped* and remains so until *brake* is de-asserted. With *accelerator* asserted, the state transitions from *S_stopped* to *S_high*. When *brake* is again asserted, the state ultimately returns to *S_stopped*.

```
initial
   begin
      clock = 0;
      forever #10 clock = ~clock;
   end

initial #400 $finish;

initial        // Test for wakeup and reset
   begin
       reset = 1'bx;
      #25 reset = 1;
      #75 reset = 0;
   end

initial begin // Test for single request
   #20 service_request = 3'b100;
   #20 service_request = 3'b010;
   #20 service_request = 3'b001;
   #20 service_request = 3'b100;
   #40 service_request = 3'b010;
   #40 service_request = 3'b001;
end

initial begin // Test for multiple requests followed by reset
   #180 service_request = 3'b111;    // Test for 3 to 2 cycle
   #60 service_request = 3'b101;     // Test for 3 to 1 cycle
   #60 service_request = 3'b011;     // Test for 2 to 1 cycle
   #60 service_request = 3'b111;     // Test for 3 to 2 cycle
   #20 reset = 1;                    // Test for reset
end
endmodule
```

The results of simulating *Polling_Circuit* are shown in Figure 7.61. The simulation demonstrates the functionality of *reset* and the polling logic. The testbench demonstrates service to a single client first, then demonstrates the response to multiple contending clients. Notice that *client_3* and *client_2* can block *client_1* from ever receiving service. (PC denotes present client.)

State transition graphs are not a convenient tool for representing complex machines. They indicate the transitions that result from inputs applied when the machine is in a particular state, but they do not directly display the evolution of states under application of input data. As an alternative, algorithmic state machine (ASM) charts present an algorithmic view of the behavior of the machine, similar to software flowcharts. ASM charts use three fundamental elements: a state box, decision box, and conditional output box. State boxes are rectangles; output boxes have round corners; and decision boxes are diamond-

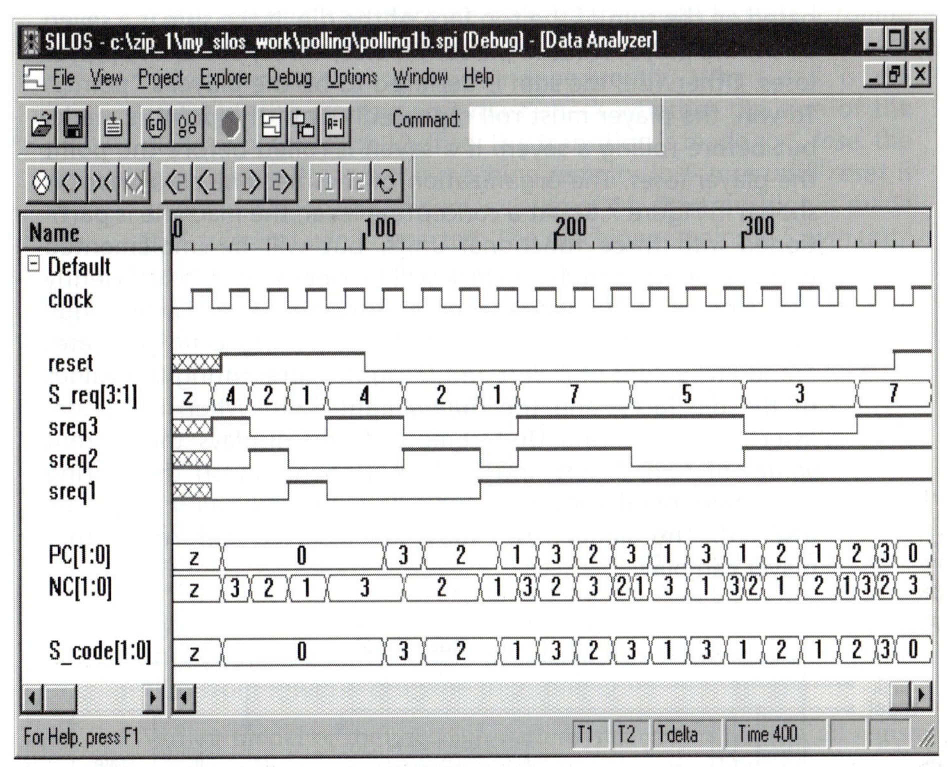

Figure 7.61 Results of simulating the polling circuit.

shaped. A chart is composed of blocks consisting of a state box and an optional configuration of decision and output boxes. The state box represents the state of the machine between synchronizing clock events. Each block has one state box; it is reached by a single entry path from some other state box in the chart. The state box is possibly connected to decision boxes, with a path to each box being associated with a particular value of the vector of inputs to the machine. If signals are asserted when a state box is entered, they are listed in the box, along with a label or state code for the box. The decision boxes determine the exit paths from a state box to another block. The outputs that depend upon both the state and an input are placed in decision boxes located along the path from a state box to an exit of the block. They contain a list of the signals asserted while the machine is in the block's state box and the inputs for the path are asserted. Outputs that depend only upon the state are listed inside the state box, following and separated from the state label by a "/". (Outputs that are not shown are implicitly de-asserted.)

Example 7.60 The dice game of craps is played with a pair of dice.[6]

A player rolls the dice, and there are three possible outcome,

```
always @ (posedge clock)
  case (speed)
    `stopped: if (brake == 1'b1)                speed <= `stopped;
                else if (accelerator == 1'b1)   speed <= `low;
    `low:     if (brake == 1'b1)                speed <= `low;
                else if (accelerator == 1'b1)   speed <= `medium;
    `medium:  if (brake == 1'b1)                speed <= `low;
                else if (accelerator == 1'b1)   speed <= `high;
    `high:    if (brake == 1'b1)                speed <= `medium;
     default:                                   speed <= `stopped;
  endcase
endmodule
```

Notice that, when the speed is high and the brake is not asserted, the speed remains at its residual value (high).

7.18.3 HANDSHAKING

Many digital machines must interact coherently with other machines. For example, a processor must interact with a peripheral device. The process of interacting coherently is referred to as "handshaking." The machine that is providing a service to another machine is called the "server," and the machine receiving the service is called the "client." The interaction of requesting and receiving service is synchronized so that the exchanged information is provided to the client while the client is expecting it. The handshake process also conserves the resources of the server by not making information available indiscriminately. Figure 7.67 shows a typical client-server configuration in which the client signals the need for service by a *client_ready* signal, and the server communicates the availability of service by the signal *server_ready*. For illustration, the server provides the client with information on a 4-bit data bus.

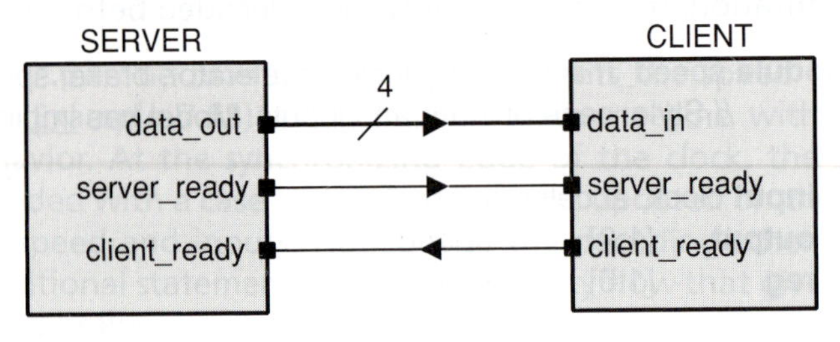

Figure 7.67 Client-server configuration.

Example 7.63 A simple handshaking process is illustrated by the ASM chart in Figure 7.68. Both client and server are shown as having an idle state in which the respective communication signals are not asserted. In this illustration, the interaction between the client and server is indicated by the dotted lines, representing the signals that are controlled and monitored to effect the exchange. The interaction need not be synchronous. Some of the edges of the graph are annotated with the delay control operator ("#") to indicate that, in a physical system, time will elapse before the transition is made. The server's idle state, *S_idle*, can correspond to an interval of time required to prepare for service before transitioning to the state *S_wait*, in which the server awaits a request for service from the client. The client's idle state, *C_idle*, can correspond to a period in which service is not desired. The server remains in *S_wait* until the client asserts *CR* (*client_ready*). When *CR* is asserted, the server enters the state *S_serve* and asserts *SR* (*server_ready*). This action signals to the client that the server has made the data available on the bus. The client then enters the state *C_client*, and after some interval removes the data from the bus. Then *CR* is de-asserted. The server then returns to *S_idle* and de-asserts *SR*. On detecting this action, the client returns to *C_idle*.

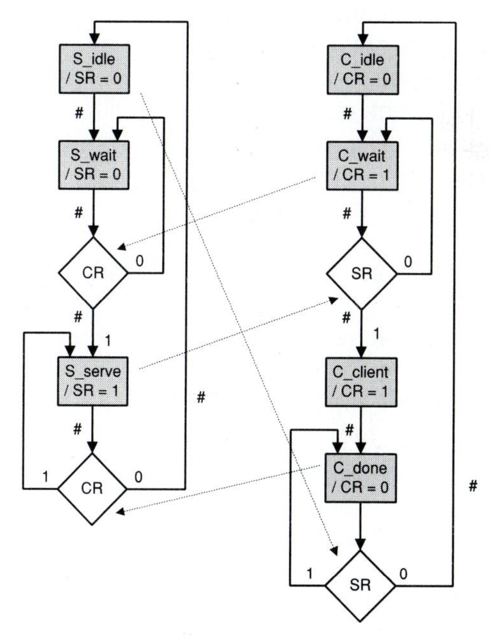

Figure 7.68 ASM chart for handshaking.

Verilog descriptions of the client and server are shown below. For simulation, and to clarify the distribution of events over the time axis, a non-zero delay in transition between states is generated randomly by the task *pause*. The server places *data_out* on the bus, and the client copies the value into *data_reg*.

```
module server (data_out, server_ready, client_ready);
  output   [3:0]   data_out;
  output           server_ready;
  input            client_ready;
  reg              server_ready;
  reg      [3:0]   data_out;

  task pause;
    reg    [3:0]   delay;
    begin
      delay = $random; if (delay == 0) delay = 1;
      #delay ; end
  endtask

  always
    forever begin
      server_ready = 0;
      pause; server_ready = 1;
        wait (client_ready)
        pause; data_out = $random;
        pause; server_ready = 0;
          wait (!client_ready)
          pause;
    end
endmodule

module client (data_in, server_ready, client_ready);
  input  [3:0]   data_in;
  input          server_ready;
  output         client_ready;
  reg            client_ready;
  reg    [3:0]   data_reg;

  task pause;
    reg  [3:0]   delay;
    begin
      delay = $random; if (delay == 0) delay = 1;
      #delay ; end
  endtask
```

```
always begin
  client_ready = 0;
  pause; client_ready = 1;
  forever begin
    wait (server_ready)
      pause; data_reg = data_in;
      pause; client_ready = 0;
      wait (!server_ready)
        pause;
        client_ready = 1;
  end
 end
endmodule
```

The waveforms in Figure 7.69 illustrate the interaction between the client and server. The delays and *data_out* have random values.

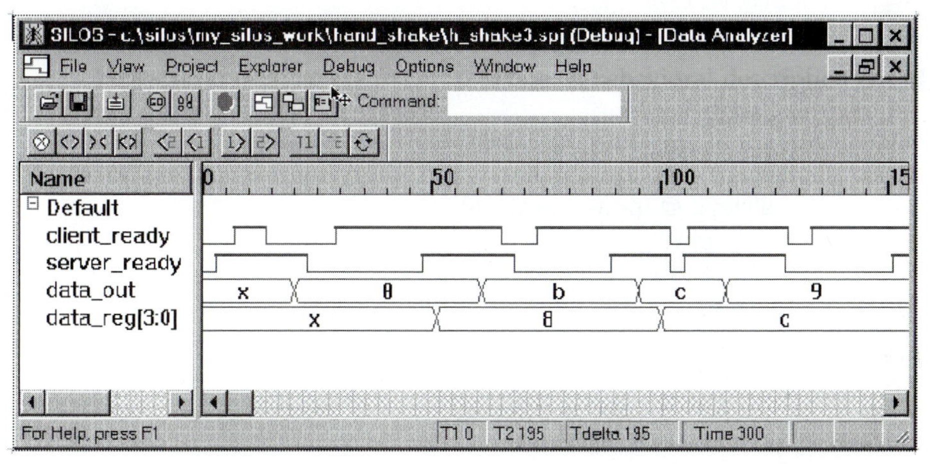

Figure 7.69 Waveforms from handshaking (simulated with random delays).

7.19 SUMMARY

The Verilog HDL contains a variety of constructs supporting behavior descriptions of combinational and sequential logic. It has two behavioral statements, which model one-shot and cyclic behavior. Within a behavior, procedural statements describe an activity flow, which may be sequential or parallel in nature. Activity flows include branching statements, looping statements, and assignments, and may include tasks and functions. The sequential constructs of the language support a variety of design styles for finite state machines and other sequential behavior.

REFERENCES

1. Sternheim, E. et al., *Digital Design and Synthesis with Verilog HDL*, Automata Publishing Co., San Jose, California, 1993.
2. Abramovici, M. et al., *Digital Systems Testing and Testable Design*, Computer Science Press, New York, 1990.
3. Smith, D. O., *HDL Chip Design*, Doone Publications, Madison, Alabama, 1996
4. Smith, Michael J. S., *Application-Specific Integrated Circuits*, Addison-Wesley, New York, 1997.
5. Katz, R.H., *Contemporary Logic Design*, Benjamin Cummings Publishing Co., Redwood City, California, 1994.
6. Roth, C.H. Jr., *Digital System Design Using VHDL.*, PWS Publishing Co., Boston, 1998

PROBLEMS

1. Sketch the waveforms of scalars *sig_a* and *sig_y*

```
always @ (posedge sig_a)
  begin
    #25 sig_y = data;
  end
initial
  begin
    #0      sig_a = 0;
    #10     sig_a = 1;
    #10     sig_a = 0;
    #10     sig_a = 1;
    #10     sig_a = 0;
    #10     sig_a = 1;
  end

initial
  begin
    #0      data = 0;
    #5      data = 1;
    #27     data = 0;
    #18     data = 1;
   end
```

2. Draw the waveforms that will be produced in simulation by the module below.

```
module Simple_Clock (clock_1, clock_2);
  output    clock_1, clock_2;
  reg       clock_1, clock_2;

  always
    begin
      #0      clock_1 = 0; clock_2 = 1;
      #100    clock_1 = ~clock_1;
      #25     clock_2 = ~clock_2;
      #75
    end
endmodule
```

3. Draw the waveforms produced by the behavior below:

```
  initial
    begin
      #0      sig_1 = 0;    sig_2 = 0;
      #300    sig_3 = 1;
      #150    sig_4 = 1;
      #280    sig_5 = 0;
    end
```

4. Compare the clock generator in Example 7.3 to the description below. Which description is more advantageous in large simulations? Why?

```
module clock_gen2 (clock);
  parameter Half_cycle = 50;
  parameter Max_time = 1000;
  output clock;
  reg clock;

  initial
    clock = 0;
  always
    begin
      # Half_cycle clock = 1;
      # Half_cycle clock = 0;
    end

  initial
    #Max_time $finish;
endmodule
```

5. Use each of the following styles to write and verify a model of a 3-input nor gate.

   a. Pre-defined primitive
   b. Continuous assignment
   c. Conditional operator
   d. UDP
   e. Cyclic behavior with procedural assignment
   f. Cyclic behavior with procedural continuous assignment

6. Write and verify descriptions of a 4-input nand gate using each of the following styles:

   a. language primitive
   b. continuous assignment with language operator
   c. continuous assignment with conditional operator
   d. user-defined primitive
   e. cyclic behavior with **casex** statement
   f. cyclic behavior with **if** statement
   g. cyclic behavior with **for** loop

7. Write and verify a cyclic behavior describing a 4 to 1 Mux using (a) a **case** statement, (b) an **if** statement, and (c) a conditional operator.

8. Write a behavioral (algorithmic) description of a D-type flip-flop synchronized to the falling edge of a clock and having active low synchronous reset.

9. a. What is the difference between an **initial** behavior and an **always** behavior?

   b. Describe Verilog's mechanisms for synchronizing procedural activity.

10. Given the following description, what are the decimal values of A, B, and C after the code has executed?

```
reg [7:0] A, B, C;
initial begin
   A = 4'd2;
   B = 3;
   C = A + B;
   A = >> A;
end
```

11. Write and simulate a Verilog description of a combinational logic circuit that will watch an 8-bit bus and detect the pattern 0110_1001. The module is to have input signals consisting of the 8-bit bus, an 8-bit matching signal, and an output signal that asserts while a match exists.

12. Explain why the module below does not model the behavior of a D-flip-flop correctly. What changes should be made to ensure correct operation? Write a testbench and produce simulation data to illustrate the behavior of both versions of the flip-flop.

```
module flop (clock, data_in, q, qbar, reset);
  input    clock, data_in, reset;
  output   q, qbar;
  reg      q;

  assign qbar = ~q;

  always @ (posedge clock or reset)
    begin
      if (reset == 0) q = 0; else
      if (clk == 1) q = data_in;
    end
endmodule
```

13. Explain why the module below does not model the behavior of a four-channel multiplexer.

```
module mux4 (a, b, c, d, select, y_out);
  input          a, b, c, d;
  input  [1:0]   select;
  output         y_out;
  reg            y_out;

  always @ (select)
    case (select)
      0: y_out = a;
      1: y_out = b;
      2: y_out = c;
      3: y_out = d;
      default y_out = 'bx;
    endcase
endmodule
```

14. Discuss what happens when the behaviors below are simulated together.

```
initial
   clock = 0;

always @ (clock)
   clock = ~clock;
```

15. Using a case statement, write an alternative model for the 2-bit comparator described in Chapter 2.

16. A 4-bit ALU is to implement the functionality shown below:

s4	s3	s2	s1	s0	C_in	Operation/Functionality
						Arithmetic Behavior
0	0	0	0	0	0	Y = A (Transfer A to Y)
0	0	0	0	0	1	Y = A + 1 (Increment A and transfer)
0	0	0	0	1	0	Y = A + B (Addition)
0	0	0	0	1	1	Y = A + B + 1 (Add with carry)
0	0	0	1	0	0	Y = A + B' (A plus 1's complement of B)
0	0	0	1	0	1	Y = A + B' + 1 (Subtraction)
0	0	0	1	1	0	Y = A -1 (Decrement A)
0	0	0	1	1	1	Y = A (Transfer A to Y)
						Logical Behavior
0	0	1	0	0	0	Y = A and B
0	0	1	0	1	0	Y = A or B
0	0	1	1	0	0	Y = A xor B
0	0	1	1	1	0	Y = A' (One's complement)
						Shift Behavior
0	0	0	0	0	0	Y = A (Transfer A to Y)
0	1	0	0	0	0	Shift left A
1	0	0	0	0	0	Shift right A
1	1	0	0	0	0	Transfer 0s to Y

a. Implement the functionality of the ALU in a module containing a single behavior using one **case** statement. Write a DUTB to verify the functionality of the design. The DUTB is to include documentation stating which features are to be tested, how they are to be tested, and any limitations to the test. Discuss the results of the test.

b. Develop an alternative implementation of the ALU using a single behavior, but with three case statements, one each implementing the arithmetic, logical, and shift behavior.

17. Write models of a 4-to-10-line decoder in the following descriptive styles:

a. structural (using Verilog primitives)

b. data flow (using Verilog language operators in continuous assignments)

c. behavioral, using a Verilog behavior incorporating a case statement.

Write a carefully documented testbench to verify your models. Use hex formatted I/O to make the output easily readable.

18. The Verilog HDL does not include an arithmetic shift operator, which has the property that it preserves the value of the MSB when shifting to the right, and the LSB when shifting to the left, as illustrated below.

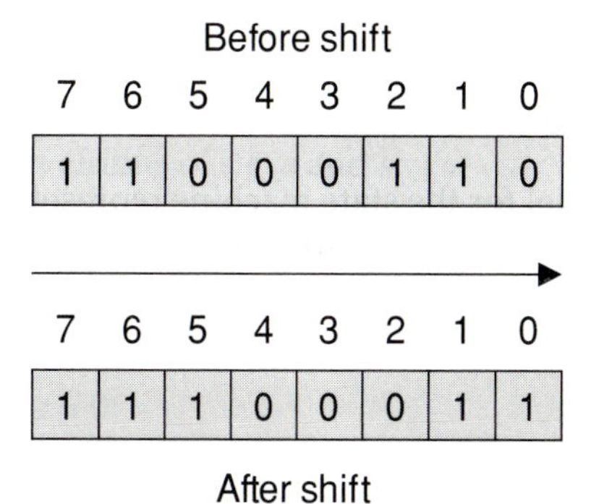

Write and verify a description that implements the arithmetic shift operation on a word of data.

19. Write a description of a counter that counts in the following sequence: 0, 3, 6, 8, 15, 24, 0.

20. Write a task that will rotate a word by a specified number of bits, in either direction.

21. Find the syntax error(s) in the following text:

```
`define stopped  2`b00;     // Compiler substitution
`define low      2`b01;
`define medium   2`b10;
`define high     2`b11;
```

22. Describe the key features of the named event construct of Verilog

23. What is the difference between an **always** behavior and a **forever** construct?

31. Which of the flip-flops below operate correctly?

a.

```
module mystery_flop_1 (q, data, clk, rst);
   input    data, clk, rst;
   output   q;

   always @ (negedge clk) q = data;
   always @ rst if (!rst) q = 0;
endmodule
```

b.

```
module mystery_flop_2 (q, data, clk, clr, set);
   input    data, clk, rst;
   output   q;

   always @ (negedge clk) q = data;
   always @ rst if (!clr) assign q = 0; else deassign q;
   always @ set if (!set) assign q = 1; else deassign q;
endmodule
```

c.

```
module mystery_flop_3 (q, data, clk, clr, set);
   input    data, clk, rst;
   output   q;

   always @ (negedge clk) q = data;
   always @ rst if (!clr) assign q = 0; else if (!set) assign q = 1;
      else deassign q;
   always @ set if (!set) assign q = 1; else if (!clr) assign q = 0;
      else deassign q;

endmodule
```

32. Discuss the following descriptions:

a. `initial assign y = a | b;`

b. `always @ (a or b) y = a | b;`

c. `always @ (a or b) #5 y = a | b;`

33. Describe the difference between blocked and nonblocked procedural assignments.

34. What are the values of *a* and *b* after the code below is compiled and simulated?

```
module order_test ();
  reg [1:0] a, b;

  initial
    begin
      a <= 3;
      a = 6;
      a = 0;
    end

  initial
    begin
      b = 0;
      b = 2;
      b <= 3;
    end
  endmodule
```

35. If the statement assigning value to *B* executes at time t_o, when does the statement assigning value to *A* execute?

```
B = 5;
fork
  #5 B1 = 2;
  #7 B2 = 3;
  begin
    #2 B3 = 4;
    #6 A2 = 3;
  end
join
#32 A = 10;
```

36. At what time does *A* change during simulation of each of the following statements?

a. A = #5 B;

b. A <= #5 B;

c. #5 A = #5 B;

d. #5 A <= #5 B;

37. Draw the waveforms produced by the following description.

```
module wave_gen1();
  reg wave1, wave2;
  initial
    begin
      #2 signal1 = 1;
      #3 signal1 = 0;
      signal2 <= #5 1;
      signal2 <= #20 0;
      signal2 <= #25 1;
      #10 signal1 = 1;
      signal1 <= #5 0;
    end
endmodule
```

38. Draw the waveforms produced by the following description.

```
module wave_gen2(wave);
  output        wave;
  reg           wave;
  reg    [2:0]  i;

  initial
     begin
       for (i =0; i <= 7; i=i+2) wave <= #(i*10) ~i[1];
     end
endmodule
```

39. Explain why the behavior *big_loop* causes a simulator to enter an endless loop, and why the behavior in *little_loop* does not.

```
always begin: big_loop
  G <= @ (A_Bus) Accumulator
end
```

```
always begin: little_loop
  G = @ (A_Bus) Accumulator
end
```

40. Explain why a simulation of the code below enters an endless loop.

```
module loop_trap ();
  reg  [15: 0] some_memory [0 : 1023];
  reg  [15: 0] a_word_register;
  reg  [7: 0] instr_register;
  reg  k;
  wire [7: 0] ir_net = instr_register;
```

```
initial
  begin
    a_word_register = some_memory [17];
  end
initial
  begin
    #5 for (k=1; 5; k = k+1)
      begin
        #5 instr_register[k] = a_word_register [k+4];
        $display ($time,, "%d %h %h", k, ir_net, instr_register [k]);
      end
      $stop;
  end
endmodule
```

41. Explain the effect of an **assign** ... **deassign** procedural continuous assignment to a register variable.

42. Explain the effect of a **force** ... **release** procedural assignment to a net.

43. Discuss the distinctions between continuous assignments, procedural assignments, and procedural continuous assignments.

44. Answer T (true) or F (false):

a. A register variable can be referenced only within a procedural statement.

b. A net may not be referenced within a procedural statement.

c. Procedural continuous assignments can implement level-sensitive and edge-sensitive behavior.

d. An **assign** ... **deassign** procedural continuous assignment statement can assign value to a **wand** net.

e. The Verilog "**case**" statement checks for a complete bitwise map between the case expression and a case item.

f. The index variable in a "**for**" loop must be a previously declared "**reg**".

g. Procedural continuous assignment establishes the binding of only a constant to a register.

h. Procedural continuous assignment establishes the binding of a RHS expression to a net.

i. The **force** ... **release** procedural continuous assignment can assign value to a register variable and to a net.

45. Determine whether the serial pattern detectors below have fatal flaws.

a.

```
module pattern (found, data, clk);
   output     found;  // 1 if pattern 0010 found
   input      data;   // serial bit stream
   input      clk;    // system clock
   reg [3:0]  sr;     // shift register

   assign found = (sr == 4'b0010);
   always @ (posedge clk) sr = (sr << 1) | data;
endmodule
```

b.

```
module pattern (found, data, clk, reset);
   output     found;  // 1 == pattern 0010 found
   input      data;   // serial bit stream
   input      clk;    // system clock
   input      reset;  // global reset
   reg [3:0]  sr;     // shift register

   assign found = (sr == 4'b0010);
   always @ (posedge clk) if (reset) sr = 0; else sr = (sr << 1) | data;
endmodule
```

46. Write and verify a model of the pipelined register structure shown below. Assume that the datapaths to *reg_a*, *reg_b*, and *reg_c* are 4 bits wide.

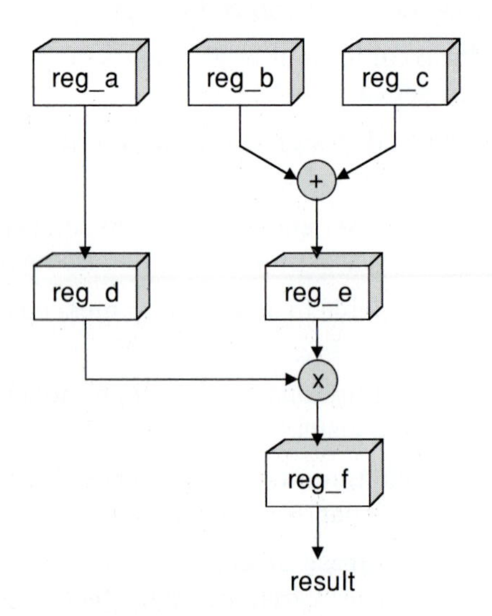

47. Write and verify a model for a 6-bit ring counter that will generate the following sequence:

100_000
010_000
001_000
000_100
000_010
000_001
100_000

48. Write and verify a model of a divide-by-8 counter.

49. Why is the description below problematic?

```
always @ (posedge clk) a = a + 1;
always @ (posedge clk) b = a + b;
```

50. Write a model of a functional unit whose output is the number of gaps exceeding 2 bits between two successive ones in a 16-bit word.

51. Write a Verilog model of a functional unit whose output is the size of the largest gap in a 16-bit word.

52. Write a parameterized model of a barrel shifter. The shifter must load a parallel word of size *w_size*, rotate the word in a selected direction, or shift it in either direction, filling in with 0s. The shifter is to have asynchronous reset.

53. Write and simulate a Verilog description of a module that will watch a synchronous serial input and detect the pattern 0100. Assume that the data is stable at the falling edge of the clock and that the rightmost-bit in this pattern arrives first. The machine is to have active-low reset.

a. Develop a state transition graph and implement the module as a state machine.

b. Implement the module using a shift register.

Caution: take measures to ensure that the machine does not issue a bogus detect after only two bits have arrived.

54. Write a parameterized model of a barrel shifter having a word of size w_size. The shifter is to synchronously shift by one bit in either direction, filling in with 0s, or take no action. The reset is to be active-low and synchronous.

55. Modify the design of the speed controller presented in Example 7.55 so that an output signal *red* is generated when the brakes are applied, and another output that is *green* when the accelerator is applied. Write and

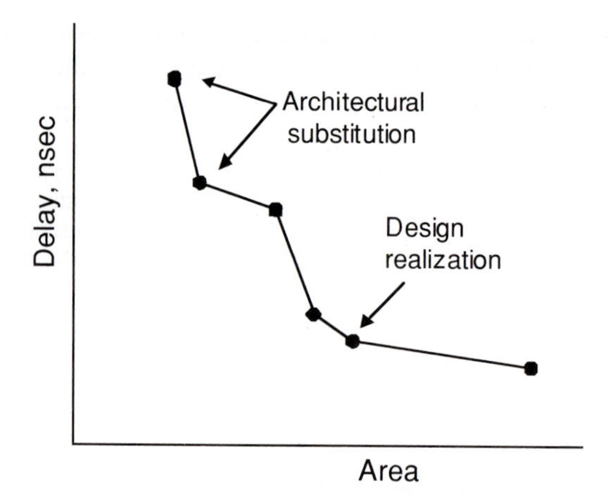

Figure 8.3 Area vs. delay tradeoff curve (banana curve).

description of a circuit; behavioral synthesis creates an RTL description from an algorithmic model of functionality.

8.1.1 LOGIC SYNTHESIS

Logic synthesis tools operate on Boolean equations and produce optimal combinational logic, which can be mapped into a physical realization. In a tool environment, logic synthesis includes circuit transformations that exploit alternative architectures and gate-level substitutions and optimizations. A logic synthesis tool has the general organization shown in Figure 8.4. The tool forms a hardware realization (technology implementation) from a Verilog behavioral description.

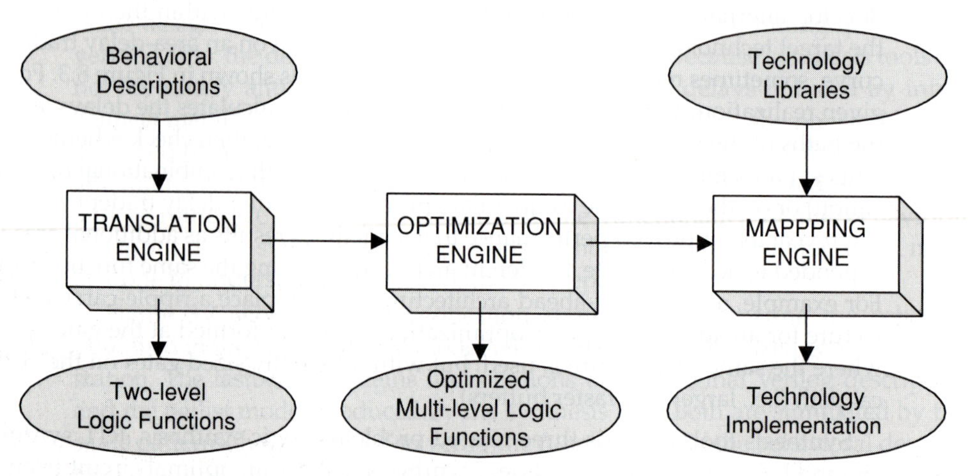

Figure 8.4 Synthesis tool organization.

The translation engine of a synthesis tool reads and translates a Verilog-based description of a circuit's input-output behavior into an intermediate internal representation of Boolean equations describing combinational logic and other representations of storage elements and synchronizing signals. The tool will translate pre-defined and user-defined combinational primitives (gates) into an equivalent Boolean expression. For example, a two-input MUX can be described by the Boolean equation below:

$$\text{y\_out} = \text{select} \cdot \text{data\_a} + \overline{\text{select}} \cdot \text{data\_b}$$

While it is convenient to write the Boolean equations of a circuit in a sum-of-products (SOP) form, the internal representation within synthesis software will be in a product-of-sums (POS) form. It is created by factoring the SOP form into expressions whose Boolean product generates the SOP form. When the POS representation for two or more outputs contain a common sub-expression the synthesis tool may minimize the internal logic needed to realize the circuit by generating the common sub-expression once, and sharing it (through fanout) among output variables. Techniques for simultaneously optimizing the set of equations will remove redundant logic, exploit "don't care" conditions, and share internal logic sub-expressions[2,3] as much as possible to produce an optimized, generic (technology-independent) multi-level logic implementation. For example, a "?" entry in the input-output table of a combinational UDP represents a "don't care" condition that can be used in Boolean minimization. However, this may lead to a mismatch between the seed code's behavior and that of the synthesis product.

The functional relationships between the inputs of a combinational Boolean circuit and their individual outputs can be expressed as a set of two-level Boolean equations in either sum-of-products or products-of-sum form, which must be optimized by the tool. A set of Boolean equations describing a multi-input, multi-output (MIMO) combinational logic circuit can always be optimized[2-4] to obtain a set of Boolean equations whose input-output behavior is equivalent, while containing the fewest literals. Software tools exist to perform this optimization and cover the resulting Boolean equations by the resources of a technology library. Consequently, *a Verilog description consisting only of a netlist of combinational primitives without feedback can always be synthesized.* Some vendors, however, choose not to implement synthesis of a UDP.

Espresso[4] is a commonly used software system developed by the University of California—Berkeley for minimizing the number of cubes in a single Boolean function. Espresso performs several transformations on a circuit to arrive at its optimal representation. *Expand* replaces cubes with prime implicants having fewer literals. *Irredundant* extracts from a cover of the function a minimal subset that also covers the function (i.e., redundant logic has been removed). *Reduce* transforms an irredundant cover into a new cover of the same function.

Espresso minimizes a single Boolean function of several Boolean variables, but does not provide a solution to the problem of optimizing a multi-input, multi-output combinational logic circuit. In general, the optimization of a set of Boolean equations is not obtained by applying Espresso separately to the individual functions in the set. Instead, a multi-level optimization program, such as MIS-II,[2] must be used to simultaneously optimize the set of equations as an aggregate.

Logic synthesis treats a set of individual Boolean input-output equations as a multi-level circuit (see Figure 8.5). By removing redundant logic, sharing internal logic, and exploiting input and output "don't care" conditions, a logic synthesis tool optimizes a multi-level set of Boolean equations, and achieves a better realization (i.e., area-efficient) than could be obtained by merely optimizing the individual input-output equations.

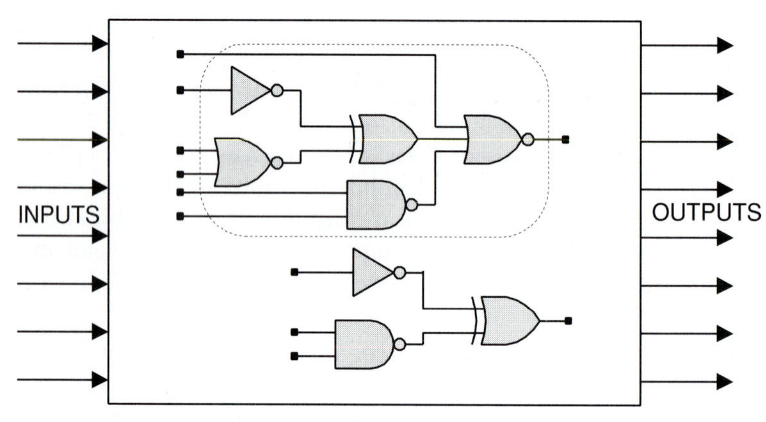

Figure 8.5 Multi-level combinational logic.

Like Espresso, the MIS-II multi-level logic optimization program performs several transformations on a logic circuit while searching for an optimal Boolean description. Four transformations play key roles in the MIS-II algorithm for logic synthesis: decomposition, factoring, substitution, and elimination.

Decomposition transforms the circuit by expressing a single Boolean function (i.e., the Boolean expression representing the logic value of a node in the circuit) in terms of new nodes.

Example 8.1 Figure 8.6 shows a function, *F*, that is to be decomposed in terms of new nodes *X* and *Y*. The original form of *F* is described by the Boolean equation:

$$F = abc + abd + \overline{a}\,\overline{c}\,\overline{d} + \overline{b}\,\overline{c}\,\overline{d}$$

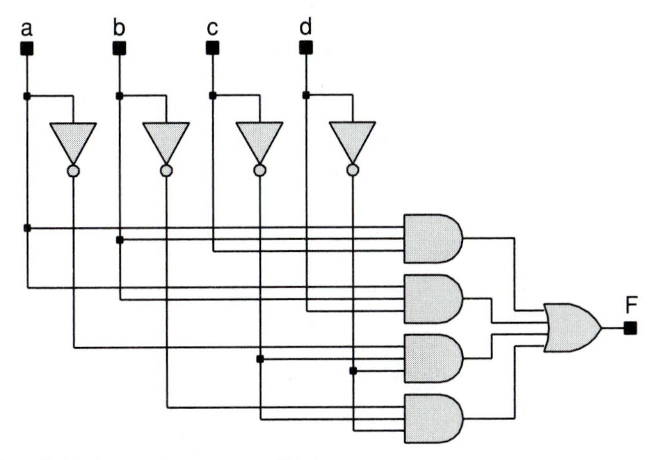

Figure 8.6 Circuit before decomposition.

The Espresso operation of *decomposition* expresses F in terms of two additional internal nodes, X and Y, to form the circuit shown in Figure 8.7. These internal nodes could then be re-used to form other expressions and thereby achieve a reduction in hardware area.

$$F = XY + \overline{X}\,\overline{Y}$$
$$X = ab$$
$$Y = c + d$$

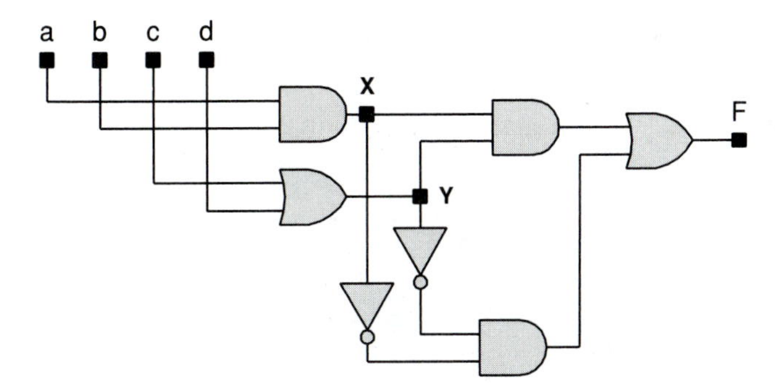

Figure 8.7 Circuit after decomposition.

The Espresso operation of *decomposition* represents an ***individual*** function in terms of intermediate nodes. *Extraction* expresses a ***set*** of functions in terms of intermediate nodes.

Example 8.2 Figure 8.8 shows a directed acyclic graph (DAG) representing the set of functions, F, G, and H, that is to be decomposed in terms of new nodes X and Y, with

$$F = (a + b)cd + e$$
$$G = (a + b)\,\overline{e}$$
$$H = cde$$

and X and Y are given by:

$$X = a + b$$
$$Y = cd$$

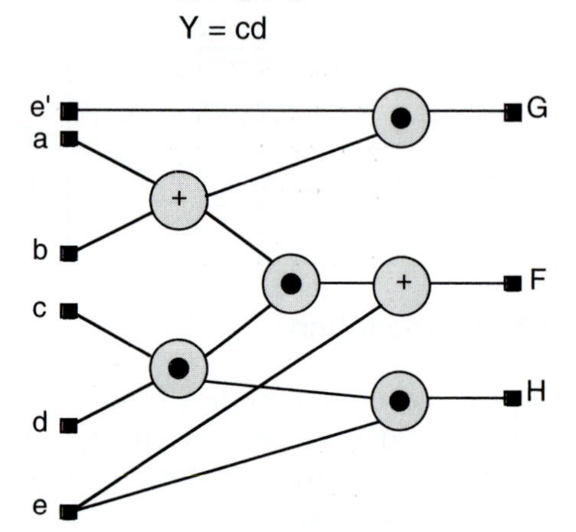

Figure 8.8 Directed acyclic graph of a set of functions before extraction.

The *extraction* process finds those members of the set of functions having the factor *(a + b)* and the factor *cd*. These factors are extracted from those functions and replaced by the new internal nodes *X* and *Y* to produce the new DAG shown in Figure 8.9.

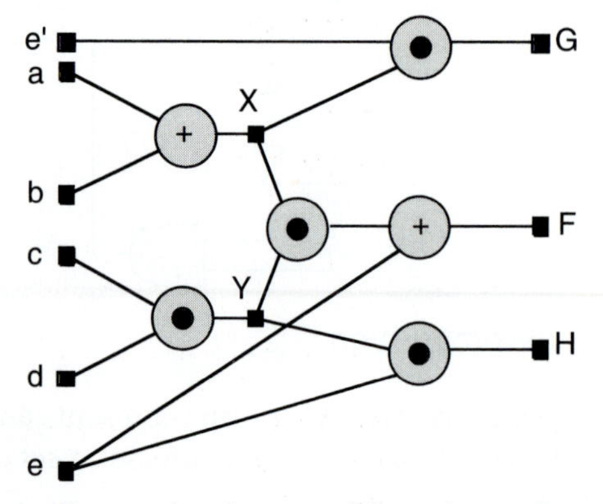

Figure 8.9 Directed acyclic graph of a set of functions after extraction.

The optimization process seeks a set of intermediate nodes to optimize the circuit's delay and area. This step may lead to significant reduction in the overall silicon area, because the intermediate nodes correspond to factors that are common to more than one Boolean function, and can therefore be shared to eliminate replicated logic. The task of finding the common factors among a set of functions is called factoring. Factoring produces a set of functions in a product of sums form. It creates a structural transformation of the circuit from a two-level realization to a multi-level realization that uses less area, but is possibly slower.

Example 8.3 Figure 8.10 shows a DAG representing a function, *F*, that is factored to identify its Boolean factors in product of sums form. The function represented by the DAG in Figure 8.10 is described by the Boolean equation:

$$F = ac + ad + bc + bd + e$$

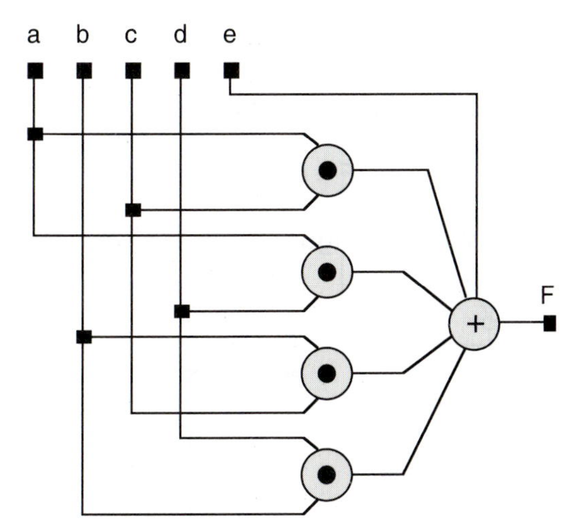

Figure 8.10 The DAG of a function before factoring.

The factored form of *F* is

$$F = (a + b)(c + d) + e$$

The *factorization* process seeks the factored representation of a function having the fewest number of literals. The DAG of the factored form of *F* is shown in Figure 8.11.

The *substitution* process expresses a Boolean function in terms of its inputs and another function. Since both functions need to be implemented, this step provides a potential reduction in replicated logic.

8.1.2 Register Transfer Level (RTL) Synthesis

Register transfer level (RTL) synthesis transforms a behavior described in terms of operations on registers, signals, and constraints, and synthesizes an optimal realization in an assumed architecture. RTL synthesis begins with the assumptions that a set of hardware resources is available, and that the scheduling and allocation of resources have been determined by a data-flow graph, subject to the constraints imposed by the resources of the architecture. The RTL description represents either a finite state machine or a more general machine that effects register transfers within the boundaries of a pre-defined clock cycle. This style of description is implicit in Verilog models using language operators and synchronous concurrent assignments to register variables (i.e., non-blocking assignments). The Verilog language operators represent a variety of register transfer operations, and are easily synthesized. Most synthesis tools operate in this domain—i.e., they lack the ability to perform tradeoffs between scheduling and allocation of resources, and use the implicit solution imposed by the writer of the description. Nonetheless, theses tools have a broad and significant scope of use.

Data-flow graphs[6] are commonly used to represent the variables that store information used and generated by the operations in an RTL description of behavior. The vertices of a data-flow graph represent the operations performed by a behavior, and the edges represent values. Data-flow graphs also display dependencies between the data. Control-flow graphs represent the sequence of branching and looping operations in an activity flow. A hierarchical sequencing graph is a special form of a combined control-data-flow graph which can be used within a synthesis tool. It uses graphs to represent the data flow in a behavior, and models control flow (i.e., branching and looping) in the hierarchy.[6] The synthesis engine must minimize and optimally encode the state of the RTL-described machine, optimize the associated combinational logic, and map the result into the target technology.

8.1.3 Finite State Machine Synthesis

Finite state machine synthesis must translate and optimally encode state transition graphs into properly sized registers and combinational logic implementing a state machine. This problem is imbedded within the general RTL synthesis problem, and state machines are easily synthesized by today's synthesis tools.

8.1.4 Behavioral Synthesis

Behavioral synthesis has the goal of creating an architecture whose resources can be scheduled and allocated to implement an algorithm, such as a DSP algorithm. A behavioral synthesis tool must synthesize datapath elements, control units, and memory. Figure 8.13a shows three non-blocking assignments, their related parse trees[7] and data flow graphs for three register assignments in a hypothetical algorithm. Consideration of scheduling leads to the data-flow graph shown in (b), which can be used to schedule activity in clock cycles using the available resources.

Behavioral synthesis is a relatively young technology, but successful tools have been developed supporting behavioral synthesis for DSP applications.[8] Behavioral synthesis poses the greatest challenges to EDA tools because there are many algorithms which cannot be synthesized easily, if at all.

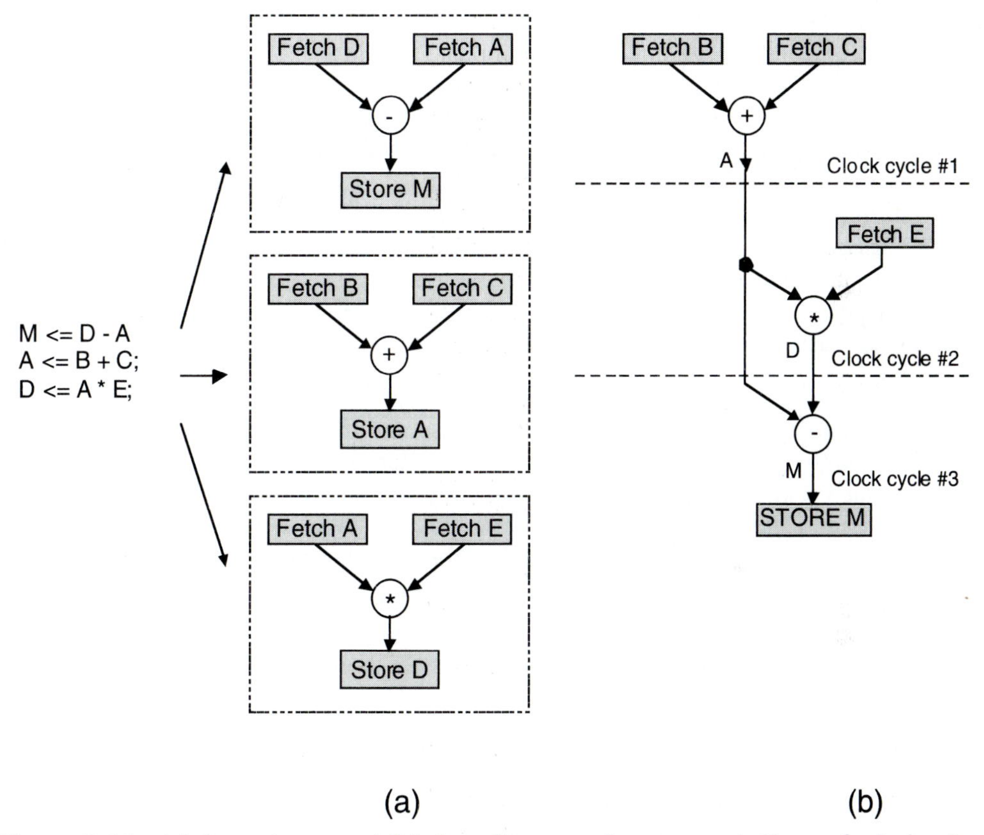

Figure 8.14 (a) Parse trees and (b) data flow graph representations of a behavior.

8.2 TECHNOLOGY-INDEPENDENT DESIGN

Technology-independent design describes only the functionality that is to be synthesized; time delays are irrelevant. What matters is that the settled signals are correct and match at the clock cycle boundaries in both the behavioral and gate-level realizations. There are some general practices that are highly recommended. First, all storage elements in the design should be controlled by an external clock and possibly a reset line. The combinational part of the design should be driven by primary inputs (through the ports) or internal storage elements. The anticipated structure is shown in Figure 8.15, where a fragment of combinational logic is situated between storage elements and primary inputs that drive it, and storage elements and primary outputs that are driven by it. The outputs themselves may or may not be registered.

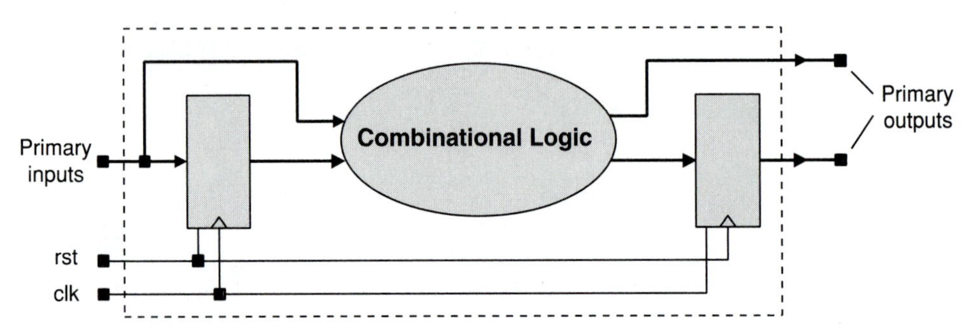

Figure 8.15 Structure for synchronous logic.

In a synchronous logic circuit, it is essential that the longest functional path through the combinational logic be shorter than the available clock cycle, which will be the clock cycle minus any margins for setup conditions of the storage elements. Under this assumption, the logic settles in one cycle of the clock, and simulation results match at the clock boundaries. Figure 8.16 depicts the match of pre- and post-synthesis behaviors.

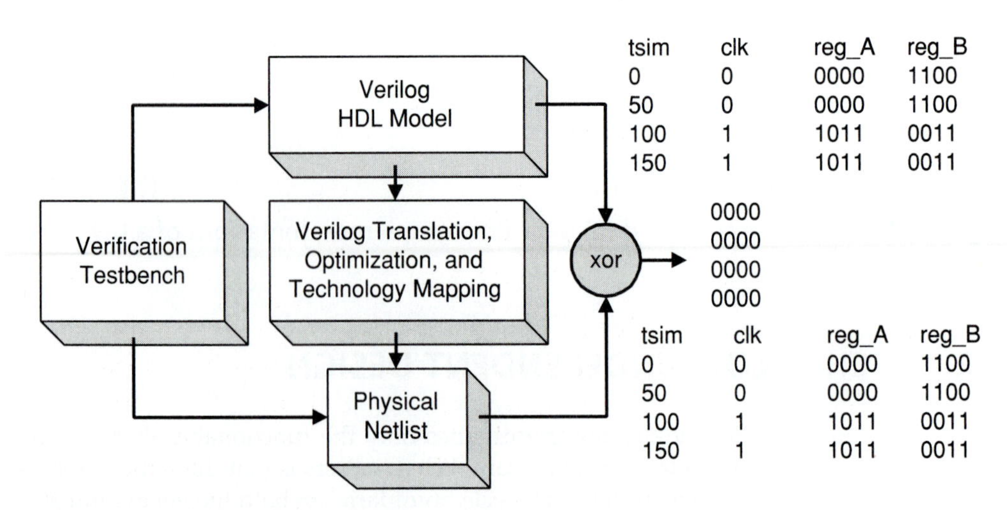

Figure 8.16 Match of pre- and post-synthesis behaviors.

8.3 BENEFITS OF SYNTHESIS

Modern EDA tools supporting a synthesis methodology provide several benefits to the design team. It takes significantly less time to write a Verilog behavioral description and synthesize a gate-level realization of a large circuit than it does to develop the gate-level realization by other means, such as bottom-up manual entry. This saves time that can be put to better use in other parts of the design cycle. The ease of writing, changing, or substituting Verilog descriptions encourages architectural exploration. Moreover, a synthesis tool itself will find alternative realizations of the same functionality and generate reports describing the attributes of the designs.

Synthesis tools create an optimal internal representation of a circuit before mapping the description into the target technology. The same internal representation can be used to re-target the design to another technology. For example, the technology mapping engine of a synthesis tool will use the internal format to migrate a design from an FPGA technology to an ASIC standard cell library, without having to re-optimize the design.

HDL-based designs are easier to debug. A behavioral description encapsulating complex behavior masks underlying gate-level detail, so there is less information to cope with in trying to isolate problems in the functionality of the design. Furthermore, if the behavioral description is functionally correct, it is a golden standard for subsequent gate-level realizations. HDLs support top-down design with mixed levels of abstraction by providing a common framework for partitioning, synthesizing, and verifying large, complex systems. Parts of large designs can be linked together easily for verification of overall functionality and performance.

HDL-based designs incorporate documentation within the design, thereby reducing the volume of documentation that must be kept in other archives. Since the language is a standard, documentation of a design can be de-coupled from a particular vendor's tools.

Effective use of synthesis tools requires that designers follow a systematic methodology, beginning with intelligent partitioning of a design, and employ a descriptive style that conforms to the vendor's requirements and anticipates the synthesis product. Vendor-specific requirements cannot be ignored. Ultimately, the equivalence of the synthesis product with the Verilog source description must be confirmed. Tools exist that automatically verify that a gate-level description consistently replicates the functionality of a behavioral description.[1] These tools use mathematical proofs to establish the correctness and equivalence of descriptions, without requiring exhaustive simulation. As an analogy, Pythagoras' theorem—that the sum of the squares of a right triangle is the square of the hypotenuse—is proved without having to demonstrate that it applies to all such triangles.

8.4 SYNTHESIS METHODOLOGY

A synthesis-oriented methodology focuses on the overall functionality of the design and creates a top-down architectural partition. Technology-independent functional unit descriptions are developed and then verified in a bottom-up sequence. Constraints on timing and area are used to synthesize a gate-level hardware realization, and post-synthesis performance verification ensures that the timing and functionality of the design meet specifications. The goal is to produce a technology-specific, optimized gate-level realization meeting timing specifications and area constraints.

The productivity and quality gains of a synthesis methodology are not free. The tools are not cheap, and a design team must learn technology and a new language. The learning curve of a synthesis tool is steep, and the results can be unpredictable. Needless to say, intelligence is still required! The restrictions of the language and the tools must be understood.

The synthesis methodology has some rules that apply to all synthesis tools, independent of the vendor. In general, avoid referencing the same variable in more than one cyclic (**always**) behavior. When variables are referenced in more than one behavior, there can be races in the software, and the post-synthesis simulated behavior may not match the pre-synthesis behavior. It is advisable, and in some cases required by vendors, to use only synchronous, re-settable flip-flops in the design. The design process with synthesis tools is not blind or beyond the influence of the designer. The designer can influence the clocking scheme, partition of the design hierarchy, and instantiation of pre-defined modules placed outside the scope of the optimization flow. The designer must explore and understand the mappings imbedded in the tool, and observe basic rules differentiating combinational and sequential logic, with an awareness that not all algorithms are synthesizable.

8.5 VENDOR SUPPORT

The degree to which the Verilog language is supported for synthesis varies among EDA vendors. There is a high level of common support, but the ultimate reference for a particular tool is the vendor's style guide, which identifies any restrictions that apply to its implementation of the language, and the language reference manual.

Hardware vendors generally strive to support the entire language so that one Verilog description serves several tools. It is strongly recommended, and in some cases required, that any technology-dependent descriptions be avoided. For example, the propagation delay of a gate should not be included in the description; it will be ignored by the synthesis tool in performing technology-independent optimization. Likewise, certain uses of language constructs are forbidden. For example, expressions which perform explicit Boolean operations on the logic values "x" and "z" are to be avoided—physical hardware operates on only two logic values: "0" and "1".

8.5.1 Commonly-Supported Verilog Constructs

A significant part of the Verilog HDL is supported (possibly with restrictions) by synthesis tools. The supported subset of the language includes constructs for structural and behavioral styles of design, as shown in Figure 8.17.

Module declaration
Port modes: **input, output, inout**
Port binding by name
Port binding by position
Parameter declaration
Connectivity nets: **wire**, **tri, wand, wor, supply0**, **supply1**
Register variables: **reg, integer**
Integer types in binary, decimal, octal, hex formats
Scalar and vector nets
Subranges of vector nets on RHS of assignment
Module and macromodule instantiation
Primitive instantiation
Continuous assignments
Shift operator
Conditional operator
Concatenation operator (including nesting)
Arithmetic, bitwise, reduction, logical, and relational operators

Procedural-continuous assignments (**assign ... deassign**)
Procedural block statements (begin ... end)
case, casex, casez, default
Branching: **if**, **if** ... **else, if** ... **else** ... **if**
disable (of procedural block)
for loops
Tasks: **task** ... **endtask***
Functions: **function** ... **endfunction**

*No timing or event control allowed.

Figure 8.17 Verilog constructs commonly supported by synthesis tools.

8.5.2 Unsupported and Ignored Constructs

Synthesis tools do not support constructs used for transistor/switch level descriptions of behavior, and other constructs that have no significance in synthesis. Figure 8.18 lists Verilog constructs that are to be avoided.

Assignment with variable used as bit select on LHS
global variables
case equaltiy , inequality (===, !==)
defparam
event
fork ... join
forever
while
wait
initial
pulldown, pullup
force ... release
repeat
cmos, rcmos, rnmos, nmos, pmos, rpmos
tran, tranif0, tranif1, rtran, rtranif0, rtranif1
primitive ... endprimitive
table ... endtable
intra-assignment timing control
delay specifications
scalared, vectored
small, medium, large
specify, endspecify
$time
weak0, weak1, high0, high1, pull0, pull1
$keyword

Figure 8.18 Verilog constructs not generally supported in synthesis.

8.6 STYLES FOR SYNTHESIS OF COMBINATIONAL LOGIC

Combinational logic describes a Boolean function whose output depends only upon the instantaneous inputs to the logic, not its history. More formally: *Logic_Outputs(t) = f(Logic_Inputs)(t)*, as shown in Figure 8.19. Combinational logic has no storage elements or memory.

Figure 8.19 Combinational logic.

The general rule that applies to combinational synthesis is that the Verilog description should not include any aspects that attempt to model specific technology. The objective is to synthesize a Boolean representation of the functionality, while ignoring the timing of the logic. Technology-dependent timing constructs, such as gate delays, are to be avoided. Although Verilog accommodates feedback loops in a description (e.g., cross-coupled nand gates), they are to be avoided in a description that is to be synthesized. Rules that ensure synthesis of combinational logic must be adhered to; otherwise the synthesis tool will either fail to operate, or will create a sequential circuit. Table 8.1 lists combinational elements commonly synthesized from a variety of Verilog descriptions. Designers can use any of the descriptive styles shown in Table 8.2.

Table 8-1 *Examples of synthesizable combinational logic elements.*

COMMONLY SYNTHESIZED COMBINATIONAL LOGIC	
Multiplexer	Adder
Decoder	Subtractor
Encoder	ALU
Comparator	Multiplier
Random Logic	PLA Structure
Lookup Table	Parity Generator

Table 8.2 lists the descriptive options available to the designer.

Table 8-2 *Options for synthesizable descriptions of combinational logic.*

OPTIONS FOR COMBINATIONAL LOGIC
Netlist of Verilog primitives
Combinational UDP
Continuous assignment
Behavioral statement
Function
Task without delay or event control
Interconnected modules of the above

8.6.1 COMBINATIONAL SYNTHESIS FROM A NETLIST OF PRIMITIVES

Combinational logic can be synthesized from a netlist of gate-level Verilog primitives. Even if a design is already expressed as a netlist of generic primitives, it is recommended that the design be synthesized to remove any redundant logic before mapping the design into a technology. The synthesis tool will automatically discover and remove redundant logic. This provides a measure of safety to the design, because most designers have difficulty discovering and removing redundant logic from any but the simplest circuits. A synthesis tool will correctly and quickly remove the redundant logic.

Example 8.6 In Figure 8.20a, *or_nand_1* consists of a set of instantiated primitives containing redundant logic. The redundant logic will be removed by a synthesis tool. This is an important step, because redundant logic presents a serious problem to tools that must generate test patterns for detecting process-induced faults in fabricated parts. The synthesized circuit is shown in Figure 8.20b.

```
module or_nand_1 (enable, x1, x2, x3, x4, y);
   input        enable, x1, x2, x3, x4;
   output       y;
   wire         w1, w2, w3;
   or (w1, x1, x2);
   or (w2, x3, x4);
   or (w3, x3, x4);    // redundant
   nand (y, w1, w2, w3, enable);
endmodule
```

A more dramatic example of Boolean optimization is illustrated by the Verilog description *boole_opt* below. Figure 8.20c shows the pre-optimized schematic of the circuit described by (generic) Verilog primitives. The circuit was then synthesized using a 1.2 micron standard cell library to obtain the optimized circuit shown in Figure 8.20d. This small circuit has a more efficient gate-level implementation (less area) than the generic circuit would have if parts from a cell library are substituted for the primitives.

```
module boole_opt (y_out1, y_out2, a, b, c, d, e);
   input a, b, c, d, e;
   output y_out1, y_out2;

   and (y1, a, c);
   and (y2, a, d);
   and (y3, a, e);
   or (y4, y1, y2);
   or (y_out1, y3, y4);
   and (y5, b, c);
   and (y6, b, d);
   and (y7, b, e);
   or (y8, y5, y6);
   or (y_out2, y7, y8);
endmodule
```

The final example, shown in Figure 8.20e, implements the Verilog bitwise operators. This generic description maps each operator to its corresponding gate. But the synthesis tool creates the circuit in Figure 8.20f, which exploits shared logic in a multi-level structure.

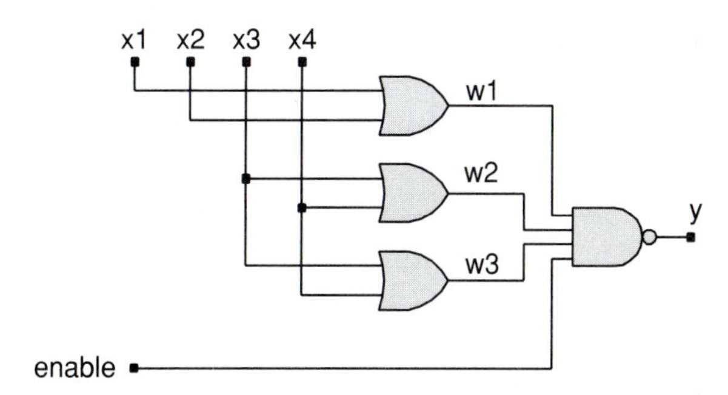

Figure 8.20 (a) Pre-optimized circuit.

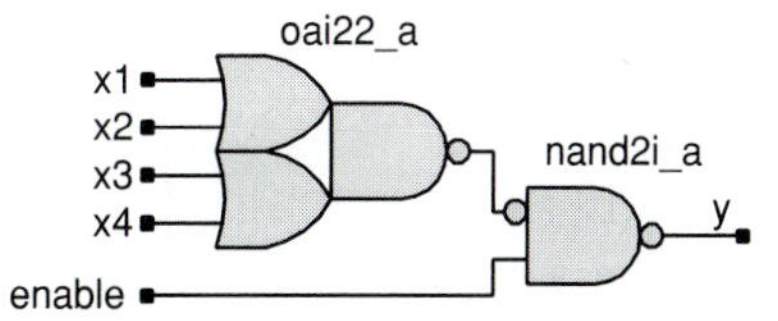

Figure 8.20 (b) Circuit synthesized from the circuit in (a).

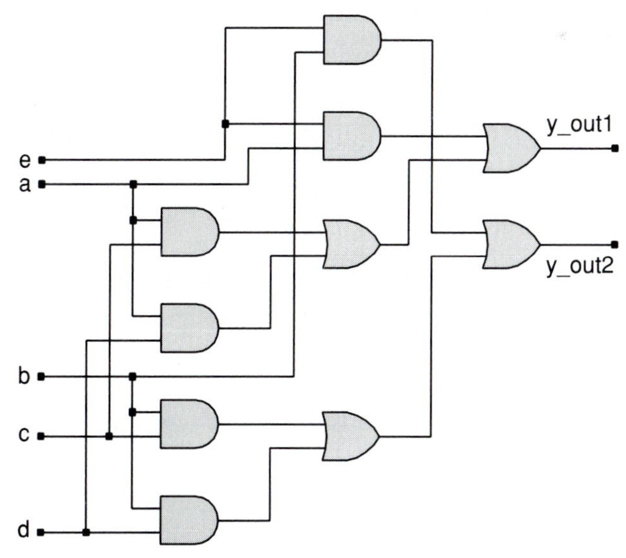

Figure 8.20 (c) Pre-optimized combinational logic circuit.

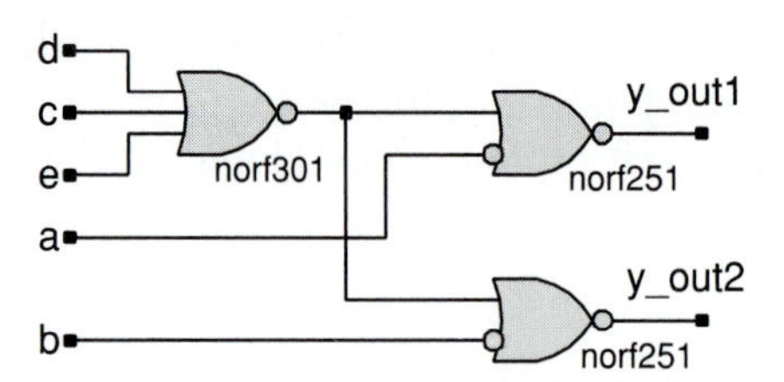

Figure 8.20 (d) Circuit synthesized from the generic circuit in Figure 8.20c.

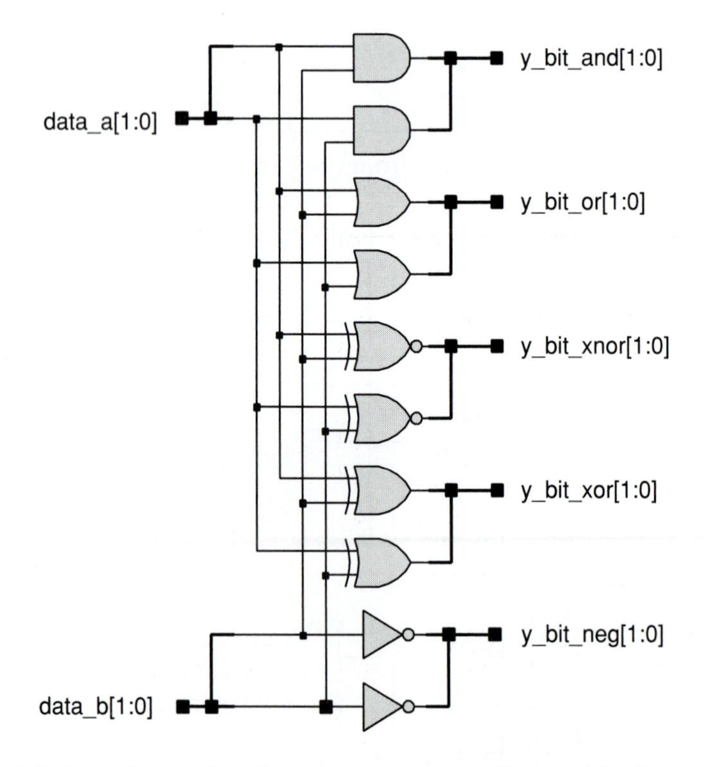

Figure 8.20 (e) Generic circuit using gates corresponding to bitwise operators.

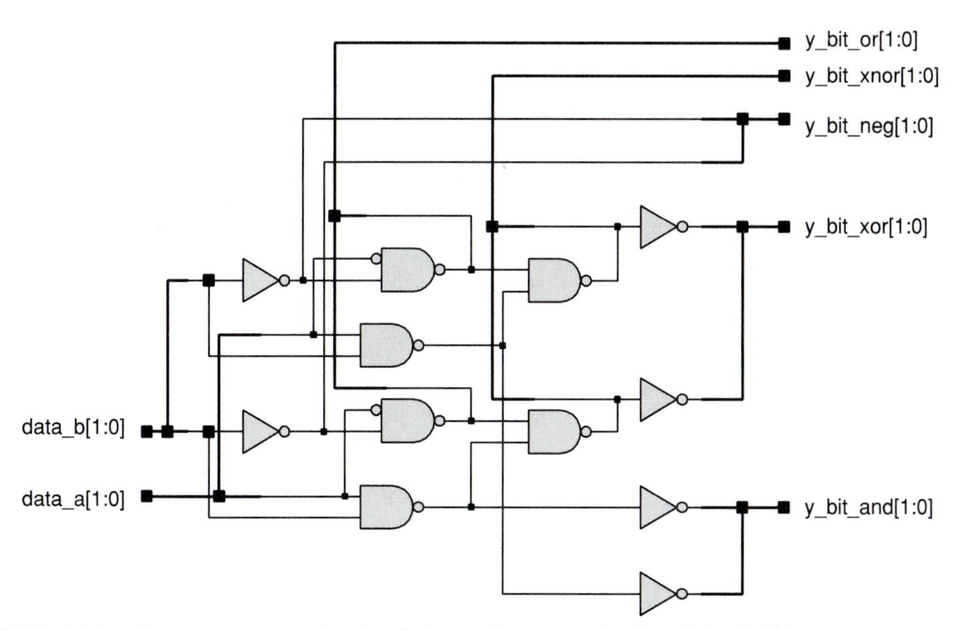

(f) Multi-level structure synthesized from the generic circuit in (e)

Figure 8.20 Examples of Boolean logic optimization.

8.6.2 COMBINATIONAL SYNTHESIS FROM UDPS.

User-defined primitives extend the set of primitives available for structural design. A synthesis tool can operate on a UDP to first obtain an equivalent representation in terms of Boolean expressions, and then optimize the logic. *Any combinational UDP can be synthesized*, but some vendors do not bother to support this construct. The synthesis tool translates the table format of the UDP into a set of Boolean equations, optimizes them, and maps them into the target technology. Table entries that have the value "x" are ignored, but the "?" entry is treated as a don't care for the purpose of optimizing the Boolean logic represented by the table. When a table has explicit uncertainty and/or "don't care"s, the post-simulation synthesis results may differ from the pre-synthesis results.

Example 8.7 The primitive below represents a Boolean function of three variables, and illustrates synthesis of a UDP. It synthesizes to the generic gate structure of a multiplexor shown in Figure 8.21.

```
primitive boolean_eqs (y, a, b, c);
   output        y;
   input         a, b, c;
```

```
table
  //    Inputs          Output
  //    a   b   c        y

        0   1   ?   :    1 ;
        0   0   ?   :    0 ;
        1   ?   1   :    1 ;
        1   ?   0   :    0 ;
endtable
endprimitive
```

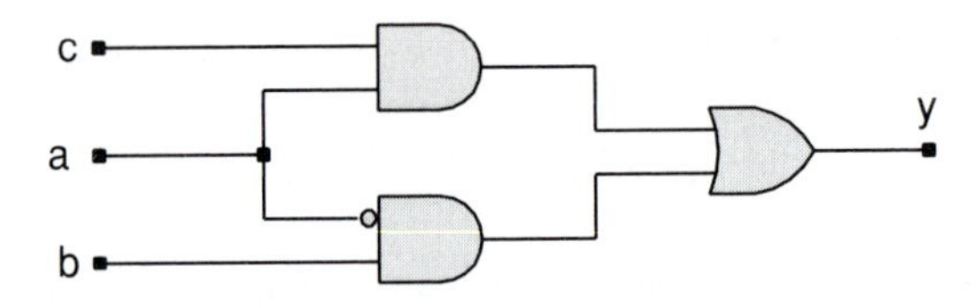

Figure 8.21 Circuit synthesized from *boolean_eqs*.

8.6.3 Combinational Synthesis from Continuous Assignments

Continuous assignments specify event-scheduling rules for nets by statically binding an expression to a net variable. The expression bound to the net is equivalent to combinational logic. A synthesis tool translates continuous assignment statements into a set of equivalent Boolean equations which can be optimized simultaneously.

Example 8.8 The module *or_nand_2* is implemented efficiently as a single continuous assignment statement. The synthesized result in Figure 8.22 is identical to that obtained from the netlist of primitives in Example 8.6.

```
module or_nand_2 (enable, x1, x2, x3, x4, y);
  input enable, x1, x2, x3, x4;
  output y;

  assign y = ~(enable & (x1 | x2) & (x3 | x4));
endmodule
```

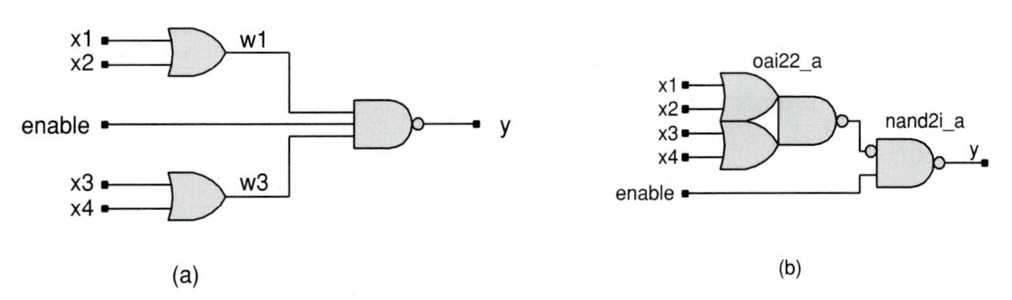

Figure 8.22 Combinational logic for a continuous assignment (a), and the circuit produced by synthesis (b).

8.6.4 COMBINATIONAL SYNTHESIS FROM A CYCLIC BEHAVIOR

Combinational logic can also be described by a cyclic behavior. The key point to remember is that the behavior must assign value to the output under all events that affect the righthand side expression of the assignment that implements the logic. **Failure to do so will produce a design with unwanted latches**. Consequently, all inputs to a behavior that is to implement combinational logic must be included in the event control expression, either explicitly or implicitly. The righthand side operands of assignments within the behavior are inputs. Likewise, any control signals whose transitions affect the assignments to the target register variables in the behavior are considered to be inputs to the behavior. Remember: **If the output of combinational logic is not completely specified for all cases of the inputs, a latch will be inferred**.

Correct synthesis of combinational logic from a Verilog behavioral description requires that the signals that appear as operands on the righthand side of any procedural assignment or procedural continuous assignment statement must not appear on the lefthand side of the expression. If this rule is not observed, the behavior has implicit feedback and will not synthesize into combinational logic. Combinational feedback is not allowed, so the target of an assignment in the behavior may not also be an operand within the behavior.

In general, it is more efficient for the purpose of simulation if a procedural continuous assignment within a behavior is used to implement combinational logic. However, not all tools support this style.

Example 8.9 The implementation of *or_nand_3* below, a combinational logic circuit, uses a cyclic behavior whose event control expression depends upon all the inputs to the circuit. The synthesized description is shown in Figure 8.23.

```
module or_nand_3 (enable, x1, x2, x3, x4, y);
  input          enable, x1, x2, x3, x4;
  output         y;
  reg            y;

  always @ (enable or x1 or x2 or x3 or x4)
    begin
      y = ~(enable & (x1 | x2) & (x3 | x4));
    end
endmodule
```

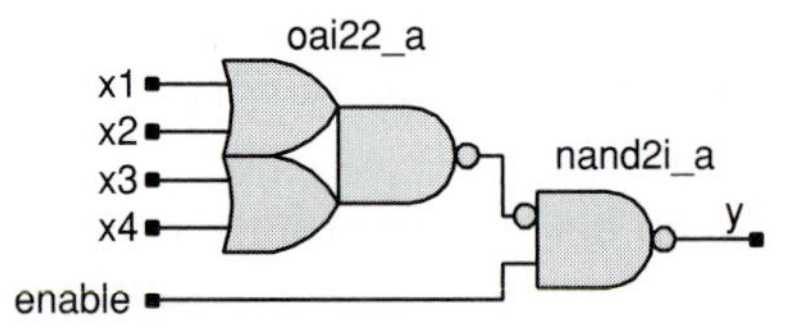

Figure 8.23 Circuit synthesized from a behavioral description of a combinational logic circuit.

We also point out that use of a register variable (e.g., **reg** y) in *or_nand_3* does not imply that the associated behavior is sequential. Nor does it imply that the synthesized hardware will have a storage device. The Verilog code in Example 8.10 uses a register variable to support the algorithm, but does not require a hardware register.

Example 8.10 A behavioral model for a parameterized-input "and" gate is described below in *and4_behav*. The event-activated **always** behavior sets a scalar register, *temp*, to "1", and then checks the bits of *x_in* in sequential order. If a bit is found to be "0", the search is discontinued and the value of *temp* is set to 0. The output of the gate *y* is formed by a continuous assignment to *temp*. The circuit synthesized from *and4_behav*, a four-input "and" gate, is shown in Figure 8.24. Chapter 10 will discuss synthesis of loops, but notice here that the index of the loop, *k*, is not apparent in the synthesized circuit, which is combinational. Also notice that the variable *y* was assigned a value before the loop, assuring that it would be assigned value within the behavior. **It is a good idea to initialize variables before a complex loop to avoid accidentally creating latches in the synthesized result.**

```
module and4_behav (y, x_in);
  parameter       word_length = 4;
  input           [word_length - 1: 0] x_in;
  output          y;
  reg             y;
  integer         k;

  always @ x_in
    begin: check_for_0
      y = 1;
      for (k = 0; k <= word_length -1; k = k+1)
        if (x_in[k] == 0)
          begin
            y = 0;
            disable check_for_0;
          end
  end
endmodule
```

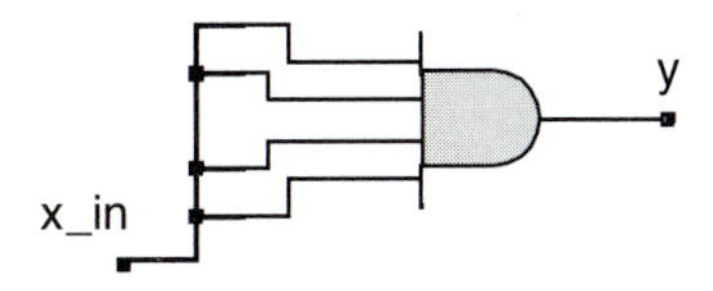

Figure 8.24 Combinational circuit synthesized from a behavioral description of an "and" gate.

Example 8.11 Chapter 2 presented four descriptions of a circuit that compares the magnitude of two two-bit words. Yet another alternative is presented below. It exploits the fact that the data words have equal magnitude if all the bits are the same in each position. Otherwise, the most significant bit at which the words differ determines their relative magnitude. This algorithm is not as simple or elegant as the one that uses the equality operators and a continuous assignment (see Example 2.24), but it serves to illustrate the power of synthesis tools to correctly synthesize combinational logic from a behavioral statement containing a loop construct. The result is shown in Figure 8.25.

```
module comparator (a, b, a_gt_b, a_lt_b, a_eq_b); // Alternative algorithm
  parameter      size = 2;
  input          [size: 1] a, b;
  output         a_gt_b, a_lt_b, a_eq_b;
  reg            a_gt_b, a_lt_b, a_eq_b;
  integer        k;

  always @ ( a or b) begin: compare_loop
    for (k = size; k > 0; k = k-1) begin
      if (a[k] != b[k]) begin
        a_gt_b = a[k];
        a_lt_b = ~a[k];
        a_eq_b = 0;
        disable compare_loop;
      end     // if
    end       // for loop
    a_gt_b = 0;
    a_lt_b = 0;
    a_eq_b = 1;
  end         // compare_loop
endmodule
```

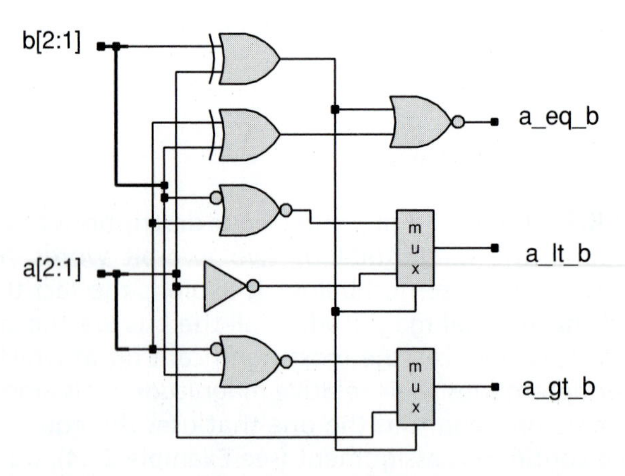

Figure 8.25 Combinational circuit synthesized from a behavioral description of a comparator.

8.6.5 COMBINATIONAL SYNTHESIS FROM A FUNCTION OR TASK

Functions and tasks can be synthesized into combinational logic with minor restrictions. Functions inherently represent combinational logic because the value produced by a function depends only upon the values of its arguments (i.e., it does not have memory). As a general rule, incomplete case statements and incomplete conditionals may not be in the description of a function that is to implement combinational logic. A task implementing combinational logic has the same restrictions as a function, but it is more general, and must be restricted so that it does not have timing control constructs imbedded within any procedural code that it contains.

Example 8.12 The module *or_nand_4* contains a function implementing the or-nand combinational logic that was implemented in the preceding examples.

```
module or_nand_4 (enable, x1, x2, x3, x4, y);
  input         enable, x1, x2, x3, x4;
  output        y;

  assign y = or_nand (enable, x1, x2, x3, x4);

  function or_nand;
    input       enable, x1, x2, x3, x4;
    begin
      or_nand = ~(enable & (x1 | x2) & (x3 | x4));
    end
  endfunction
endmodule
```

Example 8.13 The task in *or_nand_5* also implements the or-nand combinational logic that was implemented in the preceding example with a function.

```
module or_nand_5 (enable, x1, x2, x3, x4, y);
  input         enable, x1, x2, x3, x4;
  output        y;
  reg           y;

  always @ (enable or x1or x2 or x3 or x4)
    or_nand(enable, x1, x2, x3, x4, y);

  task or_nand;
    input       enable, x1, x2, x3, x4;
    output      y;
```

Example 8.18 *syn3_mux _4bits* describes the functionality of a four-channel mux using **if** branching statements, with two options for implementing the cyclic behavior.

```
module syn3_mux_4bits (y, sel, a, b, c, d);
  input        [3:0] a, b, c, d;
  input        [1:0] sel;
  output       [3:0] y;
  reg          [3:0] y;

  // Simulation efficient

  always @ (sel)
    if (sel == 0)  assign   y = a; else
      if (sel == 1) assign  y = b; else
        if (sel == 2) assign  y = c; else
          if (sel == 3) assign y = d; else
                      assign y = 4'bx;

  // Simulation and synthesis friendly

  always @ (sel or a or b or c or d)
    if (sel == 0)              y = a; else
      if (sel == 1)            y = b; else
        if (sel == 2)          y = c; else
          if (sel == 3)        y = d; else
            else               y = 4'bx;

endmodule
```

Figure 8.29 shows the result of synthesizing all three circuits when the datapaths are scalars. The circuit produced for *syn1_mux_4bits* (Figure 8.29a) differs from that produced for the "synthesis-friendly" versions of *syn2_mux_4bits* and *syn3_mux_4bits* (Figure 8.29b). The latter realization uses a 4:1 mux having individual selects at each input channel. Although the procedural continuous assignment can be used to synthesize combinational logic, a combinational behavior must not mix a procedural continuous assignment and a procedural assignment to the same variable. Such descriptions imply sequential behavior. Figure 8.29c shows a common mux structure produced for all three versions when the channels are scalars.

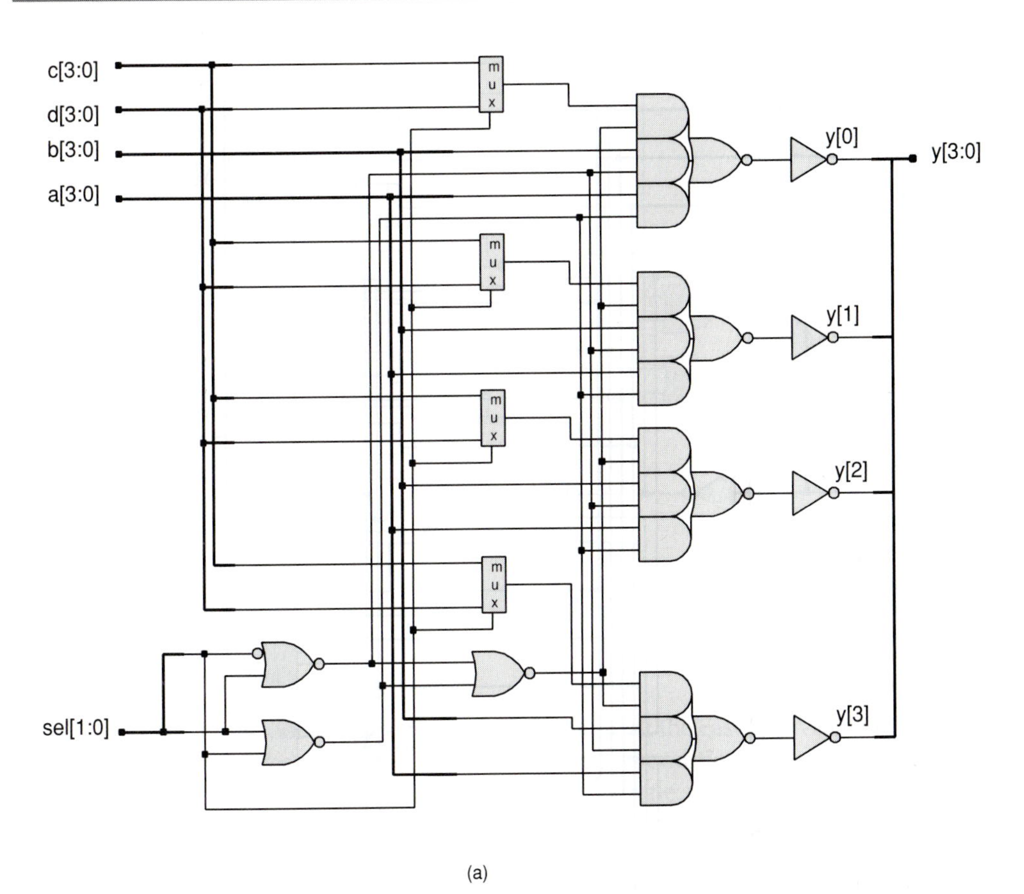

Figure 8.29 Circuits synthesized from (a) *syn1_mux_4bits*, (b) *syn2_mux_4bits* and *syn3_mux_4bits*, and (c) for all three versions with scalar channels. Continues on pages 316–317.

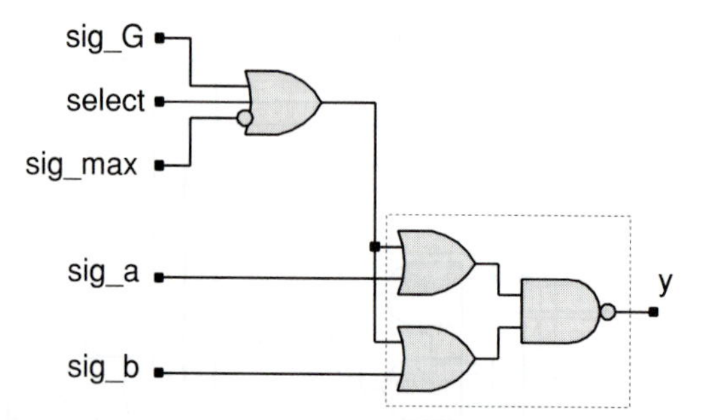

Figure 8.30 Circuit synthesized from a mux with selector logic.

case statements and conditionals may lead to the synthesis of unwanted latches in the design.

Combinational logic must specify the value of the output for all values of the input. Therefore, when a **case** statement does not specify an output for all possible inputs, the synthesis tool infers an implicit latch—i.e., the description implies that the output should retain its residual value under the conditions that were left unspecified. Caution must be taken to ensure that **case** and conditional branching (**if**) statements are complete, either explicitly or by default. Otherwise, the synthesis tool will infer latched behavior. A description using the conditional operator (? ... :) can create the functionality of a latch, but an incomplete conditional operator will cause a syntax error.

Example 8.20 When a **case** statement is incompletely decoded, a synthesis tool will infer the need for a latch to hold the residual output when the select bits take the unspecified values. The latch is enabled by the "event-or" of the cases under which the assignment is explicitly made. In this example, the latch[1] is enabled when *{sel_a, sel_b}* == *2'b10* or *{sel_a, sel_b}* == *2'b01*. Figure 8.31a shows a generic implementation, and Figure 8.32b shows an implementation using an actual library cell (*latrnb_a*). The latter uses the library's two-channel mux, and a model (esdpupd_) for an electrostatic discharge pull-up/down device. This device disables the hardware latch's active-low reset, rather than letting it float. Its internal structure consists of an instantiation of the **pullup** and **pulldown** primitives. Its outputs are the outputs of the two primitives. In this example, the output of the **pullup** primitive is connected to the latch's reset input.

[1] Appendix I contains descriptions of the various flip-flops and latches shown in the examples of synthesized circuits.

```
module mux_latch (y_out, sel_a, sel_b, data_a, data_b);
  input          sel_a, sel_b, data_a, data_b;
  output         y_out;
  reg            y_out;

  always @ ( sel_a or sel_b or data_a or data_b)
    case ({sel_a, sel_b})
      2'b10: y_out = data_a;
      2'b01: y_out = data_b;
  endcase
endmodule
```

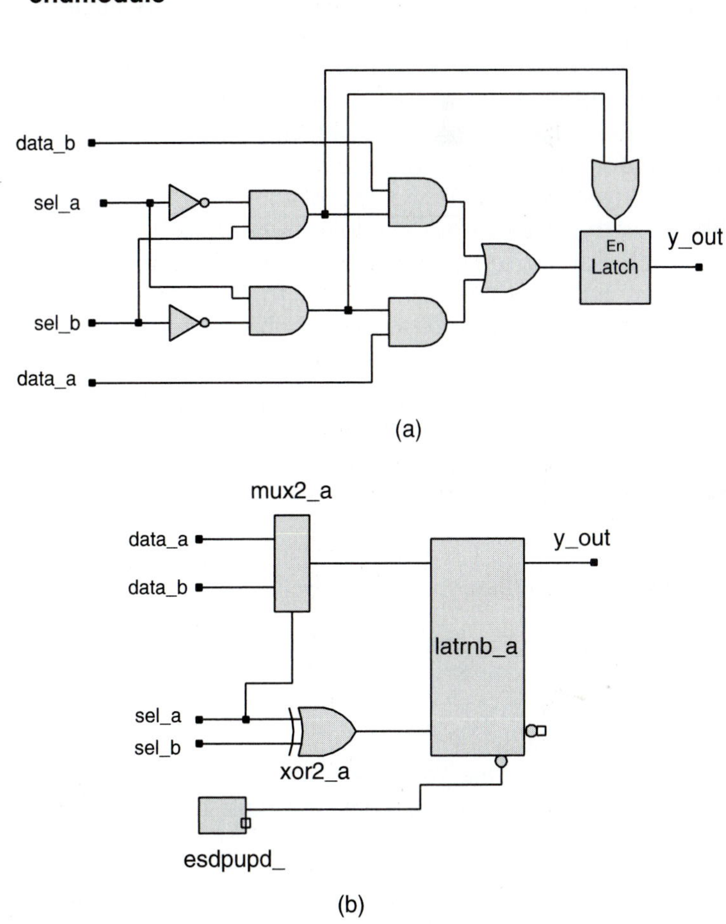

Figure 8.31 Mux with latched output synthesized from an incompletely specified **case** statement (a) generically and (b) from a cell library.

Example 8.21 Three versions of a simplified ALU are presented here to illustrate the consequences of incomplete **case** statements. The description in *alu_incomplete1a* has an incomplete event control expression and **case** statement, and lacks a default assignment. The description in *alu_incomplete2a* has a complete event control expression and is otherwise the same as *alu_incomplete1a.* The description in *alu_complete* has a complete event control expression and default *case_item*. The results of synthesis are shown in Figure 8.32. **In all three cases, only the LSB is shown in alu_out (a scalar), just for convenience of illustration**. Note that *alu_incomplete1a* and *alu_incomplete2a* synthesize to the same circuit, even though they have different event control expressions. Be careful: the event control expression in *alu_incomplete1* is sensitive to only *opcode*. The pre-synthesis behavior of *alu_incomplete1a* and *alu_incomplete2a* will be different because *alu_incomplete1* ignores events on the datapath. The post-synthesis behavior will be identical! The output of the synthesized circuit for *alu_incomplete1a* and *alu_incomplete2a* has an and-gate driven by *enable* and the output of a latch. The and-gate ensures that the output is 0 when *enable* is 0. The latch retains the value of *alu_reg* when the opcode is not a decoded *case_item*. In contrast, the circuit synthesized for *alu_complete* is combinational, because the level-sensitive behavior has a complete event control expression and complete case statement.

```
module alu_incomplete1a (alu_out, data_a, data_b, enable, opcode);
  input     [2:0]      opcode;
  input     [3:0]      data_a, data_b;
  input                enable;
  output               alu_out;      // Note: scalar
  reg       [3:0]      alu_reg;

  assign alu_out = (enable == 1) ? alu_reg : 4'b0;

  always @ (opcode)
    case (opcode)
      3'b001:   alu_reg = data_a | data_b;
      3'b010:   alu_reg = data_a ^ data_b;
      3'b110:   alu_reg = ~data_b;
  endcase
endmodule
```

```
module alu_incomplete2a (alu_out, data_a, data_b, enable, opcode);
  input   [2:0]    opcode;
  input   [3:0]    data_a, data_b;
  input            enable;
  output           alu_out;
  reg     [3:0]    alu_reg;

  assign alu_out = (enable == 1) ? alu_reg : 4'b0;

  always @ (opcode or data_a or data_b)
    case (opcode)
      3'b001:   alu_reg = data_a | data_b;
      3'b010:   alu_reg = data_a ^ data_b;
      3'b110:   alu_reg = ~data_b;
    endcase
endmodule

module alu_complete (alu_out, data_a, data_b, enable, opcode);
  input   [2:0]    opcode;
  input   [3:0]    data_a, data_b;
  input            enable;
  output           alu_out;
  reg     [3:0]    alu_reg;

  assign alu_out = (enable == 1) ? alu_reg : 4'b0;

  always @ (opcode or data_a or data_b)
    case (opcode)
      3'b001:   alu_reg = data_a | data_b;
      3'b010:   alu_reg = data_a ^ data_b;
      3'b110:   alu_reg = ~data_b;
      default:  alu_reg = 4'b0;
    endcase
endmodule
```

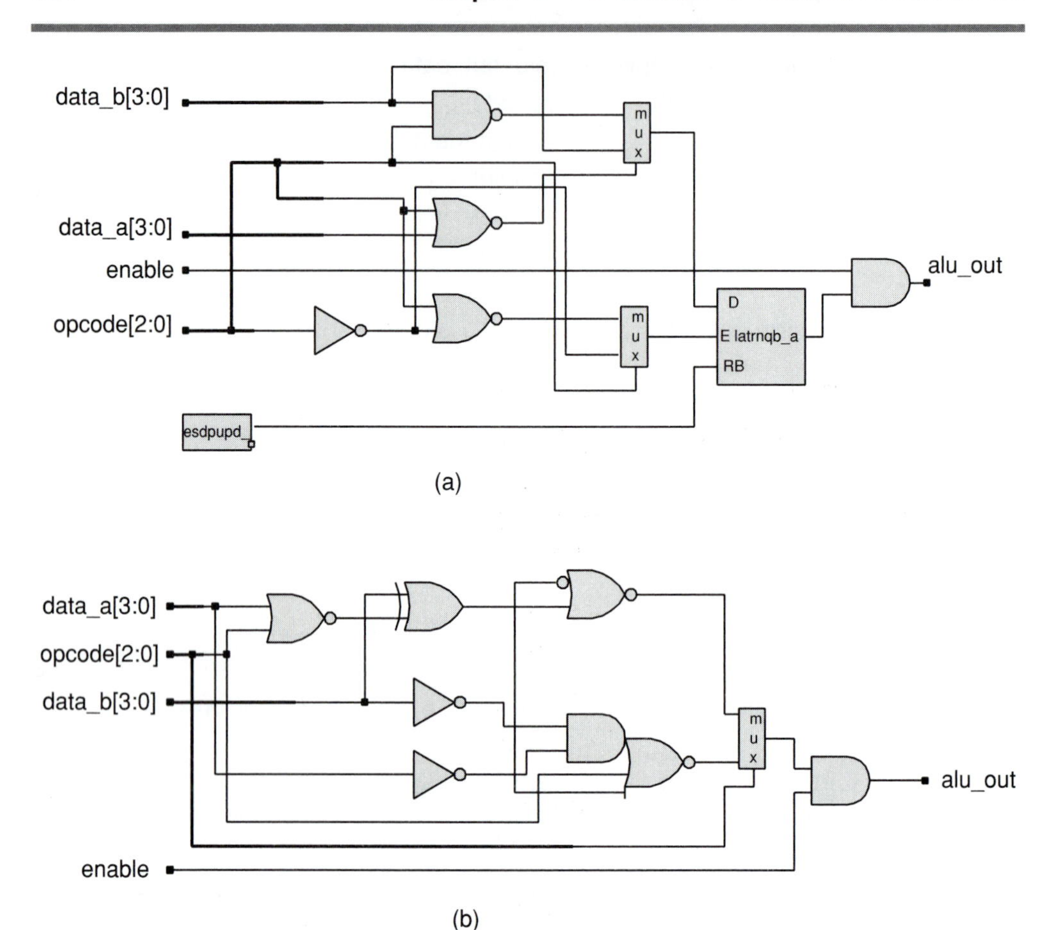

Figure 8.32 Result of synthesizing an ALU with (a) an incomplete event control expression and an incomplete **case** statement (*alu_incomplete1a*), and again with a complete event control expression and an incomplete **case** statement (*alu_incomplete2a*), and (b) a default *case_item* (*alu_complete*).

Example 8.22 The behavior implied by *incomplete_and* does not specify the output when the input pattern is (00). Under this event, the output must remain latched at its residual value. The synthesized circuit shown in Figure 8.33 shows the control logic synthesized from the incomplete conditional (**if**) statement.

```
module incomplete_and (y, a1, a2);
  input          a1, a2;
  output         y;
  reg            y;

  always @ (a1 or a2)
    begin
      if ({a2, a1} == 2'b11) y = 1; else
      if ({a2, a1} == 2'b01) y = 0; else
      if ({a2, a1} == 2'b10) y = 0;
    end
endmodule
```

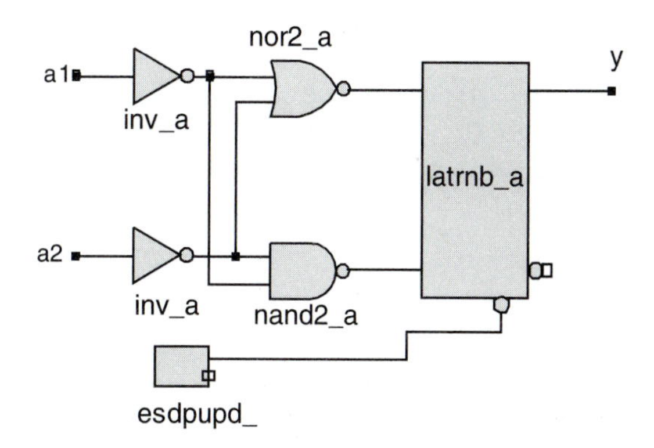

Figure 8.33 Latched "and" gate synthesized from an incomplete conditional.

A synthesis tool might allow the user to direct that an incompletely-specified case statement be treated as though it were fully specified (e.g., Synopsys *full_case* directive). The directive is included in the source code of the model and would cause a circuit to be synthesized without latches.

8.6.11 Synthesis of Priority Structures

A **case** statement implicitly attaches higher priority to the first *case_item* than to the last one, and an **if** statement implies higher priority to the first branch than to the remaining branches. A synthesis tool will determine whether or not the case items of a case statement are mutually exclusive. If they are, the synthesis tool will treat them as though they had equal priority and will synthesize a mux rather than priority structure. Even when the list of case items is not mutually exclusive, a synthesis tool might allow the user to direct that they be treated without priority (e.g., Synopsys *parallel_case* directive). This would be useful if only one *case_item* could be selected at a time in actual operation. An **if** statement will synthesize to a mux structure when the branching is specified by mutually exclusive conditions, as in Example 8.18, but when the branching is not mutually exclusive, the synthesis tool will create a priority structure.

Example 8.23 The conditional activity flow within *mux_4pri* is not governed by mutually exclusive conditions. This results in synthesis of an implied priority for datapath *a* because *sel_a* decodes *a* independently of *sel_b* or *sel_c*, shown in Figure 8.34.

```
module mux_4pri (y, a, b, c, d, sel_a, sel_b, sel_c);
  input          a, b, c, d, sel_a, sel_b, sel_c;
  output         y;
  reg            y;

  always @ (sel_a or sel_b or sel_c or a or b or c or d)
    begin
    if (sel_a == 1)  y = a; else
    if (sel_b == 0)  y = b; else
    if (sel_c == 1)  y = c; else
    y = d;
  end
endmodule
```

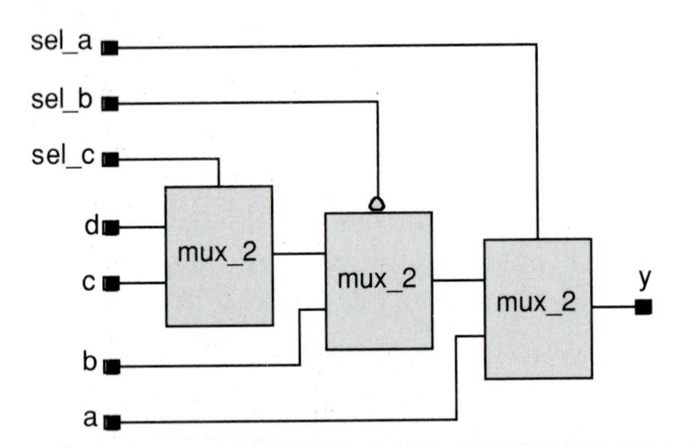

Figure 8.34 Circuit synthesized from a mux with priority decode of input conditions.

8.6.12 TREATMENT OF DEFAULT CONDITIONS

When **case**, conditional branch (**if**), or conditional assignment (**?** ... **:**) statements are used in a Verilog behavioral description of combinational logic, the seed code and synthesized netlist should produce the same simulation results (except for time-dependent behavior) if the seed code has "default" assignments that are purely "0" or "1" values (i.e., the "default" does not explicitly assign an "x" or "z" value). On the other hand, simulation results may differ if the "default" or branch statement makes an explicit assignment of an "x" or "z". A synthesis tool will treat **casex** and **casez** statements as **case** statements. Those case items that decode to explicit assignment of "x" or "z" will be treated as "don't care" conditions for the purpose of logic minimization of the equivalent Boolean expressions (i.e., it does not matter what value is assigned to the object of the **case** assignment under those input conditions). The physical hardware will propagate either a "0" or "1", while the seed behavior will propagate either an "x" or "z". This may lead to a mismatch between the results obtained by simulating the seed code and the synthesis product.

Example 8.24 Two implementations of an 8:3 encoder (without priority) are shown below. Neither fully decodes all possible patterns of *Data*, but both have **default** assignments that cover the remaining outcomes. The result of synthesis, shown in Figure 8.35, is combinational. The assumption here is that only the indicated words of *Data* occur in operation. The default assignment is necessary to prevent synthesis of a latch. The default conditions will be "don't care"s in synthesis.

```
module encoder (Data, Code);
  input     [7:0] Data;
  output    [2:0] Code;
  reg       [2:0] Code;

  always @ (Data)
    begin
      if (Data == 8'b00000001) Code = 0; else
      if (Data == 8'b00000010) Code = 1; else
      if (Data == 8'b00000100) Code = 2; else
      if (Data == 8'b00001000) Code = 3; else
      if (Data == 8'b00010000) Code = 4; else
      if (Data == 8'b00100000) Code = 5; else
      if (Data == 8'b01000000) Code = 6; else
      if (Data == 8'b10000000) Code = 7; else Code = 3'bx;
    end

/*// Alternative description is given below
```

```
always @ (Data)
  case (Data)
    8'b00000001    : Code = 0;
    8'b00000010    : Code = 1;
    8'b00000100    : Code = 2;
    8'b00001000    : Code = 3;
    8'b00010000    : Code = 4;
    8'b00100000    : Code = 5;
    8'b01000000    : Code = 6;
    8'b10000000    : Code = 7;
    default        : Code = 3'bx;
  endcase
*/
endmodule
```

The synthesized 8:3 encoder for both descriptions is shown in Figure 8.35.

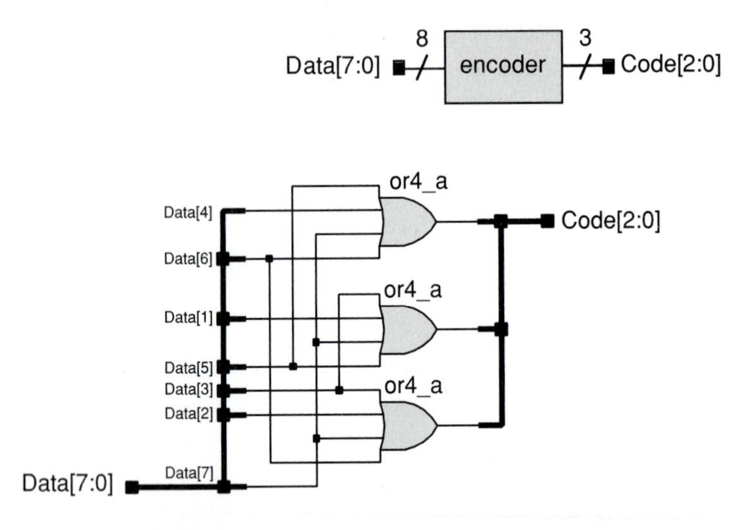

Figure 8.35 Result of synthesizing an encoder described by **if** or **case** statements.

Example 8.25 Alternative behaviors describing an 8:3 priority encoder are shown below. The result of synthesizing the circuit is shown in Figure 8.36. Notice that the branch **(if)** statement has an implied priority, and that the **casex** statement combined with "x" in the case items also implies priority. The default assignments in both styles provide flexibility to the logic optimizer.

```
module priority (Data, Code, valid_data);
  input     [7:0] Data;
  output    [2:0] Code;
  output    valid_data;
  reg       [2:0] Code;
  assign    valid_data = |Data;        // reduction or


  always @ (Data)
    begin
      if (Data[7]) Code = 7; else
      if (Data[6]) Code = 6; else
      if (Data[5]) Code = 5; else
      if (Data[4]) Code = 4; else
      if (Data[3]) Code = 3; else
      if (Data[2]) Code = 2; else
      if (Data[1]) Code = 1; else
      if (Data[0]) Code = 0; else
                   Code = 3'bx;
      end

/*// Alternative description is given below

  always @ (Data)
    casex (Data)
      8'b1xxxxxxx   : Code = 7;
      8'b01xxxxxx   : Code = 6;
      8'b001xxxxx   : Code = 5;
      8'b0001xxxx   : Code = 4;
      8'b00001xxx   : Code = 3;
      8'b000001xx   : Code = 2;
      8'b0000001x   : Code = 1;
      8'b00000001   : Code = 0;
      default       : Code = 3'bx;
    endcase
  */
endmodule
```

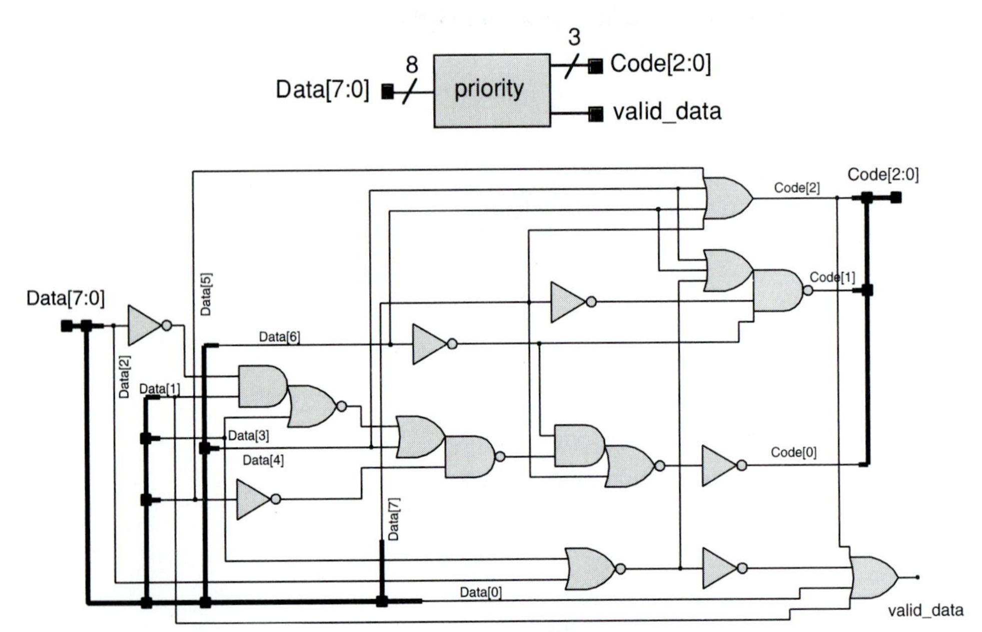

Figure 8.36 Synthesis of an 8:3 priority encoder.

Example 8.26 A 3:8 decoder is described by alternative behaviors shown below.

The decoders synthesize to the circuit in Figure 8.37.

```
module decoder (Code, Data);
  output [7:0] Data;
  input  [2:0] Code;
  reg    [7:0] Data;

  always @ (Code)
    begin
      if (Code == 0) Data = 8'b00000001; else
      if (Code == 1) Data = 8'b00000010; else
      if (Code == 2) Data = 8'b00000100; else
      if (Code == 3) Data = 8'b00001000; else
      if (Code == 4) Data = 8'b00010000; else
      if (Code == 5) Data = 8'b00100000; else
      if (Code == 6) Data = 8'b01000000; else
      if (Code == 7) Data = 8'b10000000; else
                     Data = 8'bx;
    end
```

```
/*// Alternative description is given below
  always @ (Code)
    case (Code)
      0  : Data = 8'b00000001;
      1  : Data = 8'b00000010;
      2  : Data = 8'b00000100;
      3  : Data = 8'b00001000;
      4  : Data = 8'b00010000;
      5  : Data = 8'b00100000;
      6  : Data = 8'b01000000;
      7  : Data = 8'b10000000;
      default: Data = 8'bx;
    endcase
  */
endmodule
```

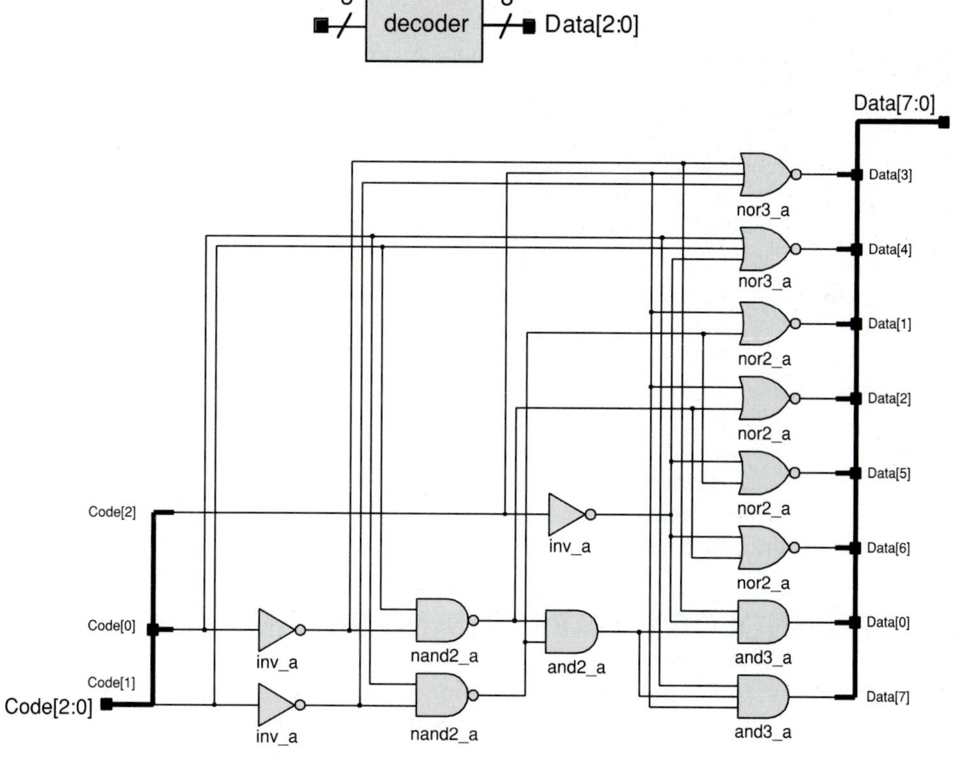

Figure 8.37 Synthesis of a 3:8 decoder.

Example 8.27 A seven-segment LED display is a useful circuit in many applications using prototyping boards. Module *Seven_seg_display* is implemented with active low outputs, and synthesizes into a combinational circuit.

```
module Seven_seg_display (data_in, data_out);
  input    [3:0]          data_in;
  output   [6:0]          data_out;
  reg      [6:0]          data_out;
  //                    abc_defg
  parameter  BLNK  = 7'b111_1111;
  parameter  ZERO  = 7'b000_0001;    // h01
  parameter  ONE   = 7'b100_1111;    // h4f
  parameter  TWO   = 7'b001_0010;    // h12
  parameter  THREE = 7'b000_0110;    // h06
  parameter  FOUR  = 7'b100_1100;    // h4c
  parameter  FIVE  = 7'b010_0100;    // h24
  parameter  SIX   = 7'b010_0000;    // h20
  parameter  SEVEN = 7'b000_1111;    // h0f
  parameter  EIGHT = 7'b000_0000;    // h00
  parameter  NINE  = 7'b000_0100;    // h04

  always @ (data_in)
    case (data_in)
      0:        data_out = ZERO;
      1:        data_out = ONE;
      2:        data_out = TWO;
      3:        data_out = THREE;
      4:        data_out = FOUR;
      5:        data_out = FIVE;
      6:        data_out = SIX;
      7:        data_out = SEVEN;
      8:        data_out = EIGHT;
      9:        data_out = NINE;
      default:  data_out = BLNK;
    endcase
endmodule
```

8.7 TECHNOLOGY MAPPING AND SHARED RESOURCES

Synthesis tools include a technology mapping engine that covers the generic, optimized multi-level Boolean description of the circuit by the physical gates in a technology library. Depending upon the sophistication of the tool, this covering may exploit basic library cells or more complex ones. Descriptions using operator-based adders (e.g., **assign** accum_out = data_a

+ accum;), subtractors, and comparators may be mapped to complex cells in the target library. If necessary, adders and subtractors are easily converted into Boolean equations and synthesized in gates. On the other hand, a tool will not recognize an adder that is implicitly represented by a hierarchical netlist structure of primitive gates. Synthesis tools provide a high degree of manual intervention to the user. For example, the user might have the freedom to direct the tool to ignore sections of a Verilog description that are already in a gate-level form.

Example 8.28 A synthesis tool mapped the addition operator in the Verilog description below to a full-adder cell in the library to create the circuit shown in Figure 8.38a. An alternative implementation shown in Figure 8.38b builds two different five-bit adder blocks out of basic library cells in a structure that depends upon the speed goal for the design (details not shown). The *esdpupd_* device provides "0" (or "1") where needed. The left-most adder forms *A[3:0]* + *B[3:0]*. The five-bit result is one of the inputs to the right-most adder block. That block has a second five-bit input formed by the *C_in* bit and four 0s. The carry-in of each block is hard-wired to ground.

```
module badd_4 (Sum, C_out, A, B, C_in);
  output  [3:0]  Sum;
  output         C_out;
  input   [3:0]  A, B;
  input          C_in;

  assign {C_out, Sum} = A + B + C_in;
endmodule
```

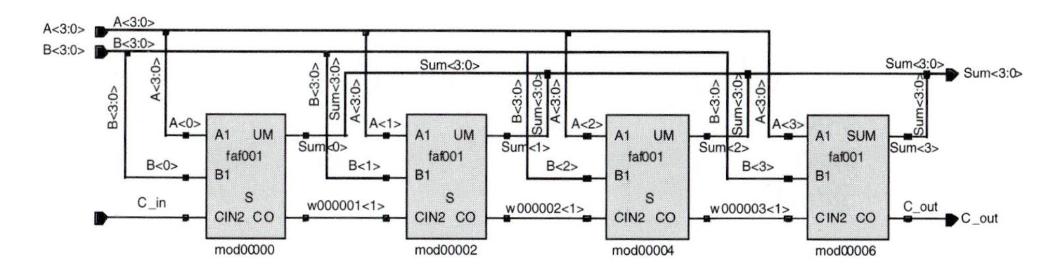

(a)

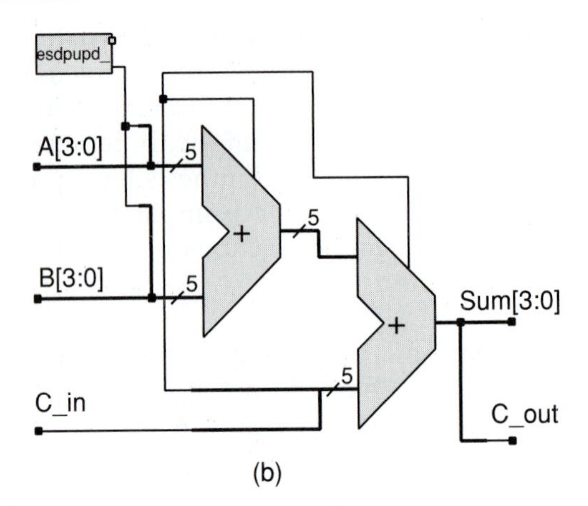

Figure 8.38 Result of synthesizing the "+" operator (a) using library full adder cells, and (b) using five-bit adder blocks.

A synthesis tool should be able to recognize whether the physical resources required to implement complex (large area!) behaviors can be shared. If the data flows within the behavior do not conflict, the resource can be shared between one or more paths. For example, resources can be shared across the mutually exclusive branches of a conditional operator. A tool may implement resource sharing automatically or by direction from the user. For example, the addition operators within the assignment described by:

```
assign y_out = sel ? data_a + accum : data_a + data_b;
```

are in mutually exclusive datapaths. Consequently, the operators can be implemented by a shared adder whose input datapaths are multiplexed. This feature is vendor-dependent. If the tool does not automatically implement resource sharing, the description must be written to force the sharing.

Example 8.29 The use of parentheses in the description in *res_share* forces the synthesis tool to multiplex the datapaths and produce the circuit shown in Figure 8.39.

```
module res_share (y_out, sel, data_a, data_b, accum);
  input    [3:0]   data_a, data_b, accum;
  input            sel;
  output   [4:0]   y_out;

  assign y_out = data_a + (sel ? accum : data_b);
endmodule
```

Failure to include the parentheses in the expression for *y_out* will synthesize to a circuit that uses two adders. The most efficient implementation multiplexes the datapaths and shares the adder between them, rather than multiplex the outputs of separate adders. The area required to implement an adder will be significantly greater than the area required to implement a mux.

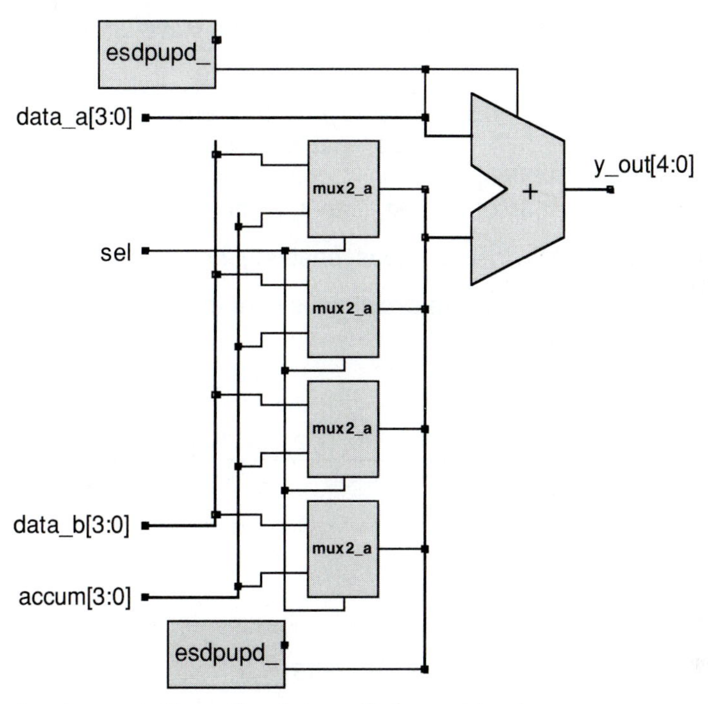

Figure 8.39 Implementation of a datapath forced to share resources.

8.8 THREE-STATE BUFFERS

The Verilog language has built-in constructs for modeling and synthesizing the functionality of three-state devices. Three-state devices are controlled by signals whose value determines whether an input signal is connected to the output terminal of the device. This functionality is important in physical circuits having multiple drivers. One common configuration is a bus, which must be shared by multiple data sources.

8.8.1 BUSES

Three-state buses are an important element of many systems. They are characterized by a "z" logic value assignment to the bus driver signal when the bus control signal is de-asserted. Otherwise, the bus is driven. This functionality is easily described by the Verilog continuous assignment using the conditional ("? ... :") operator. The assignment synthesizes to a three-state device.

Example 8.30 Figure 8.40 depicts a three-state driver of a bus, described by *stuff_to_bus1*.

```
module stuff_to_bus1 (data_to_bus, bus_enabled, clk);
  input              bus_enabled, clk;
  output    [31:0]   data_to_bus;
  reg       [31:0]   ckt_output_to_bus;

  assign data_to_bus = (bus_enabled) ? ckt_output_to_bus : 32'bz;

  // Description of core circuit goes here to drive ckt_output_to_bus

endmodule
```

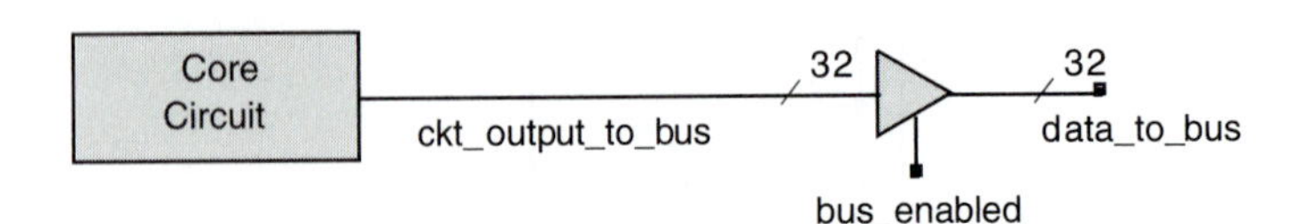

Figure 8.40 Three-state bus driver.

8.8.2 BI-DIRECTIONAL BUS DRIVERS

A bi-directional bus driver must be capable of sending and receiving data.

Example 8.31 The situation is depicted in Figure 8.41 and described below in *stuff_to_bus2*. Note that *rcv_data* and *send_data* must be non-overlapping.

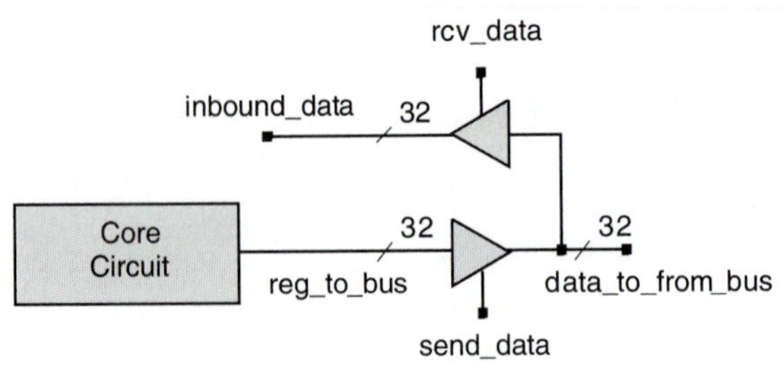

Figure 8.41 Three-state bus model.

```
module stuff_to_bus2 (data_to_from_bus, clk, send_data, rcv_data);
  input          clk, send_data, rcv_data;
  inout  [31:0]  data_to_from_bus;
  reg    [31:0]  reg_to_bus;
  wire   [31:0]  data_to_from_bus, inbound_data;

  assign inbound_data = (rcv_data) ? data_to_from_bus : 32'bz;
  assign data_to_from_bus = (send_data) ? reg_to_bus : data_to_from_bus;


// Behavior using inbound_data go here

endmodule
```

If a module having a bi-directional port (**inout**) is imbedded within a testbench, the testbench cannot drive the port directly with a register variable. To do so would erroneously imply that the module can assign value to the register. The I/O pads of an ASIC chip are typically bi-directional ports. In the simplified pad structure shown in Figure 8.42, a signal from the environment drives through the pad to the logic core. When the output is enabled, a signal from the logic core drives the output pad. The test signal applied to the pad by the testbench can be formed by a continuous assignment that assigns the value of a register variable (e.g., stimulus pattern) to a wire.

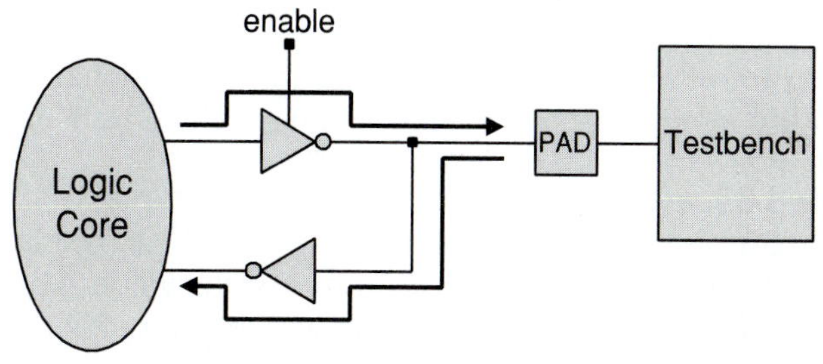

Figure 8.42 ASIC chip I/O pad structure.

8.8.3 Bus Loading

The speed at which a bus can operate is limited by the capacitive loading placed on it by driving and receiving circuits. This performance can be seriously compromised when several devices are attached to the bus. It is important that a Verilog description of a bus circuit be synthesized efficiently. The recommended practice is to multiplex the drivers of a bus so that the load is reduced to a single-driver circuit.

Example 8.32 Figure 8.43 shows a recommended style for using a mux to reduce the capacitive loading on a bus.

```
module stuff_to_bus3 (data_to_from_bus, enab_a, enab_b, clk, rcv_data);
  input              enab_a, enab_b, rcv_data, clk;
  inout  [31:0]      data_to_from_bus;
  reg    [31:0]      reg_A_to_bus, reg_B_to_bus;

  assign data_to_from_bus = (enab_a) ? reg_A_to_bus :
                                    (enab_b) ? reg_B_to_bus : 32'bz;
endmodule // Not showing other logic.
```

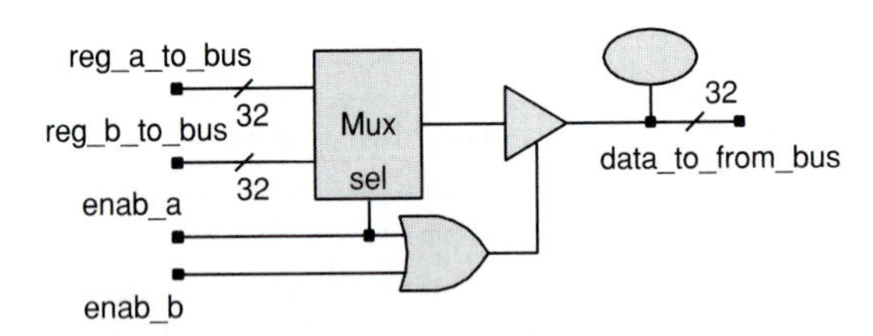

Figure 8.43 Multiplexed bus drivers.

8.9 THREE-STATE OUTPUTS AND DON'T-CARES

When the output of a module is assigned by a conditional assignment, a variety of results are possible, with some leading to three-state outputs.

Example 8.33 *alu_with_z1* describes a combinational ALU with three-state output. A second version, *alu_with_z2*, has the same description, except the default *case_item* assigns value 4'bx instead of 4'b0. The circuits that result are shown in Figure 8.43. Both circuits have three-state output inverters (which take less area than buffers), but the circuit for alu_with_z2 has a simpler realization because the synthesis tool is able to exploit the "don't-care"s implied by the default assignment of 4'bx in the **case** statement.

```
module alu_with_z1 (alu_out, data_a, data_b, enable, opcode);
  input     [2:0]   opcode;
  input     [3:0]   data_a, data_b;
  input             enable;
  output            alu_out;        // scalar for illustration
  reg       [3:0]   alu_reg;
```

```
    assign alu_out = (enable == 1) ? alu_reg : 4'bz;

    always @ (opcode or data_a or data_b)
      case (opcode)
        3'b001:    alu_reg = data_a | data_b;
        3'b010:    alu_reg = data_a ^ data_b;
        3'b110:    alu_reg = ~data_b;
        default:   alu_reg = 4'b0;
        // alu_with_z2 has default: alu_reg = 4'bx;
      endcase
endmodule
```

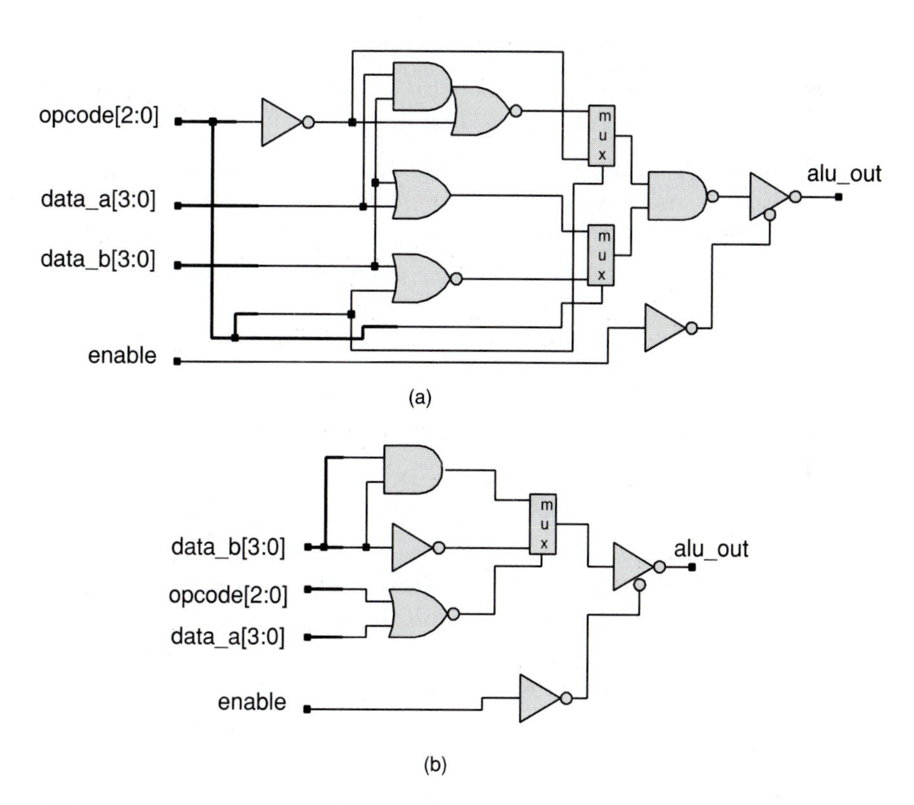

Figure 8.44 Circuit synthesized from (a) *alu_with_z1* and (b) *alu_with_z2*.

8.10 SUMMARY

Combinational logic forms its outputs from its inputs, without memory of previous values of the inputs. User defined primitives, continuous assignments, instantiated gates and behaviors that do not imply the need for memory will synthesize into combinational logic. Combinational logic is easily synthesized, but care must be taken to avoid unwanted latches that result from incompletely-specified **case** statements and conditionals.

REFERENCES

1. Smith, M. J., *Application-Specific Integrated Circuits*, Addison-Wesley Longman, Inc., Reading, Massachusetts, 1997.
2. Brayton, R. et al., *Logic Minimization Algorithms for VLSI Synthesis*, Kluwer Academic Publishers, Boston, 1984.
3. Bartlett, K. et al., *Synthesis of Multilevel Logic under Timing Constraints*, IEEE Transactions on Computer Aided Design of Integrated Circuits, CAD-7(6), pp. 582-596, October 1986.
4. Bartlett, K. et al., *Multilevel Logic Minimization Using Implicit Don't-Cares*, IEEE Transactions on Computer Aided Design of Integrated Circuits, CAD-5(4), pp. 723-740, October 1986.
5. Devidas, S. et al., *Logic Synthesis*, McGraw-Hill, Inc., New York, 1994.
6. De Micheli, G., *Synthesis and Optimization of Digital Circuits*, McGraw-Hill Inc., New York, 1994.
7. Gajski, D. et al., *High-Level Synthesis*, Kluwer Academic Publishers, Boston, 1992.
8. De Man, H. et al., *Architecture-Driven Synthesis Techniques for VLSI Implementation of DSP Algorithms*, Proc. IEEE, 78(2): 319-355, February 1990.

PROBLEMS

1. Using standard cells, write a gate-level implementation of the pre-optimized generic circuit shown below. Synthesize the circuit and compare the area/speed of the result with the area/speed of the pre-optimized circuit.

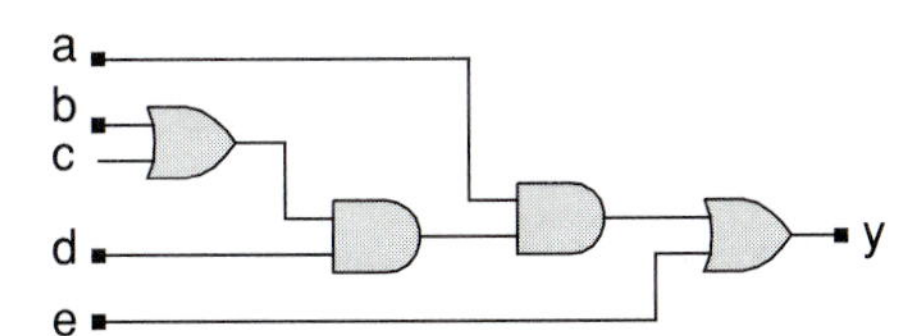

2. Using continuous assignments, form a four-bit word whose value is the number of ones in an 8-bit word. Verify and synthesize the circuit.

3. Synthesize (in a cell library) the carry-select adder from problem #14 in Chapter 4, preserving hierarchical boundaries. For word lengths of 6, 12, and 16, compare the area and speed of the carry-select adder to the area and speed of an adder synthesized directly from a continuous assignment statement.

4. Of the two versions of an ALU below, which will synthesize to the simplest circuit (fewest gates)? Note: *alu_version_2* is obtained from *alu_version_1* by changing the default assignment to 4'bx.

```
module alu_version 1 (alu_out, data_a, data_b, enable, opcode);
  input    [2:0]   opcode;
  input    [3:0]   data_a, data_b;
  input            enable;
  output           alu_out;   // scalar for illustration
  reg      [3:0]   alu_reg;

  assign alu_out = (enable == 1) ? alu_reg : 4'bx;

  always @ (opcode or data_a or data_b)
    case (opcode)
      3'b001: alu_reg = data_a | data_b;
      3'b010: alu_reg = data_a ^ data_b;
      3'b110: alu_reg = ~data_b;
      default:alu_reg = 4'b0;
                          // alu_version_2 has default: alu_reg = 4'bx;
    endcase
endmodule
```

5. Discuss and compare the results of synthesizing *circuit_1* and *circuit_2* below.

```
module circuit_1 (a, b, y_and, y_or, y_nand, y_nor, y_xor, y_xnor, y_neg);

  input  [1:0]   a, b;
  output         y_and, y_or, y_nand, y_nor,
                 y_xor, y_xnor, y_neg;
  reg            y_and, y_or, y_nand, y_nor,
                 y_xor, y_xnor, y_neg, y_band, y_bor;

  always @ (a or b) begin
    y_and    = a & b;
    y_or     = a | b;
    y_band   = a & b;
    y_nand   = ~y_band;
    y_bor    = a | b;
    y_nor    = ~y_bor;
    y_xor    = a ^ b;
    y_xnor   = a ~^ b;
    y_neg    = ~ a;
  end
endmodule

module circuit_2 (a, b, c, d, e, f, g, h, i, j, k, l, m,
    y_and, y_or, y_nand, y_nor, y_xor, y_xnor, y_neg);

  input  [1:0]   a, b, c, d, e, f, g, h, i, j, k, l, m;
  output         y_and, y_or, y_nand, y_nor,
                 y_xor, y_xnor, y_neg;
  reg            y_and, y_or, y_nand, y_nor,
                 y_xor, y_xnor, y_neg, y_band, y_bor;

  always @ (a or b or c or d or b or e or f or
    g or h or i or j or k or l or m) begin
    y_and    = a & b;
    y_or     = c | d;
    y_band     = e & f;
    y_nand     = ~y_band;
    y_bor      = g | h;
    y_nor      = ~y_bor;
    y_xor      = i ^ j;
    y_xnor     = k ~^ l;
    y_neg      = ~ m;
  end
endmodule
```

6. Explain why the comparator in Example 8.11 uses a test for $k < 3$.

7. Explain why the following implementation of a 2-bit comparator sends a simulator into a death spiral.

```
module comparator (a, b, a_gt_b, a_lt_b, a_eq_b); // Alternative algorithm
  parameter   size = 2;
  input       [size-1: 0] a, b;
  output      a_gt_b, a_lt_b, a_eq_b;
  reg         a_gt_b, a_lt_b, a_eq_b;
  reg         k;

  always @ ( a or b) begin: compare_loop
    for (k = size begin
      if (a[k] != b[k]) begin
        a_gt_b = a[k];
        a_lt_b = ~a[k];
        a_eq_b = 0;
        disable compare_loop;
      end     // if
    end       // for loop
      a_gt_b = 0;
       a_lt_b = 0;
      a_eq_b = 1;
  end         // compare_loop
endmodule
```

8. Compare the results of synthesizing *comparator* as shown in Example 8.11 with an alternative version in which *k* is declared to be an integer and the **for** loop is declared with

```
for (k = size –1; k >= 0; k = k – 1)
```

9. How will the delay control operator in the following statement influence the outcome of synthesis?

```
assign #10 y = a ^ b;
```

10. What will a synthesis tool produce from the following description? Explain.

```
module mystery1 (y, a, b, c, sel);
   input          a, b, c;
   input   [1:0]  sel;
   output         y;
   reg            y;
   always @ (sel or a or b or c) begin
      casex (sel)
         2'b00:     y = a;
         2'b01:     y = b;
         2'b10:     y = c;
      endcase
   end
endmodule
```

11. Synthesize the two implementations of the or-nand module shown. Explain the reason for any observed differences in the two realizations.

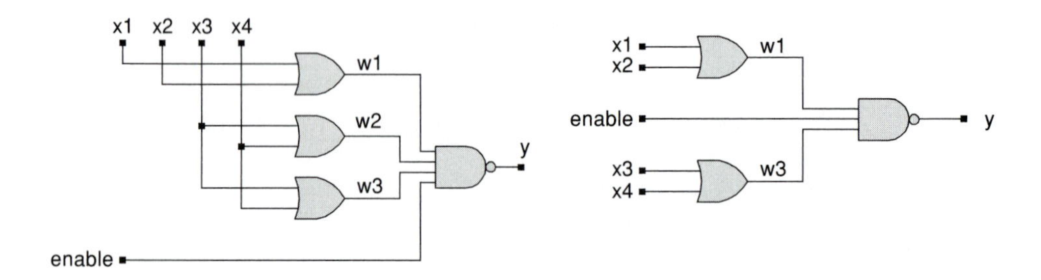

12. What will be the control signal for the latch that is inferred from the Verilog description shown below? Synthesize the circuit and confirm your answer.

```
module implicit_latch (y_out, sel_a, sel_b, data_a, data_b);
   input sel_a, sel_b, data_a, data_b;
   output y_out;
   always @ ( sel_a or sel_b or data_a or data_b)
      case ({sel_a, sel_b})
         2'b10: y_out = data_a;
         2'b01: y_out = data_b;
      endcase
endmodule
```

13. Write, verify, and synthesize a description of a function of three inputs, *sig_a*, *sig_b*, and *sig_c* that asserts if an even number of inputs are asserted.

14. Write, verify, and synthesize a Verilog description of a function of three inputs, *sig_a*, *sig_b*, and *sig_c* that asserts if an odd number of inputs are asserted.

15. Write, verify, and synthesize a Verilog description of a circuit that adds and subtracts two four-bit (unsigned) words. The module also generates a *zero_flag* indicating that the four-bit output *sum* word is 0000_2, and an *overflow_flag* to indicate the presence of an overflow condition, and an *underflow* flag.

16. Write, verify, and synthesize a Verilog description of a circuit that performs the square root function of a four-bit data word.

17. Develop, verify, and synthesize a 16-bit ripple-carry adder and a 16-bit carry look-ahead adder. Compare their area and performance.

18. Synthesize the five versions of the two-word, two-bit comparator given in Chapter 2 and this chapter.

19. Using the header shown below, write and synthesize a model of a synchronous 4-channel multiplexer using nonblocking assignments and having active-low reset to the output 4'b0101.

```
module mux_case4 (a, b, c, d, y, clock, select, reset_);
...
endmodule
```

20. Synthesize the alternative descriptions of a SN74L47 BCD-to-Seven Segment Decoder that were discussed in Problem 4.23:

a. structural (using Verilog primitives),

b. data flow (using Verilog language operators in continuous assignment statements), and

c. behavioral (using a case statement)

Compare and discuss the results. How was the wired-and logic treated by the synthesis tool?

21. Describe how latches may be inadvertently introduced into the synthesis of "combinational" logic.

22. Synthesize a circuit to multiply two 4-bit words.

23. Synthesize a circuit that generates the even-parity of a 7-bit word and forms an 8-bit word having the parity bit in the position of MSB.

24. Develop a behavioral model of a BCD-to-7 segment decoder with active-low outputs. Synthesize a gate-level realization, and verify that the simulation of the behavior and the synthesized implementation match.

25. Synthesize the following description and discuss the functionality that it implements.

```
module something (signal_1, signal_2);
  output    [7:0]   signal_2;
  input     [2:0]   signal_1;
  reg       [7:0]   dummy, signal_2;
  always @ (signal_1) begin
    dummy = 8'b0000_0001;
    signal_2 = dummy << signal_1;
  end
endmodule
```

CHAPTER 9

SYNTHESIS OF SEQUENTIAL LOGIC

A synthesis tool must distinguish sequential behavior from combinational behavior, optimize the description, and then assign hardware resources to implement the behavior. A Verilog description of sequential logic can be synthesized only if certain conditions are met. Some conditions are general; others are specific to a particular vendor or a particular construct of the language.

In general, the event control expression of a cyclic behavior describing sequential logic must be synchronized to a single edge (**posedge** or **negedge**, but not both) of a single clock (synchronizing signal). Multiple behaviors need not have the same synchronizing signal, nor the same edge of the same signal, but the optimization process considered here requires that all the synchronizing signals (clocks) have the same period. This establishes a single clock domain for optimization. A given tool may require or allow the user to specify a clock period. A timing control expression must appear at the beginning of a behavior, before any assignment statements. This constraint assures that the synthesis tool can synchronize the activity within the behavior.

Chapter 7 presented several options for describing sequential behavior in Verilog. Sequential modules are formed from sequential user-defined primitives, behaviors, sequential library cells, and nested modules implementing sequential behavior, as listed in Table 9.1. Table 9.2 lists many sequential functional units commonly synthesized from Verilog behavioral descriptions.

9.1 SYNTHESIS OF SEQUENTIAL UDPs

Verilog has no predefined sequential primitives. Chapter 5 showed how user-defined primitives can implement sequential and combinational behavior, and how a sequential UDP can implement both level-sensitive and edge-sensitive

Table 9.1 Options for implementing sequential logic.

OPTIONS FOR SEQUENTIAL LOGIC
User-defined primitive
Behavior with timing controls
Instantiated library register cell
Instantiated module

Table 9.2 Commonly synthesized sequential functional units.

COMMONLY SYNTHESIZED SEQUENTIAL LOGIC	
Data register	Gate generator
Transparent latch	Pulse generator
Shift register	Timing generator
Accumulator	Clock generator
Parallel/serial converter	Event counter
Binary counter	Memory address counter
Johnson counter	FIFO memory pointer
BCD counter	Sequencer
Gray counter	Controller
Finite state machine	Edge detector
Synchronizer	

behavior. Given a UDP, a synthesis tool must recognize the input whose value (for a latch) or edge transition (for a flip-flop) controls the datapath through the primitive. Since a sequential primitive will ultimately be mapped into logic that includes either a latch or flip-flop, the behavior described by a sequential UDP must have only one such control: either a clock signal for edge-sensitive synchronization of activity or an enable signal for level-sensitive (latch) behavior. The UDP description may also contain *asynchronous* control signals that will implement the set and reset behavior of the physical part.

A given synthesis tool might impose additional constraints on the style of Verilog code that it can synthesize into sequential logic. For example, a tool might constrain the order of the rows of a UDP table by requiring that the rows for the asynchronous controls be listed before the rows for the synchronizing or enabling signal.

The technology mapping performed by a synthesis tool is limited by the available cells in the library. A given Verilog behavior may not map directly into a library part.

Example 9.1 The J-K flip-flop described below by the UDP *jk_flop* does not map directly into a part in the scmos_12 library. Instead, it is synthesized as a D flip-flop[1] in Figure 9.1, with additional logic that creates the desired behavior in response to the *j* and *k* inputs.

```
module jk_flop (q, clk, j, k, clr);
   output      q;
   input       clk, j, k, clr;

   jk_udp (q, clk, j, k, clr);
endmodule

primitive jk_udp (clk, q, j, k, clr);
   output      q;
   input       clk, j, k, clr;
   reg         q;

   table
   // clk,  j   k   clr     q       next
      ?     ?   ?   0    :  ?   :   0;
      f     0   0   1    :  ?   :   -;
      f     0   1   1    :  ?   :   0;
      f     1   0   1    :  ?   :   1;
      f     1   1   1    :  0   :   1;
      f     1   1   1    :  1   :   0;
      r     ?   ?   ?    :  ?   :   -;
      ?     *   ?   ?    :  ?   :   -;
      ?     ?   *   ?    :  ?   :   -;
      ?     ?   ?   *    :  ?   :   -;
   endtable
endprimitive
```

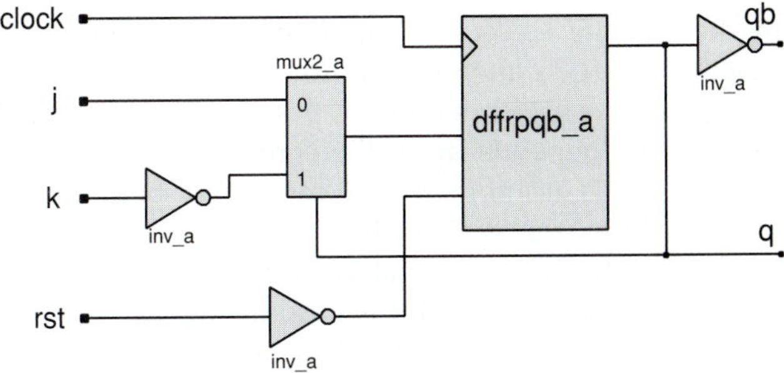

Figure 9.1 Logic synthesized for a j-k flip-flop.

[1] Appendix I contains descriptions of the various flip-flops and latches shown in synthesized circuits.

9.2 SYNTHESIS OF LATCHES

Level-sensitive, or latched, behavior is characterized by an output that is affected by the input only while a control signal is asserted. At other times, the input is ignored and the output retains its residual value (the value it had when the control signal became de-asserted). The need for a latch is inferred by the synthesis tool when it detects a level-sensitive behavior (i.e., no edge constructs) in which a register variable is assigned value in some threads of activity, but not others (e.g., an incomplete **if** statement in a behavior). Those threads of execution that do not explicitly assign a value to a register variable imply the need for the variable to retain the value it had before the behavior was activated. In general, a level-sensitive behavior may contain several such variables, and all will be synthesized as latches. The synthesis tool must identify the datapaths through the latches and their control signals. The control signal of a given latch will be the signal whose value controls the branching of the activity flow to the statements that do not assign value to the associated register variable. If the activity flow assigns value to a given register variable in all possible threads of the activity, a latch will be inferred if a path assigns a variable its own value; otherwise, in the absence of this feedback, the behavior does not imply a latch.

In synthesis, latches implement incompletely-specified assignments to register variables in **case** and conditional branch statements (**if** ...) in a level-sensitive cyclic behavior. If a **case** statement has a default assignment with feedback (i.e., the variable is explicitly assigned to itself), the synthesis tool will choose a mux structure with feedback. Likewise, if an **if** statement in a level-sensitive behavior assigns a variable to itself, the result will be a multiplexer structure with feedback. If the behavior is edge-sensitive, incomplete case and conditional statements synthesize register variables to flip-flops; if the statements are completed with feedback, the result is a register whose output is fed back through a multiplexer at its data path. (If the cell library has a cell with a gated data path, the tool may select that part. See later examples.)

The functionality of a latch is also inferred when the conditional operator (**?** ... **:**) is implemented with feedback, but the actual implementation chosen by a synthesis tool depends upon the context. If the conditional operator is used in a continuous assignment, the result will be a mux with feedback. If the conditional operator is used in an edge-sensitive cyclic behavior, the result will be a register with a gated datapath in a feedback configuration with the output of the register. If it is used in a level-sensitive cyclic behavior, the result will be a hardware latch. The reader should be aware that explicit gate-level latches (i.e., cross-coupled nand primitives) and other descriptions of combinational feedback loops will not be synthesized into hardware latches. Instead, a synthesis tool will detect the presence of combinational feedback and issue an error message. (Some FPGA tools synthesize a "latch description" into on-chip RAM.)

Example 9.2 Synthesis tools will synthesize a latch when a **case** statement has an incomplete list of case items. In this example, *latch_case_assign* decodes only five of the eight possible case items (considering only explicit 0 and 1), and the latch has asynchronous *set* and *clear* control lines. The first assigned binding to *latch_out* implements the transparent mode of behavior; the remaining assignments respond to the asynchronous control lines, and implement the latched mode of behavior (i.e., *enable* is de-asserted). This implementation is efficient in simulation, but does not synthesize with some tools; the second implementation, *latch_case1*, is less efficient in simulation but can be synthesized by all tools. A third version, *latch_case2*, is the same as *latch_case1*, except that the default assignment has been removed. This leaves a set of unspecified case items and leads to synthesis of latches. All three exhibit correct transparent behavior.

```
module latch_case_assign (latch_out, latch_in, set, clear, enable);
  input     latch_in, enable, set, clear;
  output    latch_out;
  reg       latch_out;

  always @ (enable or set or clear)
    case ({enable, set, clear})
      3'b000: assign latch_out = latch_in;  // Transparent mode
      3'b110: assign latch_out = 1'b1;      // Set
      3'b010: assign latch_out = 1'b1;      // Set
      3'b101: assign latch_out = 1'b0;      // Clear
      3'b001: assign latch_out = 1'b0;      // Clear
      default: deassign latch_out;          // Holds residual value
    endcase
endmodule

module latch_case1 (latch_out, latch_in, set, clear, enable);
  input     latch_in, set, clear, enable;
  output    latch_out;
  reg       latch_out;
```

```
  always @ (enable or set or clear or latch_in)
    case ({enable, set, clear})
      3'b000:     latch_out = latch_in;     // Transparent mode
      3'b110:     latch_out = 1'b1;         // Set
      3'b010:     latch_out = 1'b1;         // Set
      3'b101:     latch_out = 1'b0;         // Clear
      3'b001:     latch_out = 1'b0;         // Clear
      default:    latch_out = latch_out;    // Holds residual value
    endcase
endmodule

module latch_case2 (latch_out, latch_in, set, clear, enable);
  input      latch_in, set, clear, enable;
  output     latch_out;
  reg        latch_out;

  always @ (enable or set or clear or latch_in)
    case ({enable, set, clear})
      3'b000:     latch_out = latch_in; // Transparent mode
      3'b110:     latch_out = 1'b1;
      3'b010:     latch_out = 1'b1;
      3'b101:     latch_out = 1'b0;
      3'b001:     latch_out = 1'b0;
    endcase
endmodule
```

The synthesized circuit in Figure 9.2a implements *latch_case1* by connecting combinational gates with a mux in a feedback configuration, because the case statement is completed with default feedback. The description of *latch_case2*, which lacks a default assignment, synthesizes to the circuit in Figure 9.2b, using the same library that produced *latch_case1*. This circuit consists of a hardware latch and additional logic to manage the *set* and *clear* signals. As simple as this example is, it illustrates how library technology and coding style affect the outcome and overall efficiency of the implementation.

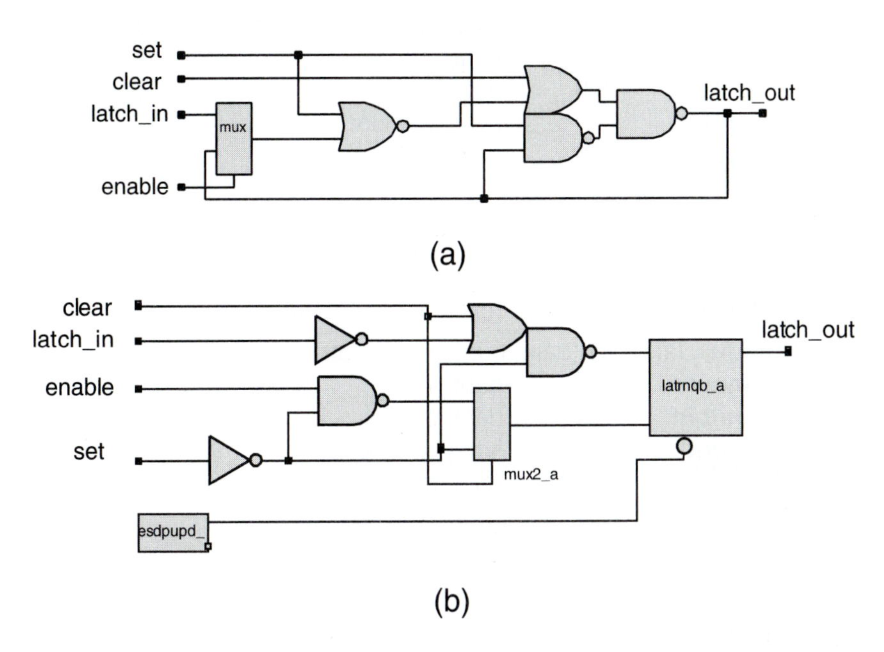

Figure 9.2 Result of synthesizing (a) *latch_case1* and (b) *latch_case2.*

Example 9.3 If a synthesis tool supports the **assign** ... **deassign** procedural continuous assignment, it will synthesize a hardware latch from a level-sensitive behavior containing an **assign** procedural continuous assignment that is also matched with the **deassign** construct. The description given by *latch_if_assign* below has a level-sensitive behavior; its event control expression is sensitive to *latch_enable*, but not to its rising or falling edge. The procedural continuous assignment binds the output of the latch to the datapath dynamically when the latch is enabled, and removes the binding when *latch_enable* is de-asserted. The output remains at its residual value until *latch_enable* is asserted again. A tool that supports this style will synthesize a circuit consisting of a bank of library latches. If the tool does not support the **assign** ... **deassign** construct, the style of *latch_if1* can be used. (The first version is more efficient in simulation because it ignores *data_in* when *latch_enable* is not asserted.) Notice that **the synthesis tool implements this descriptive style in Figure 9.3 with a mux having combinational feedback**—the **if** statement is completed with feedback.

```
module latch_if_assign (data_out, data_in, latch_enable);
  input    [3:0]    data_in;
  output   [3:0]    data_out;
  input             latch_enable;
  reg      [3:0]    data_out;

  always @ (latch_enable)
  if (latch_enable) assign data_out = data_in;
    else deassign data_out
endmodule

module latch_if1 (data_out, data_in, latch_enable);
  input    [3:0]    data_in;
  output   [3:0]    data_out;
  input             latch_enable;
  reg      [3:0]    data_out;

  always @ (latch_enable or data_in)
    if (latch_enable) data_out = data_in;
    else data_out = data_out;
endmodule
```

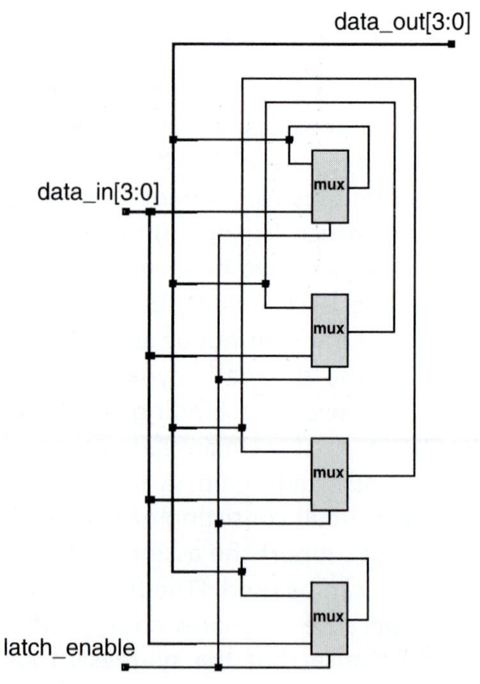

Figure 9.3 Latch behavior synthesized from *latch_if1* (as drawn by the synthesis tool).

Example 9.4 The **if** statement in Example 9.3 was completed with feedback and synthesized to a mux-feedback implementation of a latch. An **if** statement in a level-sensitive behavior will also synthesize the behavior of a latch if the statement assigns value to a register variable in some, but not all, branches (i.e., the statement is incomplete). The description given below by *latch_if2* has a level-sensitive behavior whose event control expression is sensitive to both *latch_enable* and *data_in*. The result of synthesis is shown in Figure 9.4. This style maps preferentially to a hardware latch, rather than a feedback-mux configuration.

```
module latch_if2 (data_out, data_in, latch_enable);
  input     [3:0] data_in;
  output    [3:0] data_out;
  input           latch_enable;
  reg       [3:0] data_out;

  always @ (latch_enable or data_in)
    if (latch_enable) data_out = data_in;     // Incompletely specified
endmodule
```

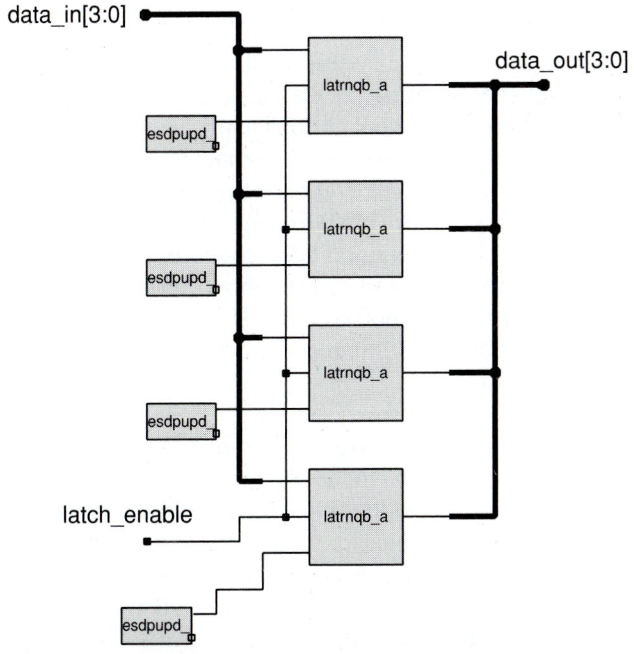

Figure 9.4 Latches synthesized from *latch_if2*, a description containing an incompletely-specified conditional branch (**if**) statement.

The difference between the synthesis results in Figures 9.3 and 9.4 illustrates how a slight change in the behavioral description can influence the outcome of synthesis. The important distinction to note is that the style in Figure 9.3 used a complete conditional branch, while the style used in Figure 9.4 used an incomplete conditional branch. The results are equivalent in simulation, but the physical circuit will have different area/speed trade-offs. The **if** statement completed with feedback equivalent is to the following conditional assignment statement:

```
assign data_out [3:0] = latch_enable ? data_in[3:0] : data_out[3:0];
```

This statement will synthesize to the same structure (see Figure 9.3), and is commonly used to describe a latch. Remember, the conditional operator must be completed.

9.3 SYNTHESIS OF EDGE-TRIGGERED FLIP-FLOPS

Sequential logic is characterized by the need for memory that stores the values of variables so that they may be referenced in other statements. A register variable in a Verilog behavior will be synthesized as a flip-flop if it is referenced outside the scope of the behavior, referenced within the behavior before it is assigned value, or assigned value in only some branches of the activity. All of these situations imply the need for memory, or residual value.

A register variable will be synthesized as the output of a flip-flop when its value is assigned synchronously with the edge of a signal. The decoding of signals immediately after the event control expression allows the synthesis tool to determine which of the edge-sensitive signals are control signals, and which is the synchronizing signal. If the event control expression is sensitive to the edge of more than one signal, an **if** statement must be the first statement in the behavior. The control signals must be decoded explicitly in the branches of the **if** statement (e.g., decode the reset condition first). The synchronizing signal is not tested explicitly in the body of the **if** statement, but by default, the last branch must describe the synchronous activity, independently of the actual names given to the signals.

Example 9.5 The non-blocking assignments to *data_a* and *data_b* in the synchronous behavior within *swap_synch* implement a synchronous swapping mechanism. Both variables are sampled (referenced) before receiving value; they are synthesized as the outputs of a flip-flop in Figure 9.5. Notice that *set1* and *set2* are explicitly decoded. The last clause of the **if** statement assigns values to *data_a* and *data_b*. Those assignments are synchronized to the rising edge of *clk*, which is not referenced explicitly in the **if** statement.

```
module swap_synch (set1, set2, clk, data_a, data_b);
  output data_a, data_b;
  input  clk, set1, set2;
  reg    data_a, data_b;

  always @ (posedge clk)
    begin
      if (set1) begin data_a <= 1; data_b <= 0; end else
        if (set2) begin data_a <= 0; data_b <= 1; end
          else
            begin
              data_b <= data_a;
              data_a <= data_b;
            end
    end
endmodule
```

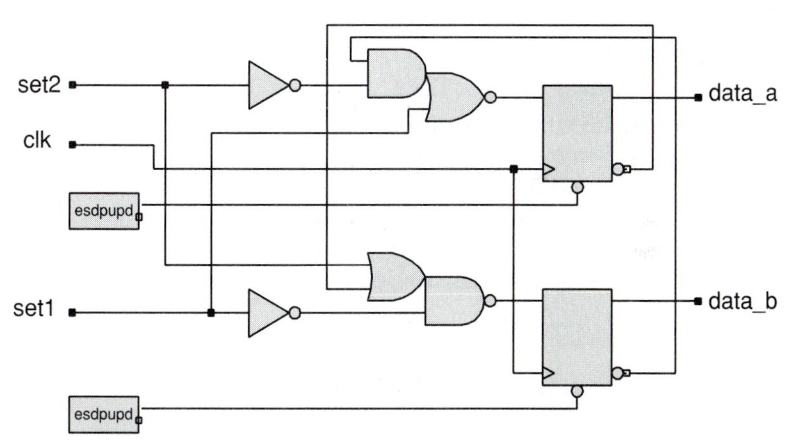

Figure 9.5 Synthesis result for variables referenced before being assigned value.

It is important to realize that not every register variable synthesizes to a hardware storage device. The four-input "and" gate that was described by a Verilog behavior in Example 8.10 has a register variable, *temp*, that is used within the behavior, but is not referenced outside the behavior. It is not referenced before it is assigned value. The variable *temp* merely supports the algorithm and does not require hardware memory—it does not have a life beyond the computation performed by the behavior. The behavior correctly synthesizes to a hardware "and" gate, without a memory element.

Example 9.6 The functionality of a four-bit parallel load data register is described by *D_reg4_a*. The positive edge of *reset* appears in the event control expression and in the first clause of the **if** statement; the positive edge of *clock* also appears in the event control expression, but is not explicitly decoded by the branch statement that follows the event control expression. This enables the synthesis tool to correctly infer the need for a re-settable flip-flop, active on the positive edge of *clock*. The value of *Data_out* is synchronized to the positive edge of *clock*, so the synthesis tool creates the four-bit array of flip-flops shown in Figure 9.6. *D_reg4_b* is an alternative (and possibly unsupported description).

```
module D_reg4_a (Data_in, clock, reset, Data_out);
  input   [3:0]  Data_in;
  input          clock, reset;
  output  [3:0]  Data_out;
  reg     [3:0]  Data_out;

  always @ (posedge clock or posedge reset)
    begin
      if (reset == 1'b1) Data_out <= 4'b0;
      else Data_out <= Data_in;
    end
endmodule

module D_reg4_b (Data_in, clock, reset, Data_out);
  input   [3:0]  Data_in;
  input          clock, reset;
  output  [3:0]  Data_out;
  reg     [3:0]  Data_out;

  always @ (posedge clock)
    begin
      Data_out <= Data_in;
    end

  always @ (reset)
    if (reset) assign Data_out <= 4'b0; else deassign Data_out;
endmodule
```

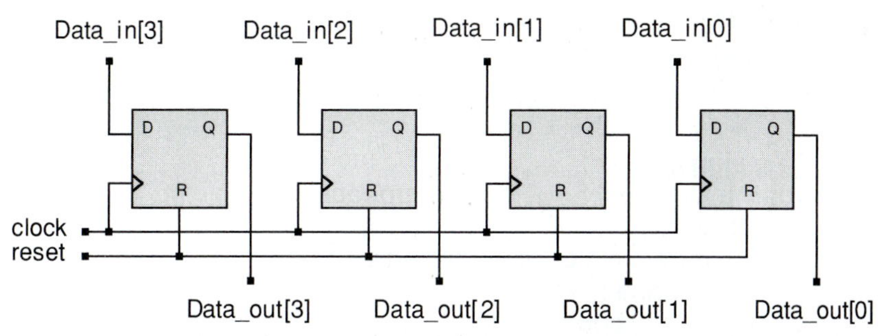

Figure 9.6 Synthesis result for a four-bit parallel-load data register.

Example 9.7 A user-defined primitive modeled a j-k flip-flop in Example 9.1. Here, two other variations are given, followed by their synthesized implementation. The first uses an **if** statement to decode the input control logic of the flip-flop, including a reset signal; the second decodes the controls with a **case** statement. Both realizations synthesize to the circuit in Figure 9.7. The first style combines the asynchronous signal (*rst*) and the synchronizing signal (*clock*) in the event control expression, and decodes the asynchronous condition in the first branch of an **if** statement. The remaining variable is decoded implicitly and is the synchronizing signal; i.e., **the order in which variables appear in the event control expression does not determine which signal synchronizes the behavior.** The second description replaces the nested **if** statements with a **case** statement. (Some vendors may allow the **assign ... deassign** procedural continuous assignment for treating asynchronous control signals.) The flip-flop selected by the synthesis tool is a D-type flip-flop active on the rising edge of the clock, with asynchronous, active-low reset.

```
module jk_flop_1 (j, k, clock, rst, q, qb);
  input       j, k, clock, rst;
  output      q, qb;
  reg         q;

  assign qb = ~q;
  always @ (posedge rst or posedge clock)
    begin
      if (rst == 1'b1) q = 1'b0; else        // decodes rst first
        if (j == 1'b0 && k == 1'b0) q = q; else
          if (j == 1'b0 && k == 1'b1) q = 1'b0; else
            if (j == 1'b1 && k == 1'b0) q = 1'b1; else
              if (j == 1'b1 && k == 1'b1) q = ~q;
    end
endmodule
```

```
module jk_flop_2 (j, k, clock, rst, q, qb);
  input       j, k, clock, rst;
  output      q, qb;
  reg         q;

  assign qb = ~q;
  always @ (posedge clock or posedge rst)
    begin
      if (rst == 1'b1) q = 1'b0; else
        case {j, k}
          2'b00:  q = q;
          2'b01:  q = 1'b0;
          2'b10:  q = 1'b1;
          2'b11:  q = ~q;
        endcase
    end
endmodule
```

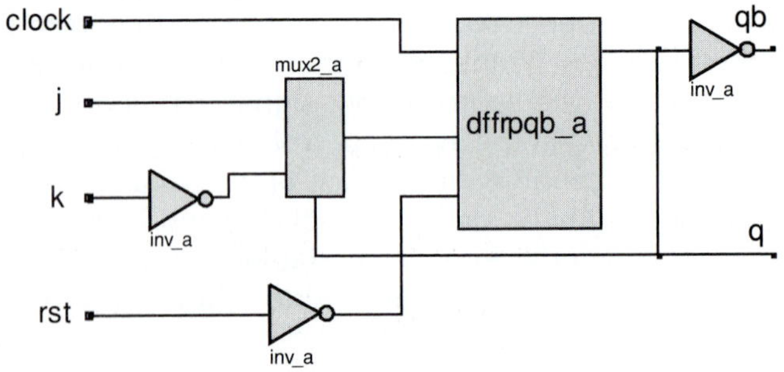

Figure 9.7 Synthesis result for alternative realizations of a j-k flip-flop.

9.4 REGISTERED COMBINATIONAL LOGIC

When a cyclic behavior that implements combinational logic is changed to be sensitive to the clock signal, without the other primary inputs, the behavior will be synthesized to implement the original combinational logic, but the outputs of the logic will be registered. The new outputs will be taken from D-type flip-flops whose inputs are the original primary outputs, and whose synchronizing signal is the clock of the new event control expression.

Example 9.8 In the Verilog description below, *reg_and* uses bitwise operators to implement a three-input "and" logic function within a behavior that is synchronized by the signal *clk*, so the register variable holding the value of *y* is updated only when the clock makes a rising-edge transition. The synthesis process produces the structure shown in Figure 9.8.

```
module reg_and (a, b, c, clk, y);
  input        a, b, c, clk;
  output       y;
  reg          y;

  always @ (posedge clk)
    begin
      y <= a & b & c;
    end
endmodule
```

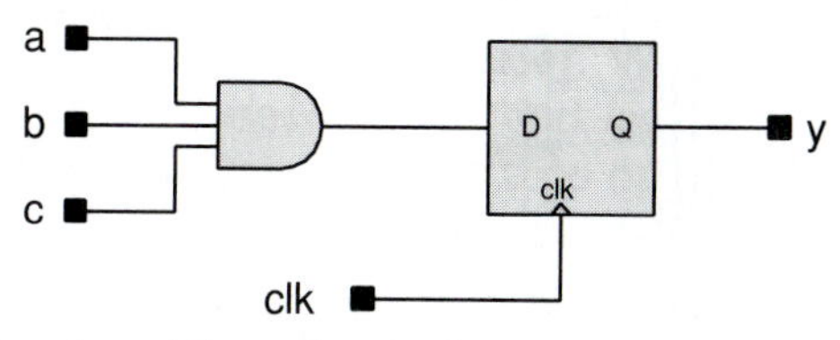

Figure 9.8 An "and" gate with registered output.

Example 9.9 A second example of registered combinational logic is given below by *mux_reg*. The output of *mux_reg* is synchronized by the rising edge of *clock*, so the synthesis tool implements the combinational logic of a four-channel multiplexer with eight-bit datapaths, but registers the outputs of the multiplexer in a bank of D-type flip-flops in Figure 9.9.

```
module mux_reg (a, b, c, d, y, select, clock);
 input        [7:0] a, b, c, d;
 output       [7:0] y;
 input        [1:0] select;
 reg          [7:0] y;

always @ (posedge clock)
  case(select)
   0: y <= a;   // non-blocking
   1: y <= b;   // same result with =
   2: y <= c;
   3: y <= d;
   default y <= 8'bx;
  endcase
endmodule
```

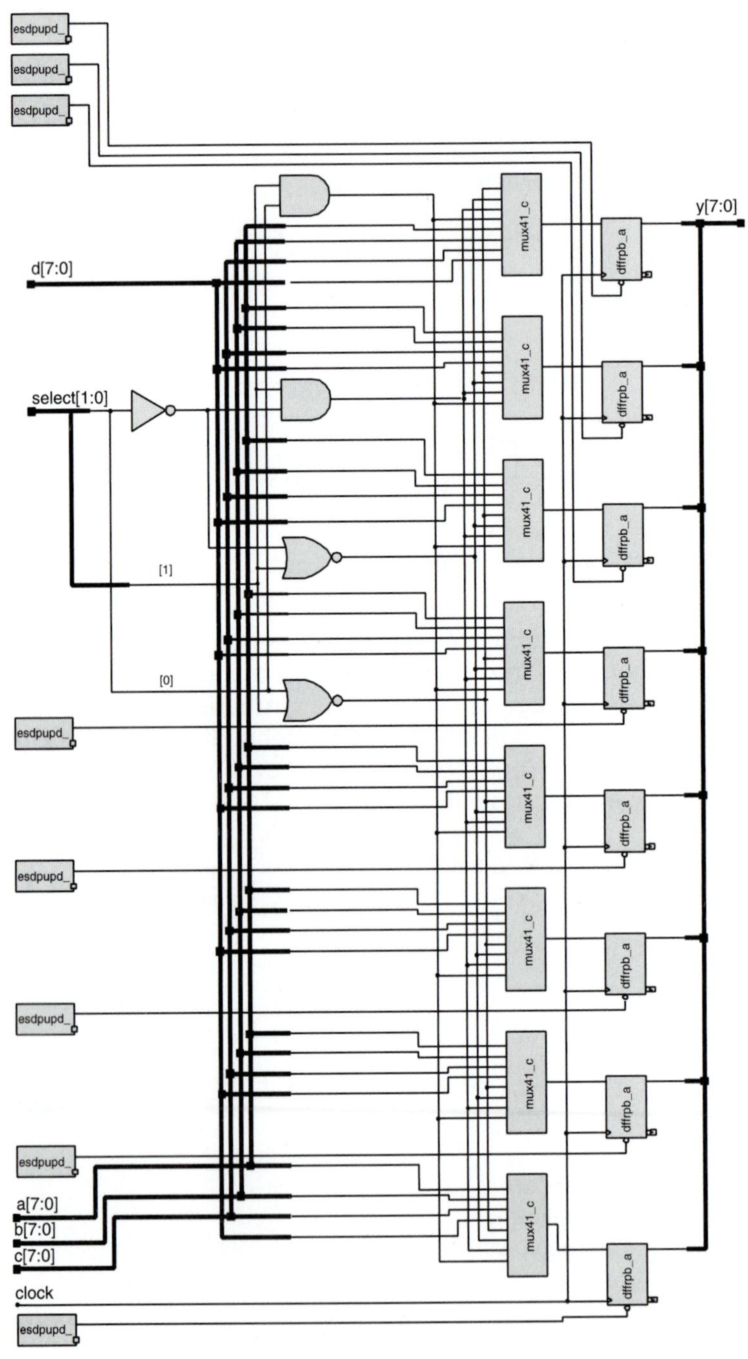

Figure 9.9 Multiplexer with registered output.

9.5 SHIFT REGISTERS AND COUNTERS

Synthesis tools easily synthesize a variety of counters and shift registers from Verilog behavioral descriptions.

Example 9.10 *Shift_reg4* below declares an internal four-bit register, *Data_reg*, creates *Data_out* by a continuous assignment to the LSB of the register, and forms the register contents synchronously from a concatenation of the scalar *Data_in* with the three leftmost bits of the register. Note that the register variable, *Data_reg*, is referenced by concatenation in a non-blocking assignment before it is assigned value in a synchronous behavior. This will synthesize to the flip-flop structure shown in Figure 9.10. Also note that the values on the right-hand side of the non-blocking assignments are the values of the variables before the active edge of the clock, and the values on the left-hand side are the values formed after the edge.

```
module Shift_reg4 (Data_in, Data_out, clock, reset);

  input    Data_in, clock, reset;
  output   Data_out;
  reg  [3:0] Data_reg;

  assign Data_out = Data_reg[0];

  always @ (negedge reset or posedge clock)
    begin
      if (reset == 1'b0) Data_reg <= 4'b0;
      else Data_reg <= {Data_in, Data_reg[3:1]};
    end
endmodule
```

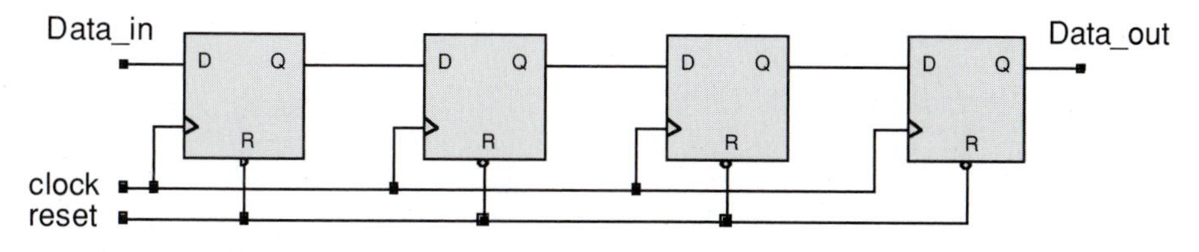

Figure 9.10 Four-bit shift register synthesized from a Verilog behavior.

Example 9.11 In this example, a register with reset and parallel load is synthesized from the Verilog description of *Par_load_reg4*. The structure of the synthesized result is shown in Figure 9.11.

```
module Par_load_reg4 (Data_out, Data_in, load, clock, reset);
  input     [3:0]    Data_in;
  input              load, clock, reset;
  output    [3:0]    Data_out;
  reg       [3:0]    Data_out;

  always @ (posedge reset or posedge clock)
    begin
      if (reset == 1'b1) Data_out = 4'b0;
        else if (load == 1'b1) Data_out = Data_in;
     end
endmodule
```

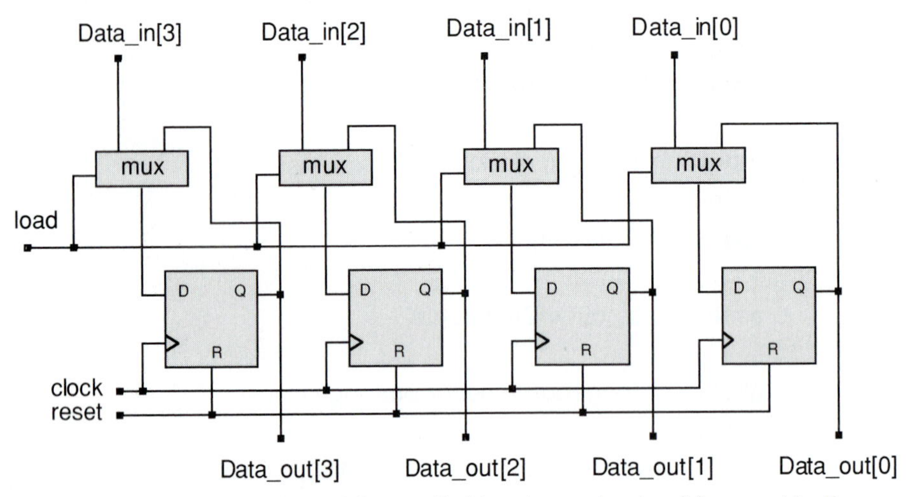

Figure 9.11 Four-bit register with parallel load, synthesized from a Verilog behavior.

Example 9.12 A barrel shifter circulates data bits in a synchronous manner. The Verilog description of *barrel_shifter* exploits concatenation to circulate the word through the register, as depicted in Figure 9.12. The top register shows the pattern before the shift, and the bottom register shows the pattern that results from the shift. The circuit synthesized from *barrel_shifter* is shown in Figure 9.13.

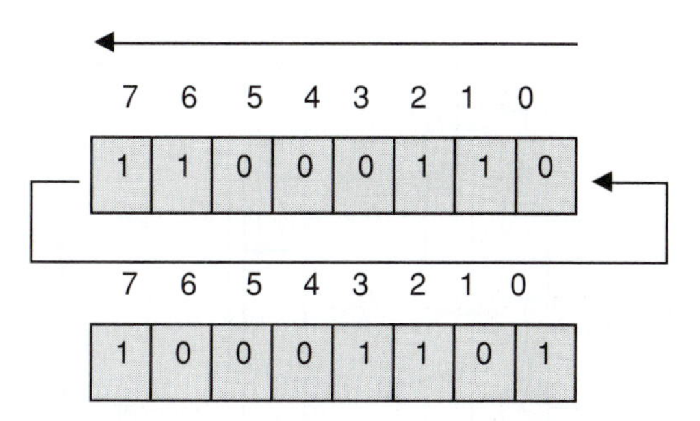

Figure 9.12 Data movement in an eight-bit barrel shifter.

```
module barrel_shifter (Data_out, Data_in, load, clock, reset);
  output [7:0] Data_out;
  input [7:0] Data_in;
  input load, clock, reset;
  reg  [7:0] Data_out;

  always @ (posedge reset or posedge clock)
    begin
      if (reset == 1'b1) Data_out <= 8'b0;
      else if (load == 1'b1) Data_out <= data_in;
      else data_out <= {Data_out[6:0], Data_out[7]};
    end
endmodule
```

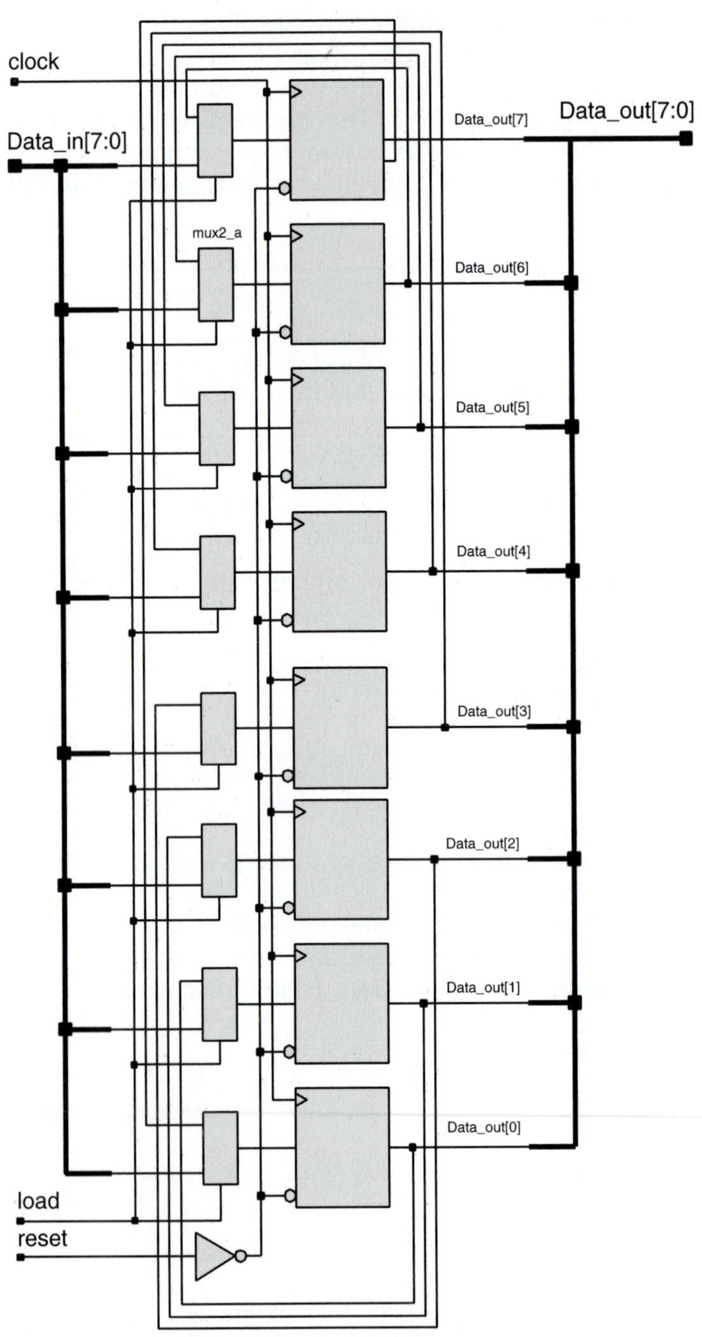

Figure 9.13 Synthesized eight-bit barrel shifter.

Example 9.13 A ripple counter can be implemented with toggle (T-type) flip-flops. This type of counter has limited practical application, because it takes excessive time to propagate changes through the cascaded chain of flip-flops, especially for long counters. The output count is also subject to glitches during the transitions.

The Verilog description of *ripple_counter* uses four behaviors to model the rippling effect, with successive stages of the counter triggered by the output of their immediately previous stage. The toggling action is controlled by the input *toggle.* The circuit simulates correctly, and synthesizes. The wires *c0, c1, c2,* and *c3* are required because the event control expression must be a simple variable (not a bit-select) to comply with the synthesis style required by the tool that produced the result (Synopsys™).

The structure of the counter, and the synthesized result are shown in Figure 9.14 and Figure 9.15, respectively.

```
module ripple_counter (clock, toggle, reset, count);
  input                 clock, toggle, reset;
  output    [3:0]       count;
  reg       [3:0]       count;
  wire                  c0, c1, c2;

  assign c0 = count[0];
  assign c1 = count[1];
  assign c2 = count[2];


  always @ (posedge reset or posedge clock)
    if (reset == 1'b1) count[0] <= 1'b0; else
      if (toggle == 1'b1) count[0] <= ~count[0];

  always @ (posedge reset or negedge c0)
    if (reset == 1'b1) count[1] <= 1'b0; else
    if (toggle == 1'b1) count[1] <= ~count[1];

  always @ (posedge reset or negedge c1)
    if (reset == 1'b1) count[2] <= 1'b0; else
    if (toggle == 1'b1) count[2] <= ~count[2];

  always @ (posedge reset or negedge c2)
    if (reset == 1'b1) count[3] <= 1'b0; else
    if (toggle == 1'b1) count[3] <= ~count[3];
endmodule
```

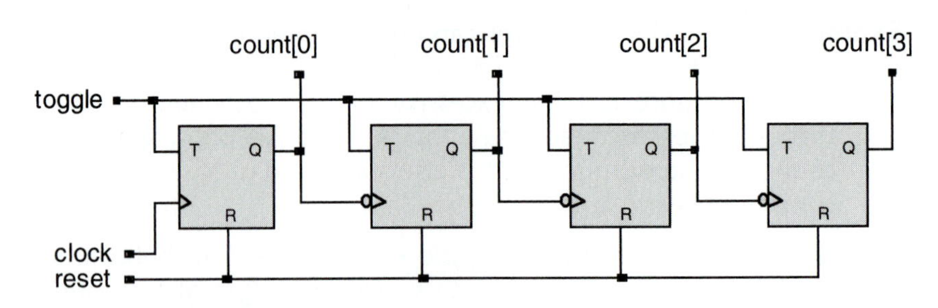

Figure 9.14 Structure of a four-bit ripple counter.

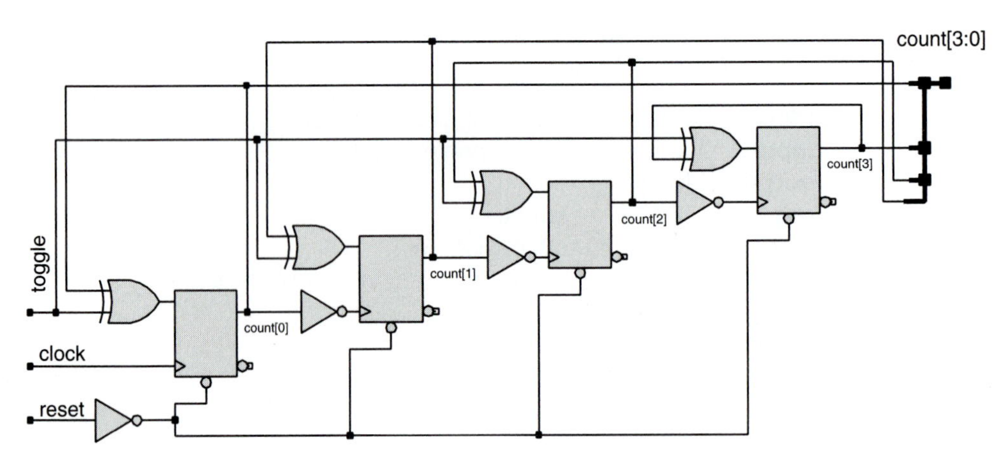

Figure 9.15 Synthesized circuit for a four-bit ripple counter.

Example 9.14 A ring counter has a single asserted bit that circulates through the counter in a synchronous manner. The movement of data in an eight-bit ring counter is illustrated in Figure 9.16. The behavior described by *ring_counter* assures the synchronous movement of the asserted bit through the register and the automatic restarting of the count at *count[0]* after *count[7]* is asserted at the end of a cycle. The synthesized circuit is shown in Figure 9.17. The D-type flip-flops (*dffgpqb_a*) in the implementation are active on the rising edge of the clock, have gated data (i.e., the internal datapath of the flip-flop is driven by the output of a multiplexer whose control signal selects between the output of the flip-flop and the external dathpath—see Section 9.8 and Appendix I) and an asynchronous active-low reset.

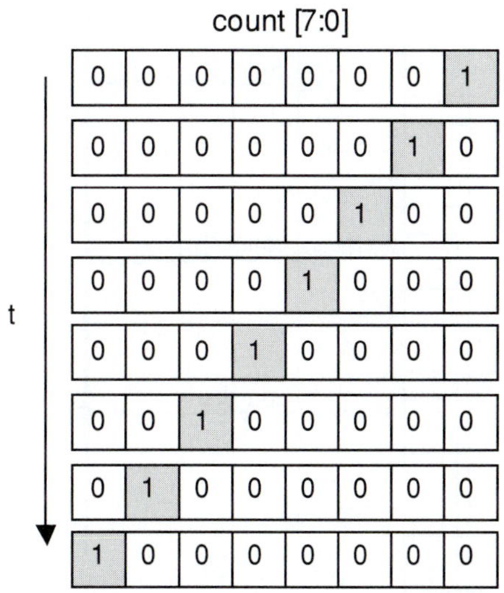

count [7:0]

0	0	0	0	0	0	0	1
0	0	0	0	0	0	1	0
0	0	0	0	0	1	0	0
0	0	0	0	1	0	0	0
0	0	0	1	0	0	0	0
0	0	1	0	0	0	0	0
0	1	0	0	0	0	0	0
1	0	0	0	0	0	0	0

t

Figure 9.16 Data movement in an eight-bit ring counter.

```
module ring_counter (enable, reset, clock, count);
  input                 enable, reset, clock;
  output     [7:0]      count;
  reg        [7:0]      count;

  always @ (posedge reset or posedge clock)
    if (reset == 1'b1) count <= 8'b0000_0001; else
    if (enable == 1'b1) count <= {count[6:0], count[7]};
endmodule
```

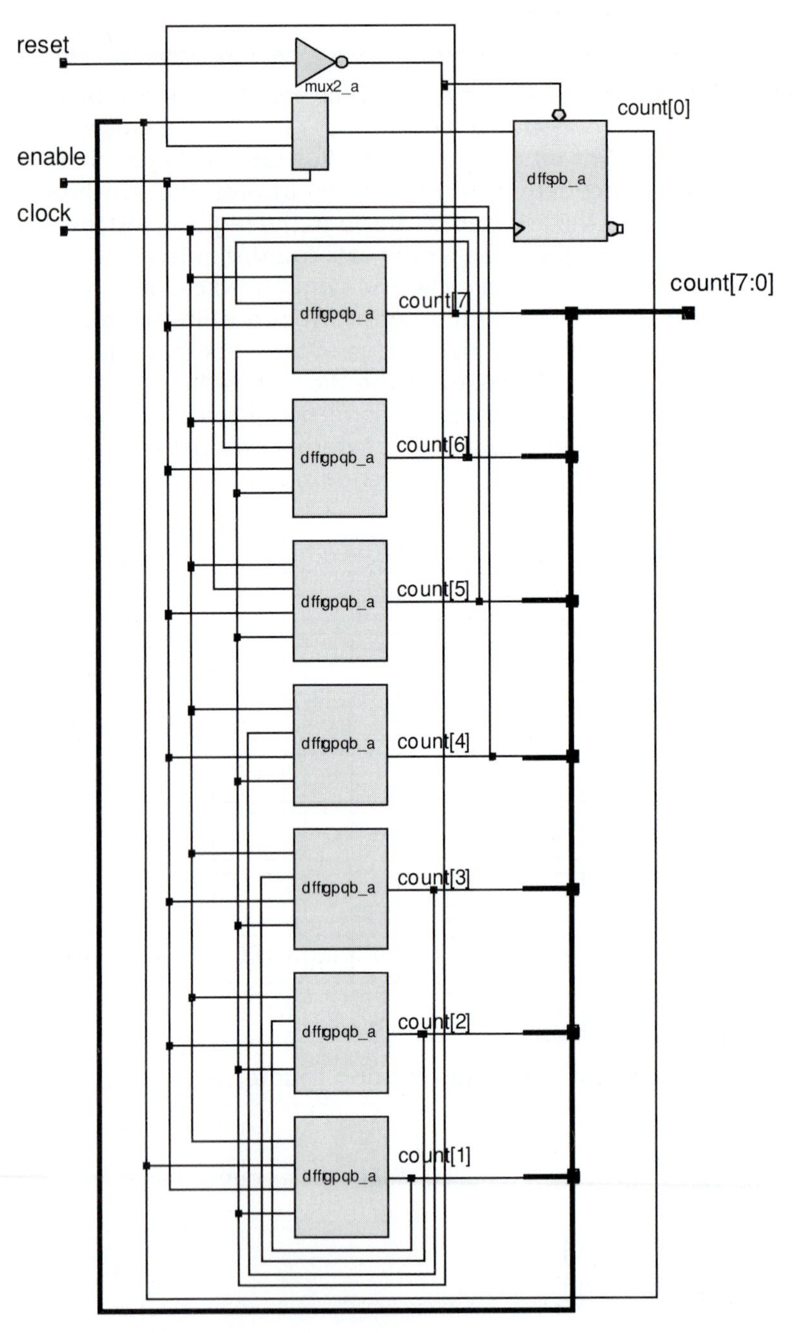

Figure 9.17 Ring counter synthesized from a Verilog behavioral description. Note: the dffgpqb_a flip-flop has (from top to bottom) rising-edge clock, data (internally gated) synchronous active-low gate control, and active-low reset inputs.

Example 9.15 Our last example of a counter is a three-bit up-down counter. The description exploits Verilog's built-in arithmetic and implements the counter with an **if** statement. The synthesized circuit is shown in Figure 9.18. In this implementation, the library cell *dffrgpqb_a* is a D-type flip-flop active on the rising edge, having gated data and asynchronous active-low reset.

```
module up_down_counter (clk, reset, load, count_up, counter_on,
                         Data_in, Count);
  input            clk, reset, load, count_up, counter_on;
  input    [2:0]   Data_in;
  output   [2:0]   Count;
  reg      [2:0]   Count;

always @ (posedge reset or posedge clk)
  if (reset == 1'b1) Count = 3'b0; else
    if (load == 1'b1) Count = Data_in; else
      if (counter_on == 1'b1) begin
        if (count_up == 1'b1) Count = Count +1;
          else Count = Count –1;
      end
endmodule
```

Data_in
3
count_up u/d D_in
load ld
reset rst
count_on cnt
clock clk count
3
Count

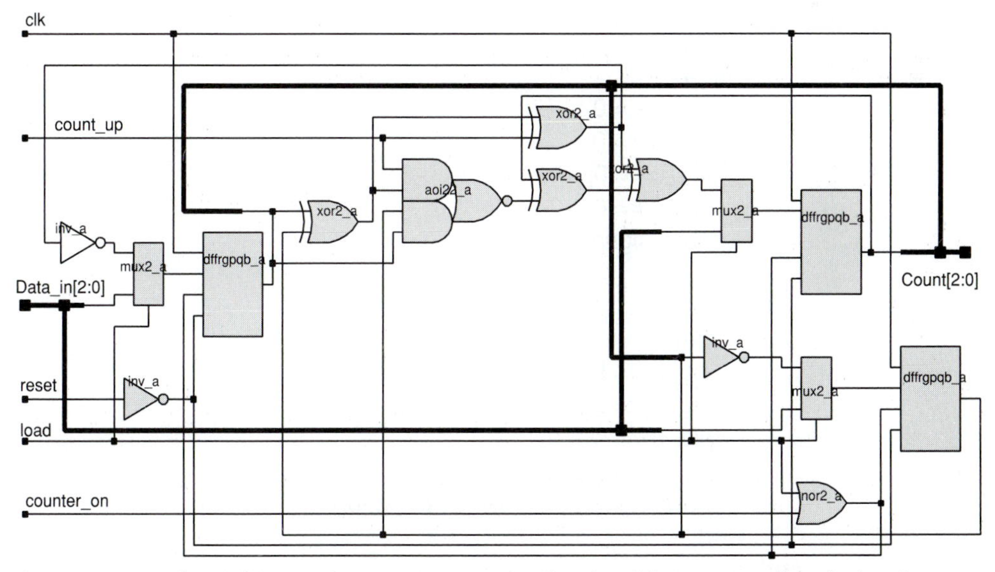

Figure 9.18 Three-bit up-down counter, synthesized from a Verilog behavior.

Register variables that are assigned values within a synchronized behavior will be synthesized as flip-flops or latches, depending upon whether the behavior is synchronized to an edge or the level of a signal. On the other hand, a signal that is assigned value outside or in a behavior that does not include a synchronizing signal in its event control expression will be synthesized as combinational logic, provided that the behavior does not have an incomplete **if**, **case** or conditional operator (? ... :).

Example 9.16 The shift register described by *shifter_1* below includes combinational logic forming the register variable *new_signal*. Since *new_signal* receives value within a synchronous behavior, it will be synthesized as the output of a flip-flop, with the structure shown in Figure 9.19. (Remember—the values of the variables on the right-hand side of a non-blocking assignment are the values before the clock; the values are on the left-hand side are assigned at the clock event.) The actual parts chosen by the synthesizer will depend upon the available technology.

```
module shifter_1 (Data_in, clock, reset, sig_d, new_signal);
  input   Data_in, clock, reset;
  output  sig_d, new_signal;
  reg     sig_a, sig_b, sig_c, sig_d, new_signal;

  always @ (posedge reset or posedge clock)
  begin
    if (reset == 1'b1)
      begin
        sig_a <= 0;
        sig_b <= 0;
        sig_c <= 0;
        sig_d <= 0;
        new_signal <= 1'b0;
      end
    else
      begin
        sig_a <= Data_in;
        sig_b <= sig_a;
        sig_c <= sig_b;
        sig_d <= sig_c;
        new_signal <= (~ sig_a) & sig_b;
      end
  end
endmodule
```

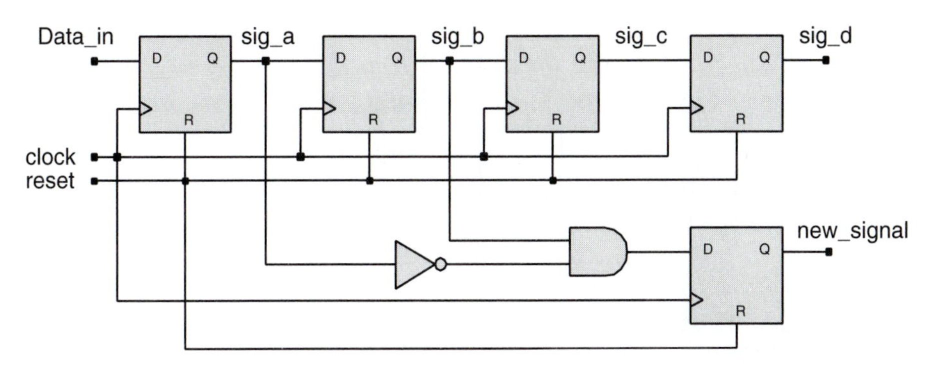

Figure 9.19 Generic structure of a shift register with registered combinational logic.

Example 9.17 In *shifter_2, new_signal* is formed outside the behavior in a continuous assignment, and is synthesized as combinational logic. The structure that will be synthesized from the behavior is shown in Figure 9.20.

```
module shifter_2 (Data_in, clock, reset, sig_d, new_signal);
  input          Data_in, clock, reset;
  output         sig_d, new_signal;
  reg            sig_a, sig_b, sig_c, sig_d;

  always @ (posedge reset or posedge clock) begin
    if (reset == 1'b1)
      begin
        sig_a <= 1'b0;
        sig_b <= 1'b0;
        sig_c <= 1'b0;
        sig_d <= 1'b0;
      end
    else
      begin
        sig_a <= shift_input;
        sig_b <= sig_a;
        sig_c <= sig_b;
        sig_d <= sig_c;
      end
  end
  assign new_signal <= (~ sig_a) & sig_b;
endmodule
```

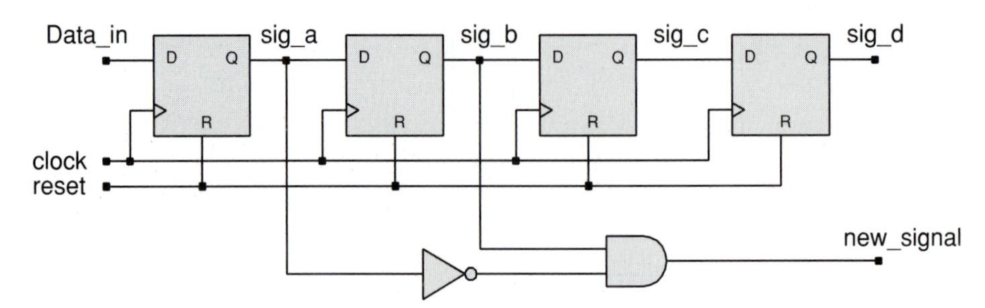

Figure 9.20 Generic structure of a shift register, with separate, unregistered combinational logic.

Example 9.18 Digital systems that operate on arrays of data sometimes contain large numbers of adders in an array structure. The processing speed in these applications is usually critical. For example, the 16-bit adder in Figure 9.21 is formed by chaining two eight-bit adders in a serial connection. If each eight-bit adder has a throughput delay of 100 ns, the worst-case delay of the configuration will be 200 ns. In a synchronous environment, this structure is organized to have all operations occur in the same clock cycle.

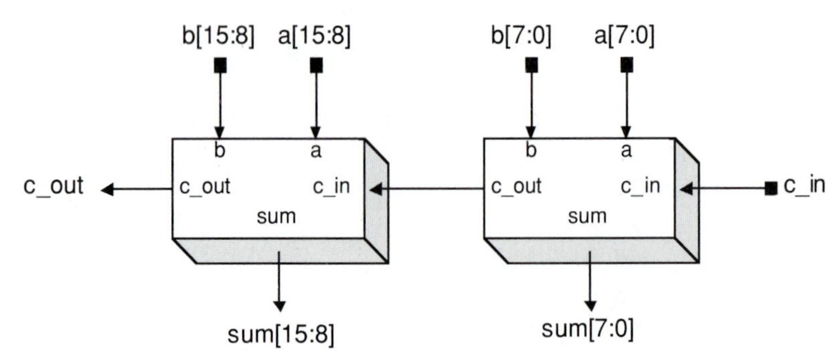

Figure 9.21 Serial connection of two eight-bit adders to form a 16-bit adder.

An alternative structure can be pipelined to operate at a higher throughput by distributing the processing over multiple cycles of the clock. The trade-off between speed and physical resources (more registers) can warrant this approach. Non-blocking assignments (see Chapter 7) are the key to modeling architectures that are pipelined to achieve high throughput on datapaths. (The pipelined structure in Figure 9.22 contains an additional register (*PR*) between the data input register (*IR*) and the data output register.) The structure sequences the data so that, in a given clock cycle, a carry bit must propagate through only half the datapath. The interface to the input datapaths still provides entire words to the unit in a synchronous manner, but the sum of only the rightmost data bytes is formed. That sum, together with the leftmost datapaths, is then stored in a 25-bit internal register. In the next clock cycle, the sum of the leftmost data bytes is formed and stored in the output register with the rightmost sum and carry from the previous cycle. With the extra internal register, the pipelined unit can operate at approximately twice the frequency of the original adder, because the longest path supported by the clock interval is through an eight-bit adder instead of a 16-bit adder. After the period of initial latency, a new sum appears at the output every 100 ns (for a 10MHZ clock).

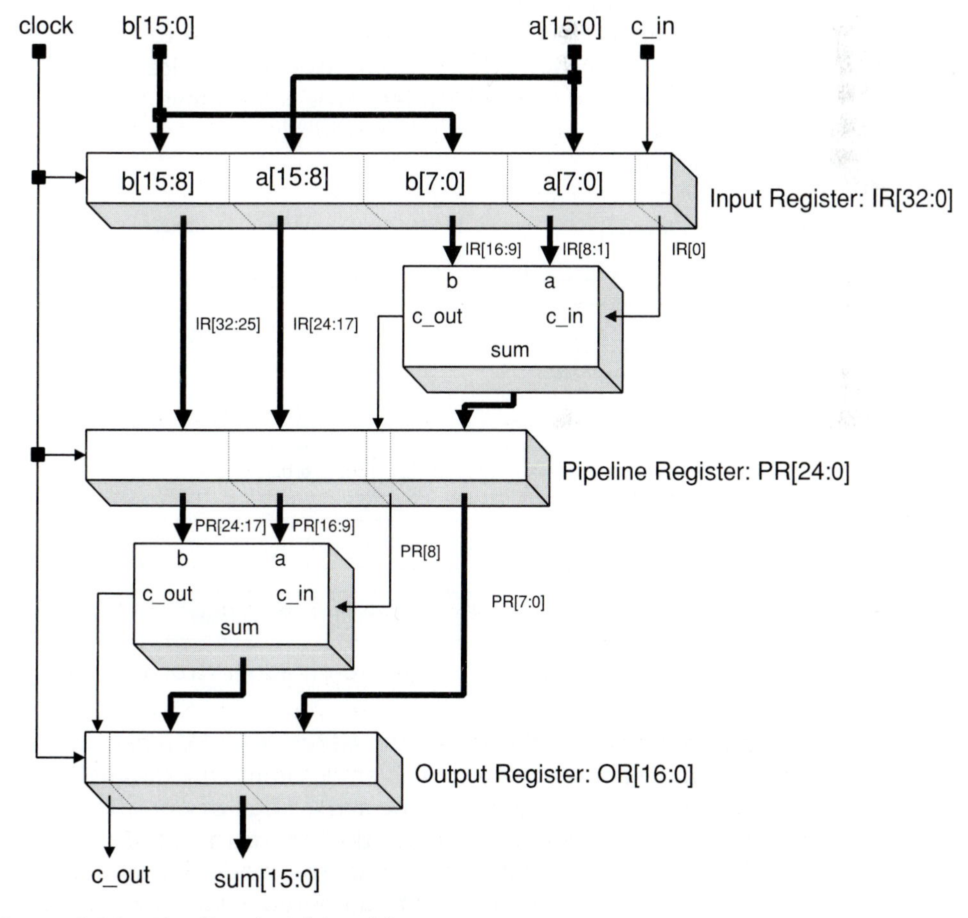

Figure 9.22 Pipelined 16-bit adder structure.

The movement of data through the pipelined adder is depicted in Figure 9.23, where $a_L a_R(1)$ denotes the first sample of the left and right bytes of input word *a*. Figure 9.24 shows the annotated result of simulating *add_16_pipe* in a testbench that uses hierarchical de-referencing to display the contents of the internal registers in a format that reveals the dataflow through the pipeline. The display outputs *IR32_17, IR16_1,* and *IR_0* show the segments of *IR*. The displayed outputs *PR24_17*, *PR16_9*, *PR_8*, and *PR7_0* show the segments of *PR*. The waveforms have been annotated to illustrate the register transfers. In the simulation results notice that the unit has a latency of two clock cycles between the application of the input data and the appearance of valid output. The first data words, 1122_h and 3344_h, are formed at t_{sim} = 100 ns, sampled and loaded into register *PR* at t_{sim} = 150 ns, partially added at t_{sim} = 250 ns (*PR7_0*), and fully added at t_{sim} = 350 ns. After the latency period, the data is correctly updated to achieve an overall maximum throughput of approximately twice that of the serially-connected eight-bit adders. (The setup times of physical registers will reduce the throughput slightly.)

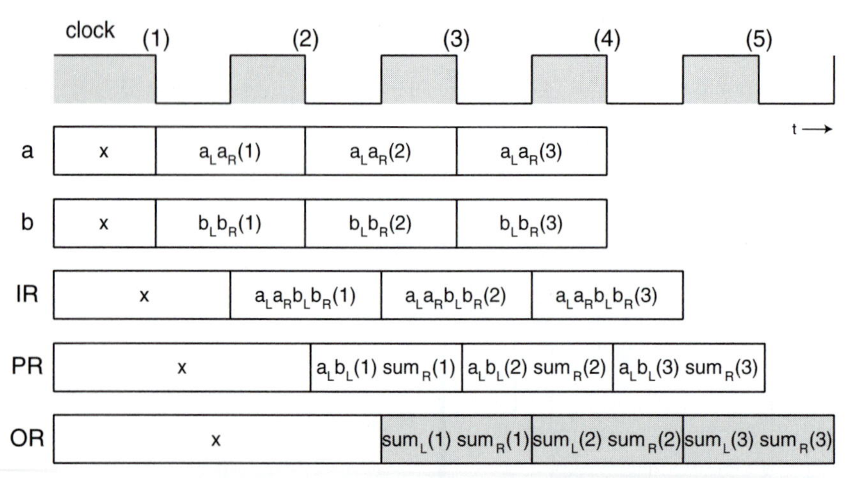

Figure 9.23 Data movement through a pipelined 16-bit adder structure.

The Verilog model of the pipelined adder, *add_16_pipe*, uses non-blocking assignments to concurrently sample the datapaths and registers immediately before the active edge of *clock*. These samples are used to form the values that will exist in the registers immediately after the clock event. The model is scaled by the value of *size*. Change the size to a desired value (it must be an even number).

```
module add_16_pipe (c_out, sum, a, b, c_in, clock);
  parameter               size   = 16;
  parameter               half   = size / 2;
  parameter               double = 2 * size;
  parameter               triple = 3 * half;

  parameter               size1 = half -1;     // 7
  parameter               size2 = size -1;     // 15
  parameter               size3 = half + 1;    // 9
  parameter               R1 = 1;              // 1
  parameter               L1 = half;
  parameter               R2 = size3;
  parameter               L2 = size;
  parameter               R3 = size + 1;
  parameter               L3 = size + half;
  parameter               R4 = double - half +1;
  parameter               L4 = double;

  input     [size2: 0]    a, b;
  input                   c_in, clock;
  output    [size2: 0]    sum;
  output                  c_out;

  reg       [double: 0]   IR;
  reg       [triple: 0]   PR;
  reg       [size: 0]     OR;

  assign {c_out, sum} = OR;

  always @ (posedge clock) begin

    // Load input register

    IR[0] <= c_in;

    IR[L1:R1] <= a[size1: 0];
    IR[L2:R2] <= b[size1: 0];

    IR[L3:R3] <= a[size2: half];
    IR[L4:R4] <= b[size2: half];

    // Load pipeline register

    PR[L3: R3] <=IR[L4: R4];
    PR[L2: R2] <=IR[L3: R3];
    PR[half:0] <= IR[L2:R2] + IR[L1:R1] + IR[0];
    OR <= {{1'b0,PR[L3: R3]} + {1'b0,PR[L2: R2]} + PR[half], PR[size1: 0]};
  end
endmodule
```

```
module seq_det_mealy_1exp (clock, reset, in_bit, out_bit);
  input                     clock, reset, in_bit;
  output                    out_bit;
  reg [2:0]                 state_reg, next_state;
  parameter                 start_state =   3'b000;
  parameter                 read_1_zero =   3'b001;
  parameter                 read_1_one =    3'b010;
  parameter                 read_2_zero =   3'b011;
  parameter                 read_2_one =    3'b100;

  always @ (posedge clock or posedge reset)
    if (reset == 1) state_reg <= start_state; else state_reg <= next_state;


  always @ (state_reg or in_bit) case (state_reg)
    start_state:    if (in_bit == 0)  next_state <= read_1_zero; else
                    if (in_bit == 1)  next_state <= read_1_one;
                    else              next_state <= start_state;
    read_1_zero:    if (in_bit == 0)  next_state <= read_2_zero; else
                    if (in_bit == 1)  next_state <= read_1_one;
                    else              next_state <= start_state;
    read_2_zero:    if (in_bit == 0)  next_state <= read_2_zero; else
                    if (in_bit == 1)  next_state <= read_1_one;
                    else              next_state <= start_state;
    read_1_one:     if (in_bit == 0)  next_state <= read_1_zero; else
                    if (in_bit == 1)  next_state <= read_2_one;
                    else              next_state <= start_state;
    read_2_one:     if (in_bit == 0)  next_state <= read_1_zero; else
                    if (in_bit == 1)  next_state <= read_2_one;
                    else              next_state <= start_state;
    default:                          next_state <= start_state;
  endcase
  assign out_bit =   (((state_reg == read_2_zero) && (in_bit == 0)) ||
                     ((state_reg == read_2_one) && (in_bit == 1))) ? 1 : 0;
  endmodule
```

The simulated response of the machine, shown in Figure 9.28, has asynchronous reset. Notice that *state_reg* enters *start_state* when *reset* is asserted; the output asserts after the second consecutive sample of a 0 or 1 is detected. The output is dependent upon the input, so this version is a Mealy machine. When *in_bit* is de-asserted after two samples, *out_bit* is de-asserted immediately, even though the clock event has not occurred.

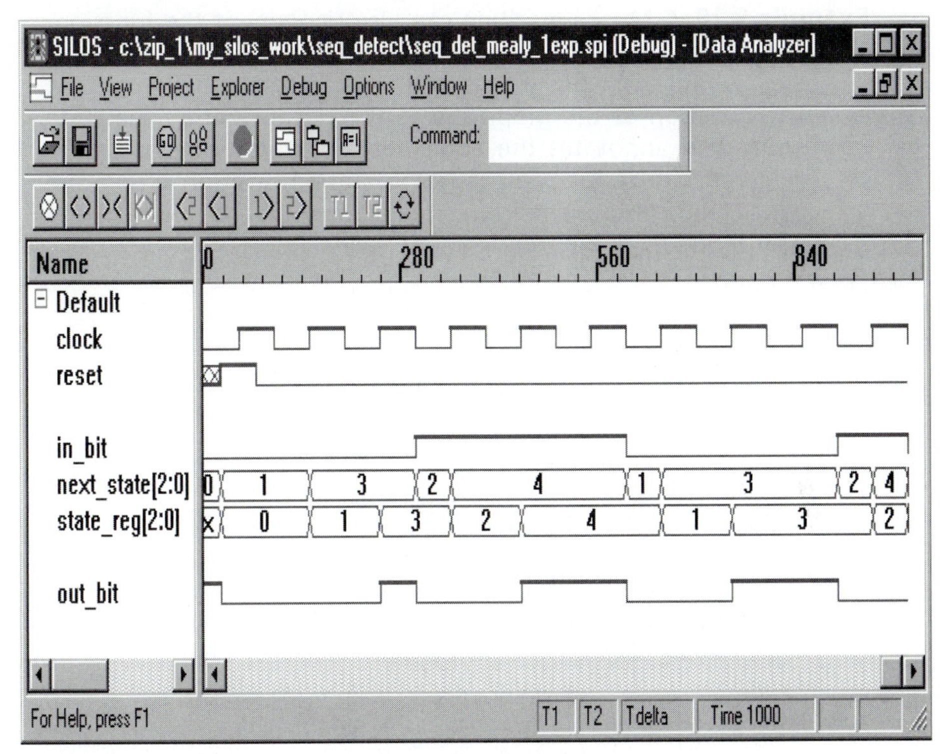

Figure 9.28 Simulation results for *seq_det_mealy_1exp*.

The sequential circuit synthesized from *seq_det_mealy_1exp* is shown in Figure 9.29. It has three flip-flops to hold the state, and the output is formed combinationally from *in_bit* and *state_reg*. The flip-flops have active-low asynchronous reset.

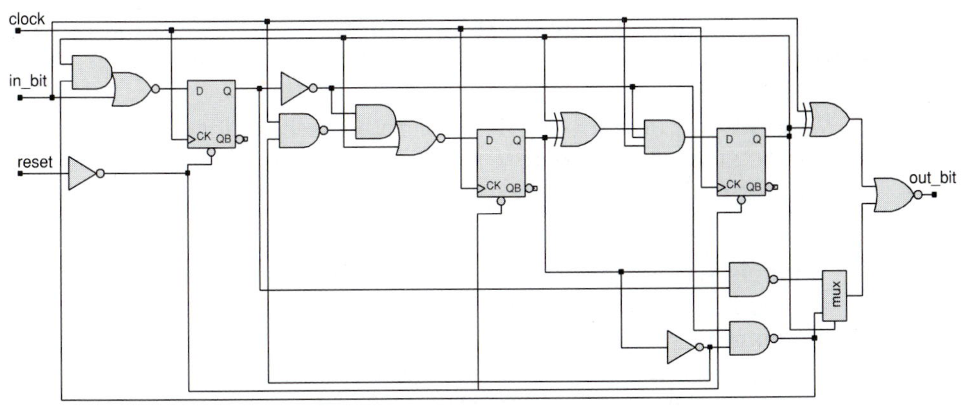

Figure 9.29 Synthesis results for *seq_det_mealy_1exp*.

Example 9.20 A Moore machine can also implement the functionality of the sequence detector described in the previous example. This machine has the state transition graph in Figure 9.30. The notation of the graph shows the output within the state node, and annotates the edge between two nodes with the value of the input that causes a state transition along the edge. In the Moore machine realization, *out_bit* can change only at the clock boundaries.

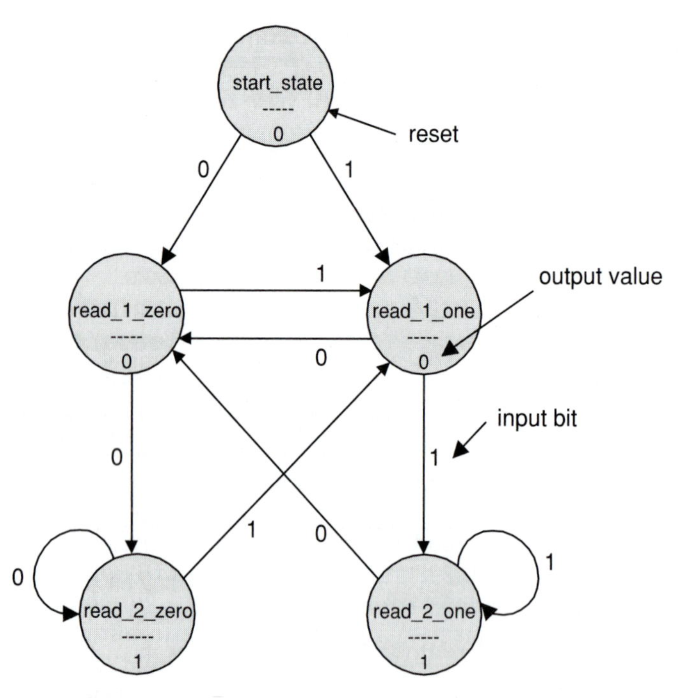

Figure 9.30 State transition graph of a Moore machine to detect two successive 1s or 0s in a serial bit stream.

The Verilog description of the Moore machine for a sequence detector is given in *moore_det_1exp*. The description is similar to *seq_det_mealy_1exp*, except that the continuous assignment to *out_bit* is a function of only *state_reg*.

```
module seq_det_moore_1exp (clock, reset, in_bit, out_bit);
  input                  clock, reset, in_bit;
  output                 out_bit;
  reg        [2:0]       state_reg, next_state;
  parameter              start_state = 3'b000;
  parameter              read_1_zero = 3'b001;
```

```
parameter          read_1_one = 3'b010;
parameter          read_2_zero = 3'b011;
parameter          read_2_one = 3'b100;

always @ (posedge clock or posedge reset)
  if (reset == 1) state_reg <= start_state; else state_reg <= next_state;

always @ (state_reg or in_bit) case (state_reg)
  start_state:   if (in_bit == 0)  next_state <= read_1_zero; else
                 if (in_bit == 1)  next_state <= read_1_one;
                 else              next_state <= start_state;
   read_1_zero:  if (in_bit == 0)  next_state <= read_2_zero; else
                 if (in_bit == 1)  next_state <= read_1_one;
                 else              next_state <= start_state;
  read_2_zero:   if (in_bit == 0)  next_state <= read_2_zero; else
                 if (in_bit == 1)  next_state <= read_1_one;
                 else              next_state <= start_state;
  read_1_one:    if (in_bit == 0)  next_state <= read_1_zero; else
                 if (in_bit == 1)  next_state <= read_2_one;
                 else              next_state <= start_state;
  read_2_one:    if (in_bit == 0)  next_state <= read_1_zero; else
                 if (in_bit == 1)  next_state <= read_2_one;
                 else              next_state <= start_state;
  default:                         next_state <= start_state;
endcase

assign out_bit = ((state_reg == read_2_zero) || (state_reg ==
                 read_2_one)) ? 1 : 0;

endmodule
```

Figure 9.31 compares the simulated outputs of *seq_det_mealy_1exp* and *seq_det_moore_1exp*, for the circuits having asynchronous reset. The machines have identical state transitions, but their outputs differ. The output of the Mealy machine makes transitions with the input and the state; the output of the Moore machine depends only on the state. The waveforms demonstrate that the reset and detection actions are functionally correct.

The circuit synthesized from *seq_det_moore_1exp* is shown in Figure 9.32. The machine has asynchronous reset. Three flip-flops implement *state_reg*, and *out_bit* is formed directly from *state_reg*. The circuit does not decode *in_bit* to form *out_bit*.

```
              flag = 0;
              disable machine; end
            else begin
              last_bit = this_bit;
              this_bit = in_bit;
              flag = 1; end
            end                                                          // 2
          end                                                            // 1
      end              // machine
    end                // wrapper_for_synthesis

    assign out_bit = (flag && (this_bit == last_bit) && (this_bit == in_bit));
endmodule
```

Figure 9.33 shows the result of simulating *seq_det_mealy_1imp*. Notice that the reset action is synchronous, and that *out_bit* de-asserts asynchronously when *in_bit* changes to a value that does not match *this_bit*, the most recently stored sample.

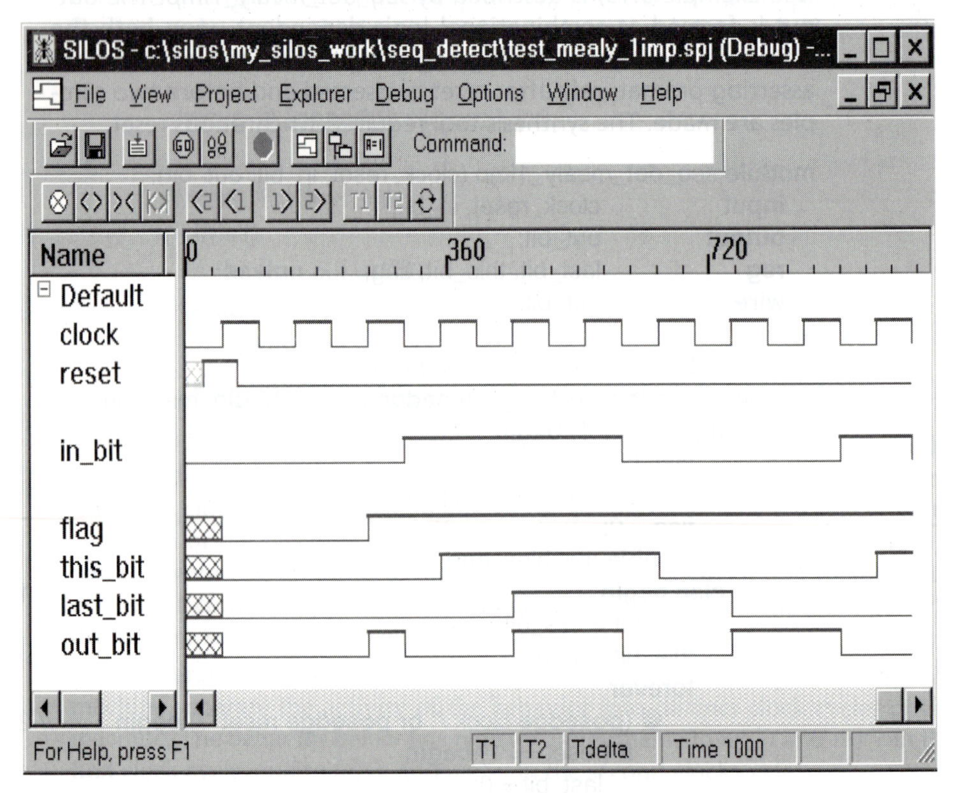

Figure 9.33 Simulation results for *seq_det_mealy_1imp*.

Example 9.22 An implicit Moore-type machine, *seq_det_moore_1imp*, also describes the sequence detector.

```
module seq_det_moore_1imp (clock, reset, in_bit, out_bit);
  input         clock, reset, in_bit;
  output        out_bit;
  reg           last_bit, this_bit, flag;
  wire          out_bit;

  always begin: wrapper_for_synthesis
    @ (posedge clock /* or posedge reset*/) begin: machine
      if (reset == 1) begin
        last_bit = 0;
        // this_bit = 0;              // see section 9.7
        // flag = 0;
        disable machine; end
      else begin                                                    // 1
        // last_bit = this_bit;       // see section 9.7
        this_bit = in_bit;
          forever
            @ (posedge clock /* or posedge reset */) begin   // 2
              if (reset == 1) begin
                // last_bit = 0;     // see section 9.7
                // this_bit = 0;     // see section 9.7
                flag = 0;
                disable machine; end
              else begin
                last_bit = this_bit;
                this_bit = in_bit;
                flag = 1; end
            end                                                     // 2
          end                                                       // 1
        end // machine
      end // wrapper_for_synthesis

  assign out_bit = (flag && (this_bit == last_bit));
endmodule
```

This version seemingly exploits the flag bit and multi-cycle nature of the machine to avoid flushing the pipeline. Also, only the first stage of the pipe is loaded at the first clock. (The implications of this will be discussed in Section 9.7.) The output is formed as a Moore output.

Figure 9.34 compares the simulated output of the two implicit machines with synchronous *reset*. With the exception of the reset action, the outputs are the same as for the explicit machines (see Figure 9.31).

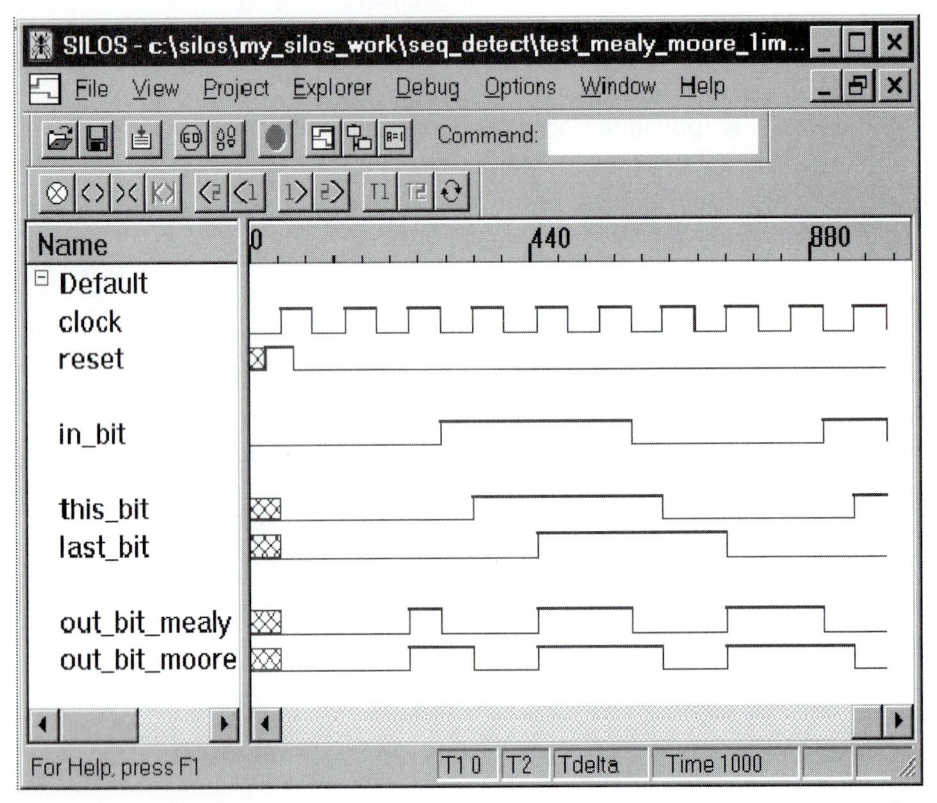

Figure 9.34 Comparison of the output of implicit Mealy and Moore state machines for the sequence detector.

The machines (with synchronous reset) synthesize to the circuits shown in Figure 9.35. In Figure 9.35b, two gated-input flip-flops (*dffrgpqb_a*) form a pipeline for *last_bit* and *this_bit*. When the gate input (*G*) is low, the *Q* output is connected to the *D* input through internal feedback, while ignoring the external *D* input.

Otherwise, the external *D* input is the input. A third gated-input flip-flop holds *flag*, which is gated together with the difference of *last_bit* and *this_bit* to form *out_bit*. The active-low input (*RB*) is disabled. The synthesis tool inserts a D-type flip-flop with multiplexed input (*dffrmpqb_a*) to hold *multiple_wait_state* (created by the synthesis tool). The active-low *RB* (*reset*) input is disabled (for synchronous operation), and the active-low *SL* (set) is wired to *reset*. The *D0* and *D1* inputs are wired to power and ground, respectively, through *esdpupd* and the active-low *SL* input, which is connected to *reset*. When *SL* is low (reset not asserted), *D0* is selected, and when *SL* is high (reset is asserted), *D1* is selected.

Now consider the action of *reset*. While *reset* is asserted, its inverted value causes *this_bit* and *last_bit* to hold their value (through internal feedback); it also drives the nand gate at the input to *flag* to get the value of its external input, which is held to 0. Thus, the reset conditions specified by the behavior description are met for *this_bit*, *last_bit*, and *flag*.

At the first clock after *reset* is de-asserted, the *multiple_wait_state* gets 1, setting up the datapath from *this_bit* to *last_bit* on subsequent clocks. Also, *this_bit* gets *in_bit* after *reset* is de-asserted.

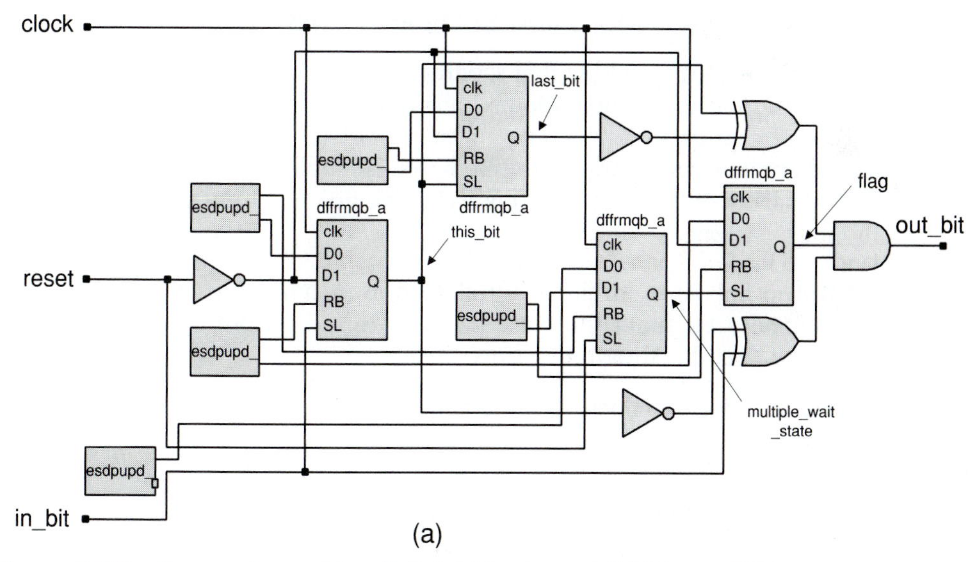

Figure 9.35 Comparison of implicit (a) Mealy and (b) Moore FSM sequence detectors. (Continues on page 392.)

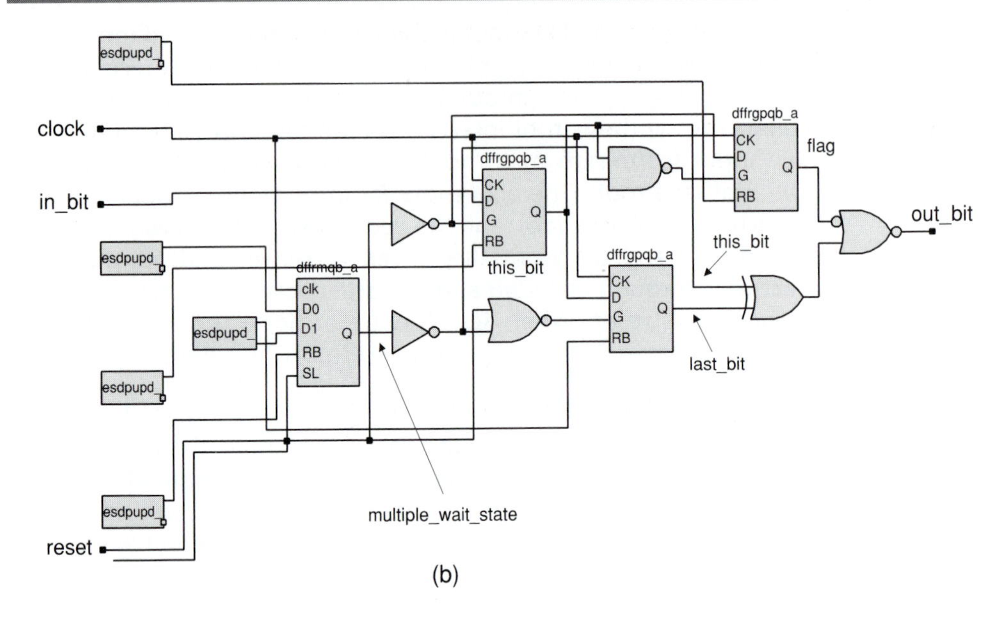

Figure 9.35 Concluded.

9.6.4 Pitfalls in Modeling State Machines

Although it is recommended that separate behaviors be written to describe the state transitions and combinational logic of explicit machines, it is possible to synthesize explicit state machines from a single description. The features of three alternative synchronous detectors (see Example 9.19) will be examined to illustrate some of the pitfalls of this style. The first will implement a fully synchronous machine.

Example 9.23 As a preliminary design, *seq_det_1a* uses a single behavior to direct the state transitions and output. The synthesized circuit is shown in Figure 9.36.

```
module seq_det_1a (clock, in_bit, out_bit);
  input       clock, in_bit;
  output      out_bit;
  reg         [1:0] state_reg;
  reg         out_bit;
  parameter   state_0 = 2'b00;
  parameter   state_1 = 2'b01;
  parameter   state_2 = 2'b10;
```

```
always @ (posedge clock) case (state_reg)
  state_0 :  if (in_bit ==0) begin state_reg <= state_1; out_bit <= 0; end else
             if (in_bit ==1)begin state_reg <= state_2; out_bit <= 0; end
  state_1 :  if (in_bit ==0) begin state_reg <= state_1; out_bit <= 1; end else
             if (in_bit ==1)begin state_reg <= state_2; out_bit <= 0; end
  state_2 :  if (in_bit ==0) begin state_reg <= state_1; out_bit <= 0; end else
             if (in_bit ==1)begin state_reg <= state_2; out_bit <= 1; end
  default:                 begin state_reg <= state_0; out_bit <= 0; end
 endcase
endmodule
```

Note that the output is registered in this implementation. Two additional flip-flops implement the state.

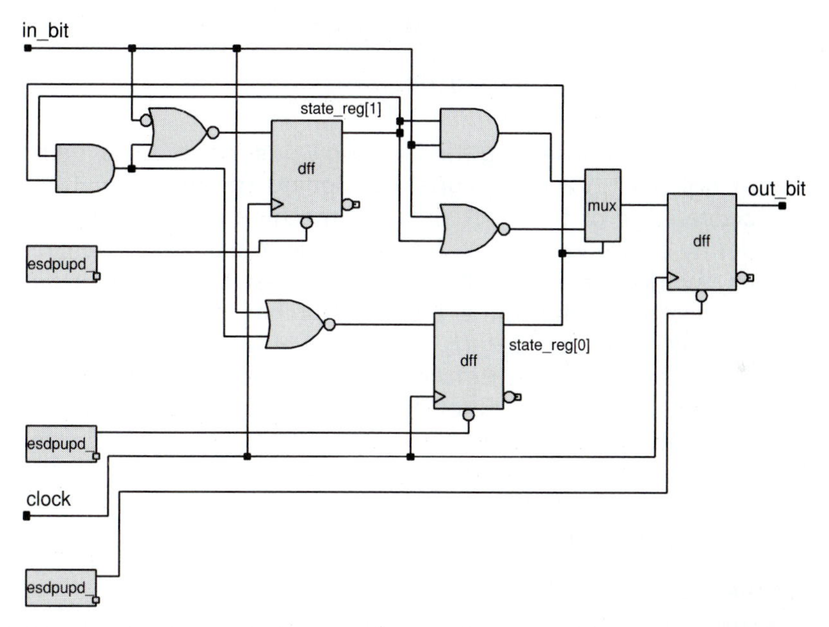

Figure 9.36 Synthesis result for *seq_det_1a*. (Resets of flip-flops are disabled.)

Notice that the synthesized circuit has no reset signal; nor does the behavioral description. Does this cause problems? Consider the simulation results shown in Figure 9.37.

Notice that *out_bit* and *state_reg* have a value of "x" until the first active edge of clock. At this first active edge, the default case item causes *state_reg* to have the value *state_0*. Also note that the value of *out_bit* is not asserted until after *in_bit* has two successive values of "1". The machine failed to assert

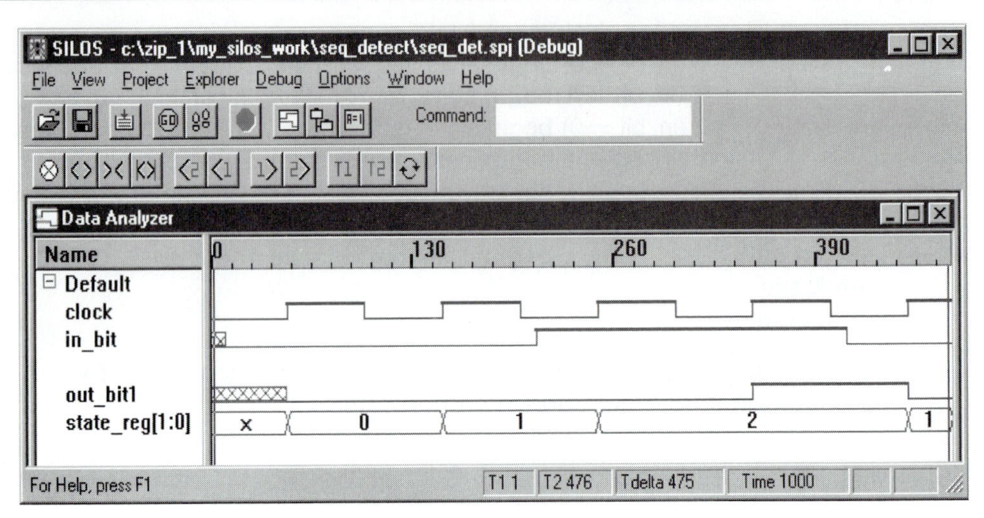

Figure 9.37 Simulation results for *seq_det_1a*.

out_bit after observing the first two values of "0". This is a result of not having a reset signal controlling the machine. The first active edge of the clock puts the machine in *state_0*, independently of the value of *in_bit*. Thus, the implementation eventually works correctly, but ignores the first sample of *in_bit*. To correct this behavior, a reset signal has been added to the description in *seq_det_1b*. This external signal can initialize the state of the machine before the first sample is made. As a general rule, an external reset should be included in the description of a state machine. This assures that the machine is initialized to a known, predictable state. Otherwise, pre-synthesis and post-synthesis simulation results might not match. Also, the physical hardware may have a random initial state.

```
module seq_det_1b (clock, reset, in_bit, out_bit);
  input             clock, reset, in_bit;
  output            out_bit;
  reg       [1:0]   state_reg;
  reg               out_bit;
  parameter         state_0 = 2'b00;
  parameter         state_1 = 2'b01;
  parameter         state_2 = 2'b10;

  always @ (posedge clock or posedge reset)
    if (reset == 1) begin state_reg <= state_0; out_bit <= 0; end else
    case (state_reg)
```

```
        state_0 :  if (in_bit ==0)   begin state_reg <= state_1; out_bit <= 0;
                   end else
                   if (in_bit ==1)   begin state_reg <= state_2; out_bit <= 0;
                   end
        state_1 :  if (in_bit ==0)   begin state_reg <= state_1; out_bit <= 1;
                   end else
                   if (in_bit ==1)   begin state_reg <= state_2; out_bit <= 0;
                   end
        state_2 :  if (in_bit ==0)   begin state_reg <= state_1; out_bit <= 0;
                   end else
                   if (in_bit ==1)   begin state_reg <= state_2; out_bit <= 1;
                   end
        default:   begin state_reg <= state_0; out_bit <= 0; end
    endcase
endmodule
```

Figure 9.38 presents simulation results for *seq_det_1a* and *seq_det_1b*, and Figure 9.39 shows the circuit synthesized from *seq_det_1b*. The latter has the same logic, but with a reset connected. Notice that *seq_det_1b* correctly detects the first occurrence of two identical bits in *in_bit*.

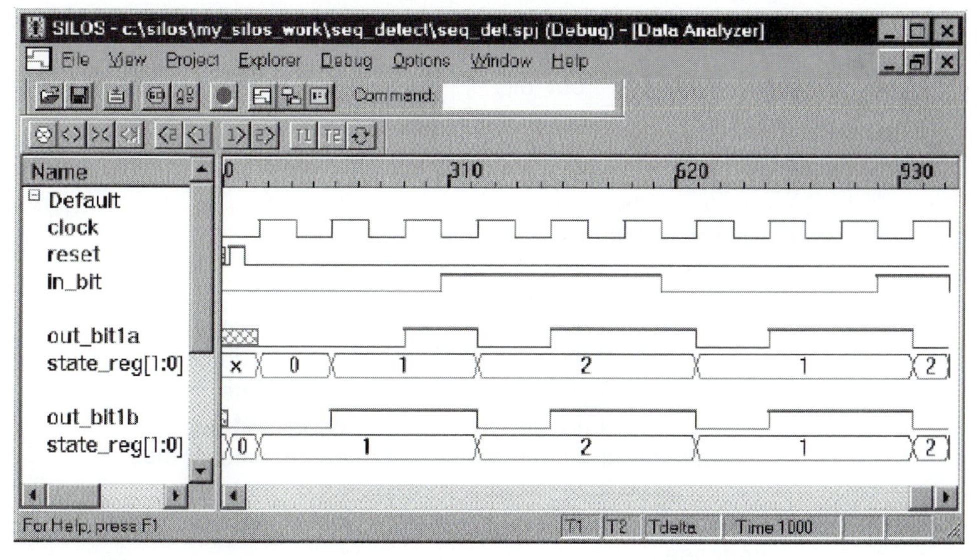

Figure 9.38 Comparison of simulation results for *seq_det_1a* and *seq_det_1b*.

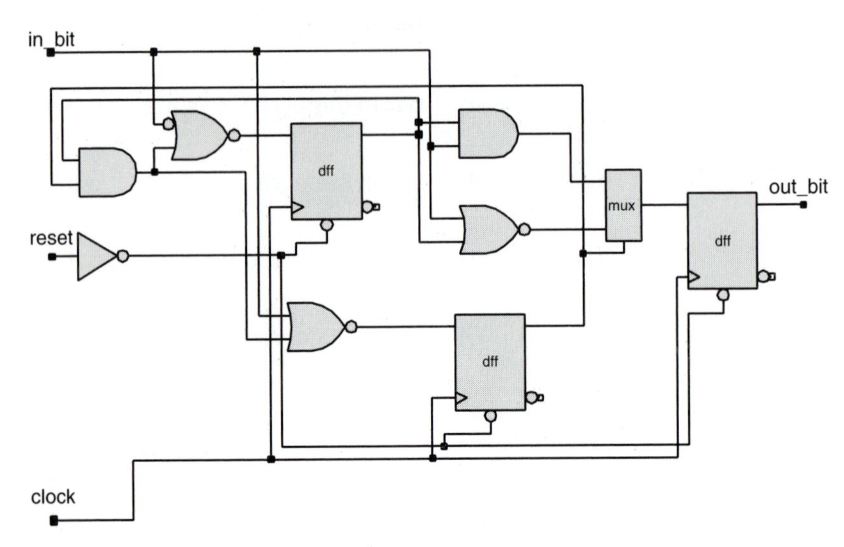

Figure 9.39 Sequence detector circuit synthesized from *seq_det_1b*.

A third variation of this machine will retain the implicit machine style, but will form the output from a behavior activated by state changes. Thus, the output of *seq_det_1c* will be derived combinationally, not registered.

```
module seq_det_1c (clock, reset, in_bit, out_bit);
  input          clock, reset, in_bit;
  output         out_bit;
  reg [1:0]      state_reg;
  reg            out_bit;
  parameter      state_0 = 2'b00;
  parameter      state_1 = 2'b01;
  parameter      state_2 = 2'b10;

  always @ (posedge clock or posedge reset)
    if (reset == 1) state_reg <= state_0; else
    case (state_reg)
      state_0 : if (in_bit ==0)   state_reg <= state_1; else
                if (in_bit ==1)   state_reg <= state_2;
      state_1 : if (in_bit ==0)   state_reg <= state_1; else
                if (in_bit ==1)   state_reg <= state_2;
      state_2 : if (in_bit ==0)   state_reg <= state_1; else
                if (in_bit ==1)   state_reg <= state_2;
      default:                    state_reg <= state_0;
    endcase
```

```
    always @ (state_reg)
      case (state_reg)
        state_0 :                       out_bit = 0;
        state_1 :  if (in_bit ==0)      out_bit <= 1; else
                   if (in_bit ==1)      out_bit <= 0;
        state_2 :  if (in_bit ==0)      out_bit <= 0; else
                   if (in_bit ==1)      out_bit <= 1;
        default:                        out_bit <= 0;
      endcase
  endmodule
```

The circuit synthesized from *seq_det_1c* is shown in Figure 9.40. The logic is the same as for *seq_det_1b*, but the output is not registered.

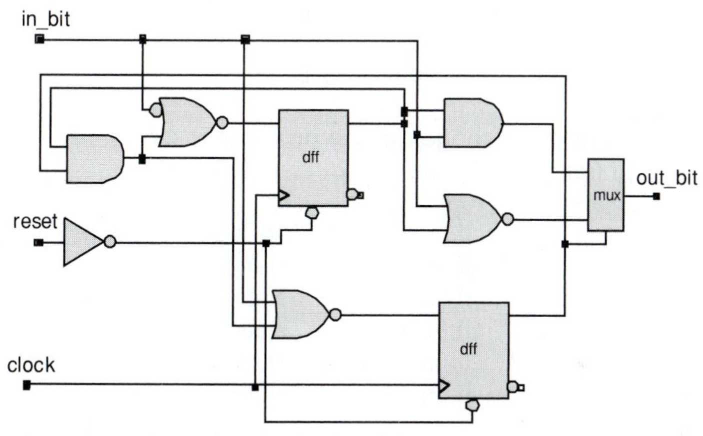

Figure 9.40 Sequence detector synthesized from *seq_det_1c*.

The presentation of simulation results for all three machines in Figure 9.41 reveals that the output of *seq_det_1c* is problematic. Because the output is formed combinationally, the machine asserts its output immediately after entering *state_1* as long as *in_bit* is "0". Likewise, when *in_bit* switches to a value of "1" the machine's output stays asserted. When the machine enters *state_1* from *state_0*, the value of *in_bit* before and after the transition was "1" and the output anticipates the next clock. Only *seq_det_1b* behaves in the desired manner. The output is formed synchronously and conditionally, based upon the value of the input *before* the clock. The output anticipates the state that will be reached at the clock transition. If the specification for behavior is that the output is to be asserted only after the second consecutive repeated sample value is found, only *seq_det_1b* is correct.

The simulation results in Figure 9.41 show that the output of *seq_det_1c* is synchronized to the change in state, and does not correctly model the behavior of the Mealy machine. A possible solution is to change the output behavior to be conditioned on changes of the state and the input signal, as shown in *seq_det_1d*. This machine synthesizes to the same circuit as *seq_det_1c*; it also asserts the output early (i.e., before the clock edge because the output is truly asynchronous).

```
module seq_det_1d (clock, reset, in_bit, out_bit);
   input          clock, reset, in_bit;
   output         out_bit;
   reg [1:0]      state_reg;
   reg            out_bit;
   parameter      state_0 = 2'b00;
   parameter      state_1 = 2'b01;
   parameter      state_2 = 2'b10;

   always @ (posedge clock or posedge reset)
      if (reset == 1) state_reg <= state_0; else
      case (state_reg)
         state_0 : if (in_bit ==0)  state_reg <= state_1; else
                   if (in_bit ==1)  state_reg <= state_2;
         state_1 : if (in_bit ==0)  state_reg <= state_1; else
                   if (in_bit ==1)  state_reg <= state_2;
         state_2 : if (in_bit ==0)  state_reg <= state_1; else
                   if (in_bit ==1)  state_reg <= state_2;
         default:                   state_reg <= state_0;
       endcase

   always @ (state_reg or in_bit)
      case (state_reg)
         state_0 :                  out_bit <= 0;
         state_1 : if (in_bit ==0)  out_bit <= 1; else
                   if (in_bit ==1)  out_bit <= 0;
         state_2 : if (in_bit ==0)  out_bit <= 0; else
                   if (in_bit ==1)  out_bit <= 1;
         default:                   out_bit <= 0;
      endcase
endmodule
```

A second style of implementation separates the next state and output forming logic from the behavior that governs the synchronous state transitions. Three variations will be considered. In the first description, *seq_det_2a*, the outputs are not registered, and

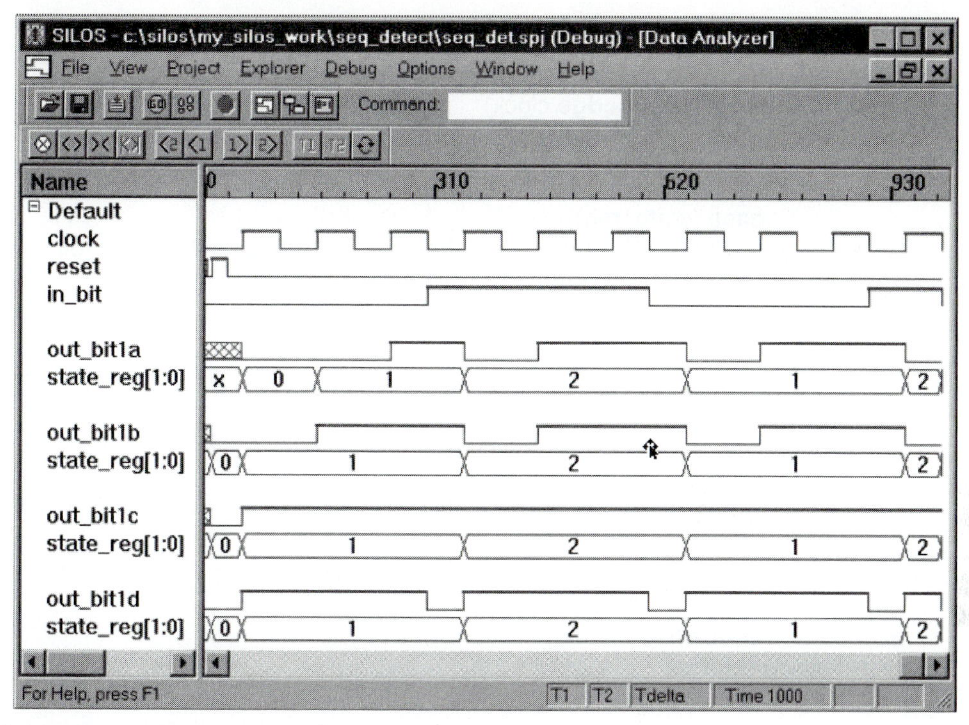

Figure 9.41 Comparison of simulation results for *seq_det_meal1a*, *seq_det_1b*, *seq_det_1c*, and *seq_det_1d*.

the combinational inputs are not covered by the decoding scheme. The unexpected synthesis result shown in Figure 9.42 has internal latches and a latched output. The latches in front of the flip-flops result from the latched description of *next_state* (due to an incomplete **case** statement), and the latch at the output results from the latched description of *out_bit* (also due to the incomplete **case** statement). The simulation results for all three versions are shown in Figure 9.43. Notice that the *seq_det_2a* description does not simulate correctly because the machine cannot recover from its initial state of 2'bx, regardless of the input.

```
module seq_det_2a (clock, in_bit, out_bit);
  input       clock, in_bit;
  output      out_bit;
  reg         [1:0] state_reg;
  reg         [1:0] next_state;
  reg         out_bit;

  parameter   state_0 = 2'b00;
```

```
parameter   state_1 = 2'b01;
parameter   state_2 = 2'b10;

always @ (posedge clock)
   state_reg <= next_state;

always @ (state_reg or in_bit)          // Next state and output logic
   case (state_reg)
      2'b00 : if (in_bit ==0)    begin   next_state <= 2'b01; out_bit <= 0;
              end else
              if (in_bit ==1)    begin   next_state <= 2'b10; out_bit <= 0;
              end
      2'b01 : if (in_bit ==0)    begin   next_state <= 2'b01; out_bit <= 1;
              end else
              if (in_bit ==1)    begin   next_state <= 2'b10; out_bit <= 0;
              end
      2'b10 : if (in_bit ==0)    begin   next_state <= 2'b01; out_bit <= 0;
              end else
              if (in_bit ==1)    begin   next_state <= 2'b10; out_bit <= 1;
              end
   endcase
endmodule
```

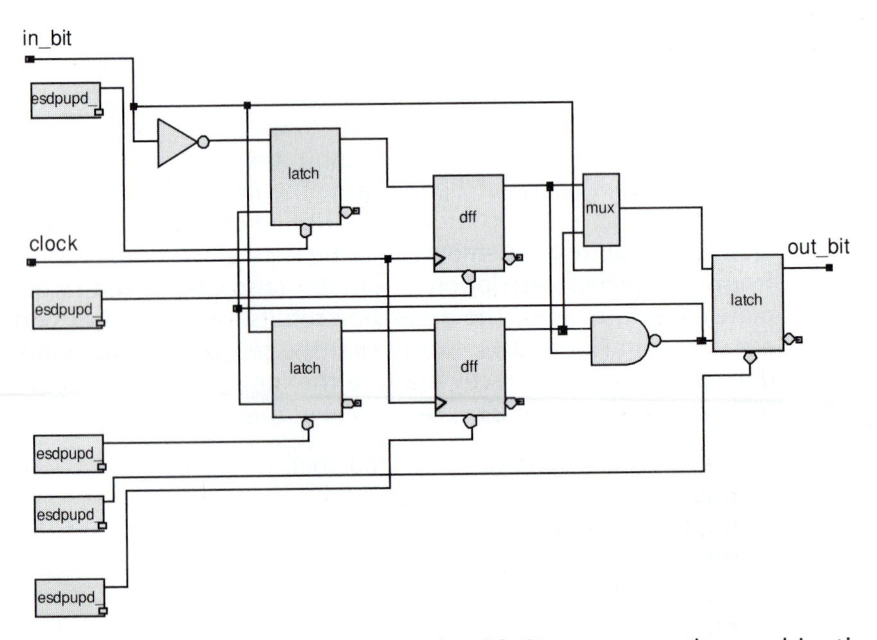

Figure 9.42 Circuit synthesized as a result of failing to cover the combinational inputs of the state machine.

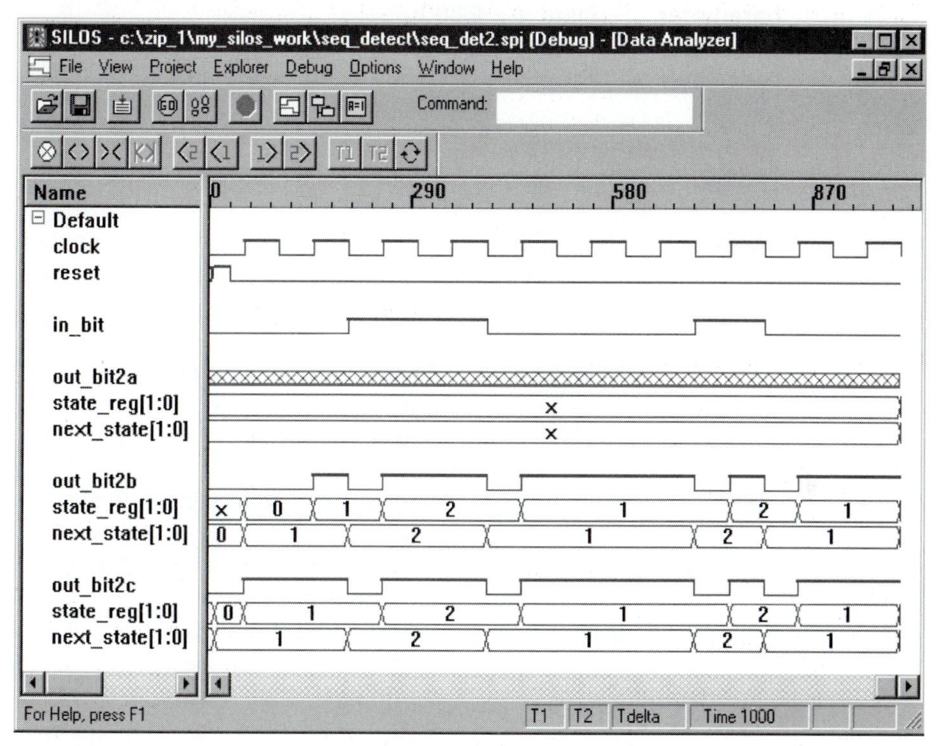

Figure 9.43 Comparison of simulation results for *seq_det_2a, seq_det_2b,* and *seq_det_2c.*

Next, a revised description, *seq_det2b*, includes a default case item, and leads to the synthesis result shown in Figure 9.44. This machine simulates correctly, even though it does not use a reset signal. The default assignment effectively resets the machine on the first active edge of the clock. Notice that the output of *seq_det2b* correctly detects the input sequence, but de-asserts after the input signal changes. This results from the unregistered combinational logic forming *out_bit* and *next_state*.

```
module seq_det_2b (clock, in_bit, out_bit);
  input          clock, in_bit;
  output         out_bit;
  reg    [1:0]   state_reg;
  reg    [1:0]   next_state;
  reg            out_bit;
```

```
parameter     state_0 = 2'b00;
parameter     state_1 = 2'b01;
parameter     state_2 = 2'b10;

always @ (posedge clock)
  state_reg = next_state;

always @ (state_reg or in_bit)           // Next state and output logic
  case (state_reg)
    state_0 :if (in_bit ==0)  begin  next_state = state_1; out_bit = 0;
                              end else
             if (in_bit ==1)  begin  next_state = state_2; out_bit = 0;
                              end
    state_1 : if (in_bit ==0) begin  next_state = state_1; out_bit = 1;
                              end else
             if (in_bit ==1)  begin  next_state = state_2; out_bit = 0;
                              end
    state_2 : if (in_bit ==0) begin  next_state = state_1; out_bit = 0;
                              end else
             if (in_bit ==1)  begin  next_state = state_2; out_bit = 1;
                              end
    default:                  begin next_state = state_0; out_bit = 0;
                              end

  endcase
endmodule
```

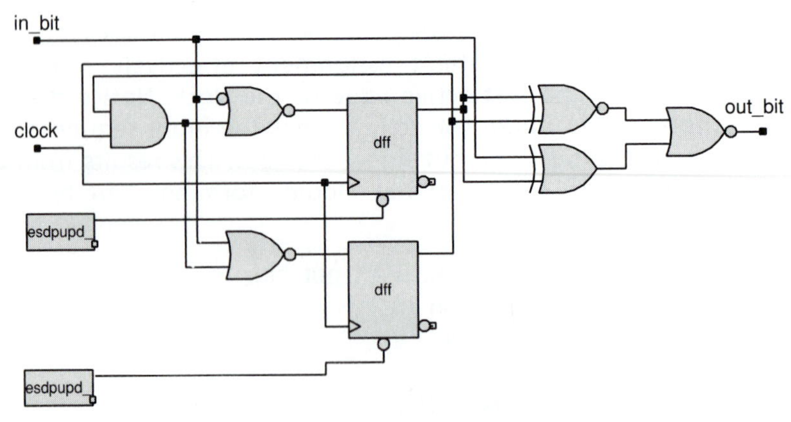

Figure 9.44 Circuit synthesized from *seq_det_2b*.

In a third version, an external reset signal is added to form the description *seq_det_2c*. The circuit synthesized from *seq_det_2c* is shown in Figure 9.45. The output of the synthesized circuit is gated by the output of an active-high latch controlled by *reset*. The input to the latch is a constant, "0". When *reset* is asserted, the latch drives *out_bit* to "0". Otherwise, *out_bit* is determined by the logic of the state machine.

```
module seq_det_2c (clock, reset, in_bit, out_bit);
  input              clock, reset, in_bit;
  output             out_bit;
  reg       [1:0]    state_reg;
  reg       [1:0]    next_state;
  reg                out_bit;

  parameter          state_0 = 2'b00;
  parameter          state_1 = 2'b01;
  parameter          state_2 = 2'b10;

  always @ (posedge clock or posedge reset)
    if (reset == 1) begin state_reg = state_0; out_bit = 0; end
      else state_reg = next_state;

  always @ (state_reg or in_bit)      // Next state and output logic
    case (state_reg)
      state_0 :  if (in_bit ==0)   begin  next_state = state_1; out_bit = 0;
                                   end else
                 if (in_bit ==1)   begin  next_state = state_2; out_bit = 0;
                                   end
      state_1 :  if (in_bit ==0)   begin  next_state = state_1; out_bit = 1;
                                   end else
                 if (in_bit ==1)   begin  next_state = state_2; out_bit = 0;
                                   end
      state_2 :  if (in_bit ==0)   begin  next_state = state_1; out_bit = 0;
                                   end else
                 if (in_bit ==1)   begin  next_state = state_2; out_bit = 1;
                                   end
      default:                     begin next_state = state_0; out_bit = 0;
                                   end

    endcase
endmodule
```

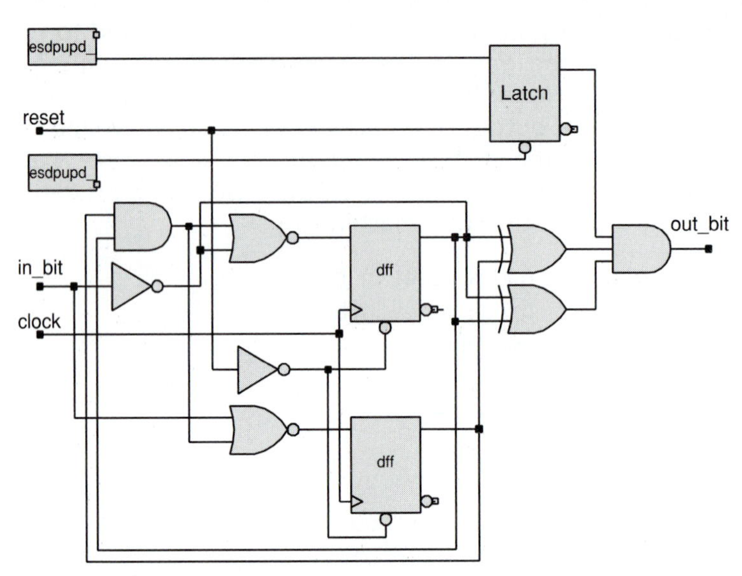

Figure 9.45 Circuit synthesized from *seq_det_2c*.

A careful examination of the simulation results in Figure 9.43 reveals interesting distinctions between the three variations of the modeling style. First, observe that *out_bit2b* correctly asserts on the second rising edge of clock, after two samples of *in_bit*, and de-asserts when *in_bit* changes to 1, but then reasserts immediately on the first sample at which *in_bit* is 1 (i.e., when the state becomes *state_2)*. This is a consequence of using an unregistered output, with *out_bit* being formed by combinational logic. The behavior of *out_bit2c* is also due to this modeling style, even though the description includes a reset.

9.6.5 Some Rules for Implicit FSMs

Synthesis tools require that the clocks (i.e., the synchronizing signals) in an implicit FSM must all be synchronized to the same clock edge (i.e., either **posedge** or **negedge**). The behavior below is valid Verilog code, but will not synthesize.

```
always @ (posedge clk)     // Synchronized event before first assignment
  begin
    reg_a = reg_b;         // Executes in first clock cycle
    reg_c = reg_d;         // Executes in first clock cycle

    @ (negedge clk)        // Illegal clock structure
```

```
    begin
      reg_g = reg_f;
      reg_m = reg_r;
    end
  end
```

Example 9.24 A model of an autonomous eight-bit linear feedback shift register was presented in Chapter 7, using a for-loop and function call. Here we re-visit the problem and present two alternative implementations: *Autonomous_LFSR2*, which uses non-blocking assignments, and *Autonomous_LFSR3*, which combines concatenation and non-blocking assignments. Both implementations are actually implicit FSMs. These two versions are less verbose, more readable and more understandable than the first (in Chapter 7). It would not be practical to write an explicit FSM model for the machine because the number of states is prohibitive. Both versions synthesize to the circuit in Figure 9.46a when modeled with *asynchronous* active-low reset, and synthesize to Figure 9.46b when modeled with *synchronous* active-low reset.

The synthesis tool selected library cells having asynchronous inputs in both cases. It added logic to create the synchronous reset in Figure 9.46b. The structure of the logic uses the bitwise exclusive-or operator, and we expect to see exclusive-or gates at the same locations as in Figure 9.46a. Instead, the circuit with a synchronous reset has a exclusive-or gate at only the input forming *Y[0]*. A mux structure is located at the other locations that would have been sites for an exclusive-or for the tap coefficients used in this example. However, the functionality of the synthesized mux structure is that of an exclusive-or gate. For example, Figure 9.46b shows that when *reset* is 1 (not low), the input to the register holding Y[6] is formed as *Y[7] (~Y[0]) + (~Y[7]) Y[0]*. The mux is controlled by *Y[7]*. If *Y[7]* is 1, *~Y[0]* is selected; if *Y[7]* is 0, *Y[0]* is selected.

Notice how the initial state of the register is formed. In the version with synchronous reset, Figure 9.46b, module *dffspqb_a* is a D-type flop-flop having active-low set; *dffrpb_a* has active-low reset. The asynchronous controls of the flip-flops are tied to power to disable them. When *reset* is 0 (asserted), the datapaths to the register cells form the appropriate values at the inputs to the D-type flip-flops for the initial state, 8'b1001_0001. Figure 9.46c shows the relevant signal values determining the value of the initial state under the condition of the active-low reset. It is apparent that imposing synchronous reset on a circuit costs area.

Decomposition is not done randomly. Skill and experience play a role, and the designer's choice of hierarchical boundaries can have a strong impact on the quality of the synthesis product and the cost of the effort. Partitioning the design into smaller functional units can improve the synthesis result, shorten the optimization cycle, and simplify the synthesis process.

Functionally-related logic should be grouped within a partition so that the synthesis tool will be able to exploit opportunities for sharing logic. If a module is used in multiple places in the design, it should be optimized separately for area, and then instantiated as needed. This strategy will result in an overall design that is very efficient in its use of area. (It also saves CPU time.)

It is also recommended that a module contain no more than one state machine. This will allow the synthesis tool to optimize the logic for a machine without the influence of extraneous logic.

The partition of a design should group registers and their logic so that their control logic may be implemented efficiently. Otherwise, splitting registers and logic across boundaries of the partition might lead to extra or duplicate control logic. Place the combinational logic driving the datapath of a register in the same module as the destination register. Likewise, any glue logic between module boundaries should be included within a module. If glue logic sits outside the modules, it cannot be absorbed by either of them. When module boundaries are preserved in synthesis, combinational logic should not be distributed between modules. Placing the logic in a single module will allow the synthesis tool to exploit common logic.

We saw in Chapter 2 that instantiating modules within modules creates a natural hierarchical organization of a design. In fact, module boundaries are the only hierarchical boundaries in Verilog. Synthesis tools have a variety of options for handling the hierarchy. Most tools can be directed to a mode in which all module instances are flattened into the parent module. This eliminates module boundaries and provides the highest degree of flexibility to the synthesis tool as it attempts to share logic through the design. Alternatively, the tool could be directed to leave a designated module untouched (i.e., the tool will not synthesize its contents). Yet another option is to synthesize the design while preserving the boundaries of the design's hierarchy. Scoping rules must be followed, and variables that are external to a module cannot be referenced either directly or indirectly (through hierarchical de-referencing), in anticipation of the flattening that will be done by the tool.

Tools can preserve the contents of a module on an instance basis or preserve the underlying hierarchical substructure, again on an instance basis. Interactive tools allow a designer to partition and re-partition a design. It is important to note that reducing the scope of the synthesis process necessarily reduces the tool's opportunities for optimizing the logic of the design. The synthesis tool cannot share logic between modules whose boundaries have been preserved. On the other hand, a massive description may not be managed very efficiently or in a reasonable amount of time by the tool.

Because a synthesis tool may eliminate a variable within a module, other modules that are to be left untouched by the operations of the synthesis tool must not reference variables outside the module (i.e., through hierarchical de-referencing).

The partition of a design should not co-locate logic having conflicting optimization objectives. Tools make a trade-off between speed and area. Logic that is to be optimized for speed should not be mixed with logic that is to be optimized for area.

Do not include clocks, I/O pads, or JTAG (i.e., boundary scan) registers in a design that is to be synthesized. Add them to the design after synthesis. I/O pads and clocks are not synthesizable. Clock trees might require special consideration of skew and other timing issues.

9.10 SUMMARY

Verilog supports a variety of descriptive styles that synthesize into sequential logic. Behaviors that do not completely decode their "inputs" will synthesize a latch. Behaviors that have no timing controls and which completely decode their inputs will synthesize to combinational logic. Register variables that are referenced before they are assigned value will synthesize in a hardware storage element. Register variables that are referenced in an assignment outside the behavior will synthesize to a storage element. Although Verilog supports a wide variety of constructs that synthesize, not all constructs do, and not all constructs that do are supported by all vendors of synthesis tools.

PROBLEMS

1. Create a model for a 16-bit counter with synchronous load and reset. Synthesize the circuit in the technology of a cell library, and determine the maximum frequency at which the counter can operate.

2. Write and verify an explicit FSM equivalent to the implicit up-down counter in Example 7.61. Compare the results of synthesizing the two realizations.

3. The state machine below is intended to implement a simple read/write controller having combinational output signals *Rd* and *Wr*. Depending on *control*, the state transitions through *s2* on the way from *s1* to *s0*. Discuss why the description is problematic. Support your discussion with results of simulation and synthesis. Develop a modified description that eliminates the problems found in this description; verify and synthesize the alternative description.

```
module controller (control, Rd, Wr, clk);
  input      control, clk;
  output     Rd, Wr;
  reg        Rd, Wr;
  parameter  s0 = 0, s1 = 1, s2 = 2;
  integer    state;
  always @ (negedge clk)
    case (state)
      s0: begin Rd = 1; Wr = 0; state = s1; end
      s1: begin Rd = 0; Wr = 1; if (control) state = s2; else state = s0;
          end
      s2: begin Rd = 0; Wr = 0; state =s0; end
    endcase
endmodule
```

4. Three versions of a speed controller FSM for a vehicle were presented in Chapter 7. Synthesize the machines and compare the circuits produced.

5. Model, verify, and synthesize a read-write controller that idles until an external start signal is asserted, then reads or writes depending on the assertion of an external read/write control signal. The operations of reading or writing terminate when an external "stop" signal asserts. The controller then enters a wait state for one cycle, before returning to the idle state. The controller has synchronous reset and is to assert individual outputs indicating that the controller is reading, writing, or waiting (done).

6. Create a 16-bit pipelined carry-look-ahead adder with 4-bit slices. Synthesize the unit in the technology of a cell library. Compare the performance of the synthesized circuit to that of a 16-bit ripple carry adder.

7. Write, verify, and synthesize a behavioral model of a synchronous serial sequence detector that is to detect a pattern of 4 identical bits. The machine is to display its output bit with a lag of three clock cycles and must not assert prematurely.

8. Write, verify and synthesize a sequence detector that is to detect where the bit stream in its serial input matches a pattern supplied by a parallel input of parameterized size.

9. Write, verify, and synthesize a parameterized model for a pipeline adder having a parameter for the word size, and a parameter for the number of stages of internal pipeline registers. For example, a pipeline adder might handle a 32-bit word and have four stages of internal pipeline registers, in addition to the input and output registers.

10. Synthesize a machine that will receive serial data and form an output word having the value of the longest string of consecutive identical bits in the most recent 64 bits that have been read.

11. Synthesize a machine that will respond to a "go" command and synchronously read an 8-bit serial input. The machine is to determine whether the received word has exactly one field of zeros or not, and set a flag to indicate whether the data is valid. The machine must also indicate the number of zeros in the field and set a flag indicating that the count is ready.

12. Synthesize a machine that will shift the bits of a 16-bit word until the MSB contains a 1.

13. Write, verify and synthesize a 8-bit bi-directional barrel shifter.

14. Write, verify, and synthesize a two-bit sequencer that will create the following pattern: (00), (11), (01), (10), (00). Include a state transition graph. Compare explicit and implicit FSM models of the sequencer.

15. Write, verify, and synthesize a three-bit sequencer that will create the following pattern: (000), (011), (101), (100), (010), (110), (001), (111). Include a state transition graph. Compare explicit and implicit FSM models of the sequencer.

16. Write, verify, and synthesize a Gray-code counter that generates the following sequence: (0000), (0001), (0011), (0010), (0110), (0111), (0101), (0100), (1100), (1101), (1111), (1110), (1010), (1011), (1001), (1000). Compare your results to an alternative implementation using a Johnson code.

17. An arithmetic shifter shifts the bits of a register, but does not fill in a trailing zero in the end cell of the register. Instead, the content of the end cell is restored after it is shifted to the adjacent cell. Write, verify and synthesize an arithmetic shifter that can load a parallel word, and shift it in either direction, with an arithmetic shift operation.

18. Synthesize the versions of an ALU described in Chapter 7, problem 16.

19. Synthesize and compare the client-polling circuits described in Chapter 7, problems 60, 61, and 62.

20. The client-server scheme in Chapter 7 uses the **wait** construct. Some tools might not synthesize this construct. Develop an explicit FSM model of a server having the behavior depicted in the ASM chart in Example 7.60. The server provides to the client the contents of a 4-bit output data port, *server_data*. Develop a testbench to verify the model, synthesize the model, then verify the synthesized result.

21. Develop an explicit FSM model of the client depicted in the ASM chart in Example 7.63. The client is to store the received data in an internal register. The server provides to the client the contents of a 4-bit input data port, *server_data*. Develop a testbench to verify the model, synthesize the model, then verify the synthesized result.

22. A new design guru claims that the Verilog model below correctly describes a pipeline adder comparable to the one discussed in Example 9.18. Prove or disprove the claim.

```
module mystery_add_16_pipe (c_out, sum, a, b, c_in, clock);
  input   [15:0]     a, b;
  input              c_in, clock;
  output  [15:0]     sum;
  output             c_out;

  reg     [15:0]     top_a, top_b, sum;
  reg     [7:0]      R1_a, R1_b, R1_sum;
  reg                R1_carry, c_out, top_c_in;

  always @ (posedge clock) begin
    top_a = a;
    top_b = b;
    top_c_in = c_in;
  end

  always @ (posedge clock) begin
    {R1_carry, R1_sum} = top_a[7:0] + top_b[7:0] + top_c_in;
    R1_a = top_a[15:8];
    R1_b = top_b[15:8];
  end

  always @ (posedge clock) begin
    {c_out, sum[15:8]} = R1_a + R1_b + R1_carry;
    sum[7:0] = R1_sum;
  end
endmodule
```

23. Synthesize a "divide-by-six" counter.

24. Consider the issue of software races and compare the results of simulating and synthesizing the two modules below.

```
module two_behav (sig_a, sig_b, sig_y1, sig_y2, clk);
  input            sig_a, sig_b, clk;
  output           sig_y1, sig_y2;
  always @ (negedge clk) sig_y1 <= sig_a;
  always @ (negedge clk) if (sig_y1) sig_y2 = sig_b else sig_y2 = 0;
endmodule
```

```
module one_behav (sig_a, sig_b, sig_y2, clk);
  input     sig_a, sig_b, clk;
  output    sig_y1, sig_y2;
  always @ (negedge clk)
    begin
      if (sig_y1) sig_y2 = sig_b; else sig_y2 = 0;
      sig_y1 <= sig_a;
    end
endmodule
```

25. A bit-serial adder forms the sum of two words by serially adding individual bits. It requires less physical area to implement the adder and to route the data path. The structure consumes less power and can operate at higher speeds than ripple adders. However, it does require a clock signal to synchronize its operation. The structure of the adder is shown below. A single full-adder sequentially adds the bits of the stored addend and augend. The unit is also called a carry-save adder because the value of the carry bit from the full-adder is stored in a flip-flop and used in the next cycle. In the configuration shown, the sum is stored in a separate register. If necessary, the *set* and *clear* signals can be used to load an initial value of *c_in*. For a 16-bit word, addition takes 16 cycles of the clock One cell of the word is added in each cycle, beginning with the LSB. The bit-serial adder can be described by the following procedural assignments:

```
sum(k) = (c_out ^ a[k] ^ b[k]) + (a[k] & b[k] & c_out);
c_out[k] = (a[k] & b[k]) + (a[k] ^ b[k]) & c_out;
```

In this description, *k* is the index of a cell in the adder, and *c_out* on the RHS of the assignment is the value passed into cell *k*. The value of *c_out* on the LHS of the assignment is the value of the carry formed at cell *k* and passed to cell *k+1* at the next clock. In practice, a control signal must govern when the addition sequence begins. The figure below also shows waveforms illustrating the movement of data in the unit. The data words (001) = 1_D and (101) = 5_D are added with c_in = 0, beginning with the least significant bit. The result is formed serially (temporally in reverse order): (110) = 6_D.

Consider implementing the adder as a state machine having the following four states: *idle*, *loading*, *adding*, and *waiting*. An active-low reset_ signal puts the machine in state idle, where it remains until *start* signal is asserted. The machine then loads the internal registers from the external busses (freeing the busses for other activity while the addition takes

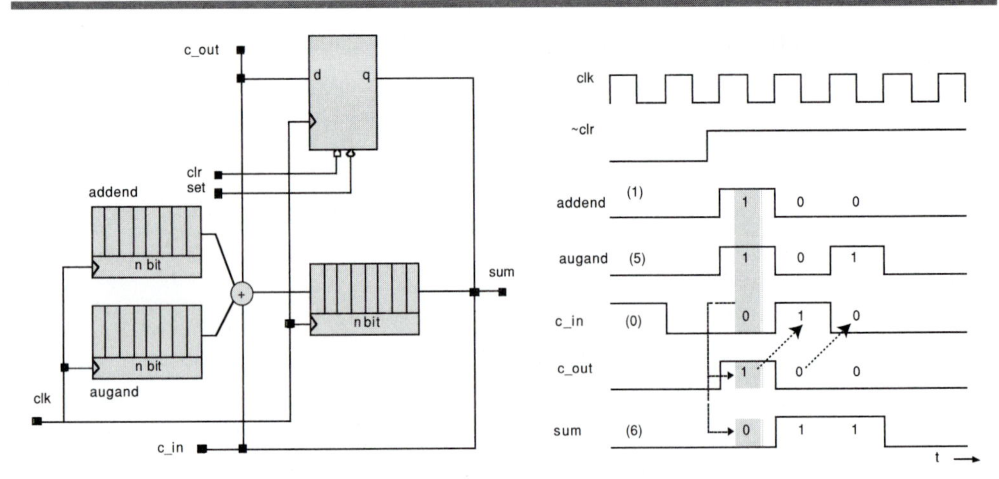

place), cycles through the addition process, and enters state *wait*, where it asserts *done* and remains until *start* is re-asserted.

Using the "shell" specified below, write and verify a description of an bit-serial adder for 8-bit datapaths.

```
module bit_serial (addend, augend, c_in, sum, c_out, start, done, clk, reset_);
  parameter word_size = 8;     // Specifies the width of the data path
  parameter reg_size = 4;      // Sized to count word_size+1
  parameter state_size = 2;    // Sized to the encoding used for the
                                  state register
  input [word_size -1: 0] addend, augend;
  input c_in, start, clk, reset_;
  output [word_size-1: 0] sum;
  output done, c_out;
  reg [word_size-1: 0] a, b;
  reg [reg_size-1:0] k;
  reg flag;                    // Internal signal that addition is complete
  reg sum, c_out, done;
  parameter idle =     4'b00;  // Serial state encoding
  parameter loading = 4'b01;
  parameter adding =  4'b10;
  parameter waiting = 4'b11;
  reg [state_size-1:0] state, next_state;
  // your code goes here
endmodule
```

The *addend* and *augend* are available when *start* is asserted, but cannot be assumed to remain throughout the addition cycle. Make any other reasonable assumptions to support your design. Verify that your adder works correctly on the following words:

alu_out while *en* is asserted; otherwise *alu_out* has a value "z". The default case assignment gives full freedom to the synthesis tool to exploit don't-cares. The synthesized circuit is shown in Figure 10.23b. Notice that the identifier *s2* is retained in the result, but is not needed. The synthesis tool retains the identity of ports.

```
module alu_reg_with_z1 (alu_out, en, a, b, s0, s1, s2, clk, reset);
  output  [1:0]  alu_out;
  input   [1:0]  a, b;
  input          en, s0, s1, s2, clk, reset;
  reg     [1:0]  out_int;

  assign alu_out = (en ==1) ? out_int : 2'bz;

  always @ (posedge clk)
    if (reset) out_int = 2'b0;
    else
      case ({s0,s1,s2})
        3'b111 :  out_int = a & b;
        3'b011 :  out_int = a | b;
        3'b001 :  out_int = a ^ b;
        default:  out_int = 2'bx;
      endcase
endmodule
```

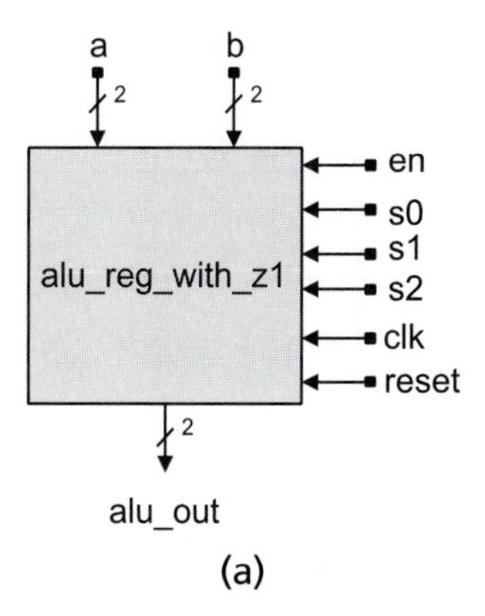

(a)

Figure 10.23 Circuit symbol (a) and circuit (b) synthesized from *alu_reg_with_z1*. Continues on page 460.

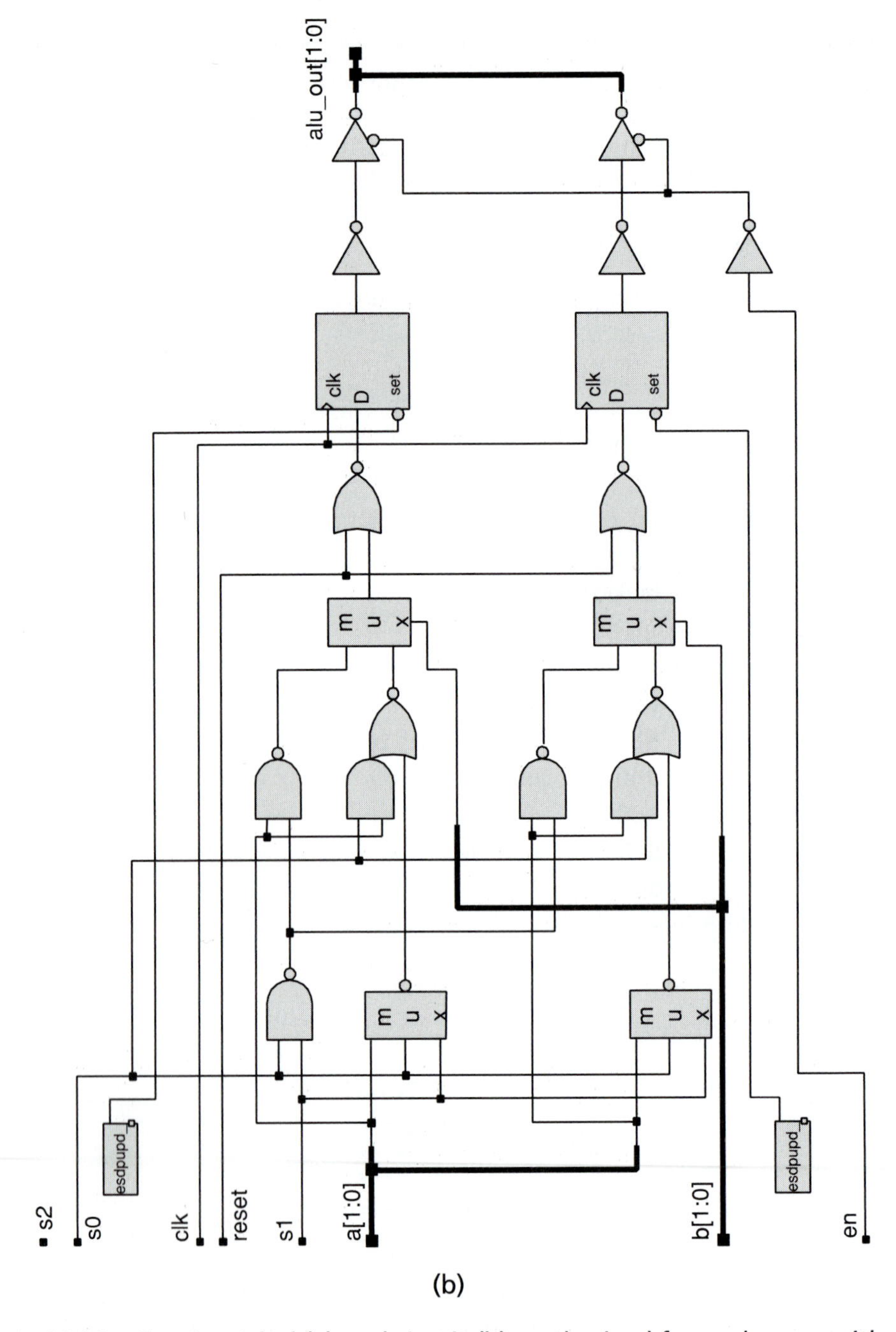

(b)

Figure 10.24 Circuit symbol (a) and circuit (b) synthesized from *alu_reg_with_z1*. Concluded.

The slightly more complex synchronous ALU in *alu_reg_with_z2* has a four-bit data path, more *case_items*, and less flexibility in its default assignments to *alu_reg*. Its realization in Figure 10.24 likewise has registered, three-state outputs.

```
module alu_reg_with_z2 (alu_out, data_a, data_b, enable, opcode, clk, reset);
  input    [2:0]   opcode;
  input    [3:0]   data_a, data_b;
  input            enable, clk, reset;
  output   [3:0]   alu_out;
  reg      [3:0]   alu_reg;

  assign alu_out = (enable ==1) ? alu_reg : 4'bz;

  always @ (posedge clk)
    if (reset) alu_reg = 0; else
      case (opcode)
        3'b001: alu_reg = data_a | data_b;
        3'b010: alu_reg = data_a ^ data_b;
        3'b101: alu_reg = data_a & data_b;
        3'b110: alu_reg = ~data_b;
        default: alu_reg = 0;
      endcase
    endmodule
```

10.7 SYNTHESIS OF RESETS

Synthesis tools support synchronous and asynchronous resets of registers and latches. Asynchronous reset activity is typically modeled within the event control expression detecting the synchronizing signal of a behavior. However, the **assign** ... **deassign** construct allows a separate behavior to be written for the asynchronous reset. The latter style is more efficient for simulation, but is not supported by all vendors. A design with synchronous resets will have additional gates to gate the D-input of a flip-flop under the reset condition. An and-gate is typically used to gate the path if the reset condition drives the output of the flip-flop to 0. If the model of a synchronous circuit does not have a global reset, the synthesis tool will tie-off unused set and/or reset lines to power/ground if it uses library cells that have those ports. FPGAs usually have a built-in global set/reset that should be used.

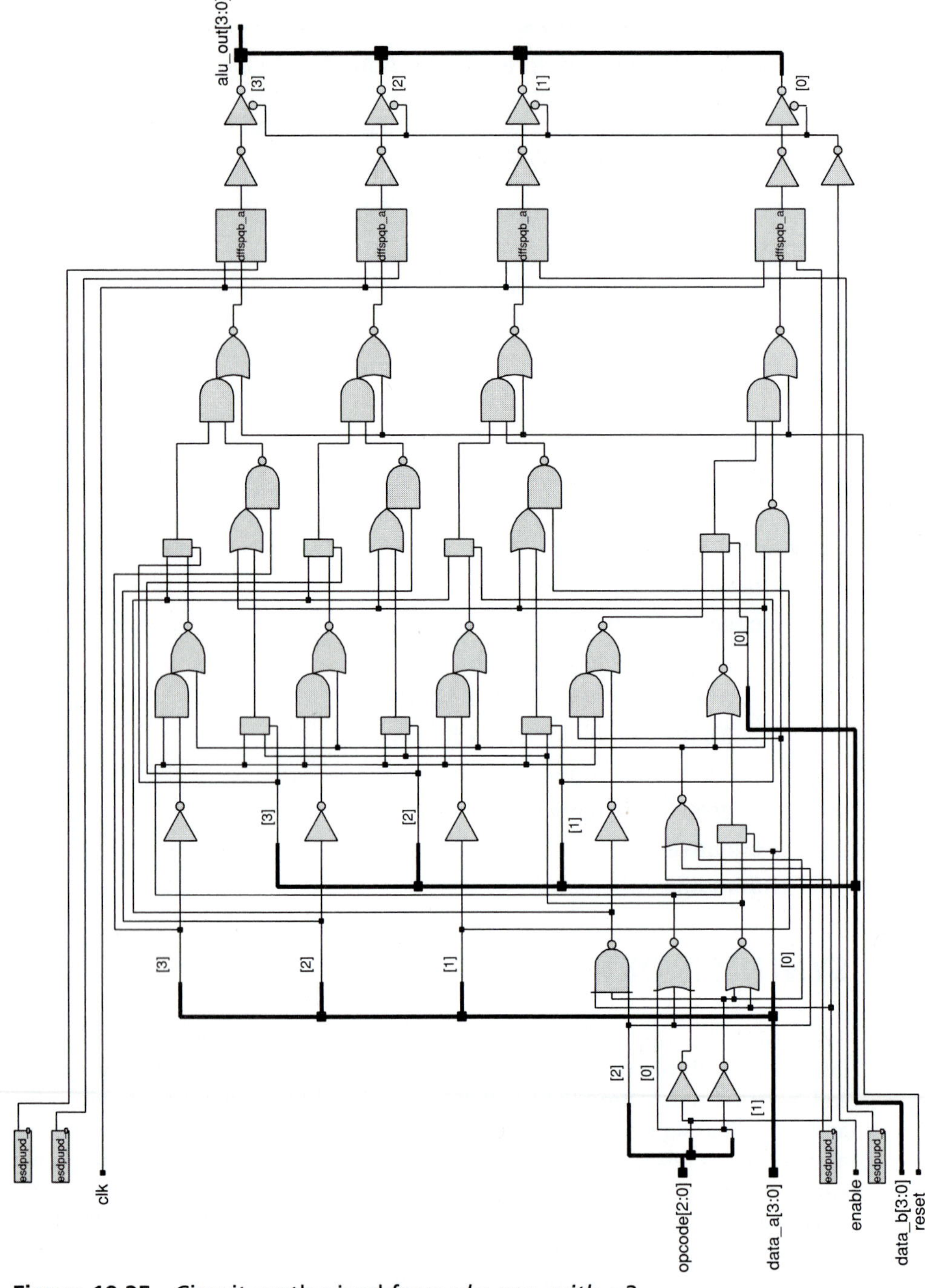

Figure 10.25 Circuit synthesized from *alu_reg_with_z2*.

10.8 TIMING CONTROLS IN SYNTHESIS

10.8.1 Delay Controls

It is not advisable to include delay controls in a Verilog behavior that is to be synthesized. In general, delay controls that are shorter than a clock period will be ignored. If a procedural assignment is associated with a delay that is longer than a clock period, some tools will place the assignment in succeeding clock cycles, corresponding to multi-cycle paths. Consult the documentation of the tool you are using.

10.8.2 Event Controls

The event control expression of a cyclic behavior that is to model combinational logic must include any signal that affects a target variable (i.e., an output of the combinational logic). The signal can appear explicitly in the event control expression, or implicitly, as it does if it is not in the event control expression but appears in a RHS expression of a procedural continuous assignment. Otherwise, a latch will be inferred by a synthesis tool. The synthesis engine will parse the event control expression to determine whether it contains a clock (synchronizing) signal (i.e., a signal associated with an edge qualifier); if one is found, the logic will be synthesized as sequential logic. Otherwise, the logic will be treated as combinational logic or a latch. The inputs of the inferred logic will be the identifiers in the RHS expressions of assignments within the behavior, and any control signals whose transitions affect assignments to the target variables.

Event control expressions determine when the procedural statements of a cyclic behavior begin to execute. If the description is to be synchronous, the expression must be synchronous with a single edge of a single clock. The same signal need not be used as the clock (synchronizing signal) in every behavior within a module, and different edges can be used in different behaviors. But all signals that are used as clocks must have the same period. (Multiple clock domains are not considered here.) It is common for the period of the clocks to be specified at the GUI of the synthesis tool.

Some restrictions commonly apply to the use of asynchronous resets (i.e., an event control expression with **posedge reset**). The **always** behavioral statement must have an event control with an expression having only one clock edge and any number of other edges connected by the event-or construct. These other signals are asynchronous control signals, and they ***must*** be accompanied individually by an edge qualifier. An event control expression may not contain a signal qualified by both **posedge** and **negedge**.

For synthesis, the statement associated with the event control expression must be a conditional branch statement (i.e., **if** ..., or **?** ... **:**). Care must be taken to separate the branches of the conditional expression into synchronous and asynchronous parts having only one synchronous branch. If a branch has no condition, or if it has a condition that does not depend upon a signal in the event control expression, the activity of the branch is considered to be synchronous. The synchronous branch must be the last branch in the statement—a style constraint that allows the synthesis tool to recognize the asynchronous control assertions, followed by defaulted synchronous activity. The style also assures that the asynchronous control signals have priority over the synchronizing signal.

10.8.3 Multiple Event Controls

The synchronous behavior of an explicit finite state machine can contain only one event control expression, but an implicit finite state machine can contain multiple event control expressions in the same behavior. The multiple event control expressions distribute the evolution of the register operations into different clock cycles. The behavior below contains an example of a cyclic behavior containing two event control expressions.

```
always @ (posedge clk)      // Synchronized event before first assignment
  begin
    reg_a = reg_b;          // Executes in first clock cycle
    reg_c = reg_d;          // Executes in first clock cycle.

    @ (posedge clk)         // Begins second clock cycle.
      begin
        reg_g = reg_f;      // Executes in second clock cycle.
        reg_m = reg_r;      // Executes in second clock cycle.
      end
  end
```

The clock edges of an implicit finite state machine define the boundaries of the machine's state transitions. It is essential for synthesis that the multiple event control expressions be synchronized to the same edge — either **posedge** or **negedge**, but not both. The behavior shown below will simulate, but will not synthesize.

```
always @ (posedge clk)      // Synchronized event before first assignment
  begin
    reg_a = reg_b;          // Executes in first clock cycle
    reg_c = reg_d;          // Executes in first clock cycle.

    @ (negedge clk)         // Illegal clock structure.
      begin
        reg_g = reg_f;
        reg_m = reg_r;
      end
  end
```

10.8.4 SYNTHESIS OF THE WAIT STATEMENT

The **wait** statement is used to introduce time delay into the execution of a sequential activity flow in a behavior. If the **wait** statement is supported, the condition that determines the duration of the delay must include at least one value that is generated by a separate concurrent behavior or continuous assignment. Otherwise, the delay will be interminable. Also, the condition that determines the delay must be held in the true state for at least one clock cycle. This guarantees that a state controller can be affected by the condition. When the **wait** condition is generated in a behavior other than the one in which the **wait** construct appears, it is necessary that the same clock be in both the block containing the **wait** and the behavior generating the condition. (Under these conditions, the simulation results of the seed behavior and the synthesized description may differ by one clock cycle.)

For some tools, the delay introduced by a **wait** statement must be synchronized with the clock. This is accomplished by inserting an event control expression immediately after the **wait** statement.

10.8.5 SYNTHESIS OF NAMED EVENTS

Named events (abstract events) provide communication between behaviors in the same or different module (see Chapter 7). Not all synthesis tools support this construct. Anticipate some general restrictions. First, two behaviors are needed to accommodate generation and triggering of a named event. Event triggering is an edge phenomenon, so do not expect a synthesis tool to build hardware that both triggers an event and references that event in the same behavior. That would be like generating a clock edge in a statement and then watching for it in the next statement, after the fact.

Two behaviors that communicate by a named event must have the same clock, or different clocks having the same period. The communicating behaviors can use the same or different edges of the same clock. If the same edge of the clock is used, the behavior that triggers (->) the named event must have a delay control of less than one clock period inserted before the triggering statement. This prevents a race between the named event and the clock edge that governs the activation of the behavior in which the event is being referenced. As a result, the behavior triggered by the named event is executed in the next clock cycle, as it will be in hardware. The general structure is illustrated below.

```
always @ (posedge clk)
  begin
    statement_1_executes;
    statement_2_executes;
    #1   // Required because both behaviors use same edge
    -> event_B_is_triggered;
  end
```

```
always
  @ event_b_is_triggered
      // edge synchronization
  @ (posedge clk)
    begin
      statement_a_executes;
      statement_b_executes;
    end
```

Example 10.20 Named events can be used to describe combinational logic, but not very efficiently. Module *and_named_event* consists of *named_event_detector*, whose event control expression contains the inputs to a hypothetical "and" gate, and *named_event_result*, whose event control expression is sensitive to event *go* inside *named_event_detector*. Both *x1* and *x2* must be passed to *named_event_detector* and *named_event_result*. *named_event_detector* has no output! This simple example can be used to test whether a synthesis tool recognizes named events in combinational logic.

```
module and_named_event (y, x1, x2);
  input     x1, x2;
  output    y;

  named_event_detector M1 (x1,x2);
  named_event_result M2 (y, x1, x2);
endmodule

module named_event_result (y, x1, x2);
  input     x1, x2;
  output    y;
  reg       y;

  always @ (and_named_event.M1.go)
    y = x1 & x2;
endmodule

module named_event_detector (x1,x2);
  input     x1, x2;
  event     go;

  always @(x1 or x2) -> go;
endmodule
```

10.9 SYNTHESIS OF MULTI-CYCLE OPERATIONS

Procedural statements within a behavior must complete execution within a single clock cycle. If an operation requires more than one clock cycle to compete, the operation must be divided into two or more parts that execute in separate, successive clock cycles, which implies that the implementation will require memory.

Example 10.21 The multi-cycle operation in *m_cycle* has the data flow graph shown in Figure 10.25. The description synthesizes into the circuit shown in Figure 10.26.

```
module m_cycle (result, data_a, data_b, data_c, clk);
  input    [3:0]  data_a, data_b, data_c;
  input           clk;
  output   [4:0]  result;
  reg      [4:0]  result, temp1;

  always @ ( posedge clk)
    begin
      temp1 = data_a + data_b;
      @ (posedge clk)
      result = temp1 + data_c;
    end
endmodule
```

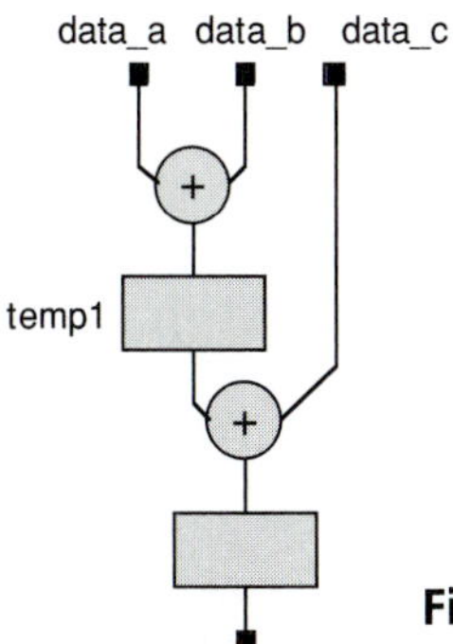

Figure 10.26 Dataflow diagram for a multi-cycle operation.

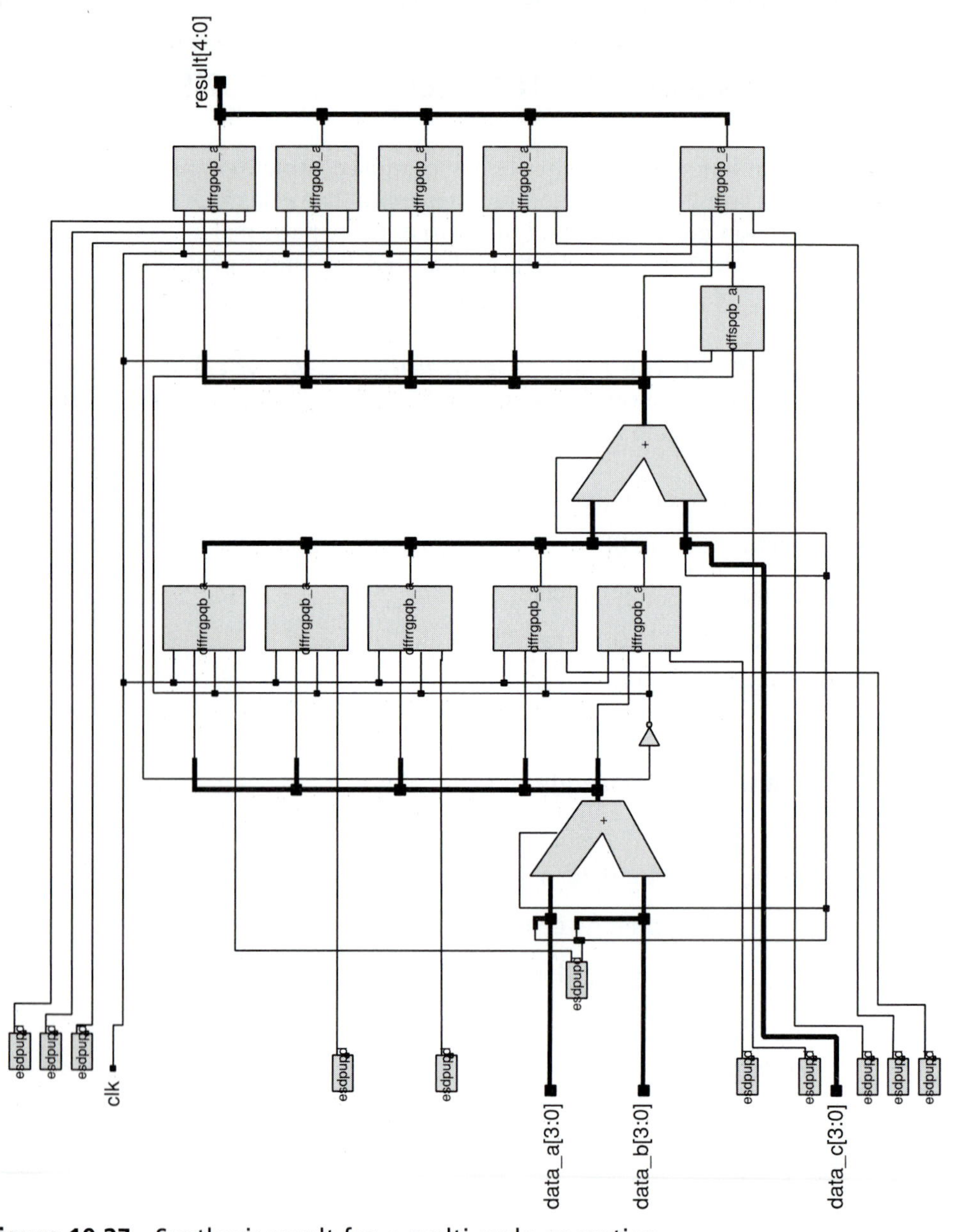

Figure 10.27 Synthesis result for a multi-cycle operation.

10.10 SYNTHESIS OF LOOPS

The support of **repeat**, **for**, **while** and **forever** loop constructs in synthesis is, of course, vendor-dependent. How and whether a synthesis tool synthesizes a loop depends upon whether the loop contains event controls and/or data dependencies. A loop is said to be static, or data-independent, if the number of iterations can be determined by the compiler before simulation. This implies that the number of its iterations is a fixed value. A loop is said to be data dependent if the number of iterations depends upon some variable during the simulation. In addition to having a dependency on data, a loop may also have a dependency on embedded timing controls. Figure 10.27 shows possible loop structures. In principle, static loops can be synthesized, but a given vendor might choose to confine the descriptive style of a static loop to a particular construct. Non-static loops that do not have internal timing controls are problematic.

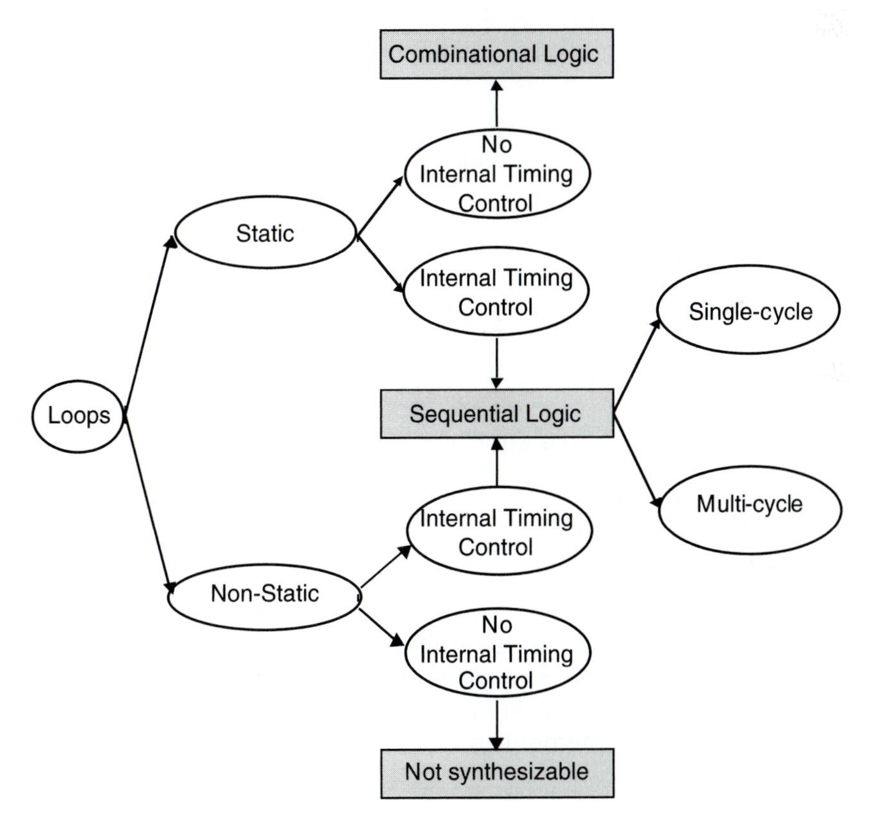

Figure 10.28 Possible loop structures.

10.10.1 STATIC LOOPS WITHOUT INTERNAL TIMING CONTROLS

When a loop in a Verilog behavior has no internal timing controls and no data dependencies, the computational activity of the loop is implicitly combinational. The mechanism of the loop is artificial—the computations of the loop can be performed without memory, instantaneously. The iterative computational sequence has a non-iterative counterpart that can be obtained by "unrolling" the loop, and the operations in the "unrolled" loop can occur at a single time step of the simulator.

Example 10.22 The number of iterations of the loop in *for_and_loop_comb* does not depend upon the data and does not have event controls. It iterates for a fixed, predetermined number of steps and then terminates. The description synthesizes to the combinational circuit with registered outputs in Figure 10.28a.

```
module for_and_loop_comb (out, a, b);
  input     [3:0]     a, b;
  output    [3:0]     out;

  reg       [2:0]     i;
  reg       [3:0]     out;

  always @ (a or b)
    begin
      for (i = 0; i <= 3; i = i+1)
        out[i] = a[i] & b[i];
    end
endmodule
```

The unrolled loop is equivalent to the following assignments:

```
out[0] = a[0] & b[0];
out[1] = a[1] & b[1];
out[2] = a[2] & b[2];
out[3] = a[3] & b[3];
```

which correspond to a bitwise-and of four-bit datapaths. The data supporting the computation at any step of the loop is independent of the data supporting other steps of the loop. So the order of the statements is inconsequential.

If the outputs are registered over a single clock cycle, as in *for_and_loop_reg*, the loop executes completely at the active edge of the clock, and the synthesized hardware must be fast enough to execute during the cycle of the clock. This description synthesizes to the circuit shown in Figure 10.28b. In the absence of any constraints, the synthesis tool selected a positive-edge-sensitive, D-type flip-flop having asynchronous reset and two multiplexed inputs, not a simple and-gate paired with a simple D-type flip-flop. The integrated cell has a more efficient layout than any combination of equivalent low-level cells, which is an advantage of a robust cell library (*dffrmpqb_a*).

When *SL* (select) is low, *D0* is clocked to *Q*, and when *SL* is high, *D1* is clocked to *Q*. The active-low reset is hard-wired to 1 (there is no reset). The connections to the *esdpupd* device provides pull to power and/or ground for fixed inputs. The *D0* inputs are hard-wired to 0, and the *D1* inputs are connected to the bits of *b[3:0]*. The *SL* inputs are the bits of *a[3:0]*. If *a[j]* is 0, the output is the 0. If *a[j]* is 1, the output is *b[j]*, the *D1* input.

```
module for_and_loop_reg (out, a, b, clk);
  input    [3:0]   a, b;
  input            clk;
  output   [3:0]   out;

  reg      [2:0]   i;
  reg      [3:0]   out;
  wire     [3:0]   a, b;

  always @ (posedge clk)
    begin
      for (i = 0; i <= 3; i = i+1)
        out[i] = a[i] & b[i];
    end
endmodule
```

A tool that supports synthesis of a static **repeat** loop having no internal timing controls will replace it by equivalent, synthesized combinational logic. The behavior of a **for** loop with static range is equivalent to a **repeat** loop with the same range, so some tools support only the **for** loop.

Example 10.23 The **repeat** loop in the fragment of code shown below executes a fixed number of times and has no event controls. The computations are ordered, because the data supporting the computation of *product* at a given step of the loop depends on the value of *product* from the previous step. But the computations do not require memory.

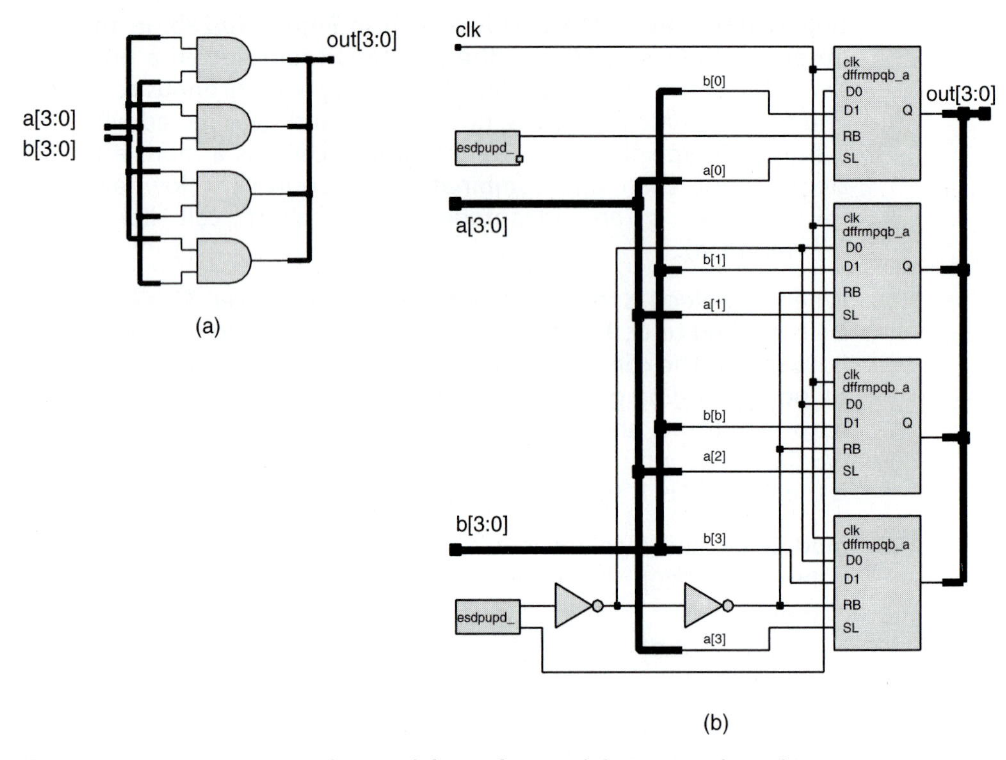

Figure 10.29 Circuits synthesized from *for_and_loop_comb* and *for_and_loop_reg*.

```
repeat (32)
  begin
    if (multiplier[0])
      begin product = product + {multiplicand, 32'h0000}; end
        product = product >> 1;
         multiplier = multiplier >> 1;
  end
```

As Example 10.22 showed, a single-cycle static loop may be associated with sequential logic. If so, the synthesis tool will eliminate register variables whose lifetime is only for the duration of the loop itself (i.e., does not cross the boundary between cycles of the clock). It will also eliminate a register variable that is not referenced outside the host module.

Example 10.24 Suppose a machine has the task of receiving a word of data synchronously, and then asserting an output that encodes the number of 1s in the word. If the hardware is fast enough, the functionality can be implemented by combinational logic in one clock cycle. The loop in *count_ones_a* has no internal timing controls and is static—it does not depend upon data.

Notice that *bit_count* asserts after the loop has expired, within the same clock cycle that the loop executes. Simulation results are shown in Figure 10.29. Be careful when interpreting the results, for the loop executes in one time step of simulation. Consequently, the displayed values of *temp*, *count* and *bit_count* are final values. Intermediate values are overwritten and not apparent. This version synthesizes into clock-compatible combinational logic (i.e., stable within one cycle of the clock) with registered outputs.

The synthesized circuit is shown in Figure 10.30. The contents of register variables *count*, *index*, and *temp* do not have a lifetime outside of the behavior (i.e., they are not referenced elsewhere). So both are eliminated by the synthesis tool. The only synthesized register is *bit_count*, which is a port. The reset action is synchronous, and the reset inputs of the selected D-type flip-flops are hard-wired to 1 (by *esdpupd_*) to disable them.

```
module count_ones_a (bit_count, data, clk, reset);
  parameter                                data_width = 4;
  parameter                                count_width = 3;
  output    [count_width-1:0]              bit_count;
  input     [data_width-1:0]               data;
  input                                    clk, reset;
  reg       [count_width-1:0]              count, bit_count, index;
  reg       [data_width-1:0]               temp;

  always @ (posedge clk)
    if (reset) begin count = 0; bit_count = 0; end
    else begin
      count = 0;
      bit_count = 0;
      temp = data;
      for (index = 0; index < data_width; index = index + 1) begin
        count = count + temp[0];
        temp = temp >> 1;
      end
      bit_count = count;
    end
endmodule
```

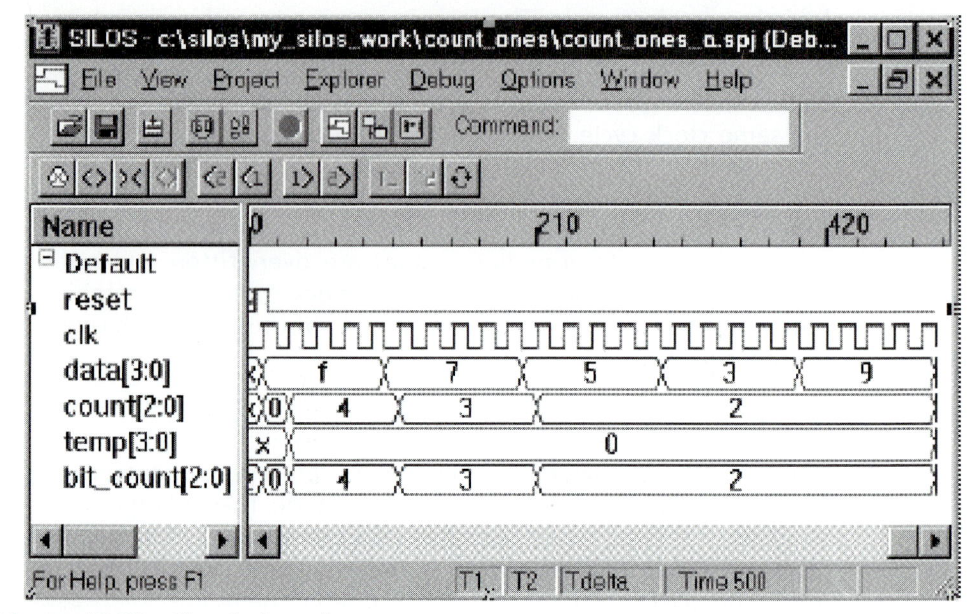

Figure 10.30 Simulation of *count_ones_a*.

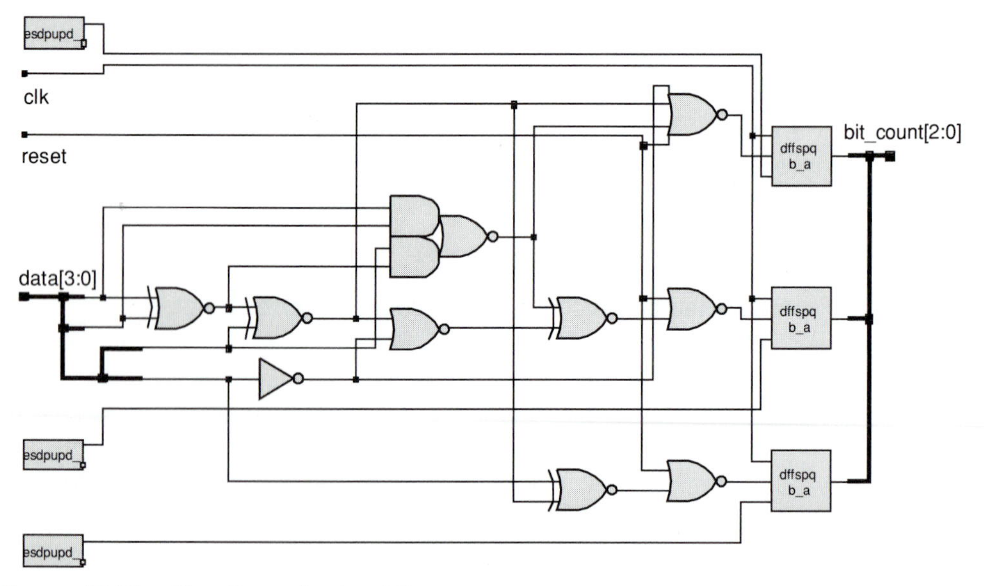

Figure 10.31 Circuit synthesized from *count_ones_a*.

10.10.2 STATIC LOOPS WITH INTERNAL TIMING CONTROLS

If a static loop has an internal edge-sensitive event control expression, the computational activity of the loop is synchronized and distributed over one or more cycles of the clock. As a result, the behavior is that of an implicit state machine in which each iteration of the loop occurs after a clock edge. The behavior may include additional computational activity that is placed in the cycle immediately following the loop's expiration.

Example 10.25 The loop in *count_ones_b* has an embedded event control expression that is synchronized to an external clock signal. The event control expression after the loop forces the assignment to *bit_count* to occur after the loop has completely executed, rather than display intermediate values. The loop always executes for a fixed number of clock cycles, independently of the data. The loop can be unrolled and controlled by a finite state machine (sequential logic) whose state transitions correspond to the iterant count in the loop. Simulation results are presented in Figure 10.31 for a four-bit data path. Register *temp* loads at the first active edge after *reset* is de-asserted. It is apparent that the loop iterates four times, that *temp* is shifting towards the LSB, that *count* iterates after a "1" is detected in *temp[0]*, and that *bit_count* is updated at the first active edge of the clock after the loop expires. *bit_count* holds its residual value until it is updated.

```
module count_ones_b (bit_count, data, clk, reset);
  parameter                                data_width = 4;
  parameter                                count_width = 3;
  output      [count_width-1:0]            bit_count;
  input       [data_width-1:0]             data;
  input                                    clk, reset;
  reg         [count_width-1:0]            count, bit_count, index;
  reg         [data_width-1:0]             temp;

  always @ (posedge clk)
    if (reset) begin count = 0; bit_count = 0;end
    else begin
      count = 0;
      temp = data;
      for (index = 0; index < data_width; index = index + 1)
        @ (posedge clk)
          begin
            if(temp[0]) count = count + 1;
            temp = temp >> 1;
          end
```

```
      @ (posedge clk)
        bit_count = count;
    end
endmodule
```

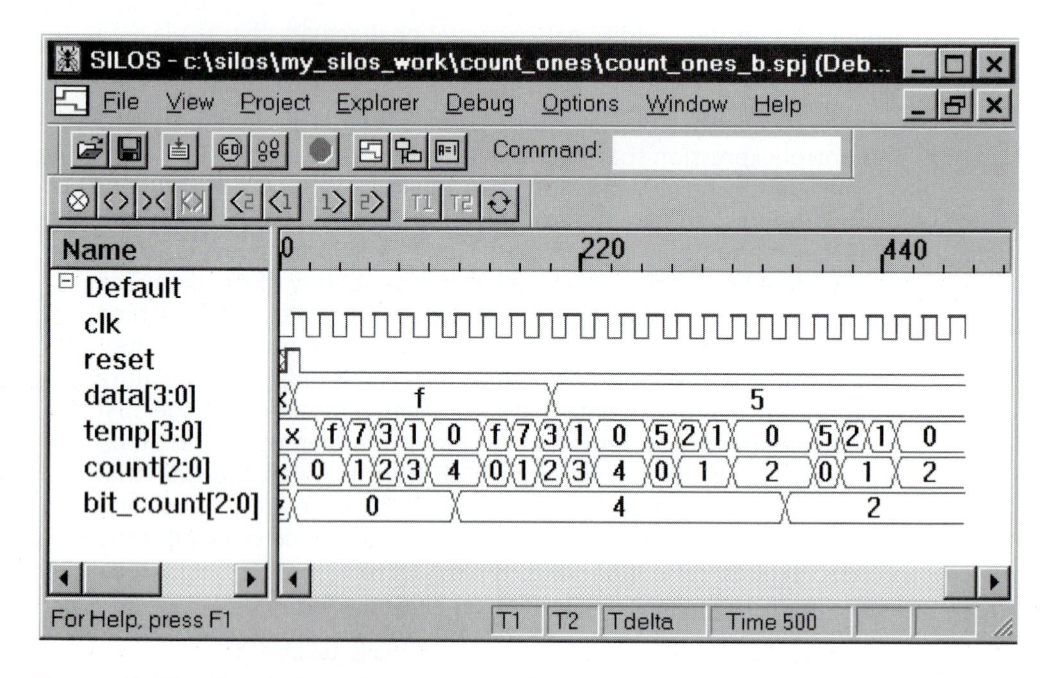

Figure 10.32 Simulation of *count_ones_b*.

10.10.3 NON-STATIC LOOPS WITHOUT INTERNAL TIMING CONTROLS.

The number of iterations to be performed by a loop having a data dependency cannot be determined before simulation. If the loop does not have internal timing control, the behavior can be simulated, but cannot be synthesized. Under the action of a simulator, the behavior is virtually sequential and can be simulated. But hardware cannot execute the computation of the loop in a single cycle of the clock.

Example 10.26 The computational activity of counting the 1s in a word of data is wasted after the last 1 is found. The data-dependent loop in *count_ones_c* implements a more efficient machine than *count_ones_b*. At an active edge of the clock, the machine loads *data* into *temp*, and counts the number of 1s in *temp* by adding the value of the LSB of *temp* to count, and shifting the word. This continues as long as the word is not empty of 1s (i.e., until the "reduction or" of *temp*, |*temp*, is false). So the data word 0001_2 will execute in fewer iterations than 1000_2. The computational activity occurs in a single cycle of the clock, so the efficiency would be apparent in simulation with long words of data. The simulation results in Figure 10.32 show the value of *index* at the end of the loop. Notice that final results are displayed (i.e., those at the end of the clock cycle of computation). ***While this behavior is attractive for simulation, it behavior cannot be synthesized.*** The task of counting the 1s in a word is fundamentally combinational. There is no combinational logic that can perform the sequential steps of the loop within one cycle of the clock, and at the same time terminate the activity if the word becomes empty of 1s. The loop cannot be unrolled because its length is data-dependent.

```
module count_ones_c (bit_count, data, clk, reset);
  parameter                          data_width = 4;
  parameter                          count_width = 3;
  output    [count_width-1:0]        bit_count;
  input     [data_width-1:0]         data;
  input                              clk, reset;
  reg       [count_width-1:0]        count, bit_count, index;
  reg       [data_width-1:0]         temp;

  always @ (posedge clk)
    if (reset) begin count = 0; bit_count = 0; end
    else begin
      count = 0;
      temp = data;
      for (index = 0; | temp; index = index + 1) begin
        if(temp[0]) count = count + 1;
        temp = temp >> 1;
      end
      bit_count = count;
    end
endmodule
```

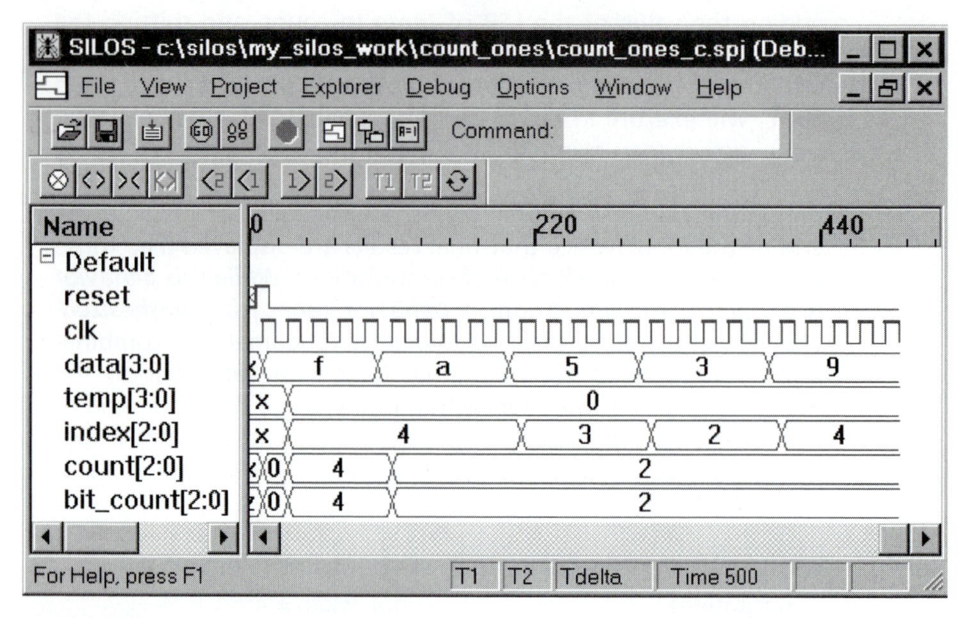

Figure 10.33 Simulation of *count_ones_c*, which has a data dependent loop, and cannot be synthesized.

10.10.4 Non-Static Loops With Internal Timing Controls

A non-static loop may implement a multi-cycle operation. The data dependency alone is not a barrier to synthesis, because the activity of the loop can be distributed over multiple cycles of the clock. The iterations of a non-static loop must be separated by a synchronizing edge-sensitive control expression in order to be synthesized.

Example 10.27 The cyclic behavior in *count_ones_d* has edge-sensitive timing controls within a non-static **while** loop. The sequential activity of the loop is distributed over multiple cycles of the clock. First, the data is loaded into a shift register. Then it shifts the data through the register on successive clock cycles. After all the data has been shifted, one more cycle elapses before *bit_count* is ready. Simulation results are presented in Figure 10.33. Notice that, when *data = 3*, the loop terminates after the second cycle.

```
module count_ones_d (bit_count, data, clk, reset);
  parameter                          data_width = 4;
  parameter                          count_width = 3;
  output      [count_width-1:0]      bit_count;
  input       [data_width-1:0]       data;
```

```
input                          clk, reset;
reg     [count_width-1:0]     count, bit_count;
reg     [data_width-1:0]      temp;

always @ (posedge clk)
  if (reset) begin count = 0; bit_count = 0; end
  else begin: bit_counter
    count = 0;
    temp = data;
    while (temp)
    @ (posedge clk)
      if (reset) begin
        count = 2'b0;
        disable bit_counter; end
      else begin
        count = count + temp[0]; // Notice missing if
        temp = temp >> 1;
      end
    @ (posedge clk);
      if (reset) begin
        count = 0;
        disable bit_counter; end
      else bit_count = count;
  end
endmodule
```

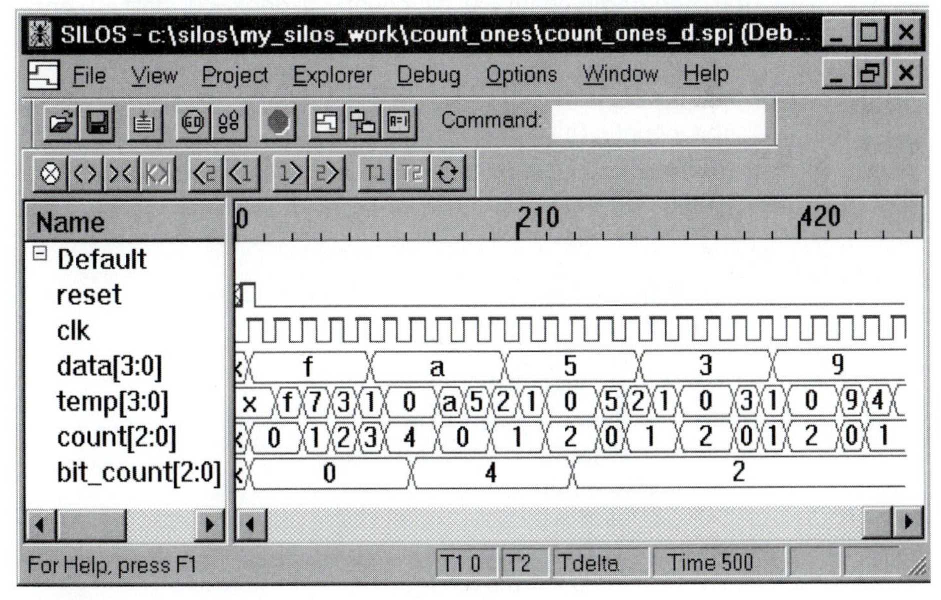

Figure 10.34 Simulation of *count_ones_d*.

A state machine can implement this functionality and include control logic that will terminate the loop if the shifted word is detected to be empty of 1s. One version of the machine includes 12 flip-flops—four to hold the data (*temp*), four to hold the value of count during the loop operation, and four to hold the value of *bit_count* data. (The practicality of the implementation is not the point here, just the illustration of style.)

The next version, *count_ones_HS*, adds *start* and *done* handshake signals to the port structure, and eliminates the data dependency in the loop. It simulates correctly.

```
module count_ones_HS (bit_count, start, done, data, clk, reset);
  parameter                          data_width = 4;
  parameter                          count_width = 3;
  output     [count_width-1:0]       bit_count;
  output                             start, done;
  input      [data_width-1:0]        data;
  input                              clk, reset;
  reg        [count_width-1:0]       count, bit_count, index;
  reg        [data_width-1:0]        temp;
  reg                                done, start;


  always @ (posedge clk) begin: bit_counter
    if (reset) begin count = 0; bit_count = 0; done = 0; start = 0;end
    else begin
      done = 0;
      count = 0;
      bit_count = 0;
      start = 1;
      temp = data;
      for (index = 0; index < data_width; index = index + 1)

        @ (posedge clk)              // Synchronize
          if (reset) begin
            count = 0;
            bit_count = 0;
            done = 0; start = 0;
            disable bit_counter; end
          else begin
            start = 0;
            count = count + temp[0];
            temp = temp >> 1;
          end
```

```
            @ (posedge clk) // Required for synchronization
              if (reset) begin count = 0; bit_count = 0; done = 0; start = 0;
                disable bit_counter; end
              else begin
                bit_count = count;
                done = 1; end
          end
        end
endmodule
```

Notice, in the simulation results in Figure 10.34, that *start* asserts at the first active edge of the clock for one cycle. *index* increments at each edge of the clock until all bits have been counted. Then *done* is asserted. When *reset* is asserted in the middle of a counting sequence, the machines re-initializes the registers and re-starts the sequence.

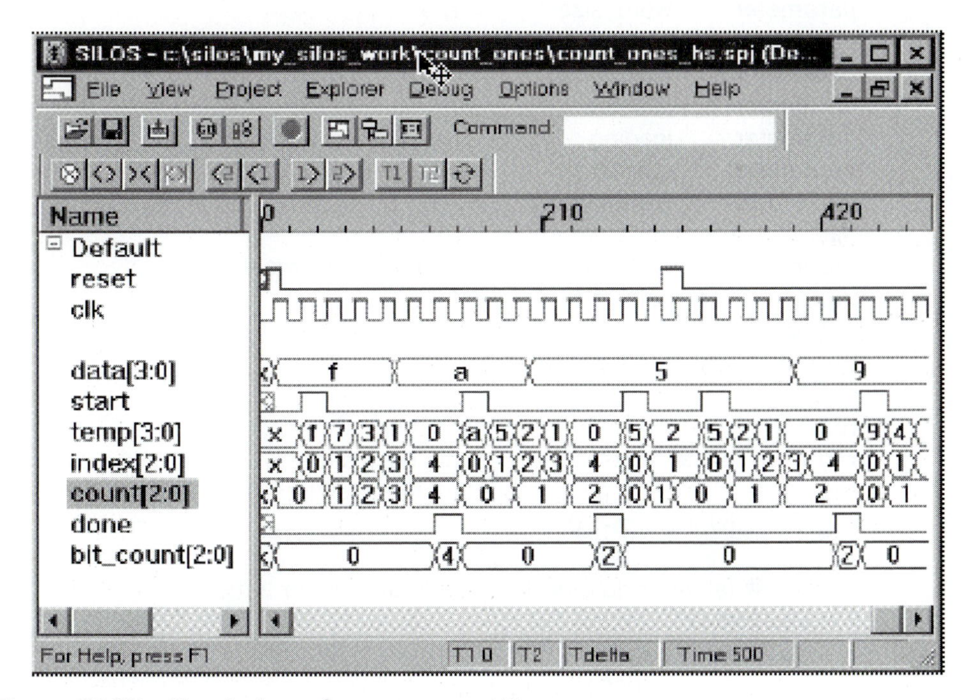

Figure 10.35 Simulation of *count_ones_HS*.

10.10.5 STATE MACHINE REPLACEMENTS FOR LOOPS

The previous examples have shown that many, but not all, forms of loops can be synthesized. Not all loop constructs are supported uniformly by vendors. The problem of limited support by a particular vendor can be avoided by replacing a loop structure with a finite state machine.

Example 10.28 The description in *count_ones_SM* is that of a state machine having the functionality of the ones counter. It circumvents the issue of whether a tool will synthesize a data-dependent loop. Also, a signal has been included, *data_ready,* to allow external control over whether the counting process begins. The ASM chart for the machine is shown in Figure 10.35, and simulation results are shown in Figure 10.36. The description *count_ones_SM* includes register transfers of *bit_count*, in addition to the state machine. Notice how counting does not begin until after register *temp* is loaded, and how the state remains in *waiting* until *data_ready* is de-asserted. (See Problem #14.)

```
module count_ones_SM (bit_count, data_ready, start, done, data, clk, reset);
  parameter     word_size      = 4;
  parameter     counter_size   = 3;
  parameter     state_size     = 2;
  parameter     idle           = 0;
  parameter     loading        = 1;
  parameter     counting       = 2;
  parameter     waiting        = 3;

  input                             data_ready, clk, reset;
  input     [word_size-1:0]         data;
  output                            start, done;
  output    [counter_size -1 : 0]   bit_count;

  reg                               bit_count;
  reg       [state_size-1 :0]       state, next_state;
  reg                               start, done, clear;
  reg       [word_size-1:0]         temp;

  always @ (state or data or data_ready or reset or temp)
    case (state)
      idle:       begin
                    done <= 0;
                    start <= 0;
                    clear <= 1;
```

```
                if (data_ready) next_state <= loading;
                else next_state <= idle; end

      loading:  begin
                start <= 1;
                clear <= 0;
                temp <= data;
                next_state <= counting; end

      counting: begin
                start <= 0;
                if (temp) next_state <= counting;
                else next_state <= waiting; end

      waiting:  begin
                done <= 1;
                if (data_ready == 0) next_state <= idle;
                else next_state <= waiting; end

      default:  begin
                done <= 0;
                start <= 0;
                clear <= 1;
                next_state <= idle; end
    endcase

  always @ (posedge clk) // state and register transfers
    if (reset) begin
      state <= idle;
      bit_count <= 0;
      temp <= 0; end
    else begin
      state <= next_state;
      bit_count <= clear ? 0 : bit_count + temp[0];
      temp <= temp >> 1;
    end
endmodule
```

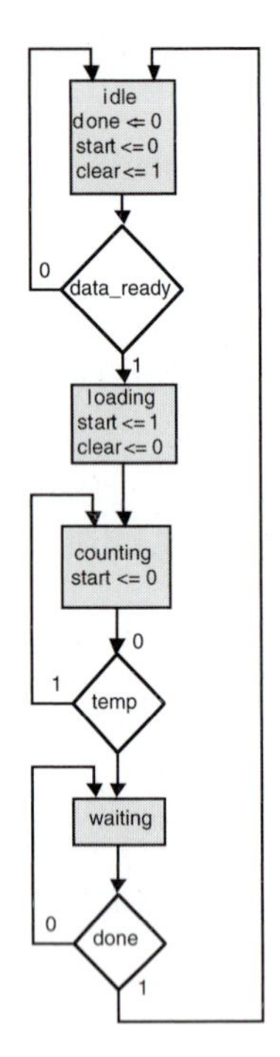

Figure 10.36 ASM chart of *count_ones_SM.*

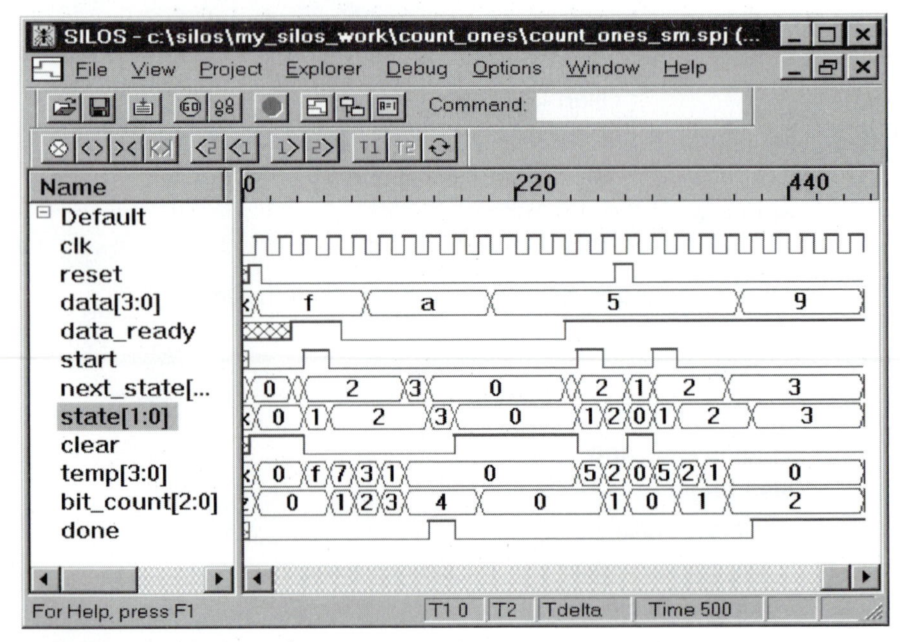

Figure 10.37 Simulation of *count_ones_SM.*

Notice that all the outputs were assigned value in state *idle*. Then only changes were assigned in subsequent states. This style makes the code more readable. Explicit state machines synthesize readily.

10.11 SYNTHESIS OF fork ... join BLOCKS

A Verilog **fork ... join** block describes parallel threads of activity in a behavior. Synthesis tools may either fail to support this construct, or require that it not contain event and delay controls that are equal to or longer than a clock cycle. These limitations effectively require that all the threads of activity complete within a single clock cycle. A synthesizable **fork ... join** description is equivalent to a behavior containing non-blocking assignments to register variables. The latter usage is recommended.

10.12 SYNTHESIS OF THE disable STATEMENT

The **disable** statement is used to disable a function, task or named block. There are two kinds of disables. External disables are located outside the task, function, or named block that they disable. Synthesis tools infer sequential logic when an external disable is used. An internal disable resides within the function, task, or named block that it disables. An internal disable is used to model the reset condition in an implicit finite state machine, or to model the external interrupt of a machine. A function that incorporates an external disable can have a race condition between the function itself and the external disable.

If a task is invoked concurrently from different behaviors, it may be disabled externally. In such a case, the synthesized state machine disables only the first invocation of the task. On the other hand, a simulator will disable all concurrent invocations of the task. This will lead to a mismatch between the simulation of the seed behavior and the synthesized state machine. If a task is invoked from different procedural blocks, self-disable should be used to ensure that all of the calls of the task are disabled.

Example 10.29 The implicit state machine *add_4cycle* adds four successive samples of a datapath. The machine implements a multicycle operation with internal disable.

```
module add_4cycle (sum, data, clk, reset);
  input    [3:0]  data;
  input           clk, reset;
  output   [5:0]  sum;
  reg             sum;

  always @ (posedge clk) begin: add_loop
```

```
        if (reset) disable add_loop; else
          sum = data;
            @ (posedge clk) if (reset) disable add_loop; else
              sum = sum + data;
                @ (posedge clk) if (reset) disable add_loop; else
                  sum= sum + data;
                    @ (posedge clk) if (reset) disable add_loop; else
                      sum = sum + data;

      end
    endmodule
```

For an additional example of internal disable, see Example 8.11.

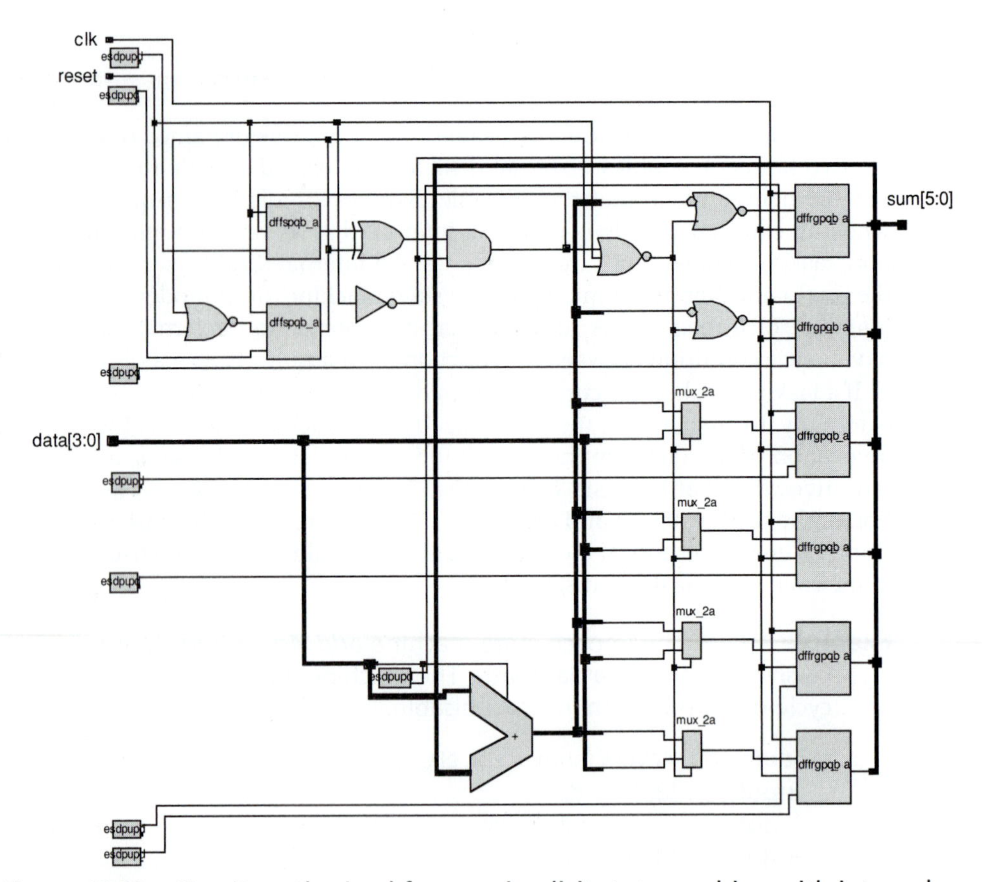

Figure 10.38 Circuit synthesized from an implicit state machine with internal disable.

10.13 SYNTHESIS OF USER-DEFINED TASKS

A user-defined task is defined within a module, and may reference variables within the module in which it is defined. Synthesis tools expand the task in-line. If multiple calls are made to the task, the synthesis tool may implement duplicate control logic. There is no mechanism for synchronizing multiple calls to the same task. A task that is to be synthesized may not contain **inout** ports. A task that does not contain event controls will synthesize to combinational logic.

Example 10.30 Module *adder_task* contains a user-defined task that adds two four-bit words and a carry bit. The circuit produced by the synthesis tool is shown in Figure 10.38.

```
module adder_task (c_out, sum, clk, reset, c_in, data_a, data_b, clk);
   input      [3:0]    data_a, data_b;
   input               clk, reset;
   input               c_in;
   output     [3:0]    sum;
   output              c_out;
   reg                 sum;
   reg                 c_out;

   always @ (posedge clk or posedge reset)
      if (reset) {c_out, sum} = 0; else
         add_values (sum, c_out, data_a, data_b, c_in);

   task add_values;
      output [3:0]     sum;
      output           c_out;
      input  [3:0]     data_a, data_b;
      input            c_in;

      reg              sum;
      reg              c_out;

      begin
         {c_out, sum} = data_a + (data_b + c_in);
      end
   endtask
endmodule
```

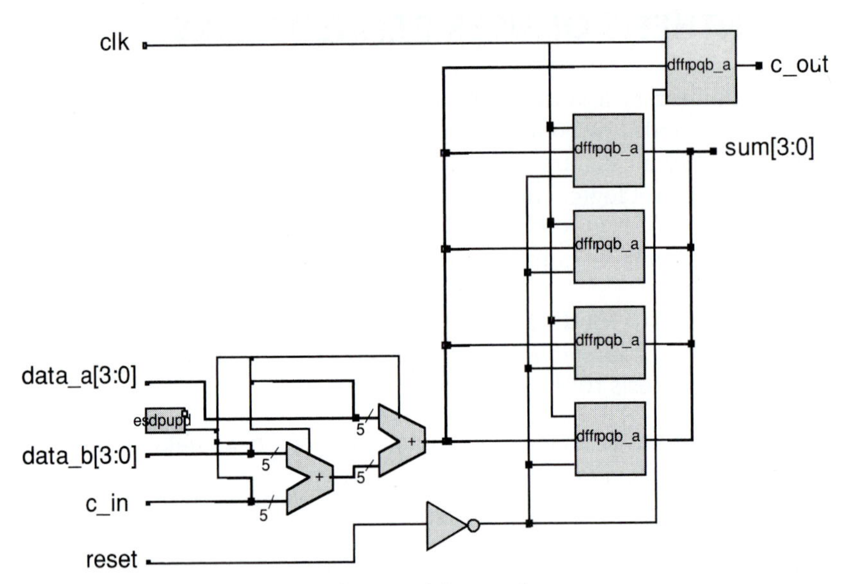

Figure 10.39 Circuit synthesized from *adder_task*.

10.14 SYNTHESIS OF USER-DEFINED FUNCTIONS

Functions are declared within modules. They bear names and may be scalars or vectors. The value of a function is returned through its name at the calling site (see Chapter 7). User-defined functions are always expanded at the site of their call. The code of a function will synthesize into combinational logic. Incomplete **case** and conditional (**if** ...) statements not are allowed within a function. This restricts user-defined functions to those that can be synthesized by combinational logic. Functions are ordinarily called in the RHS of an expression (e.g., in a continuous assignment or procedural statement). It is possible to preserve the hierarchy of a function and thereby make its implementation available for use in other areas of the design. If the hierarchy of a function is not preserved, the function can reference variables and make assignments to variables in the module from which it is called.

Example 10.31 The description of *arithmetic_unit* uses functions to describe a set of arithmetic operations. In general, function names tend to make the source code more readable. The combinational circuit synthesized from *arithmetic_unit* is shown in Figure 10.39.

```
module arithmetic_unit (operand_1, operand_2, result_1, result_2);
  input     [3:0]   operand_1, operand_2;
  output    [4:0]   result_1;
  output    [3:0]   result_2;

  assign result_1 = sum_of_operands (operand_1, operand_2);
  assign result_2 = largest_operand (operand_1, operand_2);

  function [4:0] sum_of_operands;
    input [3:0] operand_1, operand_2;
      sum_of_operands = operand_1 + operand_2;
  endfunction

  function [3:0] largest_operand;
    input [3:0] operand_1, operand_2;
    largest_operand = (operand_1 >= operand_2) ? operand_1 :
      operand_2;
  endfunction
endmodule
```

10.15 SYNTHESIS OF SPECIFY BLOCKS

Synthesis tools accomplish technology-independent optimization of the Boolean description of a behavior. Any **specify ... endspecify** block contained within a module is ignored by the synthesis tool during Boolean optimization.

10.16 SYNTHESIS OF COMPILER DIRECTIVES

Compiler directives are generally synthesized. In fact, tools use them to guide the synthesis engine. The vendor's documentation should be consulted to determine whether restrictions apply to certain directives.

10.17 SUMMARY

Most Verilog constructs can be synthesized. By exploring how a synthesis tool infers logic from a construct, a designer can adopt a style that is synthesizable and anticipate synthesis results. It is important that the mechanisms by which a tool distinguishes combinational from sequential logic be understood in order to obtain desired results. When a description implies the need for memory, a hardware latch or flip-flop will be synthesized. Incompletely specified case and conditionals lead to latches too. A priority decoder will be synthesized from an **if** statement or conditional assignment that does not have mutually exclusive conditions.

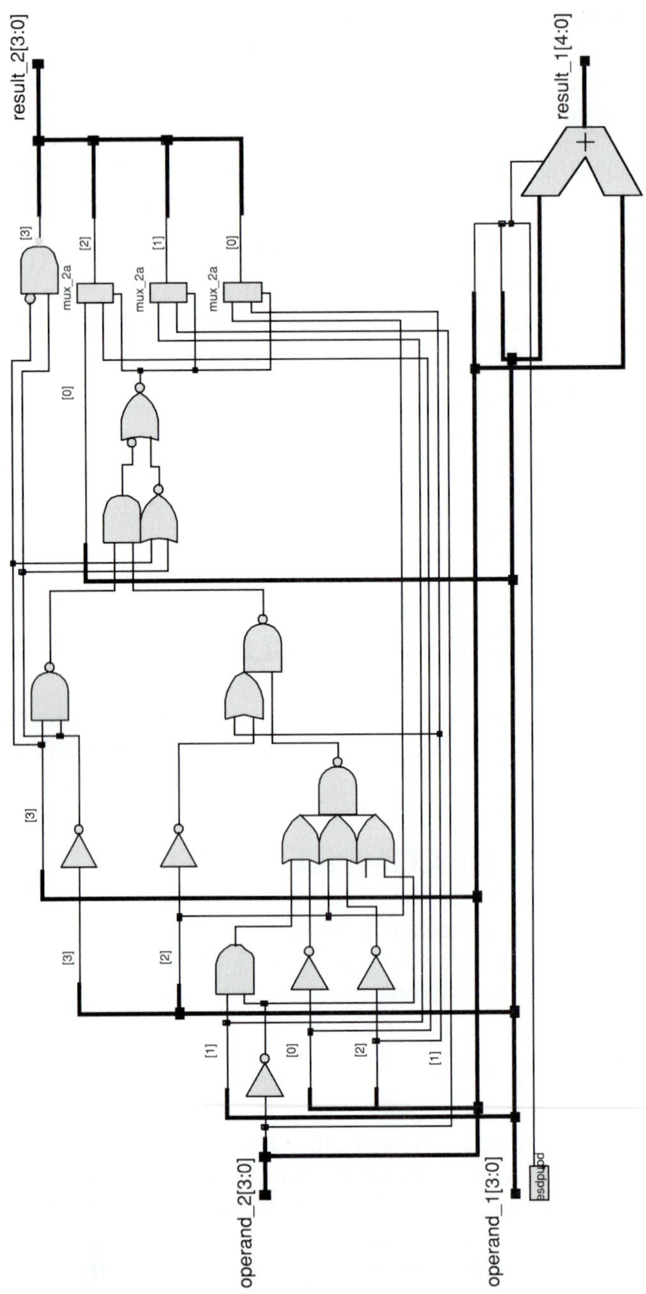

Figure 10.40 Circuit synthesized from an arithmetic unit having two declared functions.

REFERENCES

1. Smith, D.J., *HDL Chip Design*, Doone Publications, Madison, Alabama, 1996.

2. Weste, N.H.E. and Eshraghian, K., *Principles of CMOS VLSI Design*, Addison Wesley, New York, 1993.

PROBLEMS

1. A 8-bit serial adder was designed and synthesized as an explicit state machine in Problem 25 in Chapter 9. Write, verify, and synthesize (if possible) an implicit state machine version of the adder.

2. A pseudo code (not Verilog) description of an algorithm that will sort over a set of N_key 16-bit positive integers and sort them into ascending order is given below [See Knuth, *Sorting and Searching*, Addison Wesley, 1973]:

```
for i=2 to N_key
   begin
      for j = N_key downto i do
         if a[j-1] > a[j] then
         begin
            temp= a[j-1];
            a[j-1]=a[j];
            a[j]=temp
         end
end
```

   a. Develop a model of a sequential machine that will implement the algorithm.

   b. Develop a suitable testbench to verify that your sorter is implemented correctly, and demonstrate a sort on the set of integers below:

   703,765,677,612,509,154,426,653,275,897,170,908,061,512,087,503

   c. Synthesize your implementation.

   d. Now model, verify, and synthesize a sorting engine that will search over the N_key integers to find the largest, then remove it and repeat the search over the remaining N_key-1 words, etc., until the list is sorted. Demonstrate a sort on the same set of integers. Produce simulation results. Compare and discuss the relative merits of the two implementations.

3. Formulate a pipelining strategy for a magnitude comparator that compares two 8-bit words and forms outputs indicating whether the signals are equal, and, if not, indicates which word has the larger magnitude. Compare the performance of the comparators synthesized with and without pipelining.

4. Write, verify and synthesize a 16-bit adder-subtractor.

5. Explain why the code fragment below enters an endless loop when the simulation begins.

```
reg [3:0] k;
initial begin
for (k=0; k < 16; k = k+1)
   ab_data[k] = k;
end
```

6. Example 7.41 presented a model, *Add_prop_gen*, implementing carry look-ahead algorithm. What type of loop (e.g., non-static) does the algorithm implement? Synthesize the algorithm.

7. What will a synthesis tool produce from the following descriptions of behavior?

a.

```
always @ (posedge clock) begin
   b = a;
   c = b;
   d = e;
   f = e;
end
```

b.

```
always @ (posedge clock) begin
   b <= a;
   c <= b;
   d <= e;
   f <= e;
end
```

8. Draw the dataflow graphs corresponding to the pipeline structures below:

```
always @ (posedge clk)
  begin
    regc <= regb + 1;
    rega = data;
    regb = rega + 1;
  end
```

```
always @ (posedge clk)
  begin
    regc <= regb + 1;
    rega = data;
    regb <= rega + 1;
  end
```

9. What results can you anticipate from synthesizing the description below?

```
module alu_bad (alu_out, data_a,data_b, enable, opcode);
input     [2:0]   opcode;
input     [3:0]   data_a, data_b;
input             enable, clk;
output            alu_out;
reg       [3:0]   alu_reg

assign alu_out = (enable ==1) ? alu_reg : 4'b0;
  always @ (opcode)
  case (opcode)
    3'b001: alu_reg = data_a | data_b;
    3'b010: alu_reg = data_a ^ data_b;
    3'b101: alu_reg = data_a & data_b;
    3'b110: alu_reg = ~data_b;
  endcase
endmodule
```

10. Explain why a synthesis tool synthesizes *pattern_match* in *bogus_detector* as the output of a D-type flip-flop with its "data" input grounded.

```
module bogus_detector (serial_input, clk, pattern_match);
  input               serial_input, clk;
  output              pattern_match;
  reg                 pattern_match;
  reg       [2:0]     data_reg;
  parameter           pattern = 4'b0010;

  always @ (posedge clk) begin
    data_reg = data_reg << 1;
    data_reg [0] = serial_input;
    if ({data_reg, serial_input} == pattern)
      pattern_match = 1;
    else pattern_match = 0;
  end
endmodule
```

11. Write, verify, and synthesize a behavioral model for an ALU having the following opcodes:

Opcode	Operation	Operand1	Operand2
000	Pass	accumulator	
001	Pass	data	
010	Add	data	accumulator
011	Bitwise and	data	accumulator
100	Bitwise or	data	accumulator
101	Bitwise xor	data	accumulator
110	Bitwise xnor	data	accumulator
111	Bitwise complement		accumulator

The ALU has 8-bit inputs *data* and *accum*, 3-bit input *opcode*, input *clock*, an 8-bit output, and a *zero* bit that is asserted when *accum* is zero. The operation of the ALU is to be synchronized to the positive edge of *clock*. Partition your design into two behaviors: one that makes a synchronous assignment to the output, and a second that asynchronously determines the values that will be transferred to the output.

12. Model and synthesize a synchronous up-counter that is to assert at the count of 17. Compare the area and performance of two implementations: one with a serial configuration of registers in a one-hot configuration using 17 flip-flops, and another using a sequential counter having much fewer flip-flops.

13. Synthesize the following continuous assignments:

```
assign y1 = a + b + c + d;
assign y2 = ((a+b) + (c +d));
```

Compare the area and speed of the results for four-bit data paths.

14. Synthesize *count_ones_SM* in Example 10.28. Explain the role of each flip-flop and latch, and discuss its origin in the source code.

15. Using non-blocking assignments, write, verify, and synthesize a model equivalent to *m_cycle* in Example 10.21, but having a single event control expression.

CHAPTER 11

SWITCH-LEVEL MODELS IN VERILOG

The previous chapters presented gate-level and behavioral models describing the functionality of digital circuits. These models encapsulate functionality, but suppress the underlying transistor-level detail of the circuit. Although synthesis methodology exploits such models, the Verilog language also has constructs for modeling functionality at the switch level. Switch-level models of circuits are more accurate than gate-level models, but are less accurate than analog transistor models used in analog simulation.[1] Switch-level models also simulate faster than analog models. Switch-level models can be used to represent and simulate the behavior of individual transistors in a digital circuit. Table 11.1 lists and classifies the predefined structural primitives supported by Verilog. The multi-input, multi-output, and three-state gates have been discussed in previous chapters. Synthesis tools do not accommodate transistor switches and bi-directional switch models. The "pull" gates are used by synthesis tools to tie-off control signals of flip-flops to power or ground. This chapter will briefly examine the underlying MOS technology supporting the remaining primitives, and then consider their functionality in circuit models using the switch-level primitives in Table 11.1.

11.1 MOS TRANSISTOR TECHNOLOGY

The dominant technology in many of today's digital circuits is one that uses a pair of complementary metal-oxide-silicon (CMOS) transistors to implement combinational and sequential devices in logic circuits. CMOS logic circuits use two types of metal-oxide-silicon (MOS) transistors: n-channel (nmos) and p-channel (pmos). Both types of transistors are fabricated on a common silicon

Table 11.1 Gate- and switch-level primitives supported by Verilog.

COMBINATIONAL LOGIC GATES			SWITCH-LEVEL PRIMITIVES		
Multi-input gates	Multi-output gates	Three-state gates	MOS transistor switches	MOS Pull gates	MOS Bi-directional switches
and	buf	bufif0	nmos	pullup	tran
nand	not	bufif1	pmos	pulldown	tranif0
or		notif0	cmos		tranif1
nor		notif0	rnmos		rtran
xor			rpmos		rtranif0
xnor			rcmos		rtranif1

substrate, allowing a single chip to contain in excess of a million transistors isolated and interconnected to implement complex digital functionality. Figure 11.1 shows a simplified side view of the physical structure of an n-channel and p-channel transistor fabricated in a common p-type silicon substrate. In this technology, the p-channel transistor is formed within an implanted n-well of material doped with an n-type dopant serving as a surrogate background replacing the substrate.

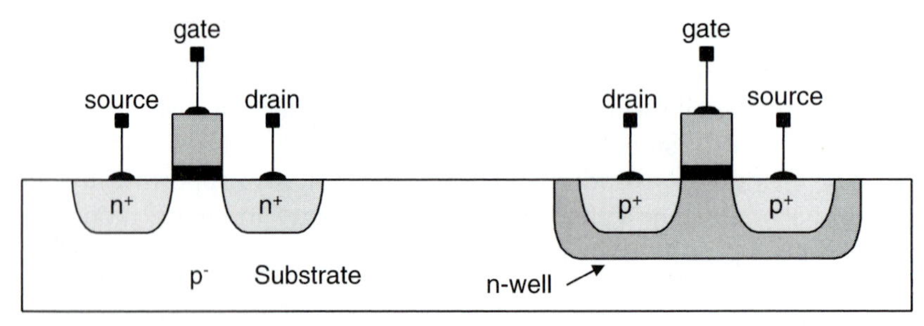

Figure 11.1 Simplified side views of n-channel and p-channel MOS transistors.

Unlike bipolar transistors,[2] MOS transistors are majority carrier devices—they have only one type of charge carrier for current. A charge "hole" is the majority carrier in a p-channel device, and an electron is the majority carrier in an n-channel device. A hole has an effective positive charge of $-q_e$, where q_e is the charge of an electron. Figure 11.2 shows schematic symbols for nmos and pmos transistors. The three terminals of the device are labeled S (source), D (drain), and G (gate). The region directly under the gate of the transistor lies between two doped source and drain regions, and is called the "channel" of the transistor. In the absence of proper voltages applied to the terminals of the device, the channel does not provide a path for current between the doped regions. Under proper biasing conditions, the gate terminal controls the flow of current through the device.

The doping of each type of transistor during fabrication determines whether it operates as an "enhancement-mode" or "depletion-mode" device.[3] An n-channel enhancement-mode device is normally biased to be "off" when the voltage between the gate and source is "0". On the other hand, a properly-biased depletion-mode device is normally "on" when the gate signal is "0". In fact, a negative voltage must be applied to the gate to suppress current. Similarly, an enhancement-mode p-channel transistor is normally off when the gate-to-source voltage is "0"; a sufficiently negative gate-to-source voltage turns the transistor "on". A depletion-mode p-channel transistor is normally on when the gate-to-source voltage is "0", and a sufficiently positive voltage must be applied to turn the transistor "off". The transistor symbols in Figures 11.1 and 11.2 are for enhancement-mode devices.

An enhancement mode MOS transistor conducts when the its gate-to-source voltage is sufficiently large and of the proper polarity relative to the device's threshold voltage. For n-channel devices, the condition can be expressed as $v_{gs} > v_{tn} > 0$ with $v_{ds} > 0$. An enhancement mode p-channel device has a negative threshold voltage, and the gate-to-source voltage must satisfy the condition $v_{gs} < v_{tp} < 0$ with $v_{ds} < 0$ (v_{tn} and v_{tp} are device threshold voltages). The gate terminal controls the operation of the device. When the gate-source voltage exceeds the transistor's threshold voltage (for nmos), the source-drain path becomes a conducting channel for current between the source and drain. In this condition, the channel is said to be "inverted" because a conducting sheet of majority carriers is drawn to the surface from the substrate by a sufficiently strong, properly polarized gate voltage. The majority carrier of either type of device flows from the source terminal to the drain terminal when the transistor conducts. The gate is insulated from the source and drain terminals by the gate dielectric material, so the only current is between the source and drain terminals. Both nmos and pmos types of transistors are used in CMOS logic circuits.

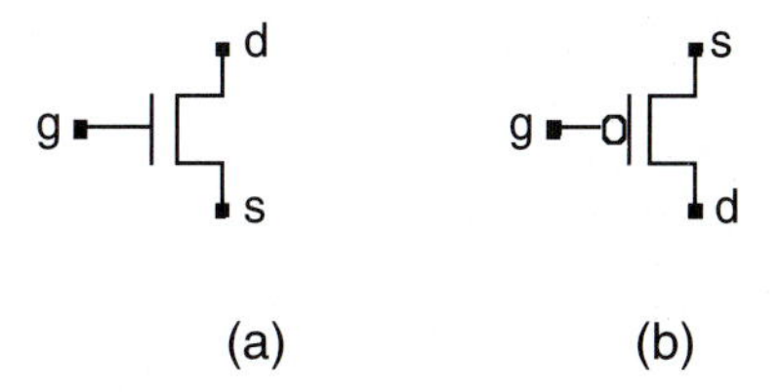

Figure 11.2 Schematic symbols for (a) nmos and (b) pmos transistors.

11.2 SWITCH-LEVEL MODELS OF MOS TRANSISTORS

Four Verilog primitives model the behaviors of individual p-channel and n-channel MOS transistors: **pmos**, **nmos**, **rnmos**, and **rpmos**. Each primitive models a unidirectional signal flow from an input terminal to an output terminal, but **rnmos** and **rpmos** primitives model physical devices having a significantly higher resistance when conducting, compared to the **pmos** and **nmos** devices. The effect of the higher resistance will be considered in a later section discussing signal strength reduction.

The input-output logic tables for the **nmos**, **pmos**, **rnmos**, and **rpmos** primitives are shown in Table 11.2. These switch-level primitives follow the syntax convention of gate-level primitives by having the output as the first terminal element, followed by the input and the control signal. When the control input is "x" or "z", the output of the primitive is either "L" or "H", depending upon whether the input is "0" or "1", respectively. In these cases, the control input is either turning the device on, turning it off, or is itself disconnected ("z"). The output could be the value of the data, or it could be "z". For example, when the input is "0" and the control is "x", the output is "L" (i.e., "0" or "z"). The symbol "L" represents "0" or "z", and the symbol "H" represents "1" or "z". Simulation cannot resolve this ambiguity of the output.

Table 11.2 Input-output relationships for switch-level MOS transistor models.

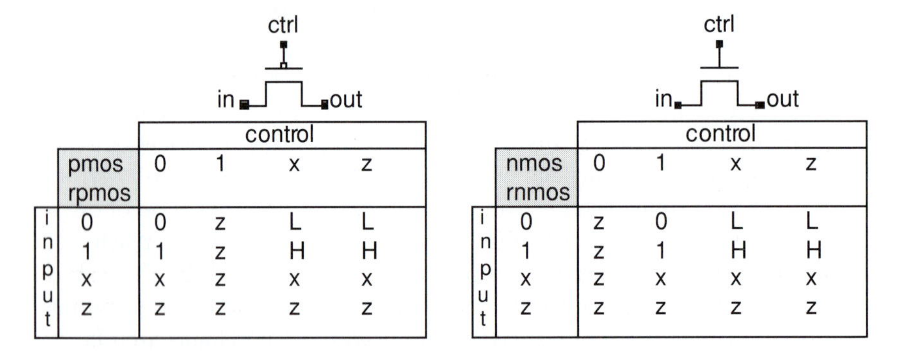

input	pmos / rpmos	control 0	control 1	control x	control z
	0	0	z	L	L
	1	1	z	H	H
	x	x	z	x	x
	z	z	z	z	z

input	nmos / rnmos	control 0	control 1	control x	control z
	0	z	0	L	L
	1	z	1	H	H
	x	z	x	x	x
	z	z	z	z	z

11.3 SWITCH-LEVEL MODELS OF STATIC CMOS CIRCUITS

Static CMOS circuits[4, 5] implement pull-up logic with p-channel transistors and pull-down logic with n-channel transistors. The simplest static CMOS circuit is the inverter shown in Figure 11.3a. A single p-channel transistor forms the pull-up logic, and a single n-channel transistor forms the pull-down logic. The drains of the two transistors are attached together in a "common drain" configuration at the output node. When the gate signal is a logical "zero", the p-channel transistor is turned on (conducting, because $v_{gsp} < v_{tp} < 0$), and the n-channel transistor is turned off (because $v_{gsn} < v_{tn}$). Charge is driven from V_{dd} to the output node through the p-channel pull-up path in Figure 11.3b. When the input signal is "1", the reverse happens. The n-channel transistor is turned "on" and the p-channel transistor is turned "off". Charge stored on the output node (capacitor) is discharged through the n-channel transistor to ground via the pull-down path shown in Figure 11.3c.

In static CMOS technology, the p-channel and n-channel transistors are turned on by complementary signal values. Therefore, the output has no DC path between the power and ground rails, and consequently the static power dissipation of the circuit is zero. When either transistor is turned on, it has a very low source-to-drain resistance, and acts like a short circuit. When the transistor is turned off, its source-to-drain resistance is very high, and the device acts like an open circuit. Current flows only during switching of the input signal, when the load capacitor is either charging or discharging. Thus, static CMOS circuits only consume power during switching (i.e., transient power).

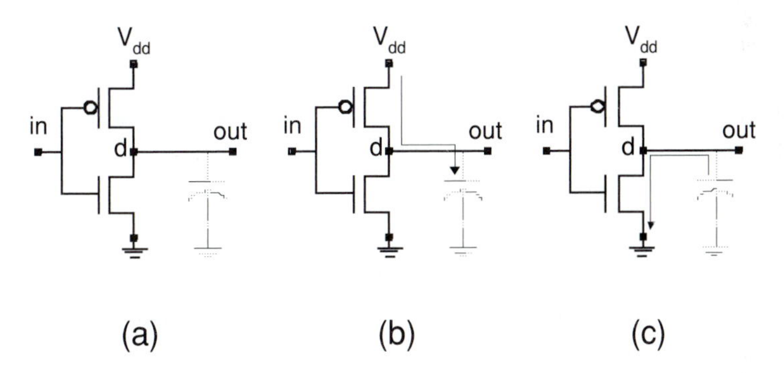

Figure 11.3 Transistor level schematic of (a) a CMOS inverter, (b) with pull-up and (c) pull-down paths for current shown.

Example 11.1 The Verilog model of the static CMOS inverter shown in Figure 11.3 is given below. The model instantiates an **nmos** and **pmos** primitive. Note that *inverter_input* is common to both devices, but the source of the **pmos** transistor is connected to *PWR* and the source of the **nmos** transistor is connected to *GND*. The common-drain output signal is *inverter_output*.

```
module cmos_inverter (inverter_output, inverter_input, gate_signal);
  output     inverter_output;
  input      inverter_input;
  supply0    GND;
  supply1    PWR;
```

```
    pmos        (inverter_output, PWR, inverter_input);
    nmos        (inverter_output, GND, inverter_input);
endmodule
```

Note that the port order is drain, source, gate. The common-drain net has two drivers, but the complementary action of the transistors assures that the net will not be driven by conflicting logic values.

Example 11.2 A three-input switch-level static CMOS nand gate is described below. The schematic in Figure 11.4 has three p-channel transistors connected in parallel to form the pull-up logic, and three connected in series to form the pull-down logic.

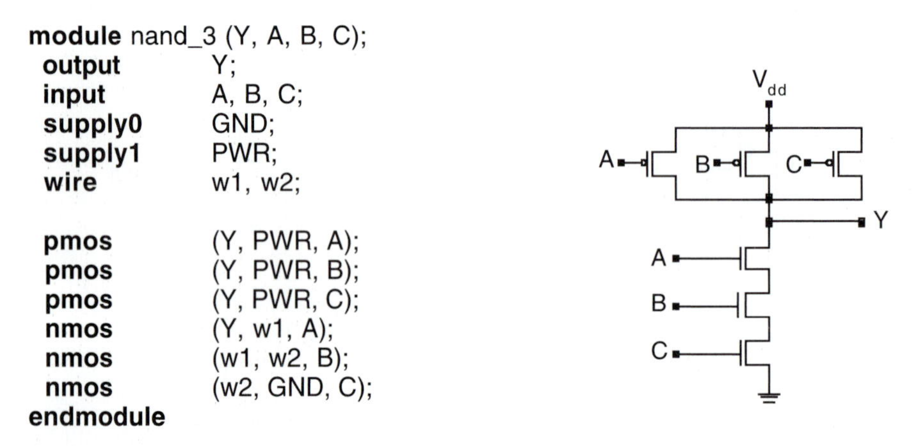


```
module nand_3 (Y, A, B, C);
 output     Y;
 input      A, B, C;
 supply0    GND;
 supply1    PWR;
 wire       w1, w2;

 pmos       (Y, PWR, A);
 pmos       (Y, PWR, B);
 pmos       (Y, PWR, C);
 nmos       (Y, w1, A);
 nmos       (w1, w2, B);
 nmos       (w2, GND, C);
endmodule
```

Figure 11.4 Verilog model and transistor level schematic of a three-input static CMOS nand gate.

11.4 ALTERNATIVE LOADS AND PULL GATES

The silicon area required to implement a given combinational logic function can be reduced by replacing the transistors of the pull-up logic by a single transistor whose gate is connected to its drain to keep the transistor turned "on". This allows logic to be fabricated with a single type of transistor (e.g., nMOS). Fabricating all of the transistors on a chip to be a single type not only reduces the silicon area; it also simplifies the process to fabricate the transistors. However, these alternative technologies suffer the penalty of having DC power consumption, which eliminates them from applications that must rely on a battery for the primary source of power. There are, however, other important applications where this technology is appropriate, as in programmable logic arrays (PLAs).[5]

In nMOS technology, n-channel devices form the pulldown logic, and a single nMOS transistor forms the pull-up. If the pull-up transistor is a depletion-mode transistor, its gate is connected to its source (of majority carriers, i.e., electrons), as shown in Figure 11.5a. If it is an enhancement-mode transistor, its gate is connected to its drain, as in Figure 11.5b. In either configuration, the pull-up transistor acts like a resistor (see Figure 11.5b). Such circuits have reduced area and switch quickly, but dissipate a significant amount of DC power.

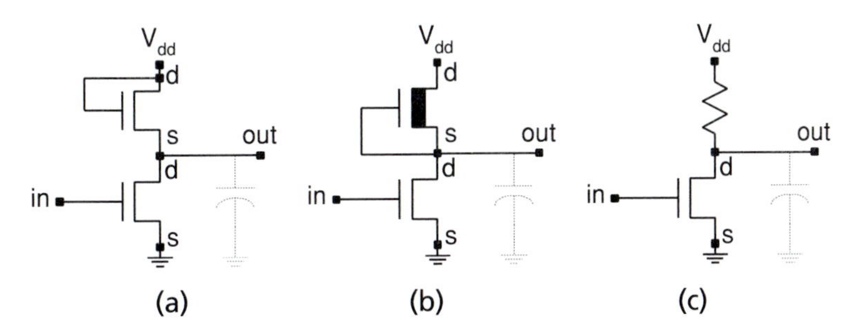

Figure 11.5 (a) Enhancement-mode, (b) depletion-mode, (c) and resistive pull-up devices in nMOS technology.

Two Verilog primitives model pull-up and pull-down functionality: **pullup** and **pulldown**. Both primitives have a single port element—the net that is to be pulled to "1" or "0". Depletion/enhancement pull-up and pull-down devices (pull gates) are also referred to as sources because they place a logic "1" (**pullup**) or a logic "0" (**pulldown**) on the single net in their terminal list, provided that the net has no other driver. The pull-up (pull-down) gate has a predefined strength of **pull1** (**pull0**). Note: The **pulldown** and **pullup** gates are not to be confused with **tri0** and **tr1** nets. The latter provide structural connectivity; the former are functional elements in the design. Synthesis tools sometimes use pull-up and pull-down gates to tie off unused set/reset inputs on flip-flops.

Example 11.3 Figure 11.6 shows the transistor-level schematic and the Verilog switch-level model of a two-input and gate using nmos pull-down logic, and a depletion load pull-up device. The instantiation of the pullup device has a single port, the net that is pulled to the power rail. The other terminals of the physical transistor are implicitly connected

11.5 CMOS TRANSMISSION GATES

The static CMOS inverter connects a p-channel transistor in *series* with an n-channel transistor in a common-drain configuration. The *parallel* configuration of a p-channel and n-channel transistor forms a transmission gate. Two Verilog

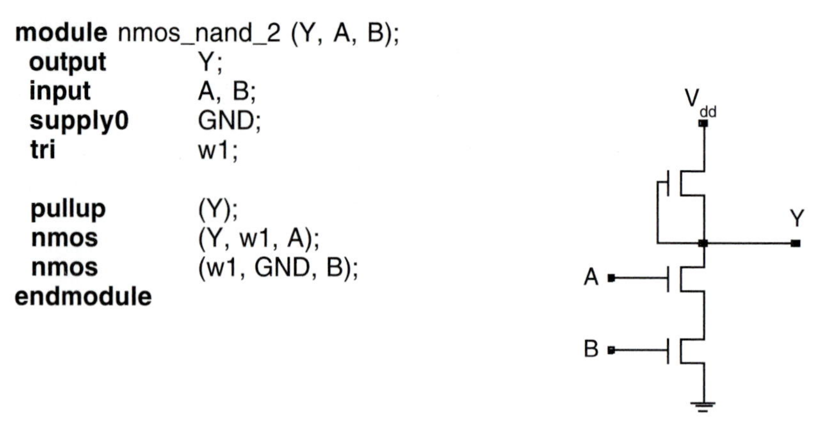

```
module nmos_nand_2 (Y, A, B);
  output       Y;
  input        A, B;
  supply0      GND;
  tri          w1;

  pullup       (Y);
  nmos         (Y, w1, A);
  nmos         (w1, GND, B);
endmodule
```


Figure 11.6 nMOS implementation of a two-input nand gate with a depletion-mode pull-up device.

primitives implement the functionality of a CMOS transmission gate: **cmos** and **rcmos**. The **rcmos** primitive is a resistive version of the **cmos** primitive.

A CMOS transmission gate passes a signal data value when it is turned on, and acts like an open circuit when it is turned off. Figure 11.7 shows the circuit symbol and transistor level schematic for a CMOS transmission gate. A complementary pair of control signals (*enable* and *~enable*) are connected to the gates of the transistor pair. This ensures that both transistors will be turned "off", or both will be turned "on". The gate has four terminal elements: the first connects to the data output, the second connects to the data input, the third connects to the gate of the n-channel transistor (enable), and the last element (~enable) connects to the gate of the p-channel transistor.

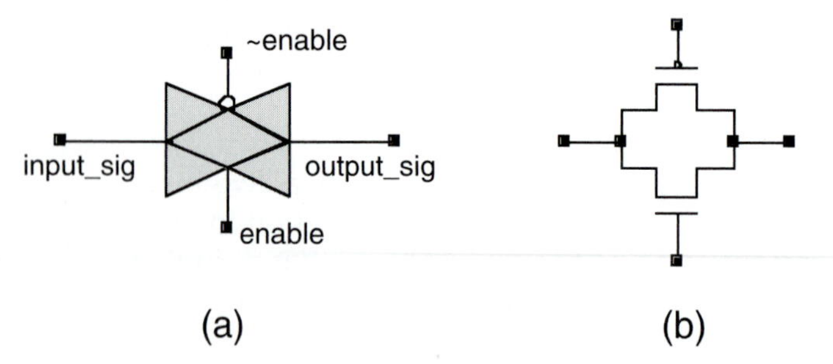

Figure 11.7 (a) Transistor schematic and (b) schematic symbol for a CMOS transmission gate.

The signal transmission characteristic of a CMOS transmission gate is symmetrical. A logic "0" and a logic "1" are passed in either direction without compromising the voltage levels and noise margins of the circuit.[4]

Example 11.4 The functionality of a CMOS transmission gate can be implemented directly with the **cmos** primitive, or by a switch-level structure with the **nmos** and **pmos** primitives. The Verilog description *Tgate_str* below implements a transmission gate with **nmos** and **pmos** primitives, and matches the terminal structure of the **cmos** primitive.

```
module Tgate_str (data_in, data_out, n_enable, p_enable);
  input data_in, n_enable, p_enable;
  output data_out;

  pmos (data_out, data_in, p_enable);
  nmos (data_out, data_in, n_enable);
endmodule
```

In practice, *n_enable* and *p_enable* are a pair of complementary signals, so both transistors are either "on" or "off". An alternative implementation of a CMOS transmission gate would use the **rpmos** and **rnmos** primitives to implement the parallel transistors.

Example 11.5 A master-slave D flip-flop with active-low clear can be modeled at the switch-level with transmission gates connected in the configuration shown in Figure 11.8.

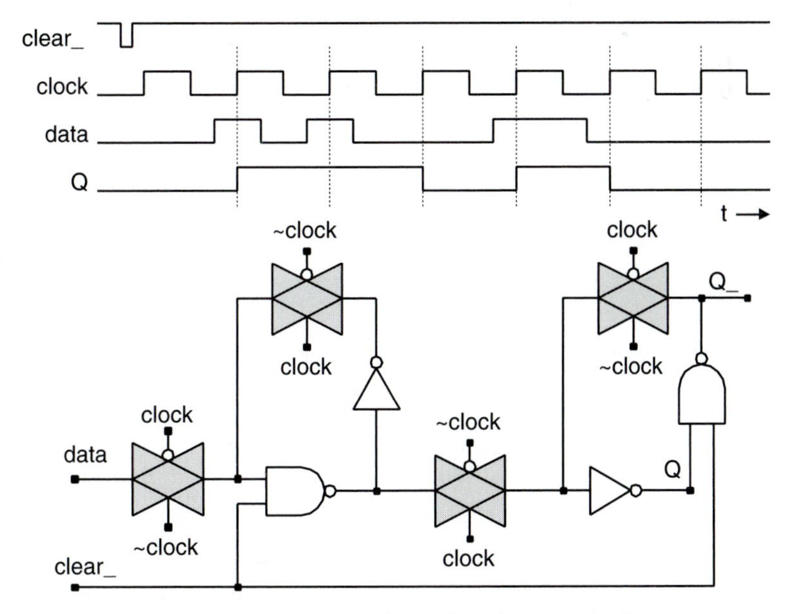

Figure 11.8 Transmission gate model of a D flip-flop with clear.

The action of the flip flop is illustrated in Figures 11.9a and b; a Verilog description and testbench are given in Figure 11.9c; and results of a simulation (with strengths) are shown in Figure 11.9d. The gate primitives were instantiated with unit delays to reveal the master-slave action of the two stages of the flip-flop, and additional signals were inserted to describe multiple drivers at *w1* and *w2*. In Figure 11.9a, notice the relationship between *data*, *w01*, and *w1*, and between *w2*, *w04*, and *w4*. The model of the flip-flop, *tgate_dflop*, has two drivers on the nets *w1* and *w4*. For example, observe how *w1* is driven by *data* through the instantiated forward path transmission gate *(cmos (w1, w2, clock, clock_bar))* and by the continuous assignment (**assign** *w1* = *w01*), with *w01* being the output of the transmission gate in the feedback path. Also notice the movement of data from the master stage to the slave stage.

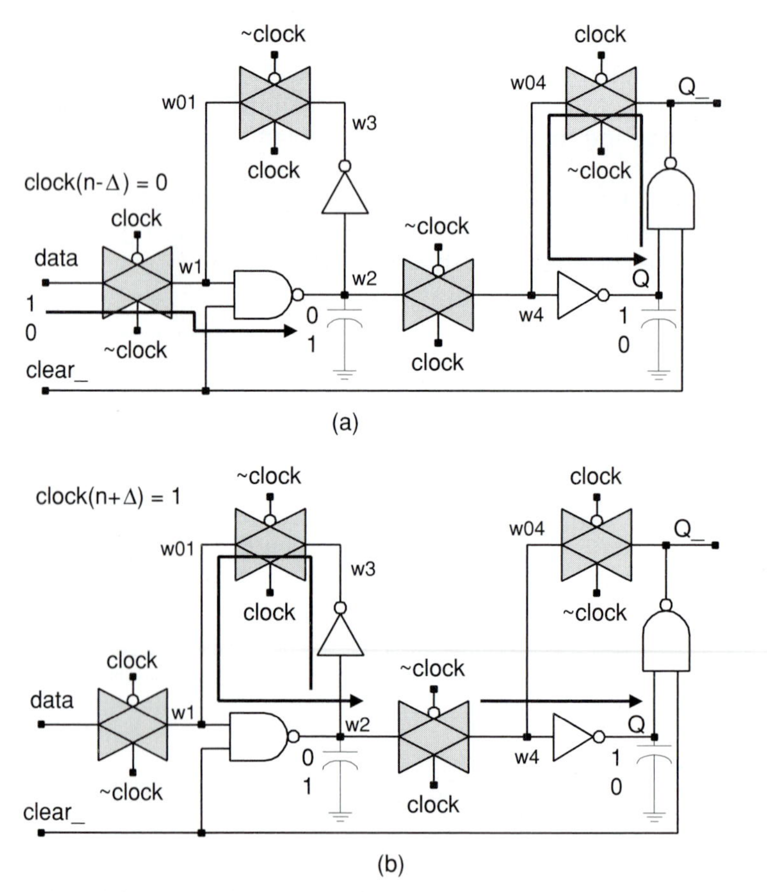

Figure 11.9 Signal flow, testbench, and simulation results for a master-slave D flip-flop. Continues on pages 505–508.

```
module test_dflop();
  reg data, clock, clear_;
  wire clock_, Q, Q_;
  parameter cl_half_cycle = 10;
  assign clock_ = ~clock;

   tgate_dflop M1 (Q, Q_, data, clock, clock_, clear_);

  task clock_gen;
    begin
      #cl_half_cycle clock = 0;
      #cl_half_cycle clock = 1;
    end
  endtask

  task setup_monitor;
    begin
      $display (...                Time   ");
      $display ("                  data   ");
      $display ("                    clear   ");
      $display ("                      clock   ");
      $display ("                        Q   ");
      $display ("                          Qbar   ");
       //$monitor ($time,, data,,,clear,,,clock,,, Q,,, Qbar);
      //$display ("Done with setup_monitor task");
    end
  endtask

  initial #500 $stop;

  initial
    begin
      setup_monitor;
    end
  initial
    begin
      $monitor ($time,",, %v,,,%v_,,%v,,, %v,,, %v_",
      data, clear_,clock,Q,Q_);  // %v displays signal strength
      #5 $display ("Done setting up monitor");
      #10 clock = 0;
      #5 clear_ = 1;
    #10 repeat(5) clock_gen;
    #10 clear_ = 0;
    #10 repeat(5) clock_gen;
     end
```

(c) continues

Figure 11.9 Continued.

```
    initial
      repeat (20) begin
        #30 data = 0;
        #30 data = 1;
      end
  endmodule

  module tgate_dflop (Q, Q_ , data, clock, clock_, clear_);
    input data, clock, clock_, clear_ ;
    output Q, Q_;
    wire w01, w1, w2, w3, w04,w4;
    assign w1 = w01;
    assign w4 = w04;
    cmos (w1, data, clock_, clock);
    nand #1 (w2, w1, clear_);
    not #1 (w3, w2);
    cmos (w01, w3, clock, clock_);
    cmos (w04, w2, clock, clock_);
    not #1 (Q, w4);
    nand #1 (Q_, Q, clear_);
    cmos (w4, Q_ , clock_, clock);
  endmodule
```

(c)

```
SILOS III  Version 97.1

Copyright (c) 1997 by SIMUCAD Inc.   All rights reserved.
No part of this program may be reproduced,   transmitted,
transcribed,   or stored in a retrieval system,    in any
form or by any means without the prior written consent of

SIMUCAD Inc., 32970 Alvarado-Niles Road, Union City,
                  California, 94587, U.S.A.
             (510)-487-9700  Fax: (510)-487-9721
          Electronic Mail Address:   "silos@ simucad.com"

input c:\silos\my_silos_work\sw_level\tgate_dflop.v
Reading "c:\silos\my_silos_work\sw_level\tgate_dflop.v"
Ready: sim to 0
  Highest level modules (that have been auto-instantiated):
    (test_dflop test_dflop
  11 total devices.
```

(d) continues

Figure 11.9 Continued.

```
  Linking ...
  13 nets total: 21 saved and 0 monitored.
  68 registers total: 68 saved.
  Done.
                Time
                    data
                          clear
                                clock
                                       Q
                                              Qbar
                0,, StX,,,StX_,,StX,,,  PuX,,, PuX_

  0 State changes on observable nets.

  Simulation stopped at the end of time 0.
Ready: sim
Done setting up monitor
               15,, StX,,,StX_,,St0,,, PuX,,, PuX_
               20,, StX,,,St1_,,St0,,, PuX,,, PuX_
               30,, St0,,,St1_,,St0,,, PuX,,, PuX_
               50,, St0,,,St1_,,St1,,, PuX,,, PuX_
               51,, St0,,,St1_,,St1,,, St0,,, PuX_
               52,, St0,,,St1_,,St1,,, St0,,, St1_
               60,, St1,,,St1_,,St0,,, St0,,, St1_
               70,, St1,,,St1_,,St1,,, St0,,, St1_
               71,, St1,,,St1_,,St1,,, St1,,, St1_
               72,, St1,,,St1_,,St1,,, St1,,, St0_
               80,, St1,,,St1_,,St0,,, St1,,, St0_
               90,, St0,,,St1_,,St1,,, St1,,, St0_
              100,, St0,,,St1_,,St0,,, St1,,, St0_
              110,, St0,,,St1_,,St1,,, St1,,, St0_
              111,, St0,,,St1_,,St1,,, St0,,, St0_
              112,, St0,,,St1_,,St1,,, St0,,, St1_
              120,, St1,,,St1_,,St0,,, St0,,, St1_
              130,, St1,,,St1_,,St1,,, St0,,, St1_
              131,, St1,,,St1_,,St1,,, St1,,, St1_
              132,, St1,,,St1_,,St1,,, St1,,, St0_
              140,, St1,,,St0_,,St1,,, St1,,, St0_
              141,, St1,,,St0_,,St1,,, St1,,, St1_
              142,, St1,,,St0_,,St1,,, St0,,, St1_
              150,, St0,,,St0_,,St1,,, St0,,, St1_
              160,, St0,,,St0_,,St0,,, St0,,, St1_
              170,, St0,,,St0_,,St1,,, St0,,, St1_
              180,, St1,,,St0_,,St0,,, St0,,, St1_
              190,, St1,,,St0_,,St1,,, St0,,, St1_
```

(d) continues

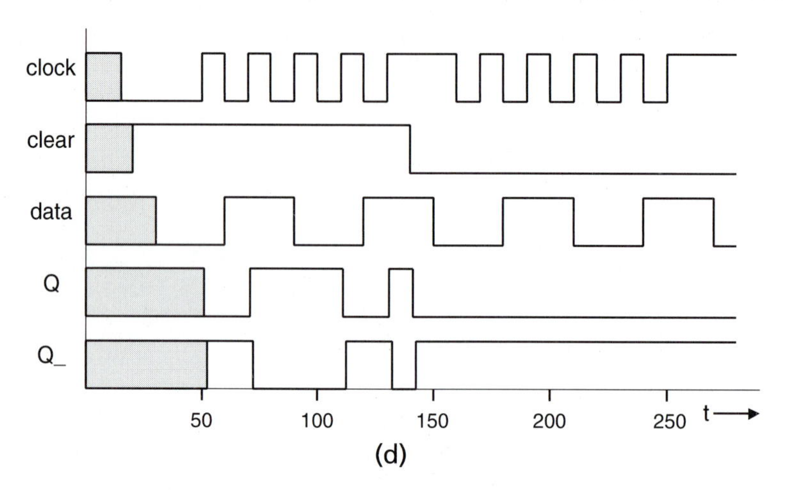

Figure 11.9 Signal flow, testbench, and simulation results for a master-slave D flip-flop. Concluded.

When *clock* is low (see Figure 11.9a), the forward-path transmission gate of the master stage charges the capacitive node output of the nand gate to hold the inverted value of *data* as *w2*. During this phase of the clock, the feedback-path transmission gate of the slave stage supports a self-sustaining loop holding the value of *Q*, while the entry transmission gate of the slave stage isolates it from the master stage. When *clock* goes high (see Figure 11.9b), the entry-path transmission gate isolates the master stage from *data*, and the feedback-path transmission gate supports a self-sustaining loop holding the inverted value of *data* at the output of the nand gate (*w2*). The entry-path transmission gate of the slave stage passes the inverted value of *data* to the inverter of the slave stage to form *Q*, and the feedback-path transmission gate is an open switch (blocking signal flow). Note that the action of *clear* is to cause *Q* = 0 and ~*Q* =1, independently of *clock*.

The listed simulation results show how *w01* and data drive *w1*. When *clock* is low, *w01* presents a high impedance to *w1*, allowing *w1* to be driven by *data*.

11.6 BI-DIRECTIONAL GATES (SWITCHES)

Verilog includes two-terminal and three-terminal bi-directional transmission gates. Bi-directional gates pass a signal value from one terminal to another. The two-terminal primitives, **tran** and **rtran**, have two bi-directional terminals. The mode of either terminal may be declared as an **input** or **inout**, and the mode of the other terminal may be declared as an **inout** or **output**. Figure 11.10 shows the circuit symbol for the **tran** and **rtran** gates. The gates act as a buffer between the two bi-directional signals.

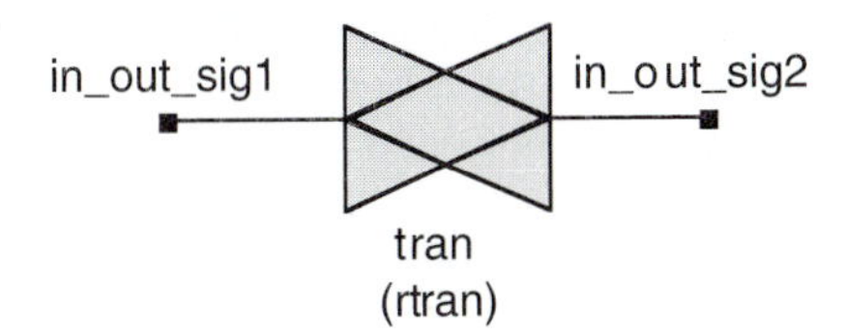

Figure 11.10 Circuit symbol for the (a) **tran** and (b) **rtran** primitive gates.

The three-terminal transmission gates, **tranif0**, **tranif1**, **rtranif0**, and **rtranif1**, likewise have two bi-directional terminals; a third terminal in the terminal list acts as a control input. The switch is closed when the control value is asserted, and open when the control value is de-asserted. The logic value of the output of this type of gate is either the logic value of the input or a high impedance. Figure 11.11 shows the circuit symbols for the three-terminal bi-directional switches implemented as predefined Verilog primitives. The actual direction of signal flow depends upon the circuit structure to which the bi-directional gate is attached.

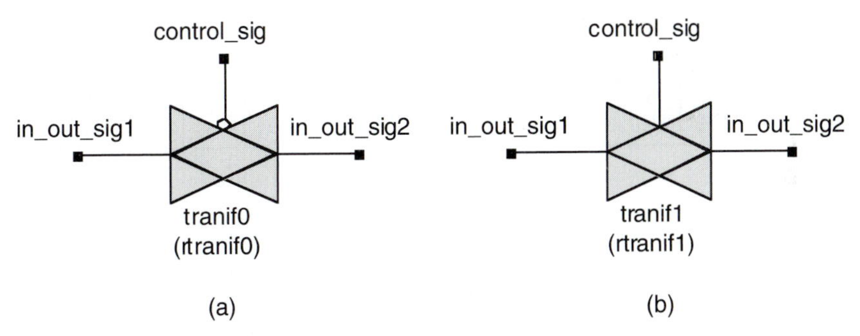

Figure 11.11 Circuit symbols for the (a) **tranif0 (rtranif0)**, and (b) **tranif1 (rtranif1)** bi-directional primitive gates.

11.7 SIGNAL STRENGTHS

All the previous examples focused on how gate- and switch-level primitives determine the logic value of the output signal from the logic value of the input signal. These models are adequate when signal contention and charge redistribution are not a factor in the circuit. A more accurate model considers both the value and strength of a signal, and the type of net it drives. The strength of a signal represents additional information that determines the result of contending drivers on a net when multiple drivers are present, or when the nets are modeled as charge storage capacitors.

Verilog's logic system has four logic values and seven logic strengths. The strength of a signal refers to the signal's ability to act as a logic driver determining the resultant logic value on a net. This is necessary when models must account for signal contention between multiple drivers of nets, or when charge redistribution[4, 5] can occur between nodes in a circuit. By default, all the signals presented in gate-level models in the previous chapters were implicitly "strong".

The range of allowed strengths for logic values "0" and "1" is represented by a "strength diagram", shown in Figure 11.12. Continuous assignments and the outputs of gates have "driving" strengths. There are four driving strengths: supply, strong (the default), pull, and weak, with keywords **supply0**, **strong0**, **pull0**, **weak0**, **highz0**, **supply1**, **strong1**, **pull1**, **weak1**, and **highz1**. There are also three charge storage strengths: **large**, **medium**, and **small,** which may be assigned only to **trireg** nets.

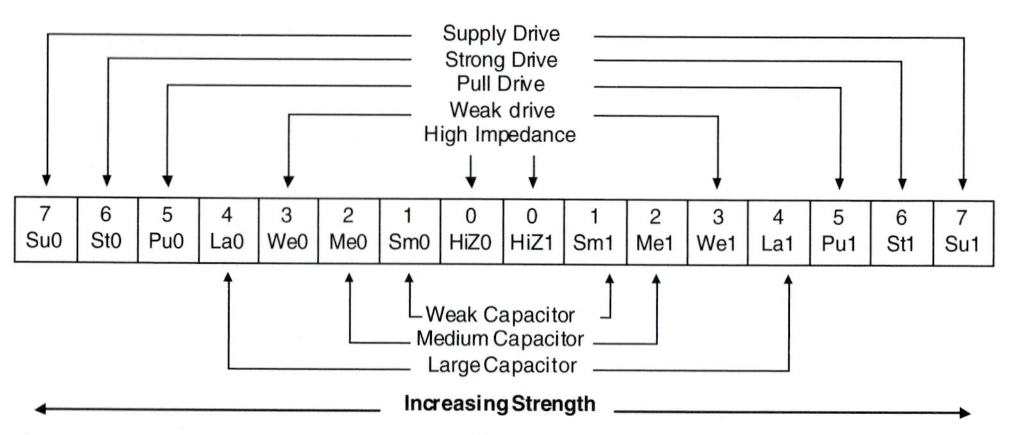

Figure 11.12 Strength diagram showing the range of signal strengths in Verilog.

A strength specification for a signal consists of two parts: a separate strength assigned to a logic value "0" and a separate strength assigned to a logic value "1". The integer associated with a strength indicates its relative strength, with "7" being the highest strength. When the logic value of a signal is known ("0" or "1"), the strength of the signal is one of the strengths (1 ... 7). If the logic value is ambiguous (x), it may have strengths in both sides of the scale. If the signal has a logic value of "z", it may be HiZ0 or HiZ1. Note that HiZ0 and HiZ1 may not be associated with the output of the combinational and pull gates. ***To see the strength of a signal, use the %v format specifier in the* $display *and* $monitor *system tasks.***

Of the indicated strengths, the capacitive strengths may be assigned only to **trireg** nets; the others may be assigned only to nets that are outputs of the combinational gates, pull gates, and continuous assignments. The strength of the output signal of a MOS transistor switch or bi-directional switch depends upon the type of gate and the strength of the input signal (see the discussion of strength reduction in Section 11.9). The following section will present the syntax for the assignment of strength to nets and continuous assignments.

11.7.1 STRENGTH OF A "DRIVEN" NET

In general, nets of type **wire, tri, wand, wor, trior,** and **triand** get their value from the output of a primitive or from a continuous assignment. Their strength depends upon the strength of the primitives or continuous assignments driving the net. Various rules for determining the value and strength of these nets are discussed in later sections of this chapter.

An optional signal *drive_strength* may be associated with the instantiation of the explicit combinational logic primitives and the MOS pull gates, according to the simplified syntax shown below. The default strength of the output of these gates is (**strong0, strong1**).

Verilog Syntax: Primitive Gate Instantiation

gate_instantiation ::= [GATE_TYPE] [drive_strength][delay] gate_instance {, gate_instance} ;

In this syntax, the GATE_TYPE is one of the following predefined keywords: **and, nand, or, nor xor, xnor, but, not, bufif0, bufif1, notif0, notif1, pullup,** or **pulldown**. The optional *drive_strength* is an unordered pair with one value from the set {**supply0, strong0, pull0, weak0, highz0**} and one from the set {**supply1, strong1, pull1, weak1, highz1**},such as (**strong1, weak0**).

A signal *drive_strength* may also be associated with continuous assignments (implicit combinational logic) to a net variable. The simplified syntax for a continuous assignment shown below includes an optional *drive_strength* item. The implicit assignment of logic with the declaration of a net uses the same syntax for the specification of strength. Only scalar nets may receive strength assignments.

Verilog Syntax: Continuous Assignment to a Net

continuous_assign ::= **assign** [drive_strength][delay] list_of_net_assignments

Example 11.6 Examples of strength declarations of nets and gates are given below.

```
nand (pull1, strong0) G1 (gate_output, input1, input2);
wire (pull0, weak1) A_wire = net_1 || net_2;
assign (pull1, weak0) A_net = Register_B;
```

11.7.2 Supply Nets

The "supply" types of nets **supply0**, **supply1**, **tri0**, and **tri1** are used to model structural connections to power ("1") and ground ("0"), and have a fixed logic value that is independent of a driver (i.e., they may not be the output port of a module, output terminal of a primitive, or target of a continuous assignment). The **supply0** and **supply1** fixed nets have a predefined strength of **supply0** and **supply1**, respectively.

The **tri0** and **tri1** fixed nets represent pull-down and pull-up resistive terminations. These nets may have multiple drivers. When such a **tri0** or **tri1** net is not driven, it is pulled to the indicated logic value and has a strength of **pull0** or **pull1**. A driver may override this value according to the rules given in Section 11.9.

11.7.3 Charge Storage Nets

The **trireg** net models physical layout in which the capacitance at a node holds a charge after the drivers to the net have been removed. The removed drivers present a high impedance (open circuit) to the net, and the net has a user-specified *charge_strength* value of **small**, **medium** (default) or **large** capacitor. The size of the capacitor determines how charge will be shared between connected storage nodes when conducting paths are established. The simplified syntax for the declaration of a charge storage net is shown below.

Verilog Syntax: Charge_Storage_Net

net_declaration ::= **trireg** [**vectored** | **scalared**][charge_strength][range] list_of_net_identifiers

11.8 AMBIGUOUS SIGNALS

The logic value "x" represents an ambiguous condition in which the signal could have a value of "0" or "1". The strength of a signal may also be ambiguous. When the strength of a signal is ambiguous, its value is a member of a set of two or more values, but the exact value is not known. Thus, a signal may have a range of logic strengths. For example, when the control input of the three-state

bufif0 gate in Figure 11.13 has a value of "0", the output of the buffer would also be St0. But if the control input is "1", the buffer is turned off and the output is high impedance. When the control input is ambiguous, "x", the output could be "0" or "z", denoted by "L". The strength-value pair of the output is StL, and includes all the shaded values shown in the strength diagram in Figure 11.13. The actual strength is not known and cannot be determined.

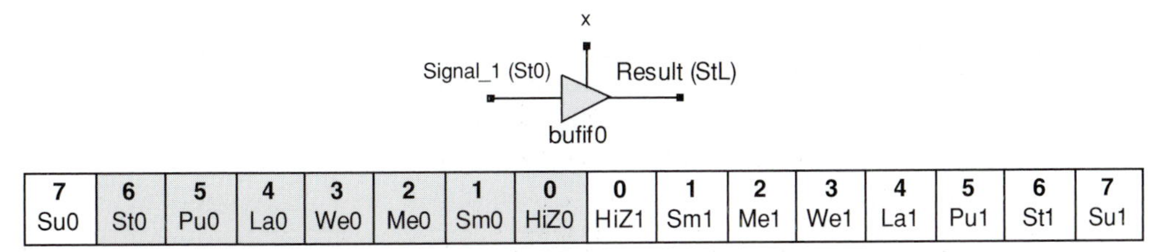

7	6	5	4	3	2	1	0	0	1	2	3	4	5	6	7
Su0	St0	Pu0	La0	We0	Me0	Sm0	HiZ0	HiZ1	Sm1	Me1	We1	La1	Pu1	St1	Su1

Figure 11.13 Three-state buffer with ambiguous control input and StL output.

In Figure 11.14, the input signal to the three-state buffer has strength-value pair (Pu1), and the output has strength-value pair (PuH).

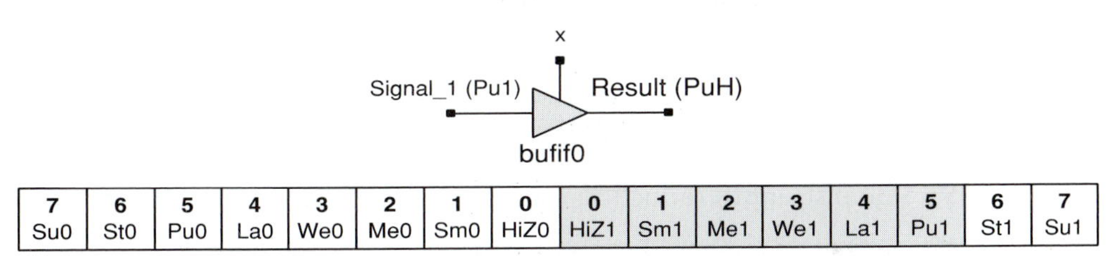

7	6	5	4	3	2	1	0	0	1	2	3	4	5	6	7
Su0	St0	Pu0	La0	We0	Me0	Sm0	HiZ0	HiZ1	Sm1	Me1	We1	La1	Pu1	St1	Su1

Figure 11.14 Three-state buffer with ambiguous control input and StL output.

11.9 STRENGTH REDUCTION BY PRIMITIVES

With the exception of the three-state gates, the strength of the output of a combinational gate primitive or pull primitive is independent of the strength of the input signal. On the other hand, the strength of the output of a MOS transistor switch or MOS bi-directional transistor switch depends upon the strength of the input signal and the type of the gate. The strength of the input and output need not be the same. In general, the strength of the output signal of this type of primitive may be less than that of the input signal (i.e., the signal undergoes strength reduction in passing through the primitive).

11.9.1 TRANSISTOR SWITCH AND BI-DIRECTIONAL SWITCHES

If the data input signal to a MOS transistor switch (**nmos**, **pmos**, and **cmos**) or the data input signal to a bi-directional transistor switch (**tran**, **tranif0**, and

tranif1) has a strength of **supply0** or **supply1**, the output signal will have a strength of **strong0** or **strong1**, respectively. Otherwise, the strength of the output signal will be the same as the strength of the input signal.

11.9.2 RESISTIVE MOS DEVICES

Resistive MOS devices (**rpmos**, **rnmos**, **rcmos**, **rtran**, **tranif0**, and **rtanif1**) form the strength of the output signal according to the rule illustrated in Figure 11.15. For example, if the input to a **rnmos** primitive has strength St0, the output of the device will have strength Pu0. If the output of resistive device is a charge storage net (**trireg**) having the strength of either a medium or small capacitor, the output signal will be a signal having the strength of a small capacitor.

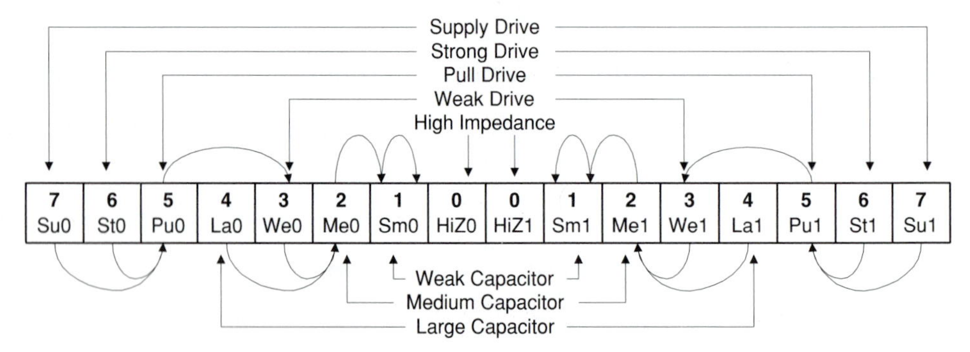

Figure 11.15 Signal strength reduction for resistive MOS devices and **trireg** nets.

Example 11.7 The **rnmos** and **rpmos** primitives reduce the strength of their input signal. Figure 11.16 shows (a) a simple inverter having a pull-up net replacing the active pull-up transistor, and output connected to an **rnmos** primitive acting as a pass transistor, (b) a Verilog model of the inverter, and (c) simulation results showing the reduction in strength (compare *rinv_out* and *inv_out*).

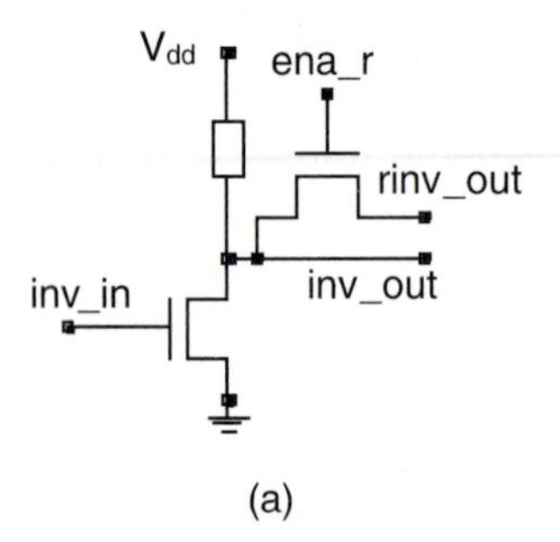

Figure 11.16 (a) Inverter with pullup net, (b) Verilog description, and (c) simulation results. Continues on pages 515–516.

```
module sw_inverter (inv_out, rinv_out, inv_in, ena_r);
   output inv_out, rinv_out;  input inv_in, ena_r;
   supply1 Vdd;
   supply0 Gnd;
   pullup (inv_out);
   nmos  (inv_out, Gnd, inv_in);
   rnmos (rinv_out, inv_out, ena_r);
endmodule

module t_sw_inverter ();
   reg inv_in;
   wire inv_out, rinv_out;

   sw_inverter M1(inv_out, rinv_out, inv_in);

   initial repeat (3) begin
      #5 inv_in = 1;
      #5 inv_in = 0;
   end

   initial #50 $stop;
   initial $monitor ($time,,"inv_in = %v inv_out = %v rinv = %v",
                              inv_in, inv_out, rinv_out);
endmodule
```

(b)

```
              S I L O S   I I I    Version 97.1

Copyright (c) 1997 by SIMUCAD Inc.   All rights reserved.
No part of this program may be reproduced,  transmitted,
transcribed,   or stored in a retrieval system,    in any
form or by any means without the prior written consent of

    SIMUCAD Inc., 32970 Alvarado-Niles Road, Union City,
                  California, 94587, U.S.A.
             (510)-487-9700  Fax: (510)-487-9721
         Electronic Mail Address:   "silos@ simucad.com"

input c:\silos\my_silos_work\sw_level\sw_inv.v
Reading "c:\silos\my_silos_work\sw_level\sw_inv.v"
```

(c) continues

Figure 11.16 Continued.

```
Ready: sim to 0
  Highest level modules (that have been auto-instantiated):
    (t_sw_inverter t_sw_inverter
  8 total devices.
  Linking ...
  6 nets total: 14 saved and 0 monitored.
  65 registers total: 65 saved.
  Done.
                0 inv_in = StX inv_out = 56X rinv = PuX

  0 State changes on observable nets.

  Simulation stopped at the end of time 0.
Ready: sim
                5 inv_in = St1 inv_out = St0 rinv = Pu0
               10 inv_in = St0 inv_out = Pu1 rinv = Pu1
               15 inv_in = St1 inv_out = St0 rinv = Pu0
               20 inv_in = St0 inv_out = Pu1 rinv = Pu1
               25 inv_in = St1 inv_out = St0 rinv = Pu0
               30 inv_in = St0 inv_out = Pu1 rinv = Pu1
Ready:
```

(c)

Figure 11.16 Concluded.

11.10 COMBINATION AND RESOLUTION OF SIGNAL STRENGTHS

Simulators use signal strengths to determine the result of driving a net with multiple drivers. The driving signals may have definite (unambiguous) known value and strength, or they may not. The logic value "x" is ambiguous; a range of strengths is said to be ambiguous. This section will treat each of four possibilities for contention between signals having possibly different values and strengths.

11.10.1 Signal Contention: Known Strength and Known Value

When a net is driven by signals having known logic values and known (unambiguous) strengths, the signal having the greater strength dominates. In the wired configuration of Figure 11.17, *Signal_1* has the strength-value pair (We0), and *Signal_2* has the pair (Pu1). Both signals have a known logic value. The pull strength dominates the weak strength, and the result has the strength-value pair (Pu1).

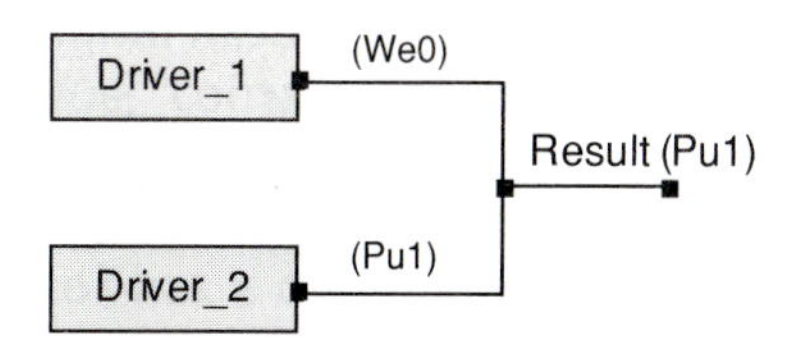

	Strength of Logic 0								Strength of Logic 1							
Signal_1	7 Su0	6 St0	5 Pu0	4 La0	3 We0	2 Me0	1 Sm0	0 HiZ0	0 HiZ1	1 Sm1	2 Me1	3 We1	4 La1	5 Pu1	6 St1	7 Su1

	Strength of Logic 0								Strength of Logic 1							
Signal_2	7 Su0	6 St0	5 Pu0	4 La0	3 We0	2 Me0	1 Sm0	0 HiZ0	0 HiZ1	1 Sm1	2 Me1	3 We1	4 La1	5 Pu1	6 St1	7 Su1

	Strength of Logic 0								Strength of Logic 1							
Result	7 Su0	6 St0	5 Pu0	4 La0	3 We0	2 Me0	1 Sm0	0 HiZ0	0 HiZ1	1 Sm1	2 Me1	3 We1	4 La1	5 Pu1	6 St1	7 Su1

Figure 11.17 Value and strength resolution for different known strengths and different logic values.

If the drivers of a net have the same value and identical strengths, the resulting signal has that value and strength. If the drivers of a net have the *same* logic value, the resulting signal has that value and the strength of the greatest of those drivers.

If the drivers of a net have the same dominating strength but different logic values, the drivers are in contention, and the resulting logic value depends upon the type of net. If the net is a **wand** (wired-and) or **wor** (wired-or), the result is determined by the rules for this form of logic (see Sections 4.3.1 and 11.10). Otherwise, the resulting logic value is ambiguous, "x", and the strength includes the range of all strengths between the greatest strength of each driver. This situation is illustrated in Figure 11.18 for a net driven by a signal having the strength-value pair St0 and a signal having the strength-value pair St1. The resulting logic value and strength are both ambiguous.

When two signals having ambiguous strengths drive the same net, the result is a signal having ambiguous strength. In the wired configuration of Figure 11.19, *Result_1* and *Result_2* have ambiguous strength. These signals combine to form *Result* having ambiguous value ("x") and strength ("53"); i.e., the output could range from a logical "0" with a pull strength ("5"), to a logical 1 with a weak strength ("3"). Note that the range of the strengths of *Result* includes all of the strengths in its component signals.

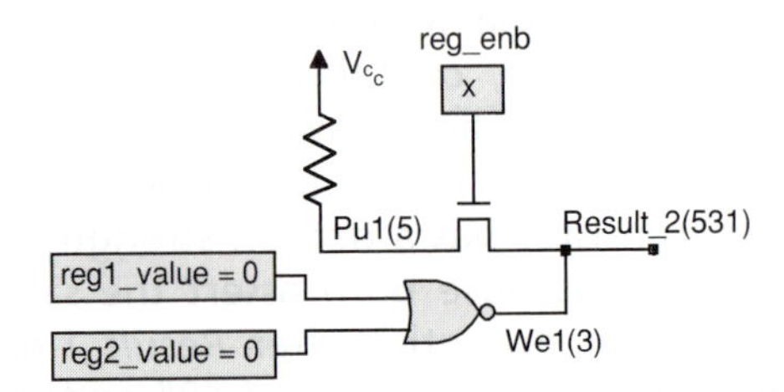

	Strength of Logic 0								Strength of Logic 1							
Result	7 Su0	6 St0	5 Pu0	4 La0	3 We0	2 Me0	1 Sm0	0 HiZ0	0 HiZ1	1 Sm1	2 Me1	3 We1	4 La1	5 Pu1	6 St1	7 Su1

Figure 11.21 Signal with ambiguous strength.

A switch-level Verilog model corresponding to the circuit in Figure 11.21 is given below.

```
reg reg1_value, reg2_value, reg_enb;
pullup (Vcc);
wire Result_2;
nmos (Result_2, Vcc, reg_enb);
nor (strong0, weak1) (Result_2, reg1_value, reg2_value);
```

Example 11.10 When the circuits of Example 11.8 and 11.9 are tied together in a wired connection of their outputs, *Result_1* and *Result 2* are in contention. In Figure 11.22, *Result_1* and *Result_2* have opposing logic values and ambiguous strengths. *Result* has a logic value of "x" and has a range of strengths ranging from the strongest strength of the "0" logic value to the strongest strength of the "1" logic value (i.e., from **strong0** to **pull1**).

The switch-level model given below can be simulated to examine the concepts of signal contention.

```
module sw3 (Result_1, Result_2, Result);
  output Result_1, Result_2, Result;
  wire Result_1, Result_2;
  reg reg1_value, reg2_value, reg_enb;
  reg reg_ena, reg_value;
  pullup (Vcc);
  pulldown (Result_1);
  nmos (Result_2, Vcc, reg_enb);
  nmos (Result_1, reg_value, reg_ena);
  nor (strong0, weak1) (Result_2, reg1_value, reg2_value);
  assign Result = Result_2;
  assign Result = Result_1;
endmodule
```

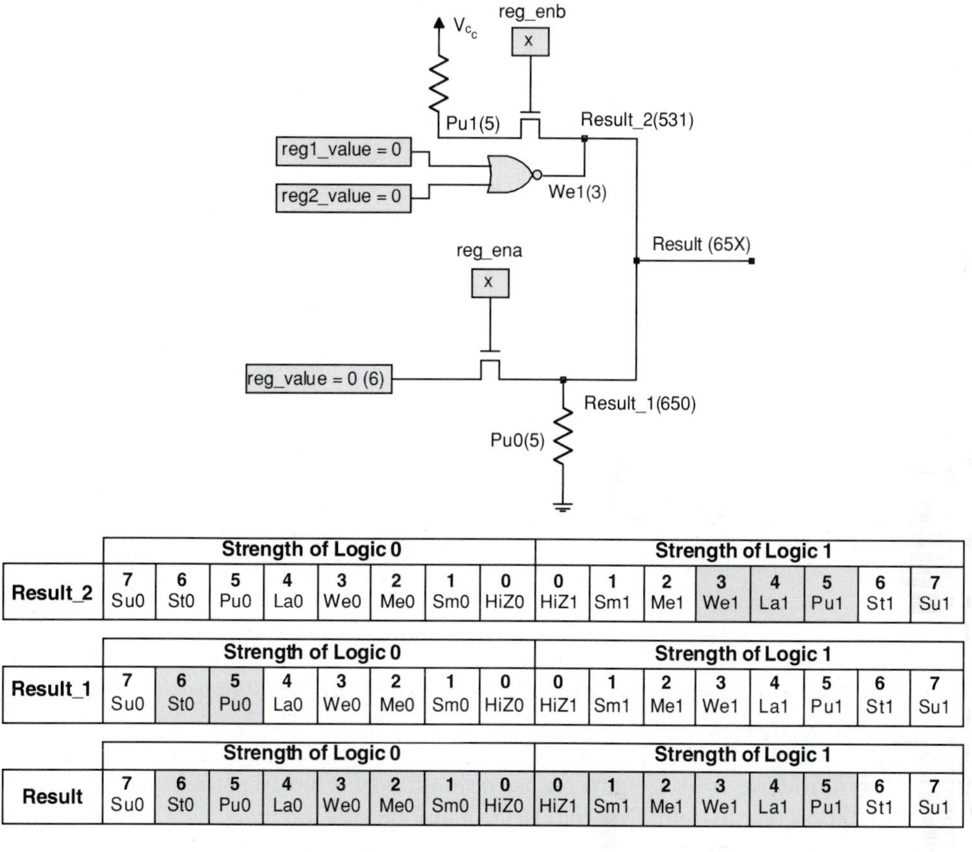

	Strength of Logic 0								Strength of Logic 1							
Result_2	7 Su0	6 St0	5 Pu0	4 La0	3 We0	2 Me0	1 Sm0	0 HiZ0	0 HiZ1	1 Sm1	2 Me1	3 We1	4 La1	5 Pu1	6 St1	7 Su1

	Strength of Logic 0								Strength of Logic 1							
Result_1	7 Su0	6 St0	5 Pu0	4 La0	3 We0	2 Me0	1 Sm0	0 HiZ0	0 HiZ1	1 Sm1	2 Me1	3 We1	4 La1	5 Pu1	6 St1	7 Su1

	Strength of Logic 0								Strength of Logic 1							
Result	7 Su0	6 St0	5 Pu0	4 La0	3 We0	2 Me0	1 Sm0	0 HiZ0	0 HiZ1	1 Sm1	2 Me1	3 We1	4 La1	5 Pu1	6 St1	7 Su1

Figure 11.22 Ambiguous signal resulting from switch network signal contention.

Example 11.11 Predefined and user-defined primitives may be instantiated with strengths. Wired connections of these gates create the possibility for signal contention. In Figure 11.23, the upper **nand** gate has a known output, "0", having **weak0** *drive_strength*, as specified by the instantiation. The lower **nand** gate has an ambiguous input, and an ambiguous output. The values of *Signal_1* and *Signal_2* are possibly in contention, and even when they are not, their strengths differ. The simulator must account for all of these possibilities. The *drive_strength* of the output, as governed by the instantiation, ranges from strong1 down to HiZ0 (i.e., the *drive_strength* is StL augmented by HiZ0). The *drive_strength* of *Result* ranges from St1 down to We0 (i.e., 36X).

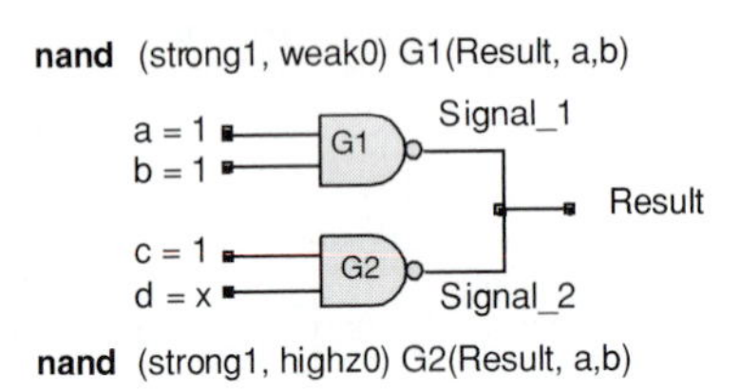

	Strength of Logic 0								Strength of Logic 1							
Signal_1	7 Su0	6 St0	5 Pu0	4 La0	3 We0	2 Me0	1 Sm0	0 HiZ0	0 HiZ1	1 Sm1	2 Me1	3 We1	4 La1	5 Pu1	6 St1	7 Su1

	Strength of Logic 0								Strength of Logic 1							
Signal_2	7 Su0	6 St0	5 Pu0	4 La0	3 We0	2 Me0	1 Sm0	0 HiZ0	0 HiZ1	1 Sm1	2 Me1	3 We1	4 La1	5 Pu1	6 St1	7 Su1

	Strength of Logic 0								Strength of Logic 1							
Result	7 Su0	6 St0	5 Pu0	4 La0	3 We0	2 Me0	1 Sm0	0 HiZ0	0 HiZ1	1 Sm1	2 Me1	3 We1	4 La1	5 Pu1	6 St1	7 Su1

Figure 11.23 Ambiguous signal resulting from combinational gate signal contention.

11.10.2 Combination of Ambiguous Strength and Known Value

The following rules govern the combination of a signal with known strength and known value with a signal having ambiguous strength.

Rule 1: The result includes those strengths of the ambiguous strength signal that are greater than the strength level of the unambiguous signal.

Rule 2: With the exception of Rule 3, the result omits the strengths of the ambiguous strength signal that are equal to or less than the strength of the unambiguous signal.

Rule 3: If the unambiguous signal and ambiguous signal have different logical value, the range of strengths of the result includes the intermediate values in the gap between the strengths taken from Rule 1 and the strength of the unambiguous signal.

Example 11.12 In the strength diagrams shown in Figure 11.24, the unambiguous signal and the signal having ambiguous strength both have the same logic value. The result excludes the strengths of the ambiguous signal that are less than the strength of the unambiguous signal. Given that the signals have the same logic value, the known strength of *Signal_1* prevents the output from ever having a strength less than We1. On the other hand, *Signal_2* could provide strengths that could exceed We1. The strength diagram for *Result* includes all of these possibilities.

	Strength of Logic 0								Strength of Logic 1							
Signal_1	**7**	**6**	**5**	**4**	**3**	**2**	**1**	**0**	**0**	**1**	**2**	**3**	**4**	**5**	**6**	**7**
	Su0	St0	Pu0	La0	We0	Me0	Sm0	HiZ0	HiZ1	Sm1	Me1	We1	La1	Pu1	St1	Su1

	Strength of Logic 0								Strength of Logic 1							
Signal_2	**7**	**6**	**5**	**4**	**3**	**2**	**1**	**0**	**0**	**1**	**2**	**3**	**4**	**5**	**6**	**7**
	Su0	St0	Pu0	La0	We0	Me0	Sm0	HiZ0	HiZ1	Sm1	Me1	We1	La1	Pu1	St1	Su1

	Strength of Logic 0								Strength of Logic 1							
Result	**7**	**6**	**5**	**4**	**3**	**2**	**1**	**0**	**0**	**1**	**2**	**3**	**4**	**5**	**6**	**7**
	Su0	St0	Pu0	La0	We0	Me0	Sm0	HiZ0	HiZ1	Sm1	Me1	We1	La1	Pu1	St1	Su1

Figure 11.24 Application of Rule 1 and Rule 2 for combining an unambiguous signal with a signal having ambiguous strength.

Example 11.13 In the strength diagrams shown in Figure 11.25, *Signal_1* has strengths for both logic values. However, the strength of *Signal_2*, the unambiguous signal, dominates all the strengths of the ambiguous whose magnitude is less than four.

	Strength of Logic 0								Strength of Logic 1							
Signal_1	**7**	**6**	**5**	**4**	**3**	**2**	**1**	**0**	**0**	**1**	**2**	**3**	**4**	**5**	**6**	**7**
	Su0	St0	Pu0	La0	We0	Me0	Sm0	HiZ0	HiZ1	Sm1	Me1	We1	La1	Pu1	St1	Su1

	Strength of Logic 0								Strength of Logic 1							
Signal_2	**7**	**6**	**5**	**4**	**3**	**2**	**1**	**0**	**0**	**1**	**2**	**3**	**4**	**5**	**6**	**7**
	Su0	St0	Pu0	La0	We0	Me0	Sm0	HiZ0	HiZ1	Sm1	Me1	We1	La1	Pu1	St1	Su1

	Strength of Logic 0								Strength of Logic 1							
Result	**7**	**6**	**5**	**4**	**3**	**2**	**1**	**0**	**0**	**1**	**2**	**3**	**4**	**5**	**6**	**7**
	Su0	St0	Pu0	La0	We0	Me0	Sm0	HiZ0	HiZ1	Sm1	Me1	We1	La1	Pu1	St1	Su1

Figure 11.25 Application of Rule 2 for combining an unambiguous signal with a signal having ambiguous strength.

Example 11.14 In the strength diagrams shown in Figure 11.26, *Signal_1* and *Signal_2* have opposite logic values, and *Signal_1* has strengths that exceed the strength of *Signal_2*. Application of Rule 2 would remove Me0 and Sm0 from the result, but because there is a gap between We1 and the remaining values, Rule 3 includes the intermediate strengths. The contention of the signals results in the range of strengths shown.

9. Write a testbench and verify the functionality of the switch-level model for the master-slave D flip-flop shown in Figure 11.8. Simulate with the flip-flop with zero and unit delay on the instantiated gates. Create waveform displays showing the evolution of signals through the circuit.

10. Write a testbench and simulate the behavior of the circuit in Figure P11.4 for various strengths of the input signals. (Hint: Use continuous assignments to represent the structure.)

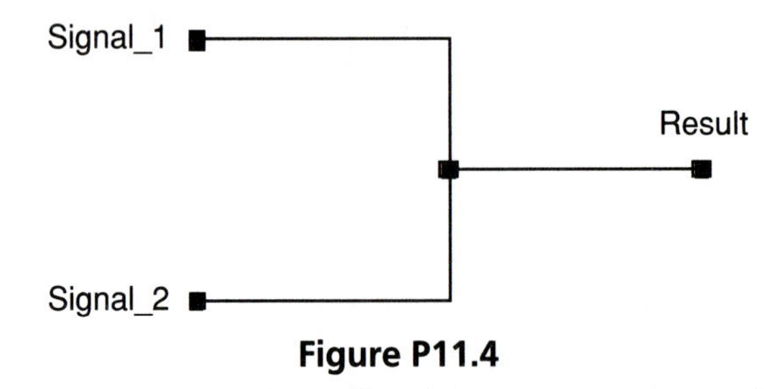

Figure P11.4

11. Write and verify a switch-level model for the linear shift register shown in Figure P11.5. The pull-up devices are depletion load nMOS transistors, and the pass transistors, N1 … N4, are enhancement mode nMOS transistors. Charge storage nets having medium strength, w1, …, w4 connect the stages of the register.

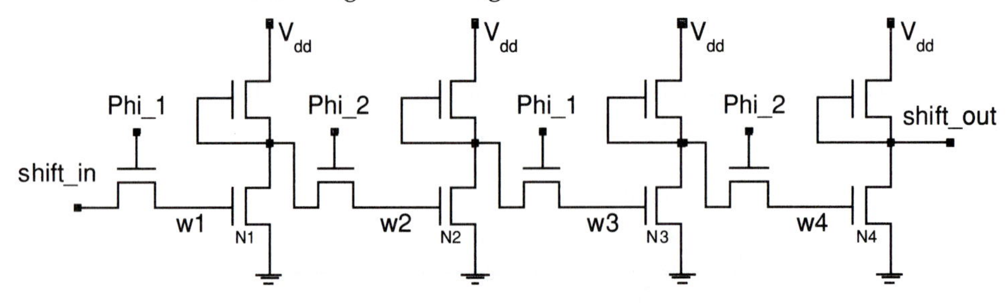

Figure P11.5

12. Indicate the value and strength of *Result* in the strength diagram in Figure P11.6.

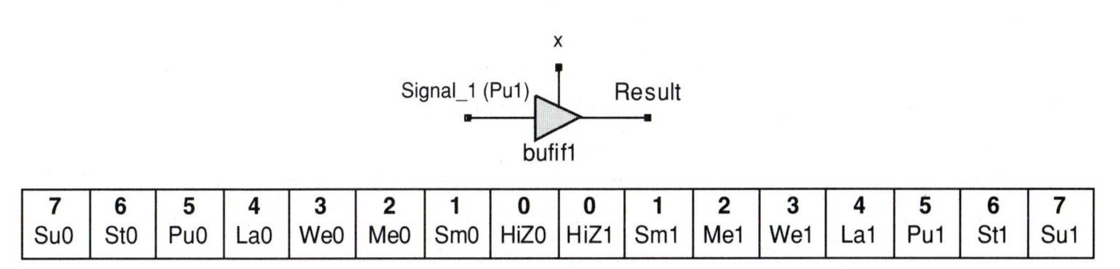

7	6	5	4	3	2	1	0	0	1	2	3	4	5	6	7
Su0	St0	Pu0	La0	We0	Me0	Sm0	HiZ0	HiZ1	Sm1	Me1	We1	La1	Pu1	St1	Su1

Figure P11.6

13. Indicate the value and strength of *Result* in the strength diagram in Figure P11.7 (assume that *Signal_1* and *Signal_2* are driving the same net).

	Strength of Logic 0								Strength of Logic 1							
Signal_1	7 Su0	6 St0	5 Pu0	4 La0	3 We0	2 Me0	1 Sm0	0 HiZ0	0 HiZ1	1 Sm1	2 Me1	3 We1	**4 La1**	5 Pu1	6 St1	7 Su1

	Strength of Logic 0								Strength of Logic 1							
Signal_2	7 Su0	**6 St0**	5 Pu0	4 La0	3 We0	2 Me0	1 Sm0	0 HiZ0	0 HiZ1	1 Sm1	2 Me1	3 We1	4 La1	5 Pu1	6 St1	7 Su1

	Strength of Logic 0								Strength of Logic 1							
Result	7 Su0	6 St0	5 Pu0	4 La0	3 We0	2 Me0	1 Sm0	0 HiZ0	0 HiZ1	1 Sm1	2 Me1	3 We1	4 La1	5 Pu1	6 St1	7 Su1

Figure P11.7

14. Indicate the value and strength of *Result* in the strength diagram in Figure P11.8 (assume that *Signal_1* and *Signal_2* are driving the same net).

	Strength of Logic 0								Strength of Logic 1							
Signal_1	7 Su0	6 St0	5 Pu0	4 La0	3 We0	2 Me0	1 Sm0	0 HiZ0	0 HiZ1	**1 Sm1**	2 Me1	3 We1	4 La1	5 Pu1	6 St1	7 Su1

	Strength of Logic 0								Strength of Logic 1							
Signal_2	7 Su0	6 St0	**5 Pu0**	4 La0	3 We0	2 Me0	1 Sm0	0 HiZ0	0 HiZ1	1 Sm1	2 Me1	3 We1	4 La1	5 Pu1	6 St1	7 Su1

	Strength of Logic 0								Strength of Logic 1							
Result	7 Su0	6 St0	5 Pu0	4 La0	3 We0	2 Me0	1 Sm0	0 HiZ0	0 HiZ1	1 Sm1	2 Me1	3 We1	4 La1	5 Pu1	6 St1	7 Su1

Figure P11.8

15. Indicate the strength of *Result_1*, *Result_2* and *Result* in the circuit in Figure P11.9.

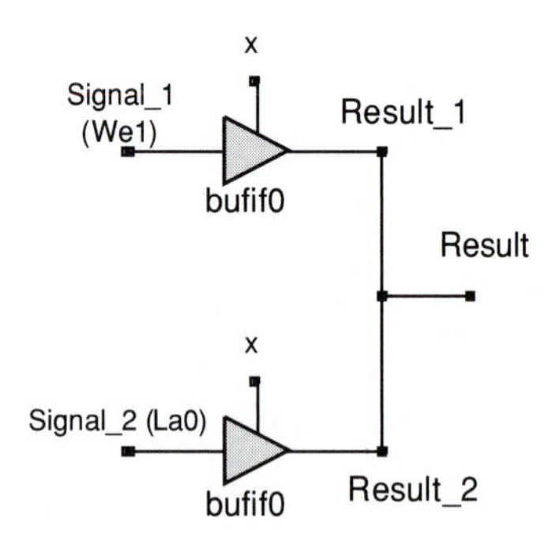

	Strength of Logic 0								Strength of Logic 1							
Result_1	7	6	5	4	3	2	1	0	0	1	2	3	4	5	6	7
	Su0	St0	Pu0	La0	We0	Me0	Sm0	HiZ0	HiZ1	Sm1	Me1	We1	La1	Pu1	St1	Su1

	Strength of Logic 0								Strength of Logic 1							
Result_2	7	6	5	4	3	2	1	0	0	1	2	3	4	5	6	7
	Su0	St0	Pu0	La0	We0	Me0	Sm0	HiZ0	HiZ1	Sm1	Me1	We1	La1	Pu1	St1	Su1

	Strength of Logic 0								Strength of Logic 1							
Result	7	6	5	4	3	2	1	0	0	1	2	3	4	5	6	7
	Su0	St0	Pu0	La0	We0	Me0	Sm0	HiZ0	HiZ1	Sm1	Me1	We1	La1	Pu1	St1	Su1

Figure P11.9

16. Indicate the strength of *Result_1, Result_2* and *Result* in the circuit in Figure P11.10.

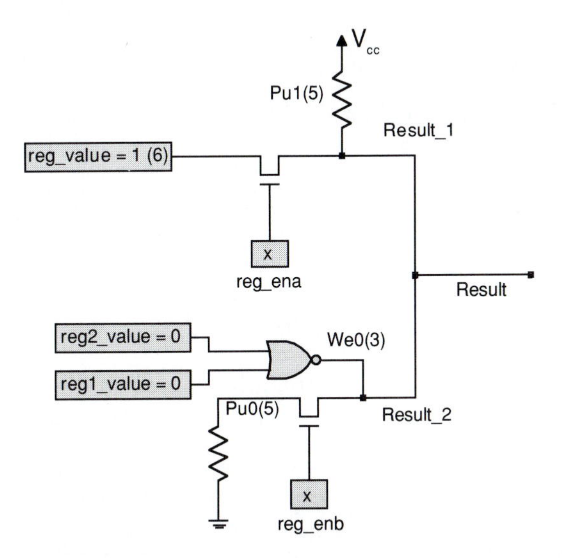

	Strength of Logic 0								Strength of Logic 1							
Result_2	7 Su0	6 St0	5 Pu0	4 La0	3 We0	2 Me0	1 Sm0	0 HiZ0	0 HiZ1	1 Sm1	2 Me1	3 We1	4 La1	5 Pu1	6 St1	7 Su1

	Strength of Logic 0								Strength of Logic 1							
Result_1	7 Su0	6 St0	5 Pu0	4 La0	3 We0	2 Me0	1 Sm0	0 HiZ0	0 HiZ1	1 Sm1	2 Me1	3 We1	4 La1	5 Pu1	6 St1	7 Su1

	Strength of Logic 0								Strength of Logic 1							
Result	7 Su0	6 St0	5 Pu0	4 La0	3 We0	2 Me0	1 Sm0	0 HiZ0	0 HiZ1	1 Sm1	2 Me1	3 We1	4 La1	5 Pu1	6 St1	7 Su1

Figure P11.10

17. Indicate the value and strength of *Result* in the strength diagram in Figure P11.11 (assume that *Signal_1* and *Signal_2* are driving the same net).

	Strength of Logic 0								Strength of Logic 1							
Signal_1	7 Su0	6 St0	5 Pu0	4 La0	3 We0	2 Me0	1 Sm0	0 HiZ0	0 HiZ1	1 Sm1	2 Me1	3 We1	4 La1	5 Pu1	6 St1	7 Su1

	Strength of Logic 0								Strength of Logic 1							
Signal_2	7 Su0	6 St0	5 Pu0	4 La0	3 We0	2 Me0	1 Sm0	0 HiZ0	0 HiZ1	1 Sm1	2 Me1	3 We1	4 La1	5 Pu1	6 St1	7 Su1

	Strength of Logic 0								Strength of Logic 1							
Result	7 Su0	6 St0	5 Pu0	4 La0	3 We0	2 Me0	1 Sm0	0 HiZ0	0 HiZ1	1 Sm1	2 Me1	3 We1	4 La1	5 Pu1	6 St1	7 Su1

Figure P11.11

18. Indicate the value and strength of *Result* in the strength diagram in Figure P11.12 (assume that *Signal_1* and *Signal_2* are driving the same net).

	Strength of Logic 0								Strength of Logic 1							
Signal_1	7 Su0	6 St0	5 Pu0	4 La0	3 We0	2 Me0	1 Sm0	0 HiZ0	0 HiZ1	1 Sm1	2 Me1	3 We1	4 La1	5 Pu1	6 St1	7 Su1

	Strength of Logic 0								Strength of Logic 1							
Signal_2	7 Su0	6 St0	5 Pu0	4 La0	3 We0	2 Me0	1 Sm0	0 HiZ0	0 HiZ1	1 Sm1	2 Me1	3 We1	4 La1	5 Pu1	6 St1	7 Su1

	Strength of Logic 0								Strength of Logic 1							
Result	7 Su0	6 St0	5 Pu0	4 La0	3 We0	2 Me0	1 Sm0	0 HiZ0	0 HiZ1	1 Sm1	2 Me1	3 We1	4 La1	5 Pu1	6 St1	7 Su1

Figure P11.12

19. Indicate the value and strength of *Result* in the strength diagram in Figure P11.13 (assume that *Signal_1* and *Signal_2* are driving the same net).

	Strength of Logic 0								Strength of Logic 1							
Signal_1	7 Su0	6 St0	5 Pu0	4 La0	3 We0	2 Me0	1 Sm0	0 HiZ0	0 HiZ1	1 Sm1	2 Me1	3 We1	4 La1	5 Pu1	6 St1	7 Su1

	Strength of Logic 0								Strength of Logic 1							
Signal_2	7 Su0	6 St0	5 Pu0	4 La0	3 We0	2 Me0	1 Sm0	0 HiZ0	0 HiZ1	1 Sm1	2 Me1	3 We1	4 La1	5 Pu1	6 St1	7 Su1

	Strength of Logic 0								Strength of Logic 1							
Result	7 Su0	6 St0	5 Pu0	4 La0	3 We0	2 Me0	1 Sm0	0 HiZ0	0 HiZ1	1 Sm1	2 Me1	3 We1	4 La1	5 Pu1	6 St1	7 Su1

Figure P11.13

20. Indicate the strength of the result of *Signal_1* and *Signal_2* driving **wand** and **wor** nets in the strength diagram in Figure P11.14.

	Strength of Logic 0								Strength of Logic 1							
Signal_1	7 Su0	6 St0	5 Pu0	4 La0	3 We0	2 Me0	1 Sm0	0 HiZ0	0 HiZ1	1 Sm1	2 Me1	3 We1	4 La1	5 Pu1	6 St1	7 Su1

	Strength of Logic 0								Strength of Logic 1							
Signal_2	7 Su0	6 St0	5 Pu0	4 La0	3 We0	2 Me0	1 Sm0	0 HiZ0	0 HiZ1	1 Sm1	2 Me1	3 We1	4 La1	5 Pu1	6 St1	7 Su1

wand	Strength of Logic 0								Strength of Logic 1							
Result	7 Su0	6 St0	5 Pu0	4 La0	3 We0	2 Me0	1 Sm0	0 HiZ0	0 HiZ1	1 Sm1	2 Me1	3 We1	4 La1	5 Pu1	6 St1	7 Su1

wor	Strength of Logic 0								Strength of Logic 1							
Result	7 Su0	6 St0	5 Pu0	4 La0	3 We0	2 Me0	1 Sm0	0 HiZ0	0 HiZ1	1 Sm1	2 Me1	3 We1	4 La1	5 Pu1	6 St1	7 Su1

Figure P11.14

CHAPTER 12

DESIGN EXAMPLES IN VERILOG

This chapter presents the following detailed examples illustrating the use of Verilog in synthesis: FIFO buffers for data acquisition, a UART, and a bit-slice micro-controller. The examples illustrate synthesis into a standard cell library with the Synopsys Design Compiler™.

12.1 FIFO—BUFFERS FOR DATA ACQUISITION

The objective of this example is to explore the role of FIFO buffers in data acquisition systems.

A first-in-first-out (FIFO) buffer (see Figure 12.1) is a dedicated memory stack consisting of a fixed, controlled array of registers. A FIFO buffer is used as an interface between a high-speed data source and a slower processor. The buffer receives and stores information, and then supplies it under the direction of a host controller. The size of the stack for a particular application is determined through simulation. In general, the processor must read from the stack often enough to prevent the loss of data under worst-case conditions.

In this example, the registers of the stack operate synchronously (rising edge) with a common clock, subject to a reset. The stack has two pointers (addresses), one pointing to the next word to which data will be written, and another pointing to the next word that will be read, subject to write and read inputs, respectively. The FIFO has input and output datapaths. Two bit-lines serve as flags to denote the status of the stack (full or empty).

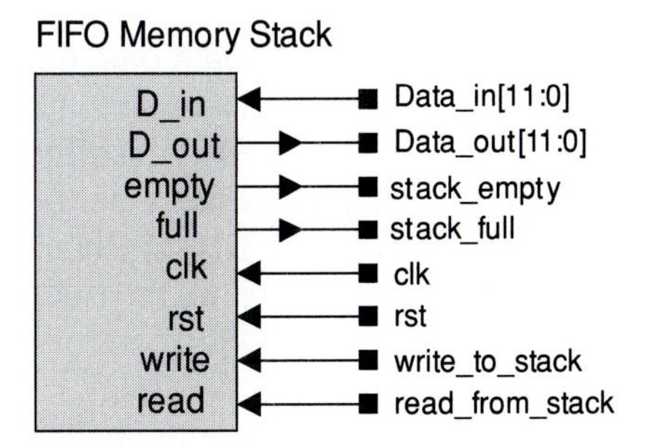

Figure 12.1 FIFO memory stack.

The FIFO writes the word at *Data_in* to memory if the stack is not full and *write_to_stack* is asserted. It reads a word from memory if the memory is not empty and *read_from_stack* is asserted. Data is inserted and retrieved in a first-in-first-out manner. Figure 12.2 shows more details of the stack. Two pointers, *write_ptr* and *read_ptr,* point to the memory cell to which data will be written next, and the cell from which memory will be read next, respectively. The stack width and height are parameterized to extend the utility of the model.

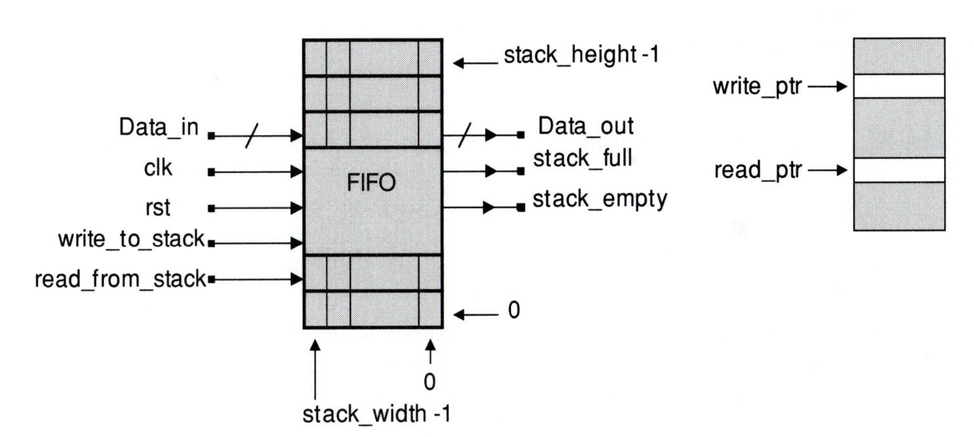

Figure 12.2 Internal organization of a FIFO stack.

We will develop a Verilog description of the FIFO for a 12-bit datapath (this choice is dictated by a later application). For convenience, we choose a stack of eight words, and need three-bit pointers to address the stack for writing and reading. Then we create the header, port declarations, parameters, and documentation shown in Figure 12.3. This much of the description can be compiled to see whether there are syntax errors before proceeding to a more complicated description. Then the work begins to describe the stack's operation.

```
module FIFO (
   Data_out,                          // Data path from FIFO
   Data_in,                           // Data path into FIFO
   stack_empty,                       // Flag asserted high for empty stack
   stack_full,                        // Flag asserted high for full stack
   clk, rst,                          // External clock and reset
   write_to_stack,                    // Flag controlling a write to the stack
   read_from_stack                    // Flag controlling a read from the stack
   );
   parameter stack_width = 12;        // Width of stack and data paths
   parameter stack_height = 8;        // Height of stack (in # of words)
   parameter stack_ptr_width = 3;     // Width of pointer to address stack

   output       [stack_width -1 : 0] Data_out;
   output       stack_empty, stack_full;
   input        [stack_width -1 : 0] Data_in;
   input        clk, rst, write_to_stack, read_from_stack;
   reg [stack_ptr_width -1 : 0]   read_ptr, write_ptr;// Pointers (addresses) for
                                                      // reading and writing
   reg [stack_ptr_width : 0]      ptr_diff;           // Register holding diff of ptrs
   reg [stack_width -1 : 0]       Data_out;
   reg [stack_width -1 : 0]       stack [stack_height -1 : 0];// The memory array
endmodule
```

Figure 12.3 Header, ports, parameters, variables, and documentation for the FIFO.

The stack consists of eight 12-bit words, addressed by three-bit pointers. The output register, pointers, and stack are declared as register variables, and will be manipulated within cyclic behaviors. There are alternative ways to model a FIFO; our model for the stack will include two status flags indicating whether the stack is full or empty. These conditions will be revealed by *ptr_diff,* a register whose value is formed from the difference between the read and write pointers to the stack. The stack is empty when *read_ptr* and *write_ptr* have the same value (i.e., *ptr_diff* = 0); it is full when *ptr_diff* reaches the stack height.

The model includes a synchronous behavior describing the data transfer to and from the stack, and a second behavior that manages the stackpointers. A read operation is performed at each active edge of the clock if three conditions are satisfied: *read_from_stack* is asserted, *write_to_stack* is not asserted, and the stack is not empty. A write operation is performed if *write_to_stack* is asserted, *read_from_stack* is not asserted, and the stack is not full. The *read_from_stack* and *write_to stack* signals must not be in contention. If they are, the *read_from_stack* command has priority. It is up to the operating environment to see that this does not happen. The complete description of *FIFO* is given in Figure 12.4; a simple testbench is given in Figure 12.5.

```
module FIFO (
  Data_out,                        // Data path from FIFO
  Data_in,                         // Data path into FIFO
  stack_empty,                     // Flag asserted high for empty stack
  stack_full,                      // Flag asserted high for full stack
   clk, rst,                       // External clock and reset
  write_to_stack,                  // Flag controlling a write to the stack
  read_from_stack                  // Flag controlling a read from the stack
  );
  parameter stack_width = 12;      // Width of stack and data paths
  parameter stack_height = 8;      // Height of stack (in # of words)
  parameter stack_ptr_width = 3;   // Width of pointer to address stack

  output [stack_width -1 : 0]   Data_out;
  output                        stack_empty, stack_full;
  input  [stack_width -1 : 0]   Data_in;
  input                         clk, rst
  input                         write_to_stack, read_from_stack;

  reg [stack_ptr_width -1 : 0]  read_ptr, write_ptr; // Pointers (addresses) for
                                                     // reading and writing
  reg [stack_ptr_width : 0]   ptr_diff;            // Gap between ptrs

  reg    [stack_width -1 : 0]   Data_out;
  reg    [stack_width -1 : 0]   stack [stack_height -1 : 0];  // memory array

  assign stack_empty = (ptr_diff == 0) ? 1'b1 : 1'b0;
  assign stack_full = (ptr_diff == stack_height) ? 'b1: 1'b0;

  always @ (posedge clk or posedge rst) begin: data_transfer
    if (rst) Data_out = 0;

    else if ((read_from_stack) && (!write_to_stack) && (!stack_empty))
      Data_out <= stack [read_ptr];
    else if ((write_to_stack) && (!read_from_stack) && (!stack_full));
      stack [write_ptr] <= Data_in;
  end     // data_transfer

  always @ (posedge clk or posedge rst) begin: update_stack _ptrs
    if (rst)
      begin
        read_ptr <= 0;
        write_ptr <= 0;
        ptr_diff <= 0;
      end
```

Figure 12.4 Verilog description of the FIFO. Continues on page 538.

```
      else
        if ((write_to_stack) && (!stack_full) && (!read_from_stack))
          begin
            write_ptr <= write_ptr + 1;      // Address for next clock edge
            ptr_diff <= ptr_diff + 1;
          end
        else if ((!write_to_stack) && (!stack_empty) && (read_from_stack))
          begin
            read_ptr <= read_ptr + 1;
            ptr_diff <= ptr_diff -1;
          end
    end // update_stack_ptrs
endmodule
```

Figure 12.4 Verilog description of the FIFO. Concluded

```
module t_FIFO ();
  parameter stack_width = 8;
  parameter stack_height = 16;
  parameter stack_ptr_width = 4;

  wire [stack_width -1 : 0] Data_out;
  wire stack_empty, stack_full;
  reg [stack_width -1 : 0] Data_in;
  reg clk, rst, write_to_stack, read_from_stack;
  wire [11:0]   stack0, stack1, stack2, stack3, stack4, stack5, stack6, stack7;

  assign stack0 = M1.stack[0];        // Probes of the stack
  assign stack1 = M1.stack[1];
  assign stack2 = M1.stack[2];
  assign stack3 = M1.stack[3];
  assign stack4 = M1.stack[4];
  assign stack5 = M1.stack[5];
  assign stack6 = M1.stack[6];
  assign stack7 = M1.stack[7];
```

Figure 12.5 Testbench to verify the Verilog description of the FIFO. Continues on page 539.

```
    FIFO M1 (Data_out, Data_in, stack_empty, stack_full, clk, rst,
    write_to_stack, read_from_stack);

    always begin
      clk = 0;  #10 clk = 1; #10;
    end

    initial #1000 $stop;

    initial b
      #5 rst = 1; #40 rst = 0;#600 rst = 1; #40 rst = 0;
    end
    initial begin
      #5 Data_in = 4'hF;
      #80 Data_in = 4'he;
    end
    initial begin
      #75 write_to_stack = 1; read_from_stack = 0;
      #310 write_to_stack = 0; #60 read_from_stack = 1;
      #400 read_from_stack = 0; write_to_stack = 1;
      #100 write_to_stack = 0; read_from_stack = 1;
      #50 read_from_stack = 0;
    end

endmodule
```

Figure 12.5 Testbench to verify the Verilog description of the FIFO. Concluded.

The model of the FIFO stack has some interesting and subtle, features. First, notice that the range of *ptr_diff* is *[stack_ptr_width : 0]*, but the range of the pointers is just *[stack_ptr_width-1 : 0]*. The extra bit in *ptr_diff* allows its value to reach 8, the *stack_height*. The value of *ptr_diff* is incremented after the clock edge; once *ptr_diff* reaches *stack_height*, the *stack_full* flag is asserted, preventing further writes. (If *ptr_diff* could only reach *stack_height-1*, the next value would be 0 and the "full" condition would not be detected.) In the second synchronous behavior, the mechanism of the *ptr_diff* allows the write and read pointers to circulate around the stack, proceeding from top to bottom on successive writes until *write_ptr* reaches the same cell being pointed to by *read_ptr*. When *rst* asserts, *write_ptr*, *read_ptr*, and *ptr_diff* are set to 0 and the output register is flushed. The stack is not flushed. Given the initialized condition of the stack pointers, the residual values in the stack will be overwritten when subsequent "writes" occur. The value of *write_ptr* and the value of *ptr_diff* are incremented when a write to the stack occurs; the values of *read_ptr* and *ptr_diff* are decremented when a read from the stack occurs. In the first synchronous behavior, in a write cycle, the value of *write_ptr* immediately before the clock edge determines the address of the cell that will receive *data_in*. In a read cycle, the value of *read_ptr* determines the address of the cell whose value is read into *data_out*. The values of *stack_empty* and *stack_full* are modeled as continuous assignments (which will lead to combinational logic in synthesis). The assignments are sensitive to the changes in *ptr_diff*.

A key feature of the model is its use of non-blocking assignments. They preclude a race between the assignment of value to *read_ptr* and *write_ptr*, and their use is at the active edge of the clock.

Some simulation results are shown in Figure 12.6. First, notice that the action of *rst* initializes the **reg** variables holding *write_ptr, read_ptr, ptr_diff,* and *Data_out*. The first set of patterns applied by the testbench was defined to step the *write_ptr* through the stack while *write_to_stack* was asserted and *read_from_stack* was de-asserted. Notice the behavior of *ptr_diff, stack_empty,* and *stack_full* as the stepping process occurs. When *ptr_diff* reaches the value 8, *stack_full* asserts and no further writing occurs. Also observe that the individual cells of the stack are written in sequential order at the rising edge of the clock, beginning with *stack[0]*. The value stored in the cell is the value of *Data_in* at the active edge of the clock. During this phase of the test, *Data_out* holds the value created by the assertion of *rst*.

The next set of patterns reads the values back from the stack. In Figure 12.7, notice the movement of *read_ptr* and *ptr_diff* on successive active edges of the clock. Reading continues while *read_from_stack* is asserted and the stack is not empty. The value read from the stack is placed in *Data_out*.

The simulation results shown in Figure 12.8 show a partial filling followed by a partial emptying of the stack. Observe the movement of *ptr_diff* and the contents of the stack. In Figure 12.9, writing is attempted while the stack is full, and in Figure 12.10, reading is attempted while the stack is empty. Figure 12.11 shows the effect of an intermediate interrupt by an assertion of *rst*.

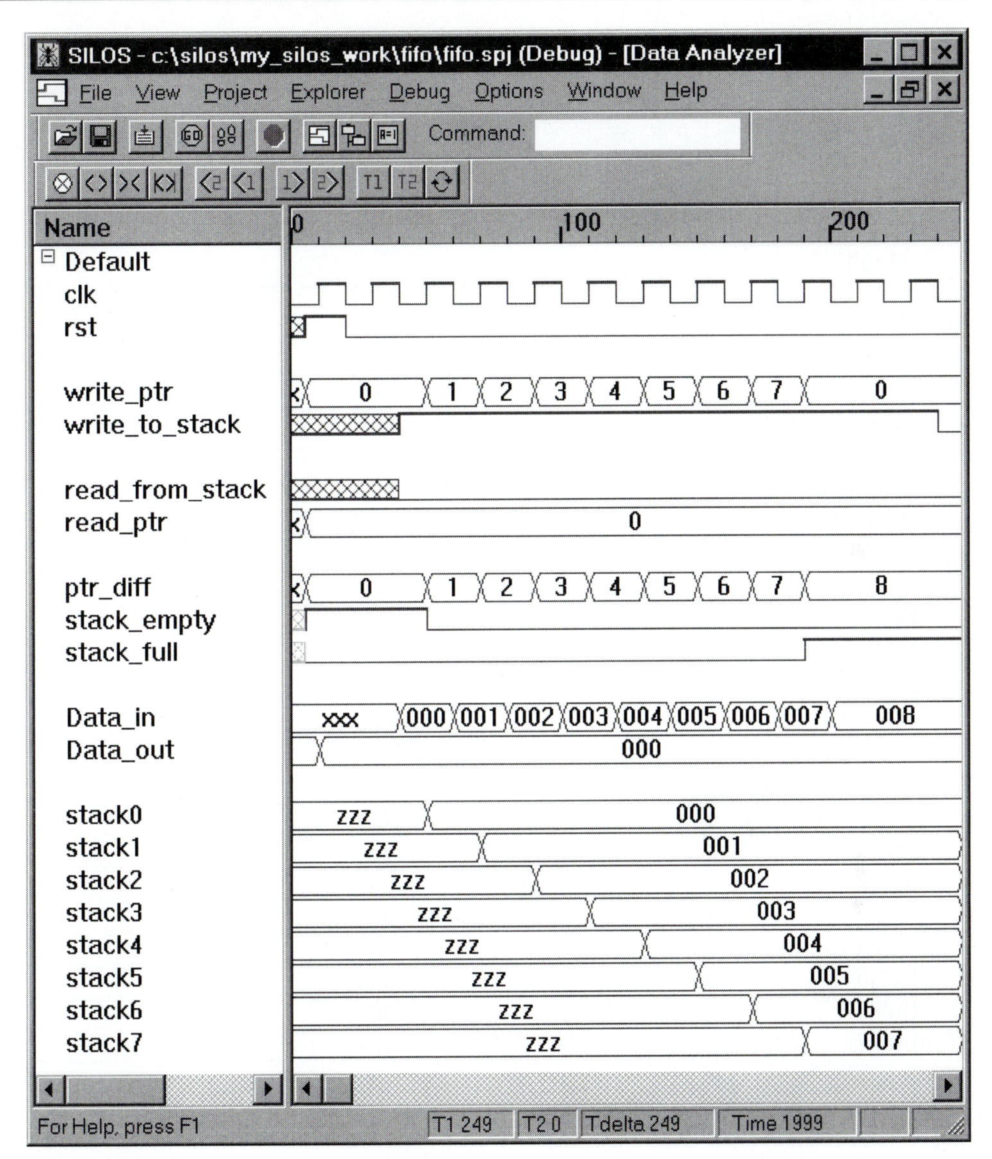

Figure 12.6 Simulation of eight-bit, eight-word FIFO: Demonstration of writing to fill the stack.

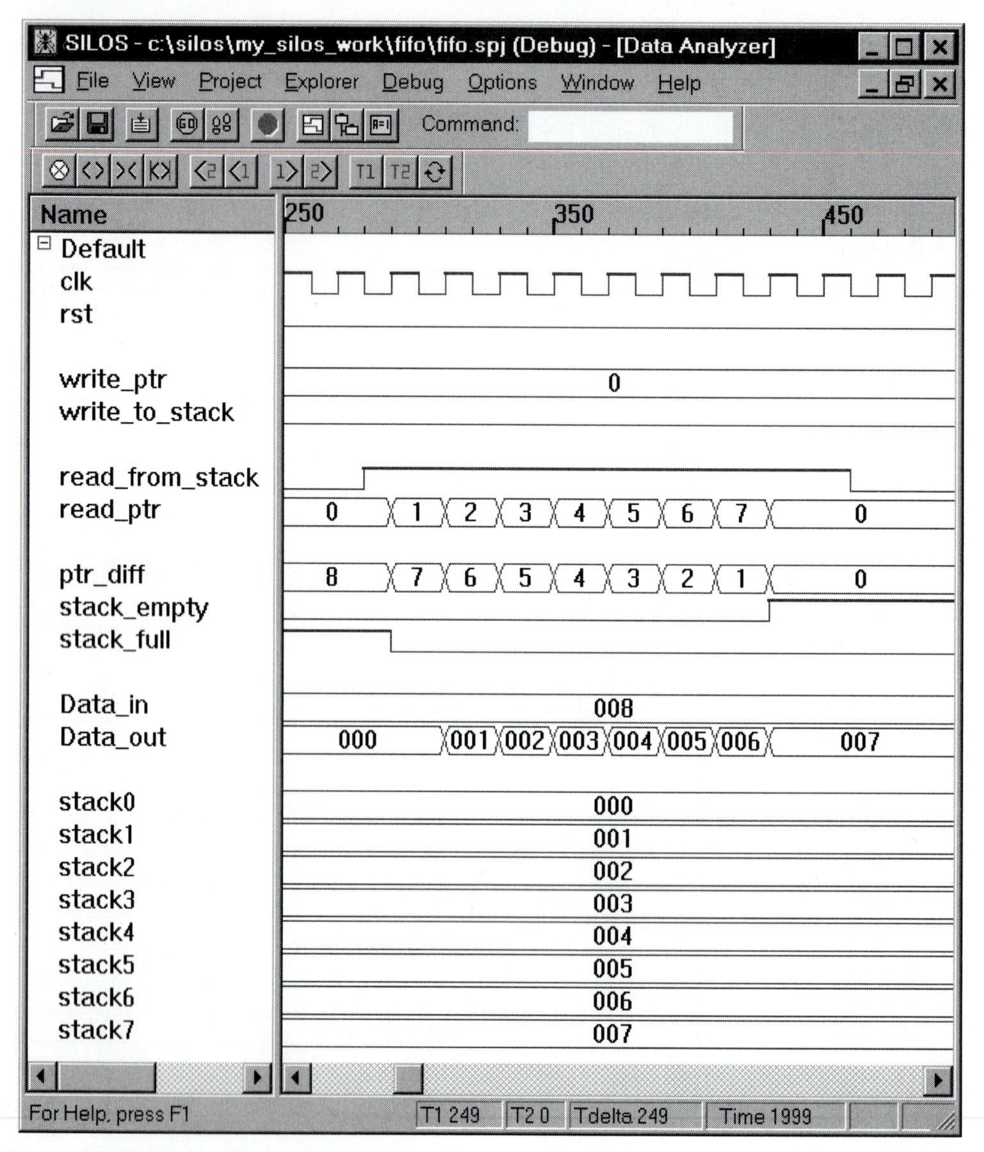

Figure 12.7 Simulation of eight-bit, eight-word FIFO: Demonstration of reading to empty the stack.

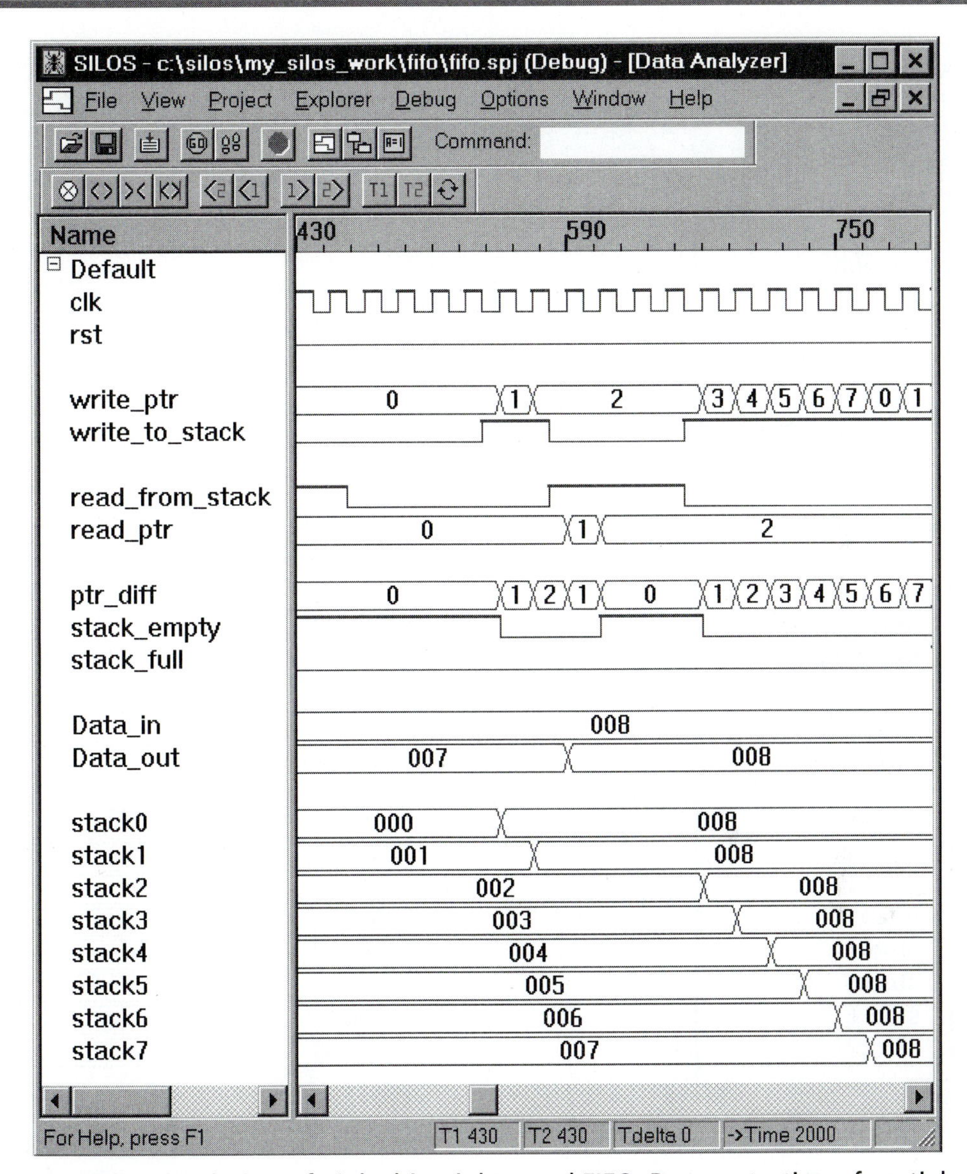

Figure 12.8 Simulation of eight-bit, eight-word FIFO: Demonstration of partial filling and emptying of the stack.

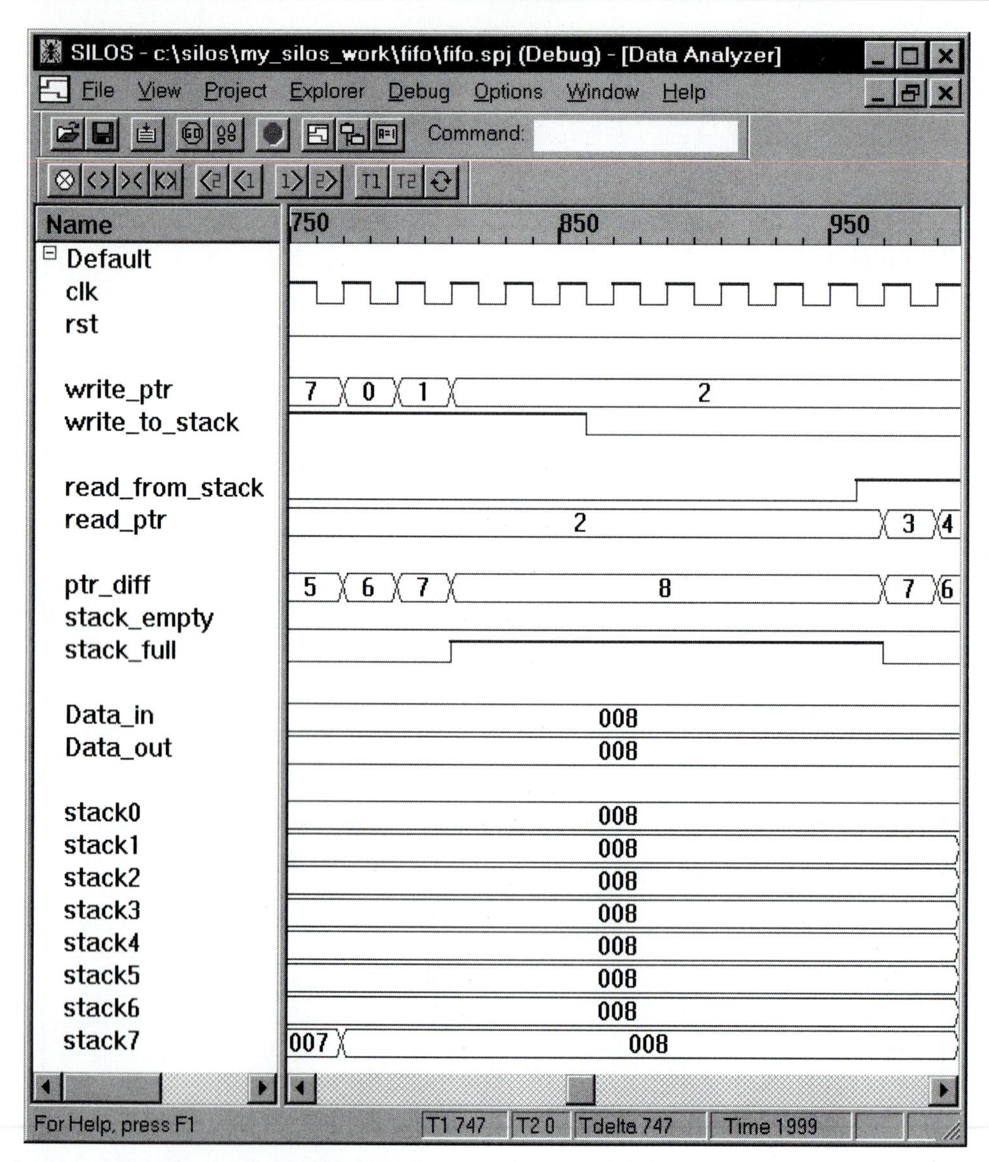

Figure 12.9 Simulation of eight-bit, eight-word FIFO: Demonstration of attempted writing while the stack is full.

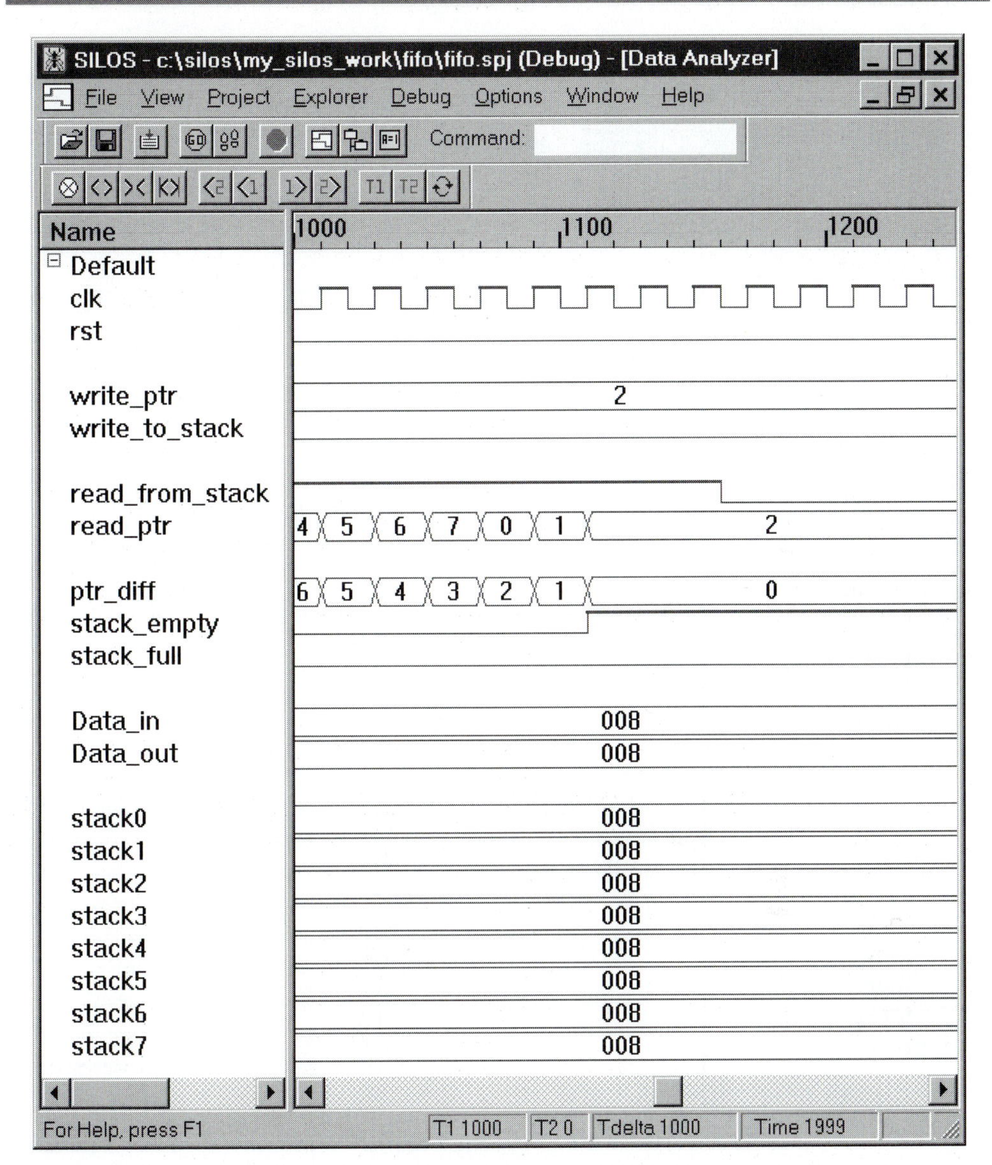

Figure 12.10 Simulation of eight-bit, eight-word FIFO: Demonstration of attempted reading while the stack is empty.

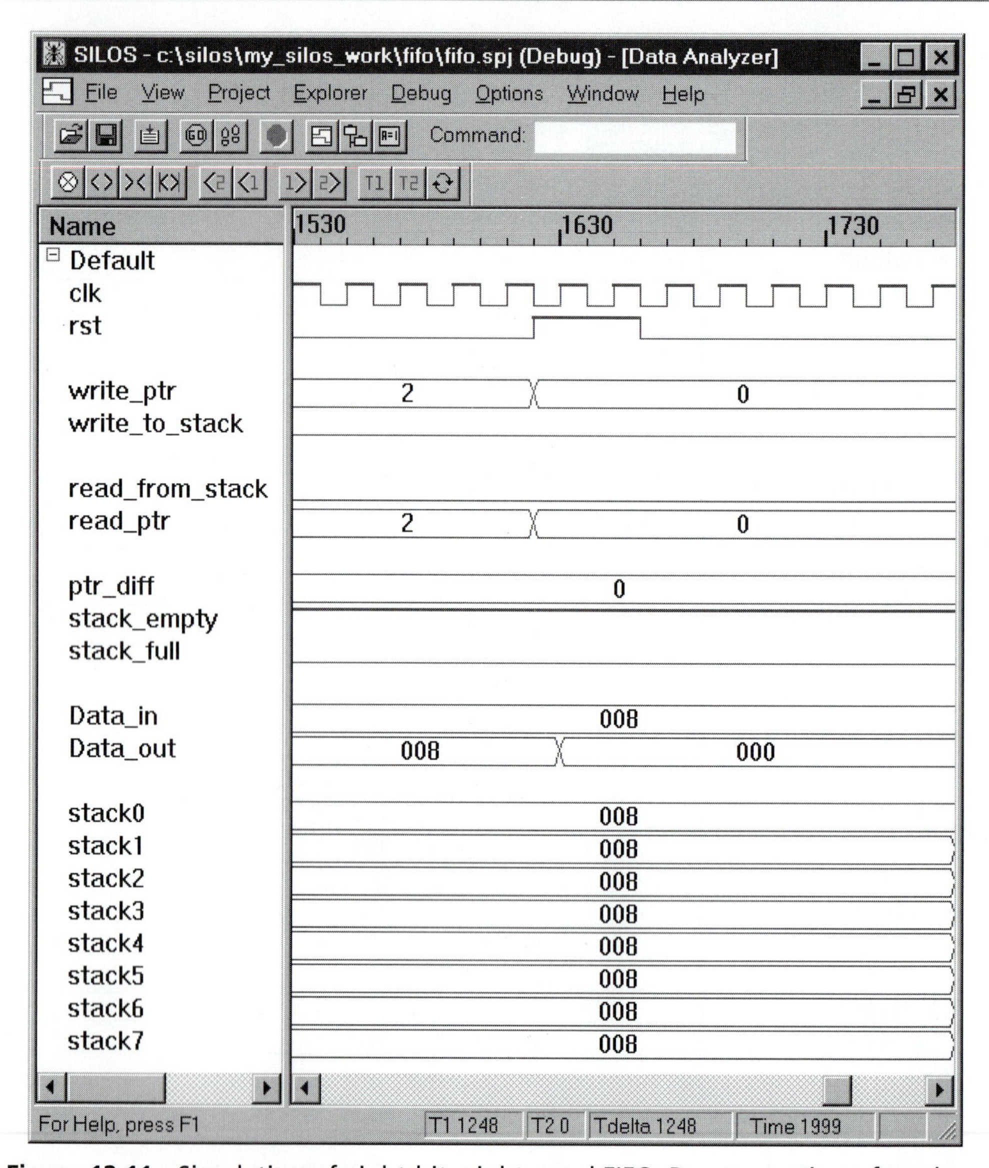

Figure 12.11 Simulation of eight-bit, eight-word FIFO: Demonstration of read to empty the stack.

Figure 12.12 shows the circuit synthesized from *FIFO* with a two-word, two-bit stack (chosen for illustration). This model has an asynchronous reset of *Data_out*, but not the stack. The loop constructs are not supported in the reset branch of an **if** statement in a cyclic behavior, so a loop could not be used to flush the stack. If it is important to flush the stack, each word must be flushed separately.

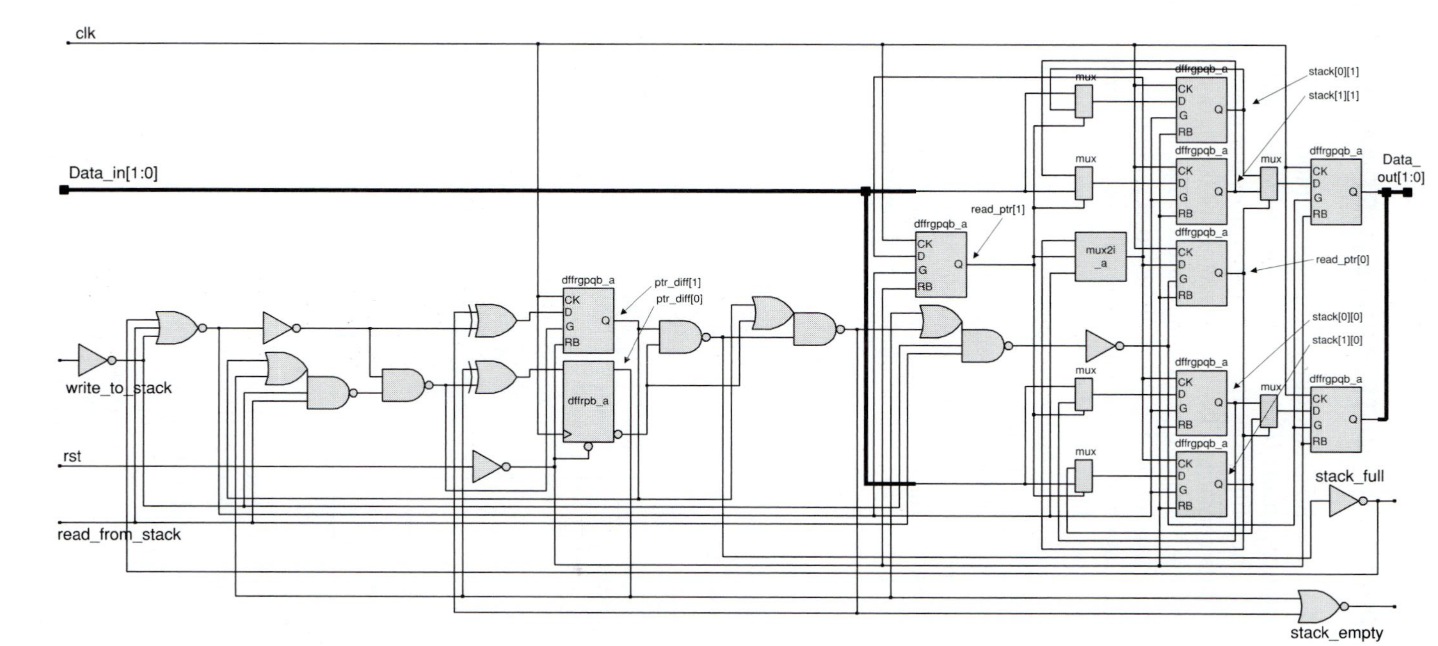

Figure 12.12 Circuit synthesized from FIFO with *stack_height* = 2 and *stack_width* = 2.

12.2 FIFO APPLICATION: TEMPERATURE MONITOR SYSTEM

Next, we'll use the FIFO stack in a temperature monitor system[1] that monitors two temperature sensors (*Data_Generator_1* and *Data_Generator_2*), as shown in Figure 12.13. The outputs of the data generators are loaded into registers and passed to a data converter. The size of a word of the data is chosen to be 12 bits, to accommodate a wide range of temperatures. The data converter is to convert temperature in units of degrees Celsius to temperature in units of degrees Fahrenheit. A data filter forms a four-sample moving average and provides the result to the input datapath of a FIFO. Given two channels of data, a FIFO controller controls the read and write operations of the two FIFOs under the direction of external signals (*read* and *select*).

The architecture of the system suggests how to proceed. First, we'll model the data register and data converter, and then verify their functionality. Then we will model and test the data filter. Next, we will connect the data register, data converter, and data filter together in tandem, and verify the integrated functionality.

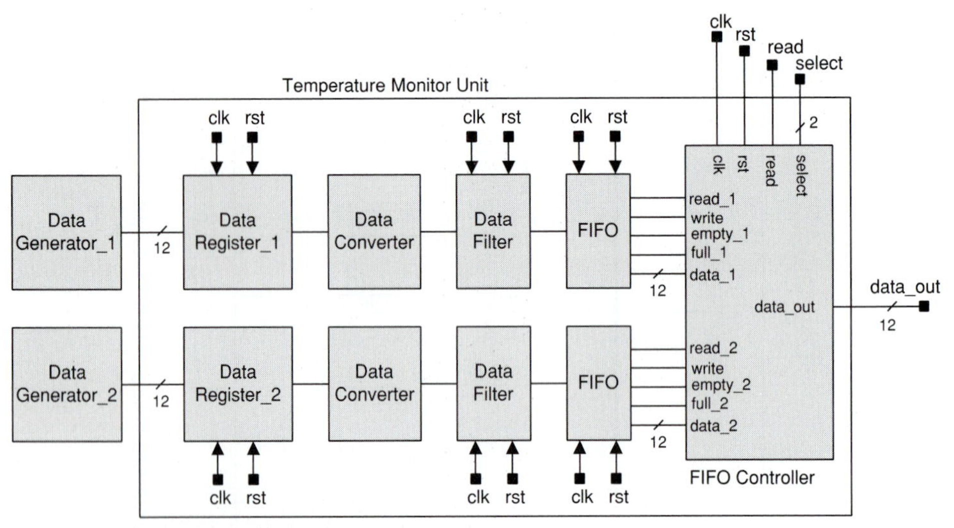

Figure 12.13 Temperature monitor system.

The description of the data register is given in Figure 12.14. The description uses asynchronous reset and synthesizes to the circuit shown in Figure 12.15.

```
module temp_register (temp_out, temp_in, clk, rst);
  output   [11:0]   temp_out;
  input    [11:0]   temp_in;
  input             clk, rst;
  reg      [11:0]   temp_out;
```

Figure 12.14 Verilog description of a temperature data register. Continues on page 549.

```
    always @ (posedge clk or posedge rst)
      begin
        if (rst == 1) temp_out = 0; else temp_out = temp_in;
      end
  endmodule
```

Figure 12.14 Verilog description of a temperature data register. Concluded.

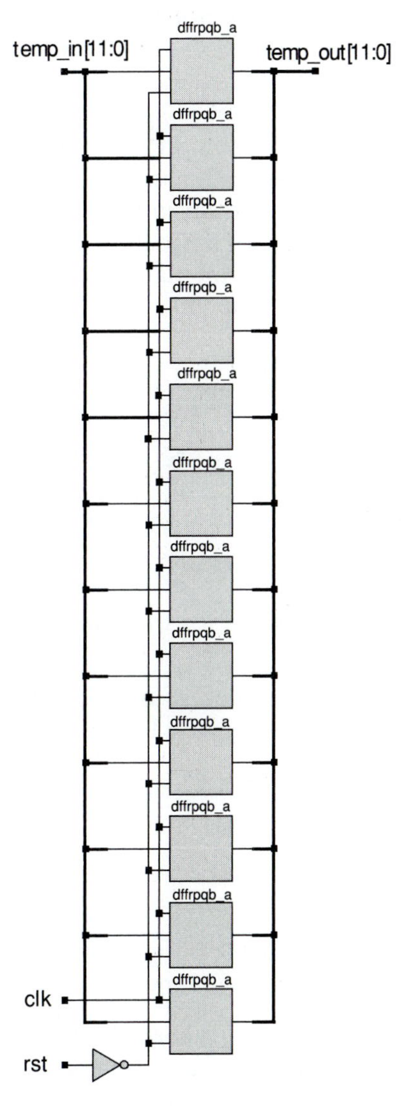

Figure 12.15 Circuit synthesized from *temp_register*.

The conversion formula used by the temperature converter is simply $T_f = (9/5)\ T_c + 32$, where T_f represents a temperature in Fahrenheit measure and T_c represents a temperature in Celsius measure. This simple formula reveals some pitfalls that can thwart the effort to synthesize a circuit. A continuous assignment statement provides the simplest implementation of the functionality of the converter:

```
assign temp_out = temp_in * 9 / 5 + 32;
```

Given the precedence of operators, the value (*temp_in *9*) is formed and divided by five. The result is added to 32. However, a synthesis tool will not synthesize the expression because the operand of the division operation is not constant. (Some tools support special compilers that can instantiate dividers.) An attempt to work around the problem by using *temp_out = (9/5) *temp_in + 32* synthesizes, but to the bogus circuit shown in Figure 12.16. The result of forming (9/5) is 1, so *temp_in* is added directly to 32. The synthesis tool is not the problem; it creates a circuit that implements the expression correctly. The expression just does not match the intended model. Another alternative form, *temp_out = (9 * temp_in) / 5 + 32*, fails to synthesize because the dividend is not a constant.

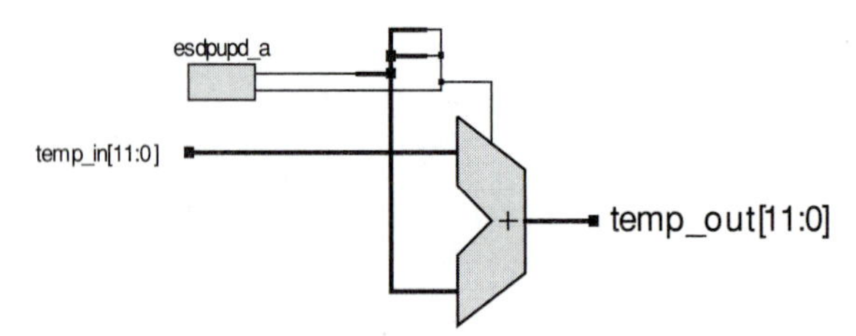

Figure 12.16 Bogus circuit synthesized for the temperature converter.

Yet another simple work-around is to approximate the value of the conversion factor 9/5 (1.8) by 1.75, which can be expressed as powers of two:

$$1.75 = 1 + .5 + .25 = 1 + 2^{-1} + 2^{-2} = 1 + 1/2 + 1/4$$

This leads to the expression:

```
parameter k1 = 2'b10;             // decimal 2
parameter k2 = 3'b100;            // decimal 4
parameter k32 = 6'b100000;        // decimal 32
assign temp_out = temp_in + temp_in / k1 + temp_in / k2 + k32;
```

This description, too, functions correctly but cannot be synthesized. The parameter *k32* is optional because numbers in Verilog are interpreted as default decimal values. The key to developing a synthesizable description is to realize that parameters having a value of a power of two can be associated with the shift operator and synthesized for division. So *temp_in /2 = temp_in >> 1*, and *temp_in / 4 = temp_in >> 2*. This leads to the synthesis-friendly description below:

```
module temp_converter (temp_out, temp_in);
  input      [11: 0]    temp_in;
  output     [11: 0]    temp_out;
  wire       [11: 0]    old_out;
  parameter k32 = 6'b100000;        // decimal 32

  // For temperature conversion, F = 9/5 C + 32
  // Approximation: 9/5 = 1.8 ~ 1 + .5 + .25

  assign temp_out = temp_in + (temp_in >> 1) + (temp_in >> 2) + k32;

endmodule
```

The data listed below was produced by simulating the temperature converter. A hand-check confirms that the model of the conversion function is correct.

```
  0 temp_in=   x temp_out =   x
 25 temp_in=   0 temp_out =  32
 50 temp_in=  20 temp_out =  67
 75 temp_in=  50 temp_out = 119
100 temp_in= 100 temp_out = 207
125 temp_in=   0 temp_out =  32
```

The temperature converter synthesizes to the circuit shown in Figure 12.17 The arrangement of bits at the inputs of the 12-bit adders accomplishes the shift operations. The last stage receives a bit set to 1 to add by 32.

An alternative description of *temp_converter* uses the same approximation formula for calculating *temp_out* from *temp_in*, but forms the approximation as

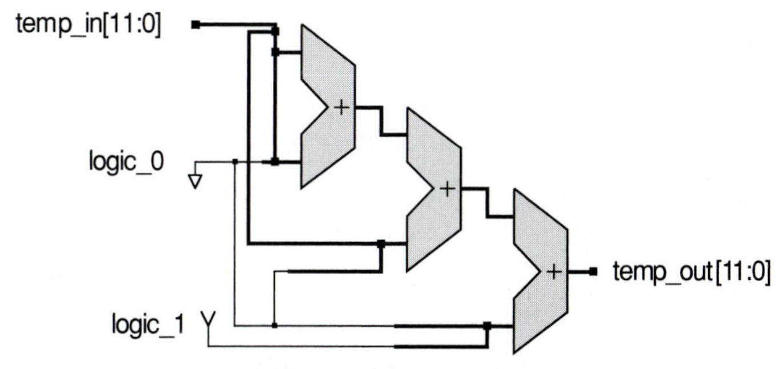

Figure 12.17 Circuit synthesized from *temp_converter*.

1.75temp_in = 2temp_in - .25temp_in. This gives the following continuous assignment:

```
assign temp_out = (temp_in << 1) – (temp_in >> 2) + k32;
```

This simple change leads to the circuit in Figure 12.18. (Note: Failure to include parens with the shift operators leads to a bogus result!) This version uses a subtractor and only one adder; the circuit uses 43% less combinational cell area than the previous version, which used three adders. Don't expect a synthesis tool to discover these alternative circuits. The quality of the results depends on your descriptive style.

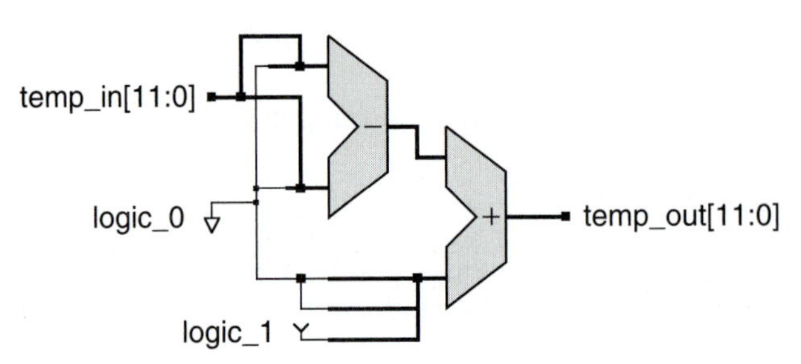

Figure 12.18 Circuit synthesized from an alternative description of *temp_converter*.

The temperature filter described in Figure 12.19 forms a four-sample moving average of the converted temperature. The synchronous behavior of *temp_filter* implements a shift register holding four samples of *temp_data*. The average temperature is formed asynchronously by a continuous assignment. The average of the four smples is obtained by a two-place shift to the right, which divides the sum by four. Figure 12.20 shows the annotated result of simulating the operation of the moving average filter. Notice how the correct value of the average temperature appears after four samples fill the pipeline of registers.

```
module temp_filter (av_temp, clk, rst, temp_data);
  output   [11:0]    av_temp;
  input               clk;
  input    [11:0]    temp_data;
  reg      [11:0]    temp [3:0];
  always @ (negedge clk or posedge rst)
    if (rst) begin
      temp[0] <= 0;
      temp[1] <= 0;
      temp[2] <= 0;
      temp[3] <= 0; end
```

Figure 12.19 Four-sample moving average temperature filter. Continues on page 553.

```
        else begin: data_pipeline
          temp[3] <= temp_data;
          temp[2] <= temp[3];
          temp[1] <= temp[2];
          temp[0] <= temp[1];
        end
      assign av_temp = (temp[3] + temp[2] + temp[1] + temp[0]) >> 2;
    endmodule
```

Figure 12.19 Four-sample moving average temperature filter. Concluded.

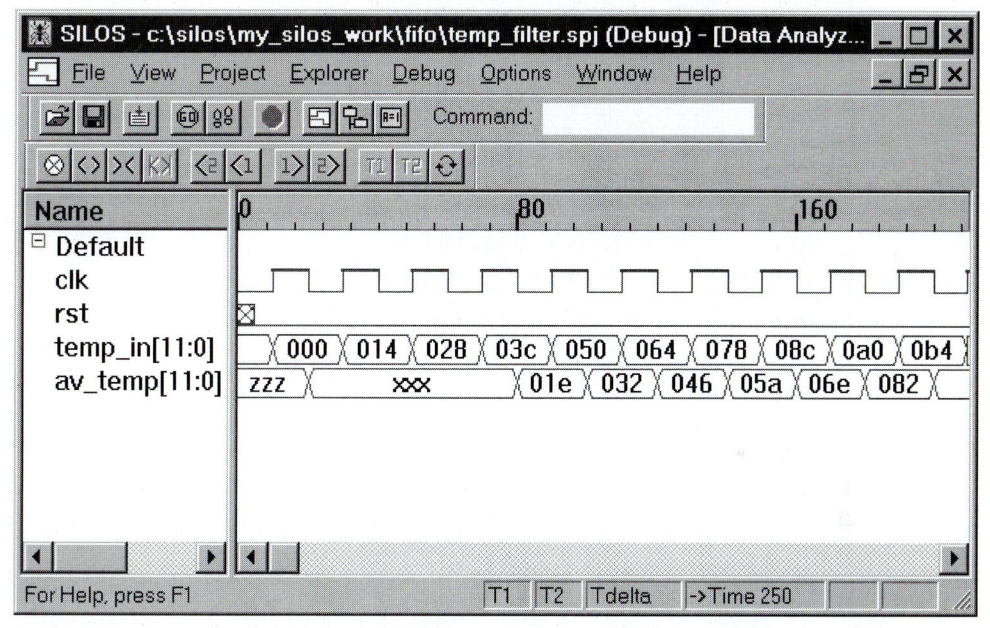

Figure 12.20 Simulation results for the moving-average temperature filter, *temp_filter*.

The circuit synthesized from *temp_filter* is shown in Figure 12.21. It consists of a four-stage pipeline adder with bits shifted to accomplish division by four. Each adder cell has a gate level representation (not shown). The carry-in to each adder is tied to 0, and the last adder stage has its two left-most bits tied to 0 for division by four. The remaining bits are formed from the left-most bits of the preceding stage.

The next step in the design is to interface the data register, temperature converter and moving-average filter together and verify that they work correctly as a unit. The result of simulating these three modules together is shown in Figure 12.22. The value of *temp_in* is registered in *temp_out* and converted to a

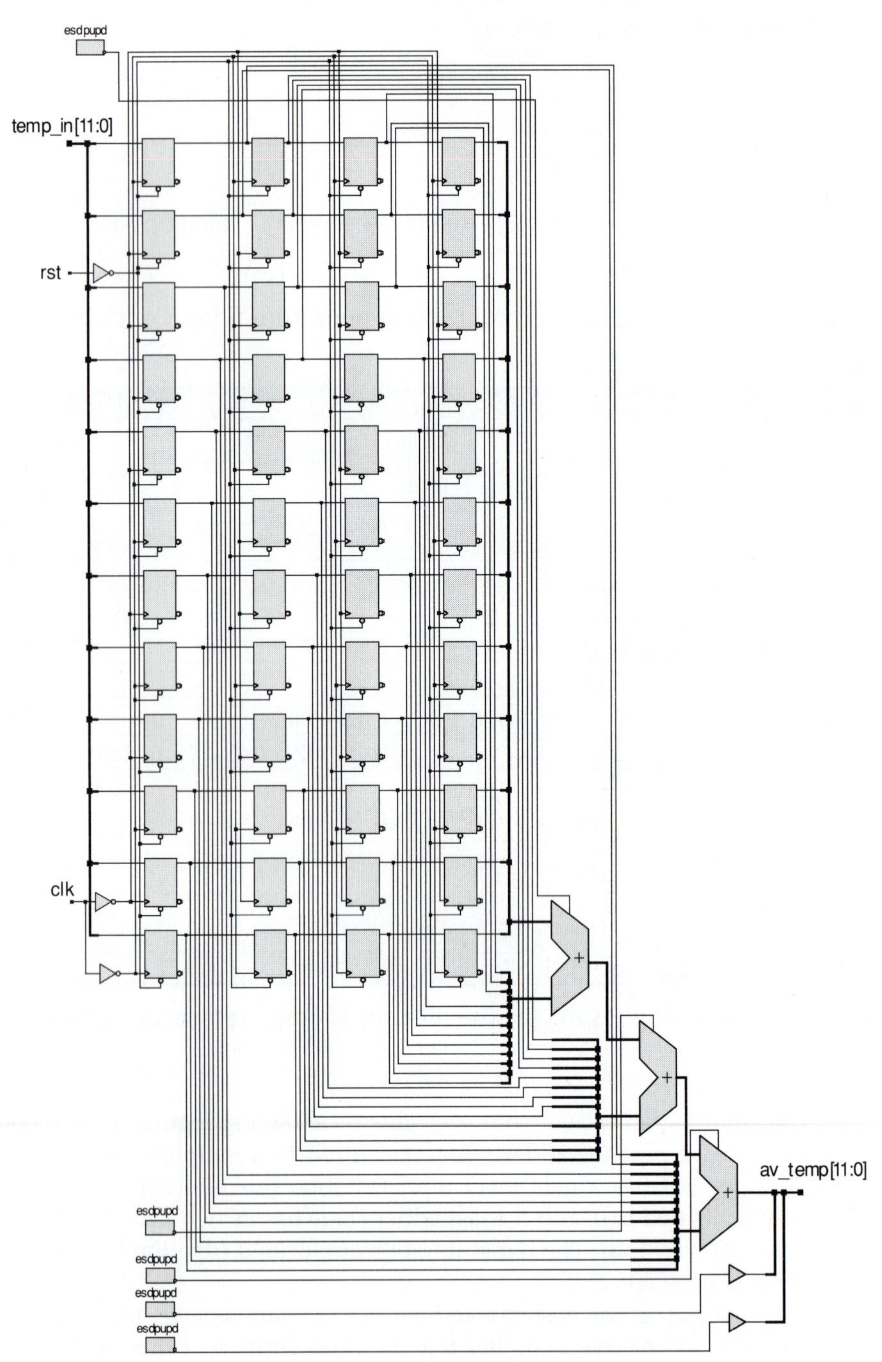

Figure 12.21 Circuit synthesized from *temp_filter*.

Fahrenheit temperature value in *temp_conv.* The movement of data through the pipeline within *temp_filter* is shown, along with the average temperature, *av_temp*. (Note: All numbers are hex values.)

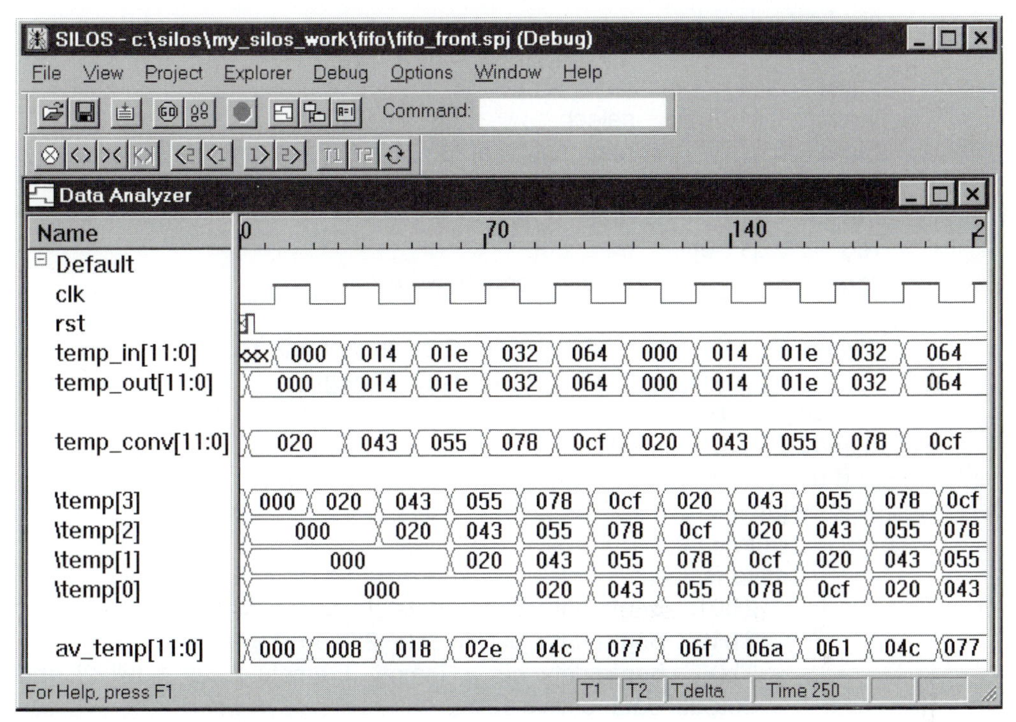

Figure 12.22 Result of simulating the temperature monitor unit's front end, consisting of *temp_register*, *temp_converter*, and *temp_filter*.

The temperature monitor system has two FIFOs, so a controller must govern the activity of the datapath for *data_out*. The signal *select* determines which controller is to be read. If *select* has the pattern 01 and *read* is asserted, the first FIFO is read (subject to the stack status); if the pattern is 10, the second FIFO is read; if the pattern is 00, the status flags of the two FIFOs are placed in the lower four bits of the output in the order {empty_1, empty_2, full_1, full_2}. Otherwise, the output is placed in the "z" state (three-stated). If *read* is not asserted, the output is put in the "z" state and *write* is asserted. Writing is done to both FIFOs simultaneously, subject to their individual status flags. The model, *fifo_controller*, is shown in Figure 12.23.

```
module fifo_controller (data_out, data_1, data_2, select, read, write, empty_1,
  full_1, read_1, empty_2, full_2, read_2);
  /* This unit selects and reads the data channels from the two FIFO units,
  directs a FIFO to write data into the stack from the sensor channel, and
  either outputs the data from a read or outputs the status bits from the two
  stacks. */

  input    [11:0]   data_1, data_2;
  input    [1:0]    select;
  input             read, full_1, empty_1, full_2, empty_2;
  output   [11:0]   data_out;
  output            read_1, read_2, write;
  reg      [11:0]   data_out;
  reg               read_1, read_2, write;

  always @ (select or data_1 or data_2 or read or full_1
           or empty_1 or full_2 or empty_2) begin
    read_1 <= 0;
    read_2 <= 0;
    if (read == 1)
      begin
        write <= 0;
        case (select)
          2'b01:  begin data_out <= data_1; read_1 <= 1; end
          2'b10:  begin data_out <= data_2; read_2 <= 1; end
          2'b00:  begin data_out <= {empty_1, empty_2, full_1, full_2}; end
          default: data_out <= 12'bz;
        endcase
      end
    else if (read == 0)
      begin
        data_out <= 12'bz;
        write <= 1;
      end
    else data_out <= 12'bz;
  end
endmodule
```

Figure 12.23 Verilog model of a controller for two FIFOs.

The results of simulating *fifo_controller* are shown in Figure 12.24. The waveforms demonstrate the operations of (1) writing to the stacks, (2) reading the status bits, and (3) reading from a selected stack. The output is three-stated when *write* is asserted.

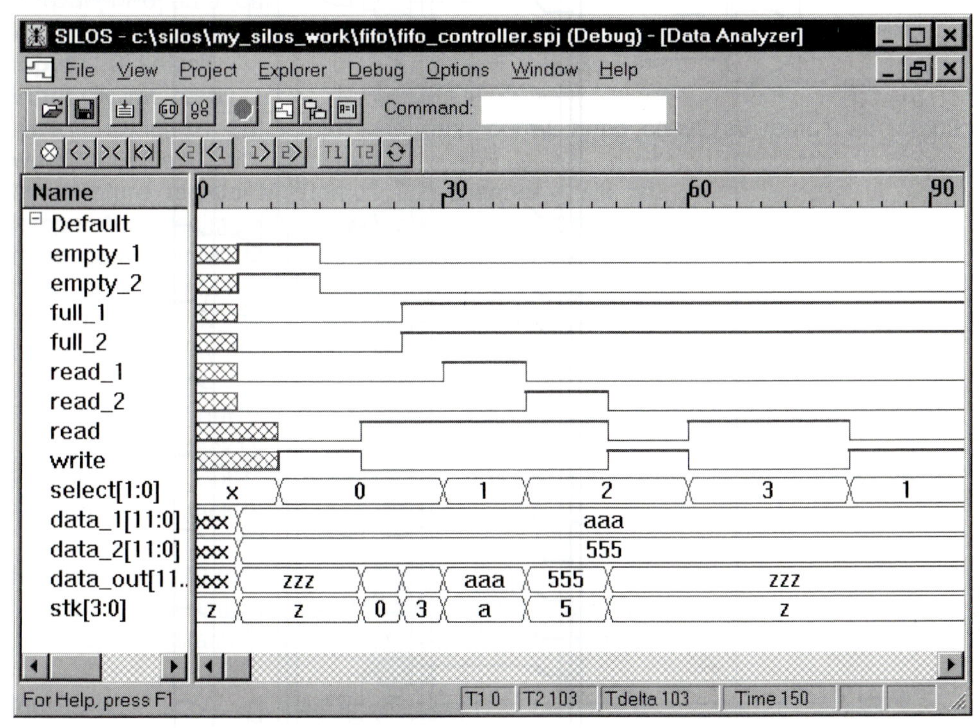

Figure 12.24 Simulation of *fifo_controller*.

The circuit synthesized from *fifo_controller* is shown in Figure 12.25. It consists of combinational logic controlling the datapaths from the two FIFOs.

12.3.1 UART—TRANSMITTER

The input-output signals of a state machine controller for the transmitter are shown in the high level block diagram in Figure 12.33. The input signals are provided by the host processor, and the output signals control the movement of data in the UART.

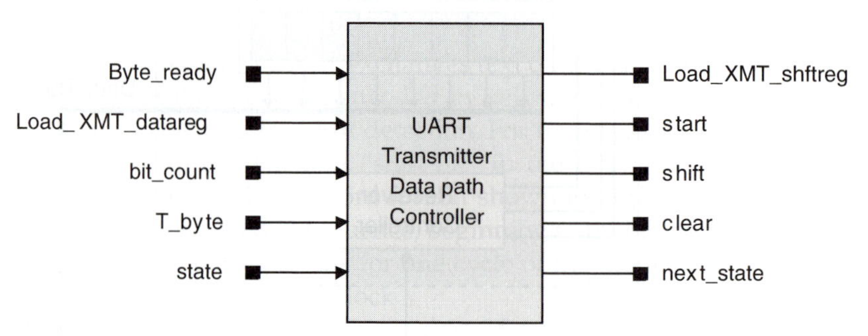

Figure 12.33 Interface signals of a state machine controller for a UART transmitter.

The controller has the following inputs:

Byte_ready	— asserted by host machine to indicate that *data_bus* has valid data
Load_XMT_datareg	— assertion transfers *Data_Bus* to the transmitter data storage register, *XMT_datareg*
T_byte	— assertion initiates transmission of a byte of data, along with the stop, start, and parity bits
bit_count	— counts bits in the word during transmission
state	— state of the transmitter controller state machine

The state machine forms the following output signals controlling the data-path of the transmitter:

Load_XMT_shftreg	— assertion loads the contents of *XMT_datareg* into *XMT_shftreg*
start	— signals the start of transmission

shift — directs *XMT_shftreg* to shift by one bit towards the LSB and backfill with a stop bit (1)

clear — clears bit_counter

next_state — the next state of the state machine controlling the data path of the transmitter

The ASM chart of the state machine controlling the transmitter is shown in Figure 12.34. The machine has three states: *idle, waiting* and *sending*. When *reset_* is asserted, the machine asynchronously enters *idle; bit_count* is flushed, *XMT_shftreg* is loaded with ones, and the control signals *clear, Load_XMT_shftreg, shift*, and *start* are driven to 0. In *idle*, if an active edge of *Clock* occurs while *Load_XMT_datareg* is asserted by the external host, the contents of *Data_Bus* will transfer to *XMT_datareg*. (This action is not part of the ASM chart because it occurs independently of the state of the machine.) The machine remains in *idle* indefinitely. When *Byte_ready* is asserted, *Load_XMT_shftreg* is asserted and *next_state* is driven to *waiting*. The assertion of *Load_XMT_shftreg* indicates that *XMT_datareg* now contains data that can be transferred to the internal shift register. At the next active edge of *Clock*, with *Load_XMT_shftreg* asserted, three activities occur: (1) *state* transfers from *idle* to *waiting*, (2) the contents of *XMT_datareg* are loaded into the left-most bits of *XMT_shftreg*, a (*word_size + 1*)-bit shift register whose LSB signals the start and stop of transmission, and (3) the LSB of *XMT_shftreg* is re-loaded with 1, the stop-bit. The machine remains in *waiting* until the external processor asserts *T_byte.* At the next active edge of *Clock*, with *T_byte* asserted, *state* enters *sending*, and the LSB of *XMT_shftreg* is set to 0 to signal the start of transmission. At the same time, *shift* is driven to 1, and *next_state* retains *sending*. At subsequent active edges of *Clock*, with *shift* asserted, *state* remains in *sending* and the contents of *XMT_shftreg* are shifted towards the LSB, which drives the external serial channel. As the data shifts occur, ones are back-filled in *XMT_shftreg*, and *bit_count* increments. With *state* in *sending, shift* asserts while *bit_count* is less than nine. The machine increments *bit_count* after each movement of data, and when *bit_count* reaches nine, *clear* asserts, indicating that all the bits of the augmented word have been shifted to the serial output. At the next active edge of *Clock*, the machine returns to *idle*.

The control signals produced by the state machine induce state-dependent register transfers in the data path. The activity of the primary inputs (*Byte_ready, Load_XMT_shftreg*, and *T_byte*) and the signals from the controller (*Load_XMT_shftreg, start, shift, clear*) are shown in Figure 12.35, along with the movement of data in the data path registers. The values of the control signals are indicated before the active edge of *Clock* and cause the register transfers shown. The sequence of output bits is also shown, with 1s filling *XMT_shftreg* under the action of *shift*.

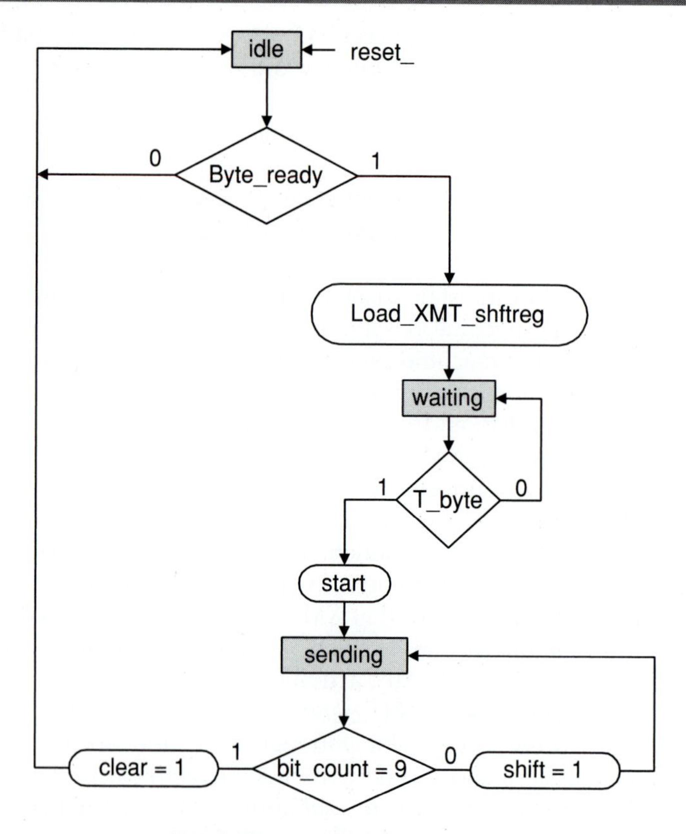

Figure 12.34 ASM chart for the state machine controller for a UART transmitter.

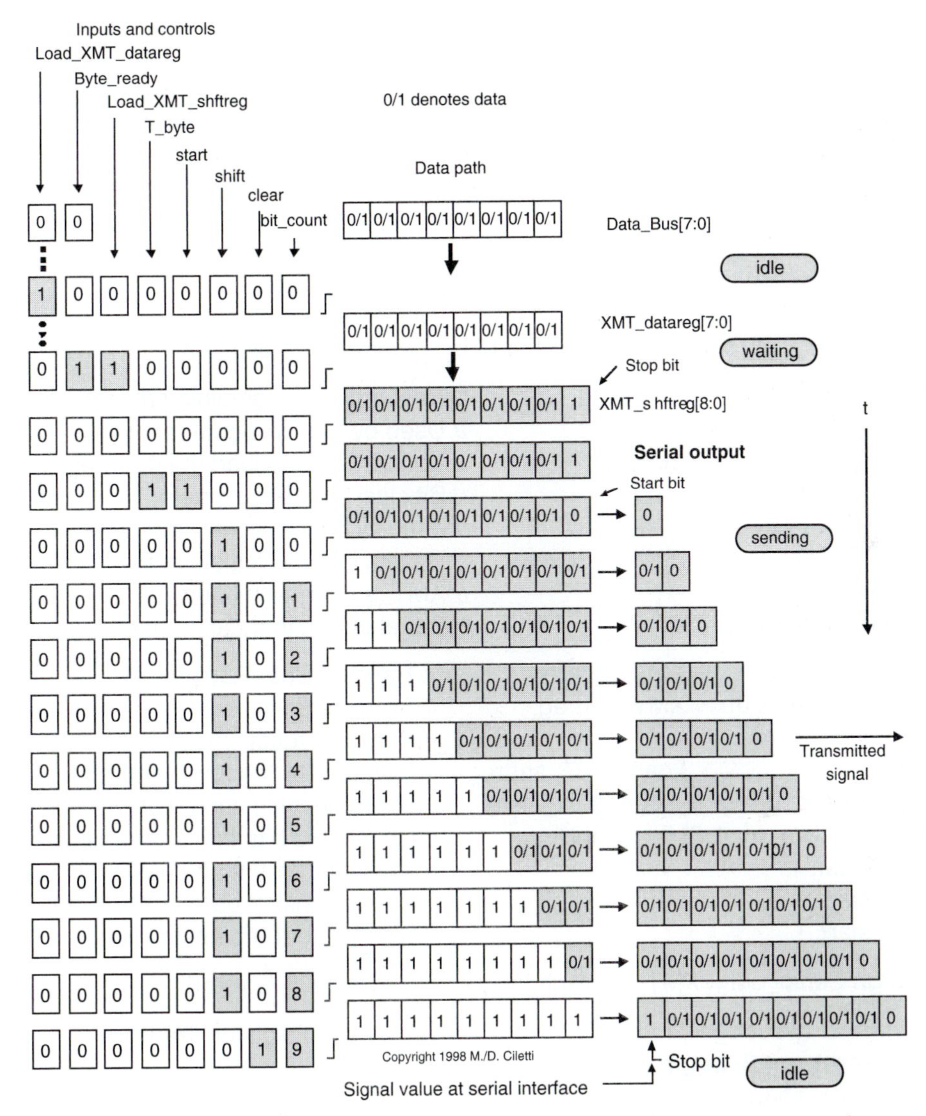

Figure 12.35 Control signals and data flow in an eight-bit UART transmitter.

The Verilog description of the transmitter shown in Figure 12.36, *UART_Transmitter*, has two cyclic behaviors—one combinational and one sequential. An edge-triggered behavior changes the state and manipulates the data path. A combinational behavior generates the control signals as a function of the external inputs and the state of the machine.

```
module UART_Transmitter
  (Data_Bus, Byte_ready, Load_XMT_datareg, Serial_out, T_byte, Clock, reset_);
  parameter   word_size = 8;                  // Size of data word, e.g., 8 bits
  parameter   one_hot_count = 3;              // Number of one-hot states
  parameter   state_count = one_hot_count;    // Number of bits in state register
  parameter   size_bit_count = 3;             // Size of the bit counter, e.g., 4
                                              // Must count to word_size + 1
  parameter   idle = 3'b001;                  // one-hot state encoding
  parameter   waiting = 3'b010;
  parameter   sending = 3'b100;
  parameter   all_ones = 9'b1_1111_1111;      // Word + 1 extra bit

  input   [word_size - 1 : 0] Data_Bus;   // Host data bus containing data word
  input   Byte_ready,                     // Used by host to signal ready
          Load_XMT_datareg,               // Used by host to load the data register
          Clock,                          // Bit clock of the transmitter
          T_byte,                         // Used by host to signal start of transmission
          reset_;                         // Resets internal registers, loads the
                                          // XMT_shftreg with ones
  output Serial_out;                      // Serial output to data channel

  reg [word_size -1: 0]  XMT_datareg;         // Transmit Data Register
  reg [word_size: 0]     XMT_shftreg;         // Transmit Shift
                                              // Register: {data, start bit}
  reg                    Load_XMT_shftreg;    // Flag to load the XMT_shftreg
  reg [state_count -1: 0] state, next_state;  // State machine controller
  reg [size_bit_count: 0] bit_count;          // Counts the bits that are transmitted
  reg  clear,                                 // Clears bit_count after last bit is sent
       shift,                                 // Causes shift of data in XMT_shftreg
       start;                                 // Signals start of transmission

  assign Serial_out = XMT_shftreg[0];         // LSB of shift register

  always @ (posedge Clock or negedge reset_) begin: Register_Transfers
    if (reset_ == 0) begin
      XMT_shftreg <= all_ones;
      state <= idle;
      bit_count <= 0;
    end
    else begin
      state <= next_state;

      if (Load_XMT_datareg == 1)
        XMT_datareg <= Data_Bus;                 // Get the data bus
```

Figure 12.36 Description of UART transmitter. Continues on page 575.

```
        if (Load_XMT_shftreg == 1)
          XMT_shftreg <= {XMT_datareg,1'b1};    // Load shift reg,
                                                // insert stop bit

        if (start == 1)
          XMT_shftreg[0] <= 0;                  // Signal start of transmission

        if (clear == 1) bit_count <= 0;
        else if (shift == 1) bit_count <= bit_count + 1;

        if (shift == 1)
          XMT_shftreg <= {1'b1, XMT_shftreg[word_size:1]}; // Shift right, fill 1's
        end
    end

    always @ (state or Byte_ready or bit_count or T_byte) begin
      clear <= 0;
      Load_XMT_shftreg <= 0;
      shift <= 0;
      start <= 0;
      case (state)
        idle:       if (Byte_ready == 1) begin
                      Load_XMT_shftreg <= 1;
                      next_state <= waiting;
                    end else next_state <= idle;

        waiting:    if (T_byte == 1) begin
                      start <= 1;
                      next_state <= sending;
                    end else begin
                      next_state <= waiting;
                    end

        sending:    if (bit_count != word_size + 1) begin
                      shift <= 1;
                      next_state <= sending;
                    end else begin
                      clear <= 1;
                      next_state <= idle;
                    end

        default:    next_state <= idle;
      endcase
    end
endmodule
```

Figure 12.36 Description of UART transmitter. Concluded.

Some simulation results are shown in Figure 12.37 for an eight-bit data word. The waveforms produced by the simulator have been annotated to indicate significant features of the machine's behavior. First, observe the values of the signals immediately after *reset_* is asserted. The state is *idle*. Note that *Data_Bus* initially contains the value $a7_h$ (1010_0111_2), a value specified by the testbench used for simulation. With *Byte_ready* low and *Load_XMT_datareg* asserted, the *Data_Bus* is loaded into *XMT_datareg*. The machine remains in *idle* until *Byte_ready* is asserted. When *Byte_ready* asserts, *Load_XMT_shftreg* asserts. This causes the state to change to *waiting* at the next active edge of *Clock*. The nine-bit *XMT_shftreg* is now loaded with the value $\{a7_h, 1\} = 1\_0100\_1111_2 = 14f_h$. Note that the LSB of *XMT_shftreg* is loaded with a one. The machine remains in *waiting* until *T_byte* is asserted. The assertion of *T_byte* asserts *start*. The machine enters *sending* at the active edge of *Clock* immediately after the host processor's assertion of *T_byte*, and the LSB of *XMT_shftreg* is loaded with a 0. The nine-bit word in *XMT_shftreg* becomes $1\_0100\_1110_2 = 14e_h$. The 0 in the LSB signals the start of transmission. Figure 12.38 shows the movement of data through *XMT_shftreg*. Notice that 1s are filled behind as the word shifts to the right. At the active edge of the clock after *bit_count* reaches 5, *clear* asserts, *bit_count* is flushed, and the machine returns to *idle*.

For diagnostic purposes, the testbench included a 10-bit shift register, *Serial_test[9:0]*, that received *Serial_out* (by hierarchical de-referencing). The inner eight-bits of this register were displayed in Figure 12.37 as *sent_word[7:0]* (skipping the start-bit and the stop-bit) to reveal the correct transmission of data, $a7_h$. The bit sequence of *Serial_out* likewise has this value. This is evident at the active edge of the clock after the assertion of *clear*. The movement of data through *XMT_shftreg* is shown in Figure 12.38

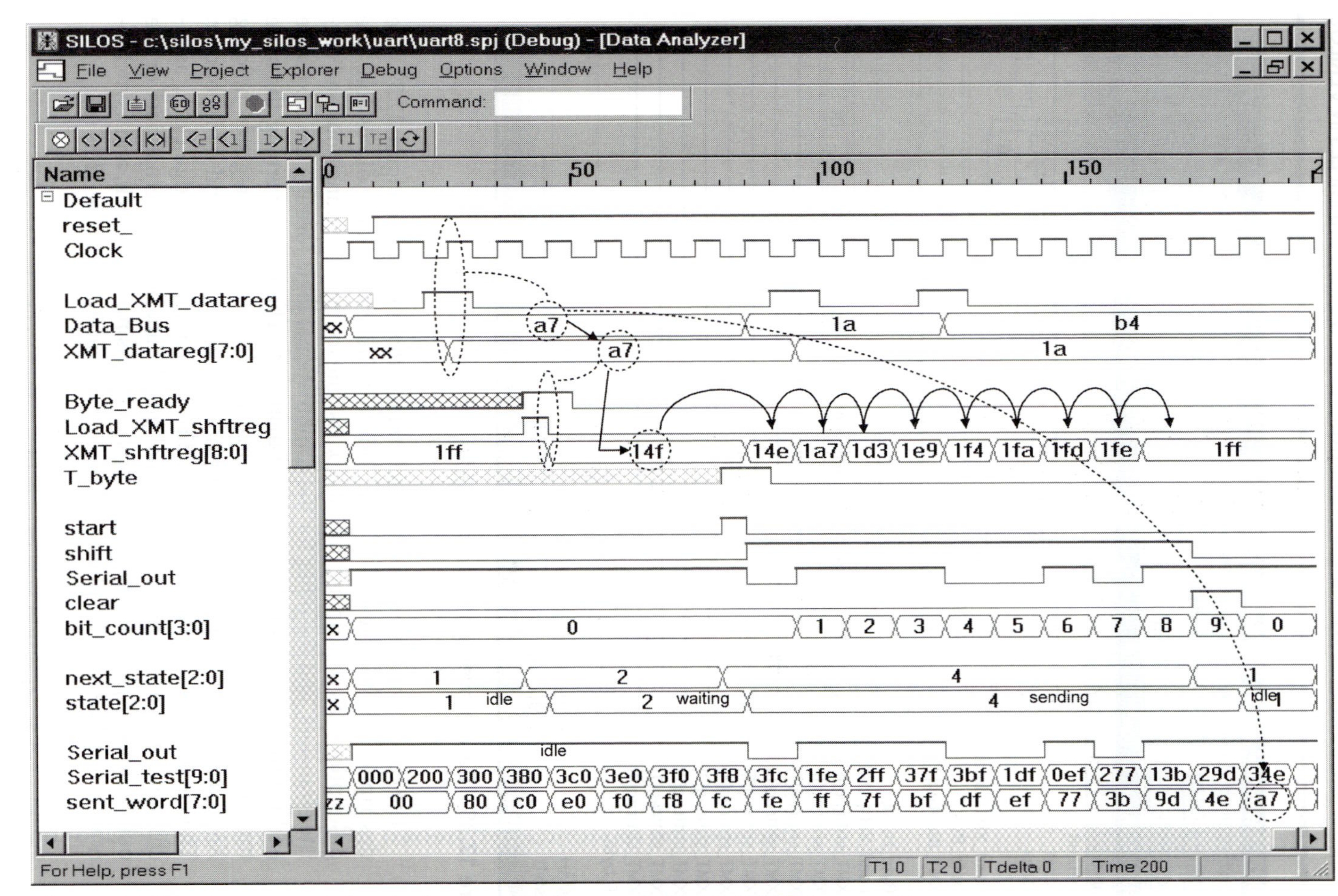

Figure 12.37 Annotated simulation results for an eight-bit UART transmitter.

```
module UART8_receiver
  (Serial_in, Sample_clk, Ready_to_read, reset_, read_not_ready_in,
  RCV_datareg, read_not_ready_out, Error1, Error2); // Sample clock is 8x Bit_clk

  parameter   word_size = 8;
  parameter   half_word = word_size / 2;
  parameter   Num_counter_bits = 4;            // Must hold count of word_size
  parameter   Num_state_bits = 2;              // Number of bits in state
  parameter   idle              = 2'b00;
  parameter   starting          = 2'b01;
  parameter   receiving         = 2'b10;

  input       Serial_in,               // Serial data input
              Sample_clk,              // Clock to sample serial data
              reset_,                  // active-low reset
              read_not_ready_in;       // status bit from host processor

  output      [word_size-1: 0]   RCV_datareg;          // Data register for the output
  output                         read_not_ready_out,   // Handshake to processor
                                 Error1,               // Host not ready error
                                 Error2;               // Data_in missing stop-bit

  parameter   idle        = 2'b00;
  parameter   starting    = 2'b01;
  parameter   receiving   = 2'b10;

  reg         RCV_datareg;
  reg         [word_size-1: 0]           RCV_shftreg;
  reg         [Num_counter_bits-1: 0]    Sample_counter;
  reg         [Num_counter_bits: 0]      Bit_counter;
  reg         [Num_state_bits-1: 0]      state, next_state;
  reg                                    inc_Sample_counter,
                                         inc_Bit_counter,
                                         clr_Bit_counter,
                                         clr_Sample_counter,
                                         shift, load,
                                         read_not_ready_out,
                                         Error1,
                                         Error2;
  //Combinational logic for next state and conditional outputs
  always @ (state or Serial_in or read_not_ready_in
             or Sample_counter or Bit_counter) begin
    read_not_ready_out <= 0;
    clr_Sample_counter <= 0;
    clr_Bit_counter <= 0;
```

Figure 12.43 Description of eight-bit UART receiver. Continues on pages 585–586.

```
inc_Sample_counter <= 0;
inc_Bit_counter <= 0;
shift <= 0;
Error1 <= 0;
Error2 <= 0;
load <= 0;

case (state)
  idle:         if (Serial_in == 0) next_state <= starting;
                else next_state <= idle;

  starting:     if (Serial_in == 1) begin
                  next_state <= idle;
                  clr_Sample_counter <= 1;
                end
                else
                  if (Sample_counter == 3) begin
                    next_state <= receiving;
                    clr_Sample_counter <= 1;
                  end
                  else begin
                    next_state <= starting;
                    inc_Sample_counter <= 1;
                  end

  receiving:      begin
                    inc_Sample_counter <= 1;
                    if (Sample_counter != word_size-1)
                      next_state <= receiving;
                    else if (Bit_counter != word_size) begin
                      next_state <= receiving;
                      shift <= 1;
                      inc_Bit_counter <= 1;
                      clr_Sample_counter <= 1;
                    end
                    else begin
                      next_state <= idle;
                      read_not_ready_out <= 1;
                      clr_Sample_counter <= 1;
                      clr_Bit_counter <= 1;
                      if (read_not_ready_in == 1) Error1 <= 1;
                      else if (Serial_in == 0) Error2 <= 1;
                      else load <= 1;
                    end
                  end
```

Figure 12.43 Description of eight-bit UART receiver. Continues on page 586.

```
      default:       next_state <= idle;

    endcase
  end

  // state_transitions_and_register_transfers

  always @ (posedge Sample_clk or negedge reset_) begin
    if (reset_ == 0) begin
      state <= idle;
      Sample_counter <= 0;
      Bit_counter <= 0;
      RCV_datareg <= 0;
      RCV_shftreg <= 0;
    end
    else begin
      state <= next_state;

      if (clr_Sample_counter == 1) Sample_counter <= 0;
      else if (inc_Sample_counter == 1) Sample_counter <= Sample_counter + 1;

      if (clr_Bit_counter == 1) Bit_counter <= 0;
      else if (inc_Bit_counter == 1) Bit_counter <= Bit_counter + 1;

      if (shift == 1) RCV_shftreg <= {Serial_in, RCV_shftreg[word_size-1: 1]};

      if (load == 1) RCV_datareg <= RCV_shftreg;
    end
  end
endmodule
```

Figure 12.43 Description of eight-bit UART receiver. Concluded.

The simulation results in Figure 12.44 are annotated to show functional features of the waveforms. The received data word is $b5_h = 1010\_1101_2$. The reception sequence is from LSB to MSB, and the data moves through the inbound shift register from MSB to LSB. The data word is preceded by a start-bit and followed by a stop-bit. With *reset_* having a value of 0, the state is *idle* and the counters are cleared. At the first active edge of *Sample_clock* after the reset condition is de-asserted, with *Serial_in* having a value of 0, the controller's state enters *starting* to determine whether a start bit is being received. Three more samples of *serial_in* are taken, and after a total of four samples have been found to be 0, the *Sample_counter* is cleared and the state enters *sending*. After the eighth sample, *shift* is asserted. The sample at the next active edge of the clock is shifted into the MSB of *Rcv_shftreg*. The value of

Rcv_shftreg becomes $80_h = 1000\_0000_2$. The sampling cycle repeats again, and a value of 0 is sampled and loaded into *Rcv_shftreg*, changing the contents of the register to $1100\_0000_2 = 40_h$.

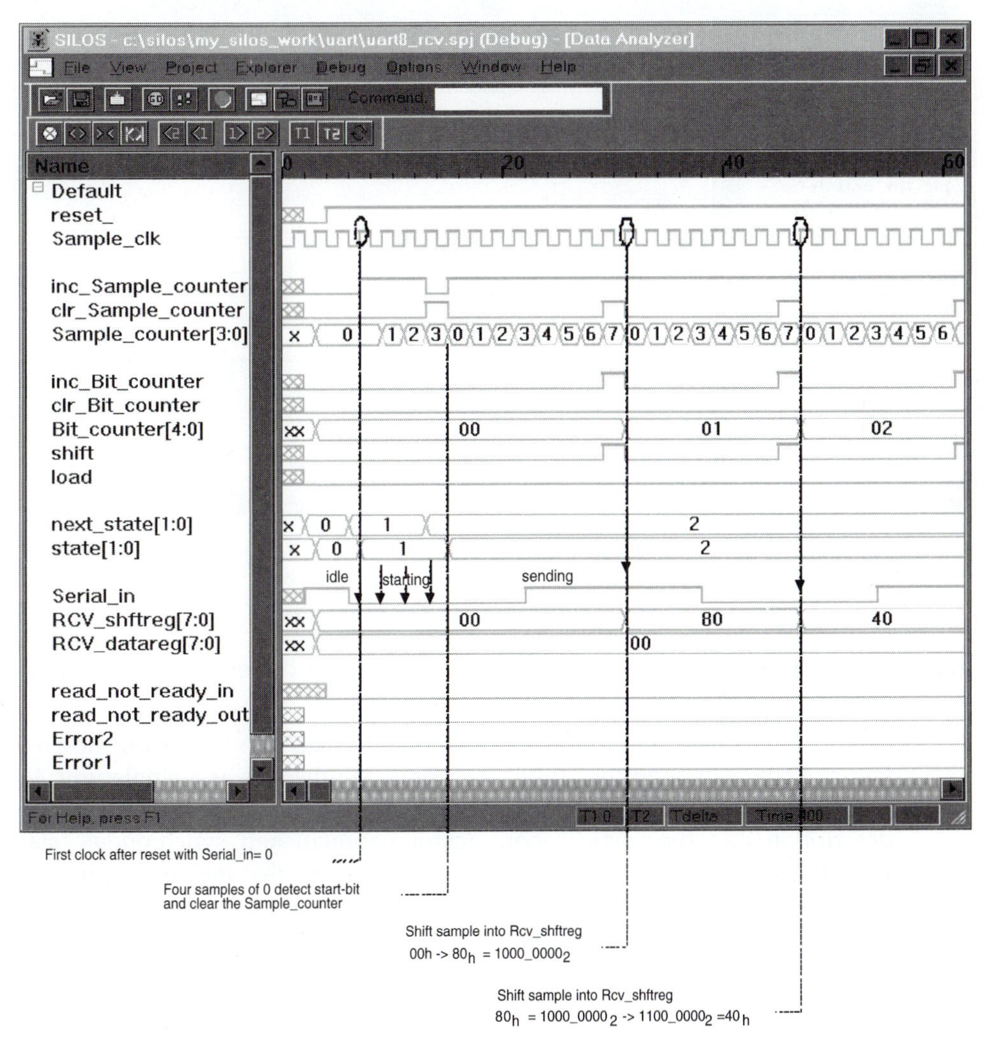

Figure 12.44 Annotated simulation results for the UART receiver.

The end of the sampling cycle of the received word is shown in Figure 12.45. After the last data bit is sampled, the machine samples once more to detect the stop-bit. In the absence of an error, the contents of *RCV_shftreg* will be loaded into *RCV_datareg*. In this example, the value $b5_h$ is finally loaded from *RCV_shftreg* into *RCV_datareg*. Other tests can be conducted to completely verify the functionality of the receiver.

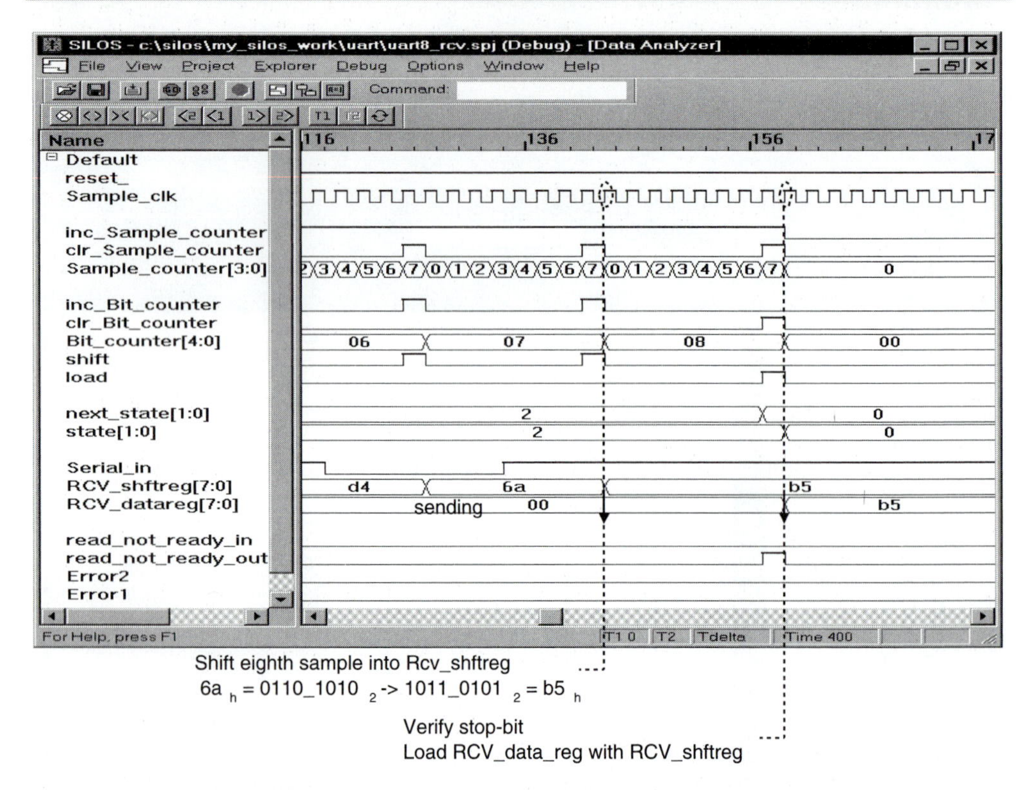

Figure 12.45 Transfer of data word into *RCV_datareg* at the end of sampling.

The Verilog description of the receiver synthesizes into the circuit shown in Figure 12.46. Although the entire description synthesizes as a single unit, the structure of the synthesized result is revealed more easily by partitioning the description into the asynchronous (combinational) and synchronous (state transition) parts shown below. It is important to note that the ports of the parent modules of a partition must be sized properly to accommodate vector ports in the child modules. Otherwise, the ports will be treated as default scalars in the scope of the parent module.

```
module UART8_rcvr_partition
  (Serial_in, Sample_clk, reset_, read_not_ready_in,
  RCV_datareg, read_not_ready_out, Error1, Error2);

  // Partitioned UART receiver

  input   Serial_in,              // Serial data input
          Sample_clk,             // Clock to sample serial data
          reset_,                 // active-low reset
          read_not_ready_in;      // status bit from host processor
```

```
output [7:0]    RCV_datareg;     // Data register for the output
output read_not_ready_out,      // Handshake to host processor
       Error1,                  // Host not ready error
       Error2;                  // Data_in missing stop-bit

wire   [7:0] RCV_shftreg;
wire   [3:0] Sample_counter;
wire   [4:0] Bit_counter;
wire   [1:0] state, next_state;
wire   inc_Sample_counter,
       inc_Bit_counter,
       clr_Bit_counter,
       clr_Sample_counter,
       shift, load,
       read_not_ready_out,
       Error1, Error2;

state_transition_part M1
  (Sample_clk, reset_, Serial_in, state, next_state, Sample_counter,
  Bit_counter, RCV_datareg, clr_Sample_counter, inc_Sample_counter,
  clr_Bit_counter, inc_Bit_counter, shift, load);

controller_part M2
  (Serial_in, read_not_ready_in, read_not_ready_out, Error1, Error2,
  state, next_state, Sample_counter, Bit_counter, inc_Sample_counter,
  inc_Bit_counter, clr_Bit_counter, clr_Sample_counter, shift, load);
endmodule


module controller_part
  (Serial_in, read_not_ready_in, read_not_ready_out, Error1, Error2,
  state, next_state, Sample_counter, Bit_counter, inc_Sample_counter,
  inc_Bit_counter, clr_Bit_counter, clr_Sample_counter, shift, load);

  input    Serial_in,                 // Serial data input
           read_not_ready_in;         // status bit from host processor
  output   read_not_ready_out,        // Handshake to processor
           Error1,                    // Host not ready error
           Error2;                    // Data_in missing stop-bit

  input    [1:0] state;
  input    [3:0] Sample_counter;
  input    [4:0] Bit_counter;
  output   [1:0] next_state;
```

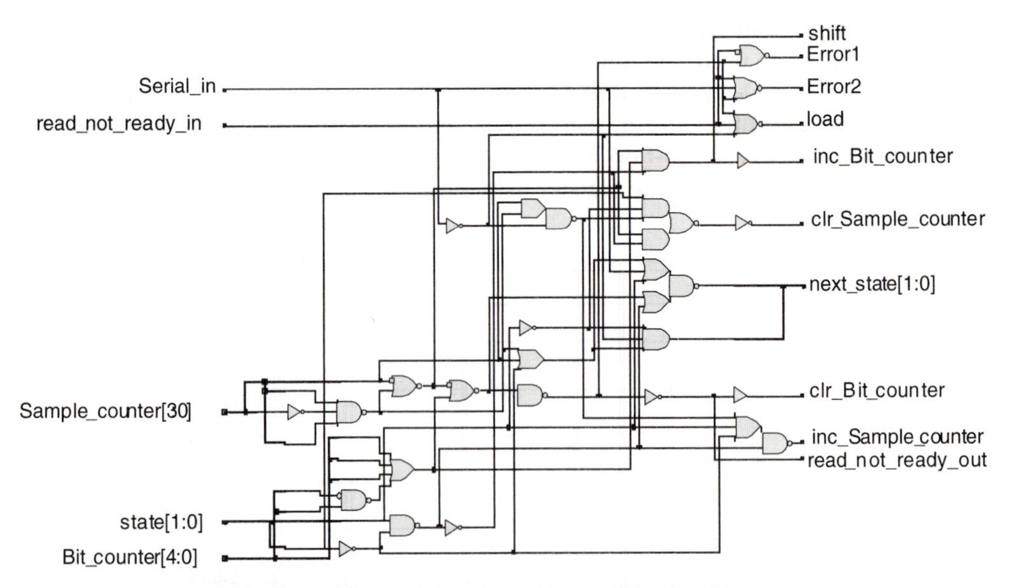

Figure 12.46 Concluded. (b) Combinational logic forming the next state, output register, and control signals for register transfers.

12.4 BIT-SLICE MICROCONTROLLER

The AMD2901 is a widely used four-bit-slice microcontroller manufactured by Advanced Micro Devices. The interface signals and datapaths of the individual device allow it to be chained with other identical devices to form an architecture having a wide datapath. The overall architecture of the AMD2901 is shown in Figure 12.47.[4] At the heart of the device an ALU performs either an arithmetic or logic operation on its operands to create an output word. This output is passed to an output selector, the register file, or the Q register. A dual-address register file (RAM array) holds 16 words that are addressed by the four-bit external words *a[3:0]* and *b[3:0]*. The outputs *ad* and *bd* hold the words addressed by *a* and *b*, respectively. The bit-lines *ram0* and *ram3* are used in shifting operations when the device is chained to other devices. A Q register acts as an intermediate register that exchanges bits with an adjacent identical device, and receives the output of the ALU. Both the register file and Q register have built-in shift registers. Under the control of a sub-word of the instruction word *i[8:6]*, the Q register can load or shift its contents. The output of the Q register is passed to the operand selector. The operand selector determines the configuration of operands for the ALU. One channel is configured from either the external data path, *d[3:0]*, or the output *ad[3:0]* output of the register file, and the other channel is formed from either the output *bd[3:0]* of the register file or the output of the Q register. Various bit lines form outputs for implementing the propagate and generate signals used in carry look-ahead addition, and outputs indicating the status of the ALU. The *p_bar* and *g_bar* outputs are inverted propagate and generate signals to the next stage; *ovr* indicates that an overflow condition has occurred; *c_out* is the carry to the next stage, *zero* indicates that the output of the ALU is zero, and *f_3* indicates that the most significant bit of the ALU is asserted.

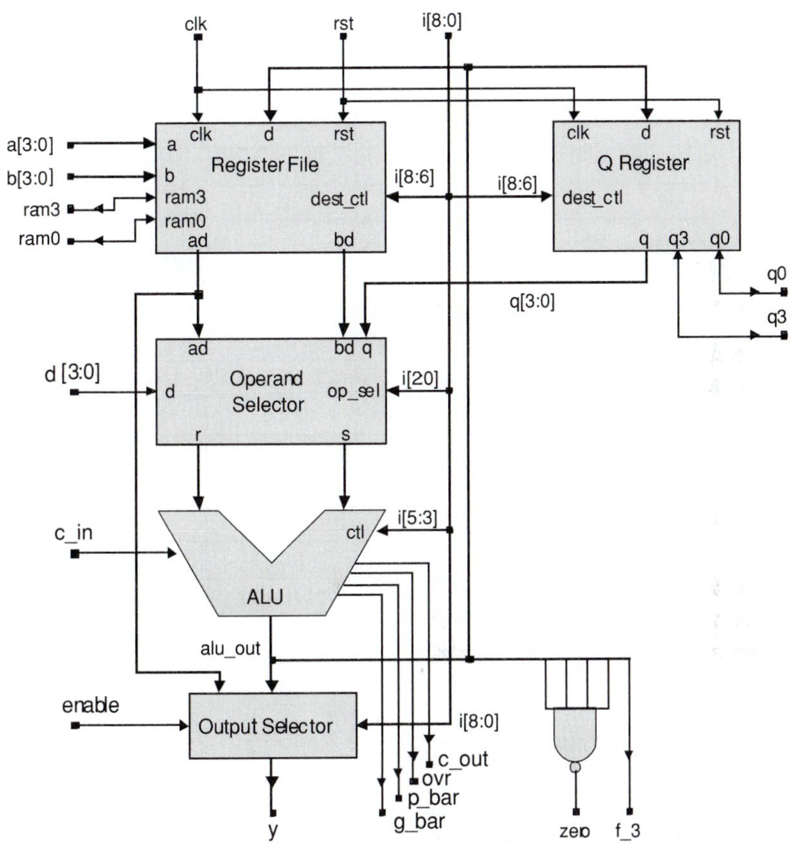

Figure 12.47 Top level architecture of the AMD2901 4-bit-slice microcontroller.

The nine-bit instruction word is partitioned into three fields determining (1) the operands for the ALU (*i[2:0]*), (2) the operation of the ALU (*i[5:3]*), and (3) the destination of the output of the ALU (*i[8:6]*)]. The mnemonics determining the operands of the ALU are shown in Table 12.1.

The operations of the AMD2901 ALU are shown in Table 12.2, along with a Verilog description of each operation.

Table 12.3 shows the destination mnemonics for the operations of the ALU, and the operations within the register file and the Q register.

Table 12.1 *Mnemonics of the AMD2901 ALU operand selector.*

Mnemonic	Octal Code	Operand Microcode i[2:0]	ALU Operands	
			r	s
aq	0	0 0 0	a	q
ab	1	0 0 1	a	b
zq	2	0 1 0	o	q
zb	3	0 1 1	o	b
za	4	1 0 0	o	a
da	5	1 0 1	d	a
dq	6	1 1 0	d	q
dz	7	1 1 1	d	o

Table 12.2 *Mnemonics of the operations of the AMD2901 ALU.*

<table>
<tr><th rowspan="2">Mnemonic</th><th rowspan="2">Octal Code</th><th rowspan="2">Operation Microcode i[5:3]</th><th colspan="2">ALU Functions</th></tr>
<tr><th>Operation</th><th>Verilog</th></tr>
<tr><td>add</td><td>0</td><td>0 0 0</td><td>r +s</td><td>r +s</td></tr>
<tr><td>subr</td><td>1</td><td>0 0 1</td><td>s-r</td><td>s-r</td></tr>
<tr><td>subs</td><td>2</td><td>0 1 0</td><td>r-s</td><td>r-s</td></tr>
<tr><td>or</td><td>3</td><td>0 1 1</td><td>ror s0</td><td>r || s</td></tr>
<tr><td>and</td><td>4</td><td>1 0 0</td><td>r and s</td><td>r && s</td></tr>
<tr><td>notrs</td><td>5</td><td>1 0 1</td><td>(not r) and s</td><td>(~r)&& s</td></tr>
<tr><td>exor</td><td>6</td><td>1 1 0</td><td>r xor s</td><td>r ^s</td></tr>
<tr><td>exnor</td><td>7</td><td>1 1 1</td><td>r xnor s</td><td>r ~^ s</td></tr>
</table>

Table 12.3 *Destination mnemonics for the AMD2901 ALU.*

Mnemonic	Octal Code	Destination Microcode i[8:6]	Output y	Register File Actions		Q Register Actions		reg_file bit shifts		Q register bit shifts	
				Shift*	Value Loaded**	Shift*	Value Loaded	bit 3	bit 0	bit 3	bit 0
qreg	0	0 0 0	alu_out	-	none	none	alu_out	-	-	-	-
nop	1	0 0 1	alu_out	-	none	-	none	-	-	-	-
rama	2	0 1 0	reg_f[a]	none	alu_out	-	none	-	-	-	-
ramf	3	0 1 1	alu_out	none	alu_out	-	none	-	-	-	-
ramqd	4	1 0 0	alu_out	down	alu_out >>1	down	Q>>1	ram3	d[1]	q3	d[1]
ramd	5	1 0 1	alu_out	down	alu_out>>1	-	none	ram3	d[1]	***	***
ramqu	6	1 1 0	alu_out	up	alu_out<<1	up	Q<<1	d[2]	ram0	d[2]	q0
ramu	7	1 1 1	alu_out	up	alu_out<<1	-	none	d[2]	ram0	***	***

*The shift action occurs before the load action.

- denotes "don't care."

**All values are loaded to the register file at the address specified by *b*.

***Denotes "not applicable."

reg_f[b] denotes the contents of the register file at the address *b*.

The Verilog modules below implement the functional units of the AMD2901. Each module has been verified and synthesized. The AMD2901 was also verified and synthesized as a top level module.

```
// amd2901 bit slice microprocessor
// operands: instruction[2:0]
  `define      aq        3'b000
  `define      ab        3'b001
  `define      zq        3'b010
  `define      zb        3'b011
  `define      za        3'b100
  `define      da        3'b101
  `define      dq        3'b110
  `define      dz        3'b111
// opcodes: instruction[5:3]
  `define      add       3'b000
  `define      subr      3'b001
  `define      subs      3'b010
  `define      orrs      3'b011
  `define      andrs     3'b100
  `define      notrs     3'b101
  `define      exor      3'b110
  `define      exnor     3'b111

// destination mnemonics: instruction[8:6]
  `define      qreg      3'b000
  `define      nop       3'b001
  `define      rama      3'b010
  `define      ramf      3'b011
  `define      ramqd     3'b100
  `define      ramd      3'b101
  `define      ramqu     3'b110
  `define      ramu      3'b111

module amd2901(clk, rst, enable, c_in, a, b, d, instruction, ram0, ram3,qs0,
  qs3, y, g_bar, p_bar, ovr, c_out, f_0, f_3);
  input              clk, rst, c_in, enable;
  input    [3:0]     a, b, d;
  input    [8:0]     instruction;
  inout              ram0, ram3;
  inout              qs0, qs3;
  output   [3:0]     y;
  output             g_bar, p_bar, ovr, c_out, f_0, f_3;
  wire     [3:0]     ad, bd, q, r, s, alu_out;
```

```
  Register_file M1 (
        clk,              // system clock
        rst,              // global reset
        a, b,             // addresses to memory
        alu_out,          // data path from ALU
        instruction[8:6], // micro instruction bits for destination
        ram0, ram3,       // LSB and MSB from memory word
        ad, bd            // memory output words (4 bits each)
  );

  ALU_amd2901 M2 (r, s, c_in, instruction [5:3], alu_out, g_bar, p_bar, ovr, c_out);
    output_selector M3 (y, ad, alu_out, instruction [8:6], enable);
    operand_selector M4 (d, ad, bd, q, instruction [2:0], r, s);
    Q_register M5 (clk, rst, alu_out, instruction [8:6], qs0, qs3, q);

  assign f_0 = (alu_out == 4'b0) ? 0 : 1'bz;
  assign f_3 = alu_out[3];
endmodule

module ALU_amd2901 ( r, s, c_in, ctl, alu_out, g_bar, p_bar, ovr, c_out );
  input  [3:0]  r, s;
  input         c_in;
  input  [2:0]  ctl;
  output [3:0]  alu_out;
  output        g_bar, p_bar, ovr, c_out;
  reg    [3:0]  alu_out;
  reg           c_out;

  always @ (r or s or c_in or ctl)
    case (ctl)
      `add:     {c_out, alu_out} = r + s + c_in;
      `subr:    {c_out, alu_out} = r + ~s + c_in;
      `subs:    {c_out, alu_out} = s + ~r + ~c_in;
      `orrs:    {c_out, alu_out} = {1'b0, r | s};
      `andrs:   {c_out, alu_out} = {1'b0, r & s};
      `notrs:   {c_out, alu_out} = {1'b0, (~r) & s};
      `exor:    {c_out, alu_out} = {1'b0, r ^ s};
      `exnor:   {c_out, alu_out} = {1'b0, r ~^ s};
      default:  {c_out, alu_out} = 5'bx;
    endcase

  assign ovr = c_out ^ alu_out[3];
  assign g_bar = ~(
    (  r[3] & s[3]) |
    (  (r[3] | s[3]) & (r[2] & s[2])  ) |
    (  (r[3] | s[3]) & (r[2] | s[2]) & (r[1] & s[1]))|
```

```
      ( (r[3] | s[3]) & (r[2] | s[2]) & (r[1] | s[1]) & (r[0] & s[0])));

  assign p_bar = ~((r[3] ^ s[3]) & (r[2] ^ s[2]) & (r[1] ^ s[1]) & (r[0] ^ s[0]));
endmodule

module reg_load_R(data, clk, reset, load, q);
  // re-verified 4-28-98
  parameter size = 8;
  input [size-1: 0]      data;
  input                  clk, reset, load;
  output [size -1: 0]    q;
  reg [size-1: 0]        q;

  always @ (posedge clk or posedge reset)
    if (reset == 1) q = 0; else if (load == 1) q = data;
endmodule

module Register_file (clk, rst, a, b, f, dest_ctl, ram0, ram3, ad, bd);
  input            clk, rst;
  input    [3:0]   a, b, f;
  input    [2:0]   dest_ctl;
  inout            ram0, ram3;
  output   [3:0]   ad, bd;
  reg      [3:0]   ab_data [15:0];      // memory
  wire     [3:0]   data;
  wire             ram_en;
  reg      [4:0]   row_index;    // Note: size of row_index is critical

  assign ram_en = ((dest_ctl == `qreg) || (dest_ctl == `nop)) ? 0 : 1;

  always @ (posedge clk or posedge rst)
    if (rst == 1) for (row_index = 0; row_index < 16; row_index = row_index + 1)
      ab_data[row_index] = 0;
    else if (ram_en) ab_data[b] = data;

  assign data = (dest_ctl == `ramqu || dest_ctl == `ramu) ? {f[2], f[1], f[0], ram0} :
    (dest_ctl == `ramqd || dest_ctl == `ramd) ? {ram3, f[3], f[2], f[1]} :
    (dest_ctl == `rama || dest_ctl == `ramf) ? f : 4'bxxxx;

  assign ad = ab_data [a];
  assign bd = ab_data [b];
  assign ram3 = (dest_ctl == `ramu || dest_ctl == `ramqu) ? f[3] : 1'bz;
  assign ram0 = (dest_ctl == `ramd || dest_ctl == `ramqd) ? f[0] : 1'bz;
endmodule
```

```
module Q_register (clk, rst, data_in, dest_ctl, qs0, qs3, q);
  // Parameterized register with shift left (up), shift right (down)
  // and transfer of interface bits to/from companion slices.

  input           clk, rst;
  input    [3:0]  data_in;
  input    [2:0]  dest_ctl;
  inout           qs0, qs3;
  output   [3:0]  q;
  reg      [3:0]  data;
  reg             q_en;

  reg_load_R #(4) M1 (data, clk, rst, q_en, q);

  bufif1 (qs3, data_in[3], (dest_ctl == `ramu || dest_ctl == `ramqu));
  bufif1 (qs0, data_in[0], (dest_ctl == `ramd || dest_ctl == `ramqd));

  always @ (dest_ctl)
    case (dest_ctl)
      `qreg, `ramqd, `ramqu:  assign q_en = 1;
      default:                assign q_en = 0;
    endcase

  always @ (dest_ctl or qs0 or qs3)
    case (dest_ctl)

      //Shift left (up)
      `ramqu:       assign data = {data_in[2], data_in[1], data_in[0], qs0};

      //Shift right (dn)
      `ramqd:       assign data = {qs3, data_in[3], data_in[2], data_in[1]};

      // Pass through
      `qreg:        assign data = data_in;

      default:      assign data = 4'bx;
    endcase

endmodule

module output_selector (mux_out, channel_1, channel_2, sel, enable);
  input  [3:0]  channel_1, channel_2;
  input  [2:0]  sel;
  input         enable;   // active low
  output [3:0]  mux_out;
```

```
    assign mux_out = (enable == 0 && sel == `rama) ? channel_1:
                     (enable == 0) ? channel_2 : 4'bz;

endmodule

module operand_selector (d, ad, bd, q, op_sel, r, s);
  input  [3:0]  d, ad, bd, q;
  input  [2:0]  op_sel;
  output [3:0]  r, s;
  reg    [3:0]  r, s;

  // operand codes

  always @ (op_sel) begin
    case (op_sel)
      `aq, `ab:        assign r = ad;
      `zq, `zb, `za:   assign r = 4'b0;
      default          assign r = d;
    endcase
    case (op_sel)
      `aq, `zq, `dq:   assign s = q;
      `ab, `zb:        assign s = bd;
      `za, `da:        assign s = ad;
      default:         assign s = 4'b0;
    endcase
  end
endmodule
```

For this example, the AMD2901 was synthesized as a hierarchically-organized unit. The top-level synthesized circuit is shown in Figure 12.48. It contains blocks representing the nested modules of the Verilog description of *amd2901* and paths for the control signals and busses. Figure 12.49 shows the result of synthesizing ALU_amd2901. Notice that the Verilog behavioral description of the circuit contains several addition operators, but the synthesized ALU contains only two adders. The synthesis tool exploits resource sharing to the extent possible to obtain an efficient realization. Figure 12.50 contains the synthesized operand selector. It consists of steering logic for the datapaths. The synthesized output selector is shown n Figure 12.51. The *invtn_c* device is a three-state inverter. Figure 12.521 contains the synthesized Q register. It uses the *dffrgpqb_a* standard cell, a D-type flip-flop having rising edge, active-low reset, and internally gated data between the external data and the output of the flip-flop. Figure 12.53 shows the circuit synthesized for the four-word, four-bit register file. It, too, uses the *dffrgpqb_a* flip-flop.

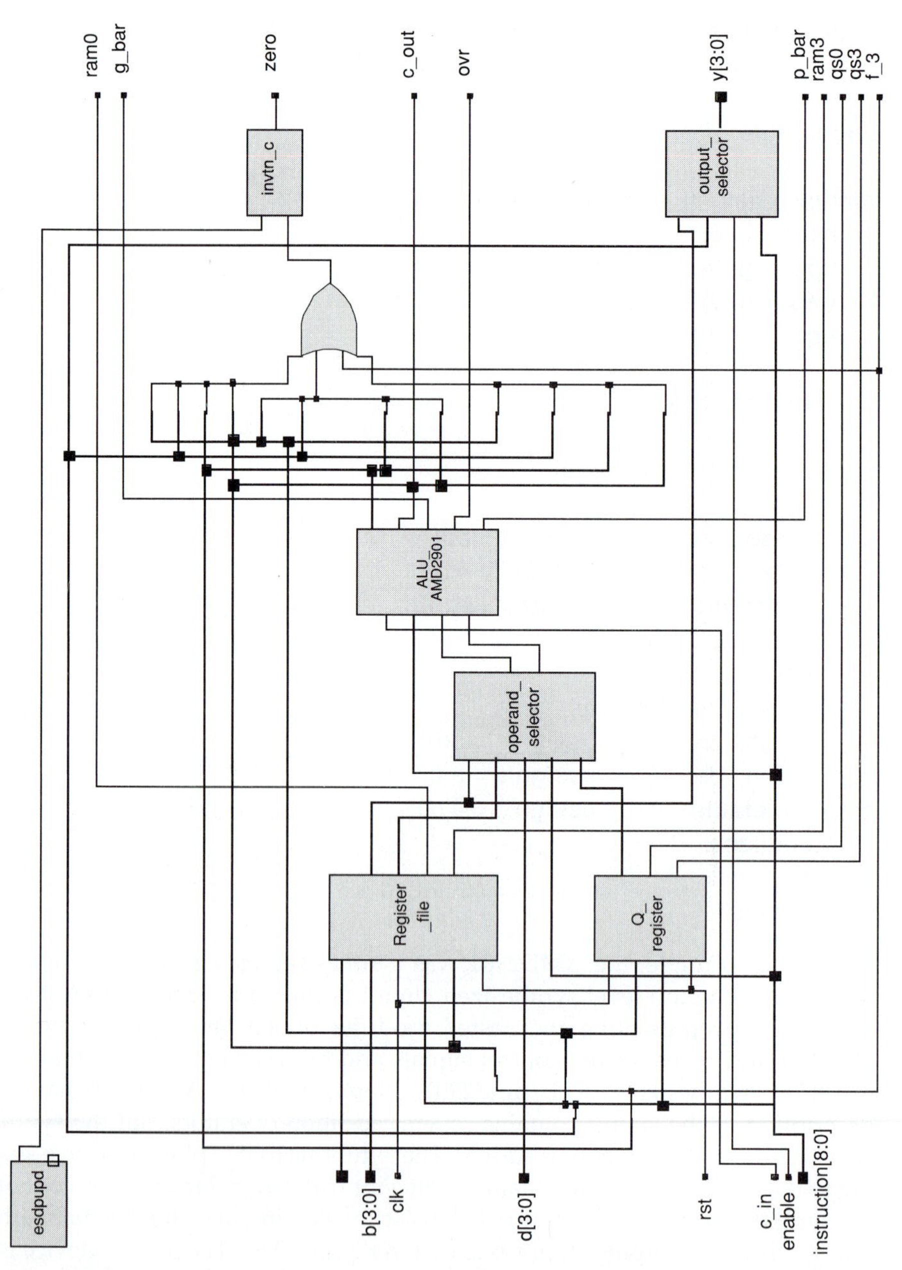

Figure 12.48 Top-level circuit synthesized from *amd2901*.

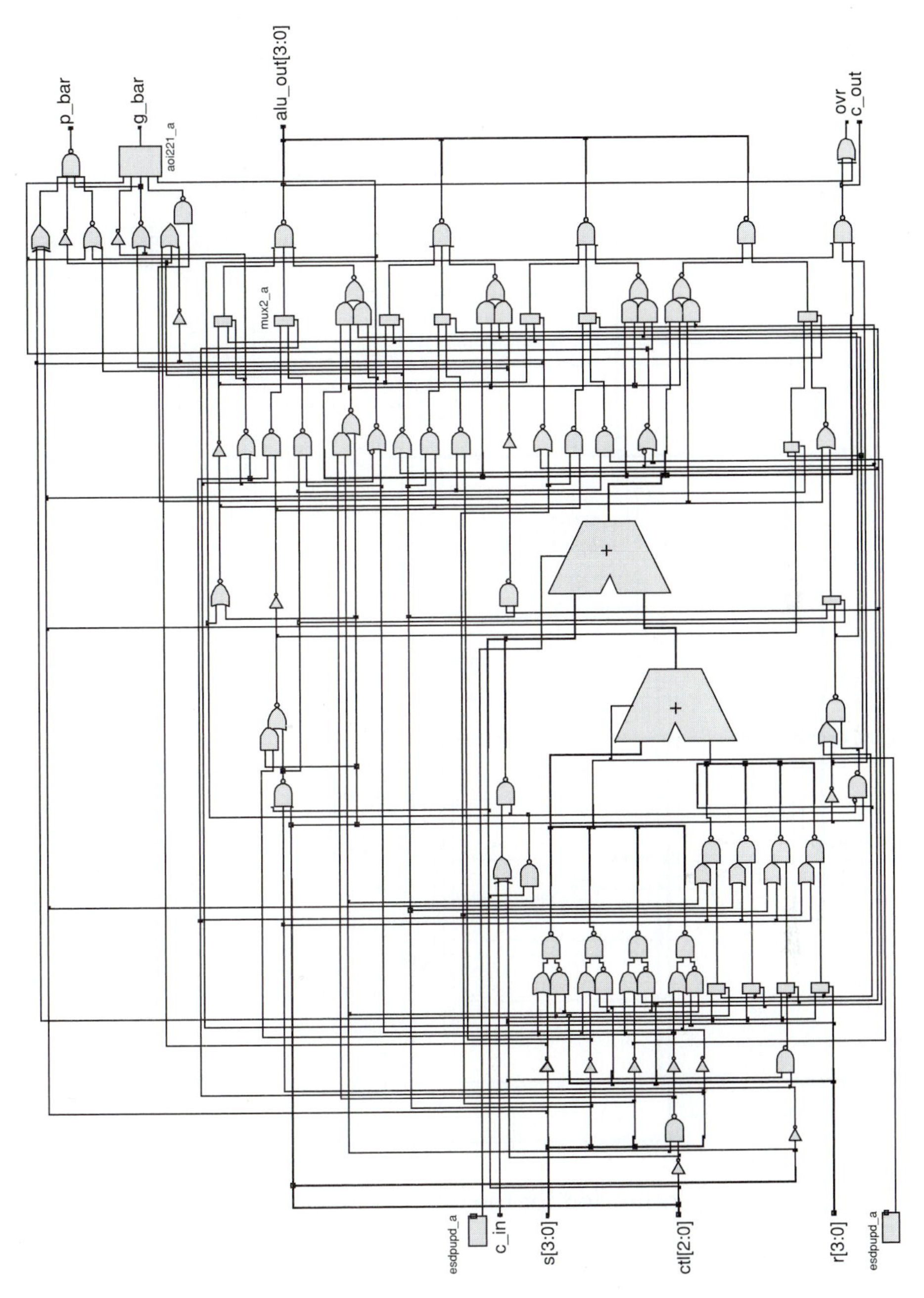

Figure 12.49 Circuit synthesized for the ALU of the AMD2901 four-bit-slice microcontroller.

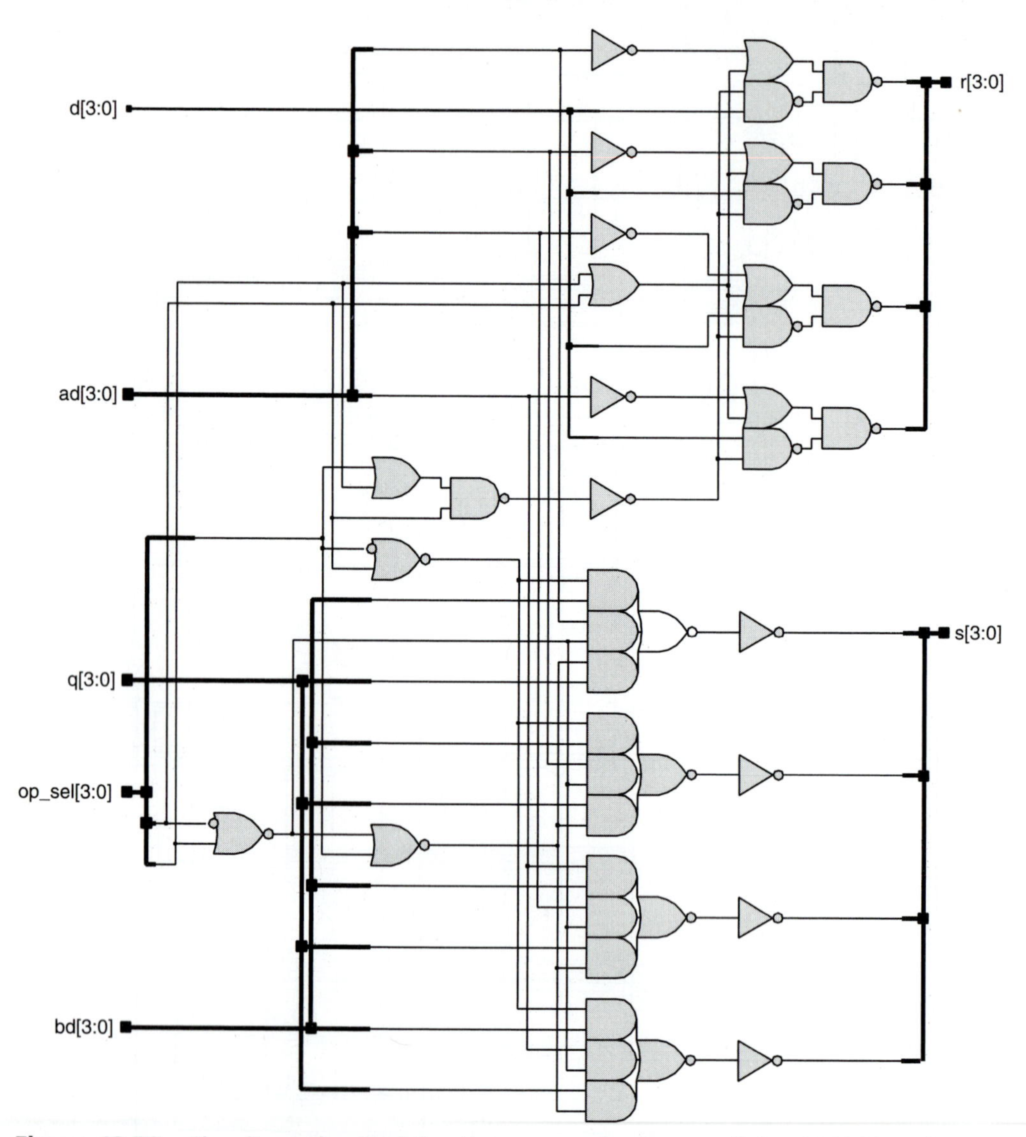

Figure 12.50 Circuit synthesized for the operand selector of the AMD2901 four-bit-slice microcontroller.

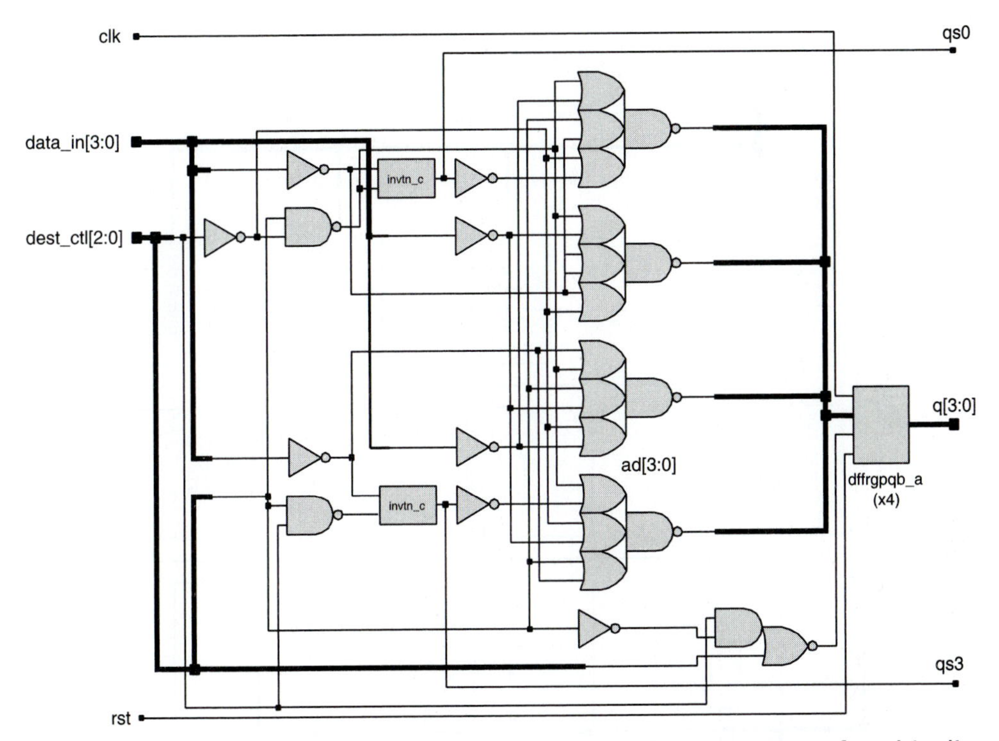

Figure 12.51 Circuit synthesized for the Q register of the AMD2901 four-bit-slice microcontroller.

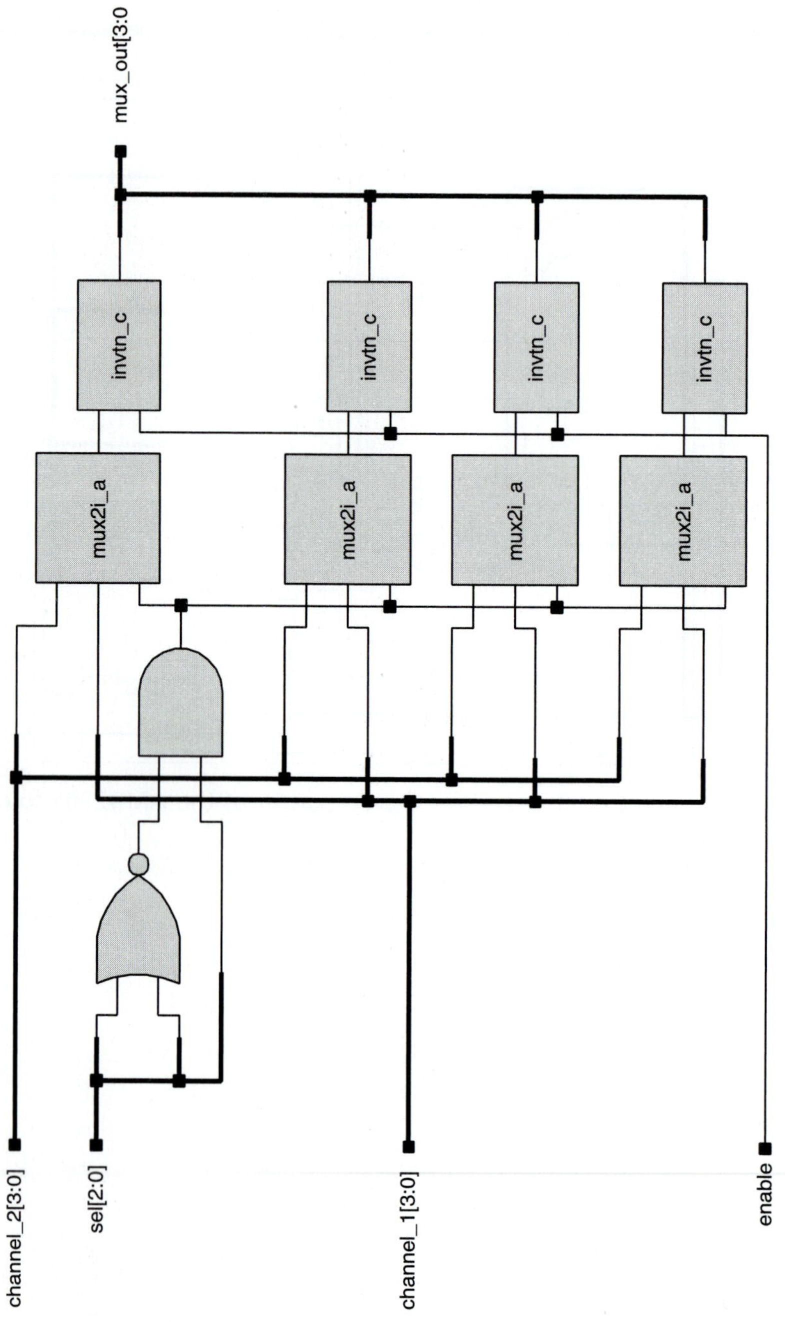

Figure 12.52 Circuit synthesized from *output_selector* of the AMD2901 four-bit-sl

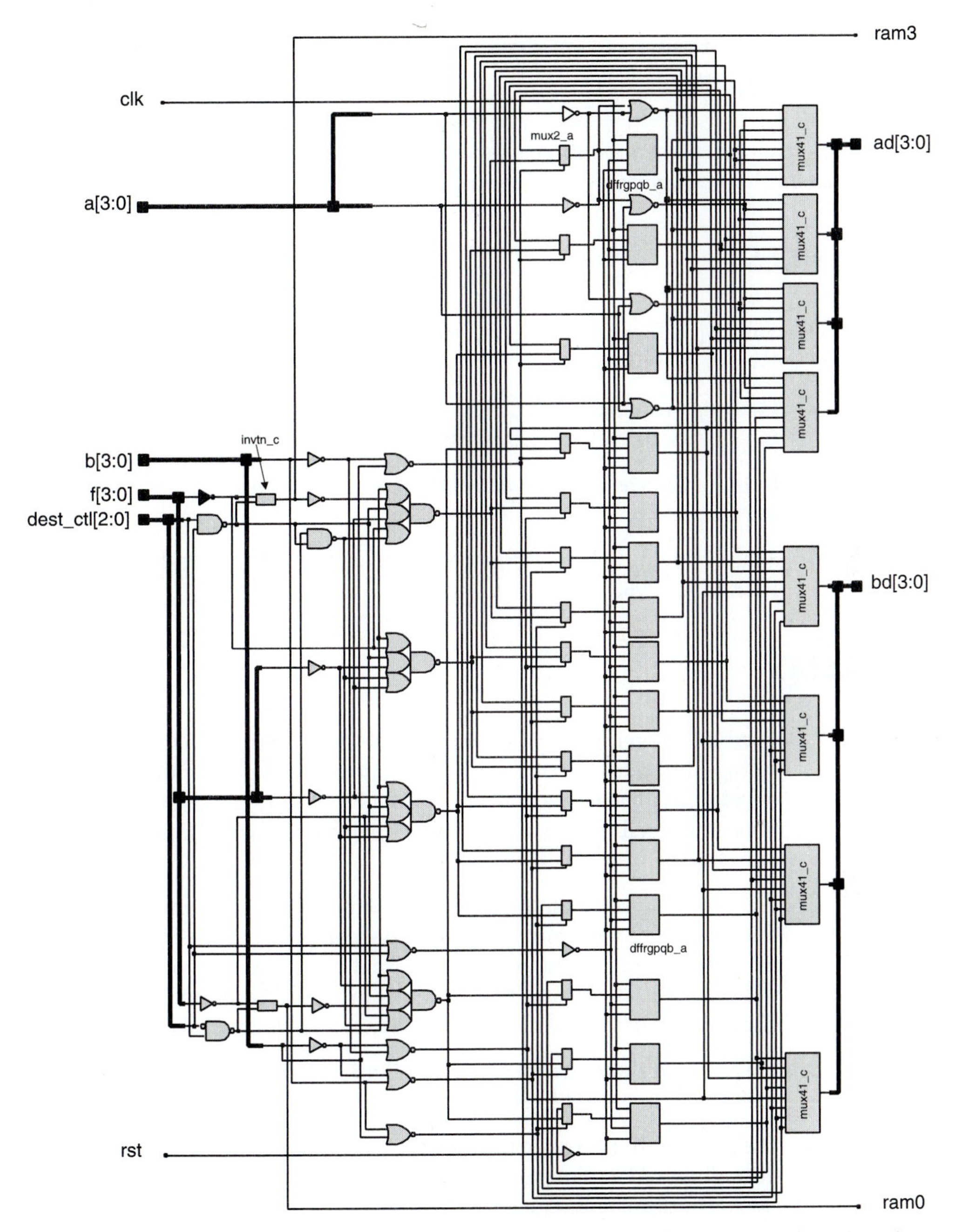

Figure 12.53 Circuit synthesized from *register_file* of the AMD2901 four-bit-slice microcontroller, modified for four words (for display purposes only).

12.5 SUMMARY

This chapter presented three elaborate examples demonstrating synthesis-friendly Verilog styles for the Synopsys Design Analyzer™.

REFERENCES

1. Smith, Michael J. S., *Application-Specific Integrated Circuits*, Addison-Wesley, New York, 1997.
2. Heuring, V.P. and Jordan, H. J., *Computer Systems Design and Architecture*, Addison Wesley Publishing. Co., Menlo Park, CA, 1997.
3. Roth Jr., Charles H., *Digital Systems Design Using VHDL*, PWS Publishing Co., Boston, 1998
4. Skahil, Kevin, *VHDL for Programmable Logic*, Addison Wesley, Menlo Park, CA, 1996.

PROBLEMS

1. The FIFO developed in this chapter uses two pointers to manage the storage and retrieval of data. Investigate other implementations and develop an alternative description of a FIFO. Synthesize both descriptions into a cell library and compare their performance (speed) and use of resources.
2. The FIFO controller in the temperature monitoring system discussed in this chapter has a common write signal controlling both FIFOs. Modify the control unit and the controller so that the FIFOs receive independent write signals. Verify and synthesize your result. Verify that the synthesized circuit has the same functionality as the behavioral description.
3. Model, verify, and synthesize a 16-bit microcontroller using the AMD2901 discussed in this chapter.
4. The AMD2901 4-bit slice microcontroller has long combinational paths through the ALU. Formulate a pipelining strategy for the improving the clock-to-output and register-to-register delays of the synthesized circuit. Compare the circuits synthesized with and without pipelining.
5. Example 7.60 described an electronic dice game. Synthesize the game and verify that the synthesized circuit matches the behavioral model of the game. (Change the source description to satisfy the style required by your synthesis tool.)

6. In the UART receiver described in this chapter, the detection of a start bit is based on the first three samples of data. Modify the description and synthesize a circuit that will abort the detection sequence if any sample within the first bit is not zero.

7. The sampling scheme in this chapter's UART receiver does not verify that the samples in a bit have the same value. Model, verify, and synthesize a receiver that verifies that the 4 inner samples of a bit have the same value. Alternatively, require that a majority of the samples have the same value.

8. Modify the UART receiver in this chapter to request retransmission of a bit of data when a transmission error is detected.

9. Develop a parity generator to accept a seven-bit format data stream and create an eight-bit data format including a parity bit. Integrate the parity generator with the UART transmitter developed in this chapter. Verify and synthesize your description.

10. Model, verify, and synthesize a behavioral description of an integrator that synchronously accumulates the sum of its data inputs. The machine has synchronous reset, and a hold signal that determines whether the integrator recycles its current value or adds the data to the current value. Use a parameterized wordlength.

CHAPTER 13

RAPID PROTOTYPING WITH XILINX FPGAs

13.1 INTRODUCTION TO FPGAs

The previous chapters focused on standard cell-based realizations of synthesizable Verilog models. At lower levels of integration and volume of product, gate arrays are a more economical technology. Mask-programmable gate arrays are fabricated in a foundry, where final layers of metal customize the wafer to the specifications of the end user. Field-programmable gate arrays (FPGAs) are sold as fully fabricated generic products. Their functionality is determined by programming done in the field by the customer and/or end user. FPGAs allow designers to turn a design into working silicon in a matter of minutes, making rapid prototyping a reality. This chapter presents an introduction to FPGA-based technology, design methodology and rapid prototyping. FPGAs are distinguished on the basis of several features: architecture, number of gates, mechanism for programming, program volatility, granularity and robustness of functional/logical unit, physical size (footprint), pinout, time-to-prototype, speed, power, and availability of resources for connectivity.[1] Some recent devices use flash memory to provide fast programmability and non-volatility, but they will not be considered here. We will focus on so-called SRAM-based FPGAs, which lose their programming when power is removed from the part.

FPGAs have a fixed architecture that is programmed in the field for a particular application. A typical architecture consists of an array of programmable functional units (FUs), as shown in Figure 13.1, together with fixed but programmable resources that establish the routing of signals and configure the functionality of the device. A functional unit has limited resources for implementing combinational and sequential logic.

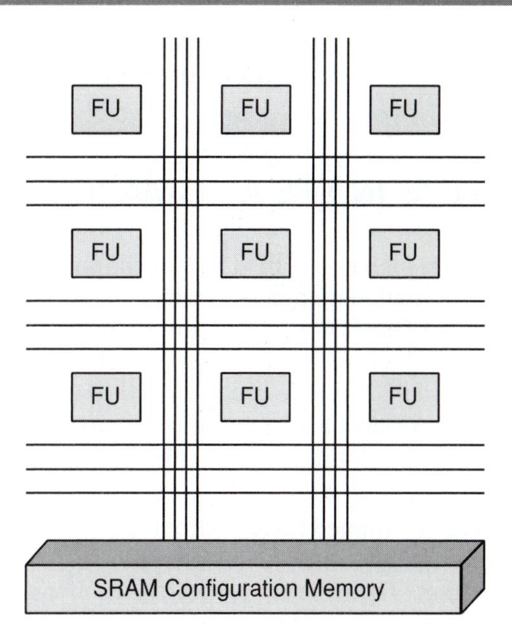

Figure 13.1 FPGA architecture.

Volatile FPGAs are configured by a program called a configuration memory that can be downloaded and stored in static CMOS memory. The contents of the static cells are applied to the control lines of CMOS transmission gates (see Chapter 11) and other devices to (1) program the functionality of the functional units, (2) customize configurable features, such as slew rate, (3) establish connectivity between functional units, and (4) establish input/output ports. The configuration program is downloaded to the FPGA from either a host machine or from an on-board PROM. When power is removed from the device the program stored in memory is lost. The device must be re-programmed before it can be used again.

The volatility of a stored-program FPGA is a double-edged sword—it presents the need for the program to be re-loaded in the event that power is disrupted, but it also allows the same generic part to serve a boundless variety of applications, and even be reconfigured on the same circuit board under the control of a processor. One of the programs that can be executed by an FPGA can even test the host system in which it is embedded. The ease of re-programming a stored-program FPGA supports rapid prototyping, enabling design teams to compete effectively in an environment characterized by narrow and ever-shrinking windows of opportunity. Time-to-market is critical in many designs, and FPGAs provide a path to early entry. In 1998, Xilinx announced a new technology called Internet Reconfigurable Logic (IRL). This technology downloads Java-based applications from the Internet, allowing designers to repair, enhance, upgrade, or completely reconfigure devices in the field.

13.2 ROLE OF FPGAs IN THE ASIC MARKET

The architectural resources of FPGAs match the general needs of digital designers for computational engines having memory, datapaths, and processors. The flexibility of an FPGA adds a dimension beyond what is available in mask-programmed devices, because mask-programmed devices cannot be reprogrammed. An FPGA can be programmed to implement a processor. The hardware resources of an FPGA can be configured for datapaths, and some device families include a limited amount of captive memory.

Figure 13.2 summarizes key distinctions between FPGA technology, cell-based and mask-programmed ASIC technology, and standard parts. The myriad applications for ASICs and FPGAs require customization to address the needs of the market. The diversity, rewards, evolving technology, and short longevity of the market preclude producing and stockpiling standard parts to meet these needs without incurring unacceptably high risks. The lower volumes demanded by individual, specialized applications provide a smaller base over which to amortize the costs of development and production, so units costs for FPGAs are higher than both for standard parts, and high-volume, mask-programmed ASICs.

TECHNOLOGY	FUNCTIONALITY	COST
STANDARD PART	Supplier-Defined	Low
FPGA	User-Defined	High
ASIC	User-Defined	Low

Figure 13.2 Comparisons of standard parts, ASICs, and FPGAs.

The early technology of mask-programmed gate arrays (MPGAs) used a fixed array of transistors and routing channels. Limited routing was a major issue in early devices, and frequently led to incomplete use of the available transistors. Today, multi-level metal routing is commonplace, with high use of resources. MPGAs are pre-processed to the point of the customizing final metal layers, which are tailored to specific applications. The customization/metallization steps connect individual transistors to form gates, and interconnect gates to implement logic. This technology provides a much quicker turnaround than do cell-based and full-custom technologies, because only the final metallization step is customized, but it is not as fast as that for FPGAs. Depending on the foundry, an MPGA can be turned around in a few days to several weeks. On the other hand, designs can be implemented, programmed, and re-programmed virtually instantaneously in an SRAM-based FPGA while the part is mounted in its target host application.

FPGAs are fully tested before they are shipped, so the designer's attention is focused on the creativity of the design, not on testing. Designers can quickly correct design flaws, and quickly reconfigure the part to a different functionality, in the field.[1] FPGAs address a market than cannot be met by mask-programmed technologies, which are one-time write. Mask-programmed technologies do not support reconfiguration, and corrections are costly. The risk of an MPGA-based design is significantly higher than for an FPGA, because a design flaw requires re-tooling of the final masks (with attendant costs) and lost time to re-enter the fab process queue.

MPGAs have a broad customer base for amortizing the NRE of most of the processing steps, compared to cell-based and full-custom solutions. Gate arrays are widely used to implement designs having a high content of random logic, such as state machine controllers.

MPGAs require the direct support of a foundry, and the completion of a design can depend upon the schedule of the foundry's other customers. FPGAs are fully manufactured and tested in anticipation of being shipped immediately to a buyer.

The software interface between the designer and FPGA technology is simple, and is now readily and cheaply available on PCs and workstations. Programmable logic technology is growing in density at exponential rates comparable to other technologies, such as DRAMs.[1] The speed of parts is growing at a linear rate.

Standard cell-based technology uses a library of pre-designed and pre-characterized cells implementing basic gates and flip-flops. The design of the individual cells in a library is labor-intensive, as efforts are made to achieve a dense, area-efficient layout. Consequently, a cell library has a high NRE cost, which a foundry must amortize over a large customer base during the lifetime of the underlying process technology. The mask set of a standard cell library is fully characterized and verified to be correct. Place and route tools select, place, and interconnect cells in rows on a chip to implement functionality. The structure is semi-regular because the cell heights are fixed, while the width of cells may vary, depending on the functionality being implemented. Placement and routing are customized for each application. Place and route are now done automatically to achieve dense configurations meeting speed and area constraints. Cell-based technology requires a fully customized mask set for each application. Consequently, the volume must be sufficient to offset high production and development costs, and ultimately drive a low unit cost.

Programmable logic device (PLD) technology is about 25 years old, with early devices offering programmable arrays to implement two-level logic in SOP form. PLDs can implement the functionality of a few hundred gates. They are programmed on-site by the end user. Programmable logic arrays (PLAs) have programmable AND-planes and programmable OR-planes to implement m Boolean functions of n variables, as shown in Figure 13.3. Programmable read-only memories have relatively small n; PLAs have a large value for n, and their architecture is fixed.

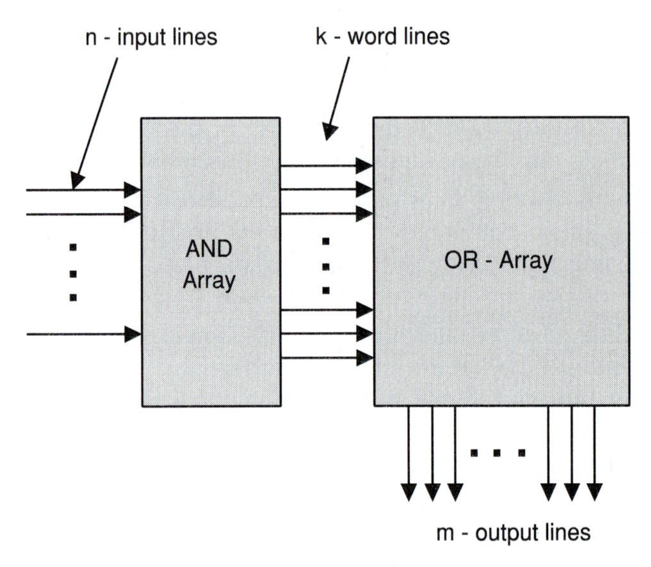

Figure 13.3 Architecture of a programmable logic array (PLA).

Programmable array logic (PAL) (circa 1978) is a special case of PLA technology. Part of the device is programmable (the AND-plane) and part of it is fixed (the OR plane). PALs are cheaper and easier to program than PLAs. Figure 13.4 shows an architecture representative of PAL devices. Inverting and non-inverting inputs are available to form product terms, which are combined as "or" terms. Dedicated output pins are three-stated; others serve as bi-directional pins having three-state inputs and outputs.

13.3 FPGA TECHNOLOGIES

State-of-the-art FPGAs can now implement the functionality of up to a million equivalent (two-input nand) gates on a single chip. Three basic types of FPGAs are available: anti-fuse, EPROM, and SRAM-based. The capacity and speed of

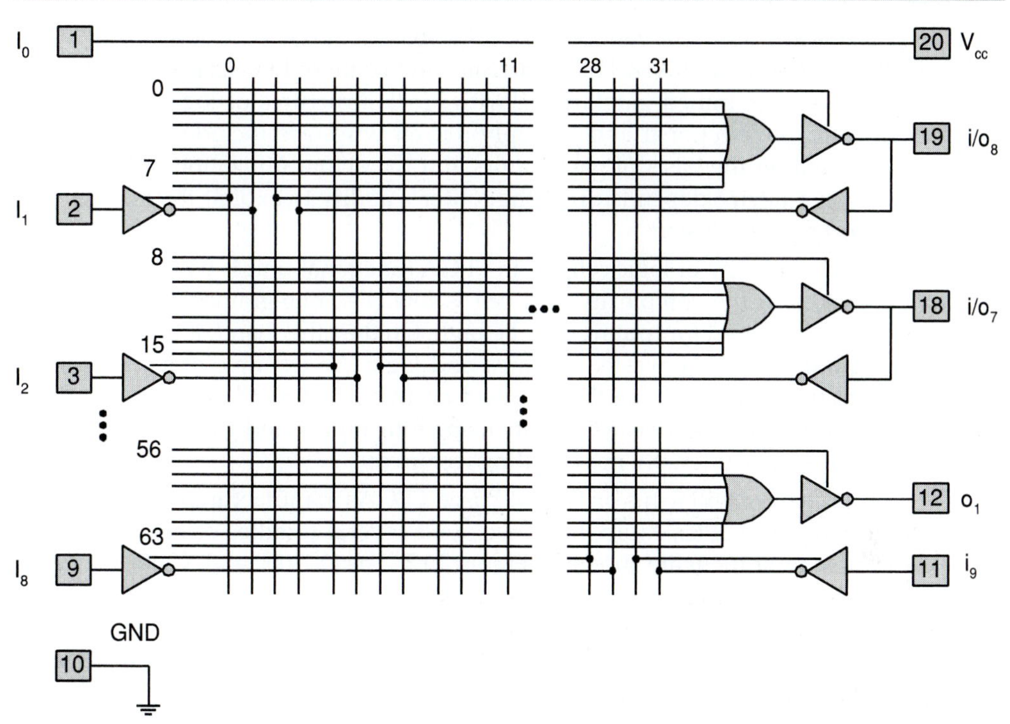

Figure 13.4 Typical PAL Architecture.

these parts continues to evolve with process improvements that shrink minimum feature sizes of the underlying transistors.

An anti-fuse device is programmed by applying a relatively high voltage between two nodes to break down a dielectric material. This eliminates the need for memory to hold a program, but the one-time write configuration is permanent. When an anti-fuse is formed, a low resistance path is irreversibly created between the terminals of the device. The anti-fuse itself is relative small, about the size of a via, and over a million devices can be distributed over a single FPGA.[2] The significant advantage of this technology is that the on-resistance and parasitic capacitance of an anti-fuse are much smaller than for transmission gates and pass transistors, which supports higher switching speeds and predictable timing delays along routed paths.

EPROM and EEPROM-based technology use charged floating gates, programmed by a high voltage. It is re-programmable and non-volatile. Such devices can be programmed off-line while embedded in the target system.

SRAM-based FPGA technology uses mainly CMOS transmission gates to establish interconnect. The status of the gates is determined by the contents of the SRAM configuration memory.

There are multiple vendors of SRAM-based FPGA products (e.g., Xilinx, Altera, ATMEL, and AT&T). The architecture of these FPGAs is similar to that of mask-programmed gate arrays, with block structures of logic and routing channels. Bi-directional and multiply-driven wires are included. Devices are advertised on the basis of gate counts, but the actual utilization of the usable gates on a device depends upon the router's ability to exploit the resources to support a given design.

The complexity of logic cells in an FPGA functional unit is based on competing factors. If the complexity of a cell is low (i.e., fine-grained), the time and resources required for routing may be high. On the other hand, if the complexity is high, there will be wasted cell area and logic. An example of a fine-grained architecture would be one that is based on two-input nand gates or muxes, as opposed to a large-grain architecture using four-input nand gates or muxes. The former uses considerably more routing resources.

This chapter will focus on SRAM-based FPGAs manufactured by Xilinx Corporation. Since 1985, Xilinx has developed several families of FPGAs; two of the most prominent are shown in Figure 13.5. The XC3000 and XC4000 families are still in widespread use, but other families have surpassed their basic capabilities. The XC4000E family offers from 2,000 to 25,000 usable gates. Its function generators are more versatile, and can be configured both as RAM with synchronous or asynchronous write addressing and dual-port RAM with concurrent read and write. This feature supports efficient and faster implementations of register files, shift registers, and FIFOs. The architecture includes a dedicated fast-carry network and companion dedicated carry logic to speed up wide arithmetic functions and counters.[3]

The XC4000EX family (see Figure 13.6a) is a high gate-count version of the XC4000E family, with devices having 28,000 to 36,000 logic gates, greatly increased routing resources (almost 2x), faster clocking options and more versatile output logic. The input-output buffers (IOBs) of the XC4000EX have a fast dedicated early clock to support fast system speeds (4 ns setup time and 6 ns clock to output delay).[4] The Spartan series (Figure 13.6b) offer 3.3 volt and 5 volt FPGAs replacing some of the earlier series and having densities from 5,000 and 40,000 system gates.

The XC4000XL family (See Figure 13.6) operates at 3.3V and includes devices having from 5,000 to over 100,000 gates. The XC4000XV family, shown in Figure 13.7 offers devices having over a half-million equivalent gates and a 92 x 92 matrix of configurable logic blocks (CLBs).[1] CLBs are the building blocks of the Xilinx technology, and they implement combinational and sequential functionality. CLBs in advanced parts can be configured to provide a limited amount of on-chip memory. These high-end parts, which are fabricated in a 0.25 micron process and can operate at a 200 MHz internal clock frequency, provide over 270,000 RAM bits (without logic).

The latest family of Xilinx parts, the Virtex™ series, announced in October 1998, includes devices having 50,000 to 1,000,000 gates at speeds of up to 160 MHz (1,728 to 27,648 logic cells). The family uses a 2.5V, 0.22 micron, five-layer metal CMOS process. The family is targeted at system-level applications, with four delay-locked loops for internal and external clock synchronization, flexible I/O technology, two configurations of on-chip memory (distributed and block), accommodations for embedded cores, and other advanced features. Some of the metrics of this family are summarized in Figure 13.8.

Feature	XC3000	XC4000
Equivalent gates*	2,000–9,000	3,000–10,000
Number of CLBs	64–320	64–400
Inputs per CLB	5	9
Flip-flops per CLB	2	2
I/O Blocks	64–144	64–240
Functions per CLB	1 of 5 variables 2 of 4 variables	1 of 5 variables 1 of 4 variables 1 of 9 variables

*Two-input Nand

(Note: The typical number of gates that can be realized in a design can be significantly less than the maximum in the part, due to limited routing resources.)

Figure 13.5 Features of Xilinx part families.

[1] (For additional information see: http://www.xilinx.com/products/xc4000XLA.html).

Device	Logic Cells	Max Logic Gates (No RAM)	Max RAM (No Logic)	Typical Gate Range (Logic and RAM)	CLB Matrix	Total CLBs	Number of Flip-Flops	Max User I/O
XC4003E	238	3,000	3,200	2–5K	10 x 10	100	360	80
XC4005E/XL	466	5,000	6,272	3–9K	14 x 14	196	616	112
XC4006E	608	6,000	8,192	4–12K	16 X 16	256	768	128
XC4008E	770	8.000	10,368	6 –15K	18 x 18	324	936	144
XC4010E/XL	950	10,000	12,800	7 –20K	20 x 20	400	1,120	160
XC4013E/XL	1368	13,000	18,432	10–30K	24 x 24	576	1,536	192
XC4020E/XL	1862	20,000	25,088	13–40K	28 x 28	784	2,016	224
XC4025E	2432	25,000	32,768	15–45K	32 x 32	1,024	2,560	256
XC4028EX/XL	2432	28,000	32,768	18–50K	32 x 32	1,024	2,560	256
XC4036EX/XL	3078	36,000	41,472	22–65K	36 x 36	1,296	3,168	288
XC4044XL	3800	44,000	51,200	27–80K	40 x 40	1,600	3,840	320
XC4052XL	4598	52,000	61,952	33–100K	44 x 44	1,936	4,576	352
XC4062XL	5472	62,000	73,728	40–130K	48 x 48	2,304	5,376	384
XC4085XL	7448	85,000	100,352	55–180K	56 x 56	3,136	7,168	448

(a)

Device	System Gates*	Max Logic Gates	Logic Cells**	Max RAM Bits	Max I/O	Number of Flip-Flops
XCS05XL	2K–5K	3,000	238	3,200	77	360
XCS10XL	3K–10K	5,000	466	6,272	112	616
XCS20XL	7K–20K	10,000	950	12,800	160	1,120
XCS30XL	10K–40K	13,000	1,368	18,432	192	1,536
XCS40XL	13K–40K	20,000	1,862	25,088	205	2,016

*20–30% of CLBs as RAM

**1 Logic Cell = 4-input Look-Up Table + Flip-Flop

(b)

Figure 13.6 Features of the (a) XC4000E, XC4000XL family, and (b) the Spartan families..

Device	Logic Cells	Max. Logic Gates (No RAM)	Max.RA MBits (No Logic)	Typical Gate Range (Logic and RAM)	CLB Matrix	Total CLBs	Number of Flip-Flops	Max Use I/O	PROM Size
XC40125XV	10,982	125,000	147,968	80-250k	68 x 68	4,624	10,336	448	2,797,040
XC40150XV	12,312	150,000	165,888	100-300k	72 x 72	5,184	11,520	448	3,373,448
XC40200XV	16,758	200,000	225,792	130-400k	84 x 84	7,056	15,456	448	4,551,056
XC40250	20,102	250,000	270,848	180-500k	92 x 92	8,464	18,400	448	5,433,888

Figure 13.7 Features of the XC4000XV family.

	XCV50	XCV100	XCV150	XCV200	XCV300	XCV400	XCV600	XCV800	XCV1000
Logic Cells	1,728	2,700	3,888	5,292	6,912	10,800	15,552	21,168	27,648
System Gates	57,906	108,904	164,674	236,666	322,970	468,252	661,111	888,439	1,124,022
CLB Array	16 x 16	20 x 30	24 x 36	28 x 42	32 x 48	40 x 60	48 x 72	56 x 84	64 x 96
Block RAM Bits	32,768	40,960	49,152	57,344	65,536	81,920	98,304	114,688	131,072
Select RAM Bits	24,576	38,400	55,296	75,264	98,304	153,600	221,184	301,056	393,216
Delay Locked Loops	4	4	4	4	4	4	4	4	4
I/O Stds Sup-ported	12	12	12	12	12	12	12	12	12
Max Avail-able I/O	180	180	260	284	316	404	500	514	514

Figure 13.8 Features of the Virtex family.

13.4 THE XILINX XC3000 FPGA FAMILY

The architecture of the XC3000 family consists of an array of 64 to 320 configurable logic blocks (CLBs), a variety of local and global routing resources, input-output (I/O) blocks (IOBs), programmable I/O buffers, and an SRAM-based control store memory. The overall structure of devices in this family is illustrated in Figure 13.9.

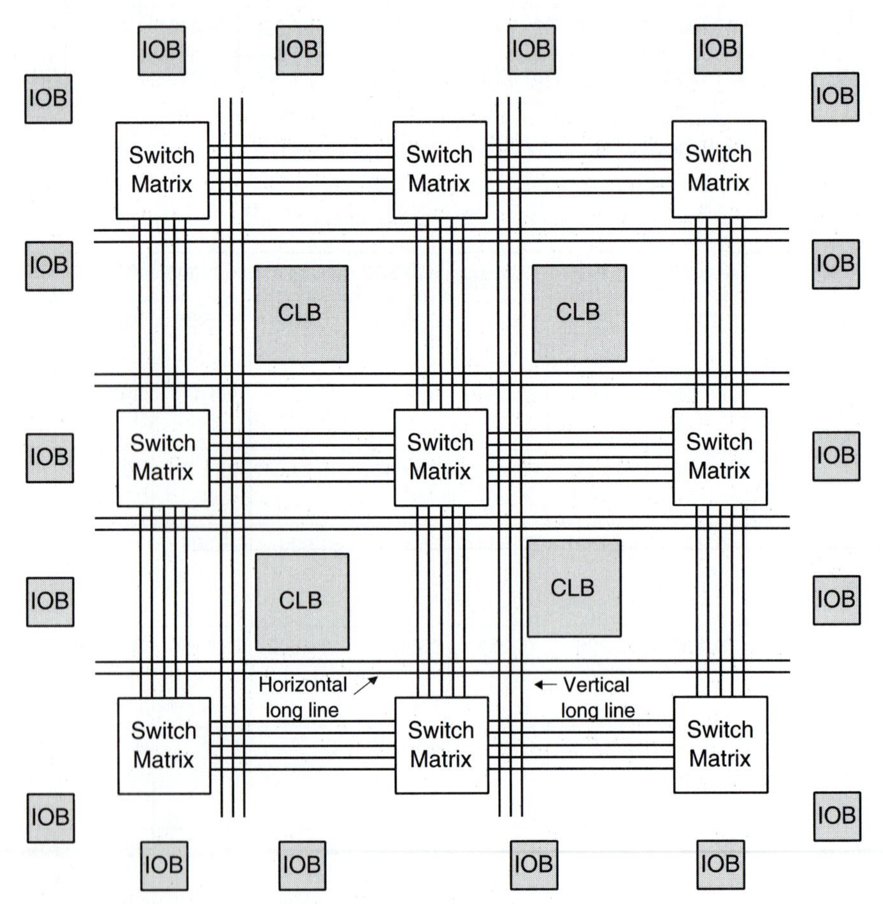

Figure 13.9 Structure of Xilinx 3000 series devices.

13.4.1 The XC3000 Configurable Logic Block (CLB)

Each CLB consists of a look-up table, multiplexers, registers, and paths for control signals. Figure 13.10 shows the architecture of a CLB in the Xilinx 3000 series. Within a CLB, programmable muxes configure the datapaths at the inputs of the flip-flops and route the control signals, such as the reset, through the CLB. The control lines of a configuration mux are connected to dedicated memory cells, whose contents are loaded by the programming software. (Note: a mux that is shown without a control line is configured by the program.) Each CLB has a look-up table (LUT) implementing two combinational logic functions of five input variables (5 ns CLB delay) and two D-type flip-flops. Each CLB has two outputs and a programmed mux selects between the output of the LUT and the output of an internal flip-flop. The outputs of the flip-flops are also routed by an internal feedback path to dedicated inputs of the LUT. Thus, a group of CLBs can support register-rich state machine and pipelining architectures. A mux at the input of each flip-flop operates under user control to gate the output of the mux back to its D-input, or to gate an external datapath (*DATA IN*) into the flip-flop. Each flip-flop can be reset individually or globally. Note that although the schematic of the XC3000 is shown with feedback paths from the flip-flops back to separate inputs (*QX* and *QY*) of the LUTs, the feedback paths are actually programmed to connect to the *.a* and *.e* inputs of the LUT, thereby reducing the number of available inputs.

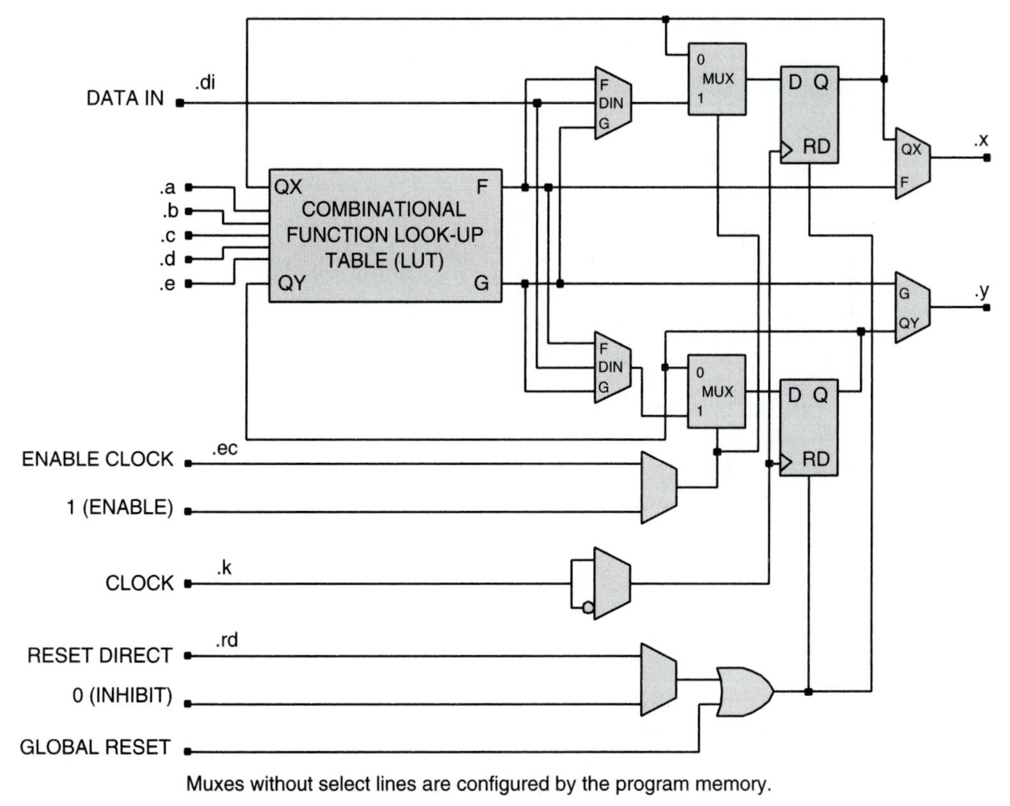

Figure 13.10 Xilinx 3000 series configurable logic block (CLB).

13.4.2 XC3000 Interconnect Resources

The Xilinx 3000 series devices have three types of routing resources: direct interconnection, long lines, and general purpose interconnect. These resources provide capability for nearest-neighbor and across-the-chip connection between CLBs. The interconnect structure of the Xilinx 3000 family is shown in Figure 13.11.

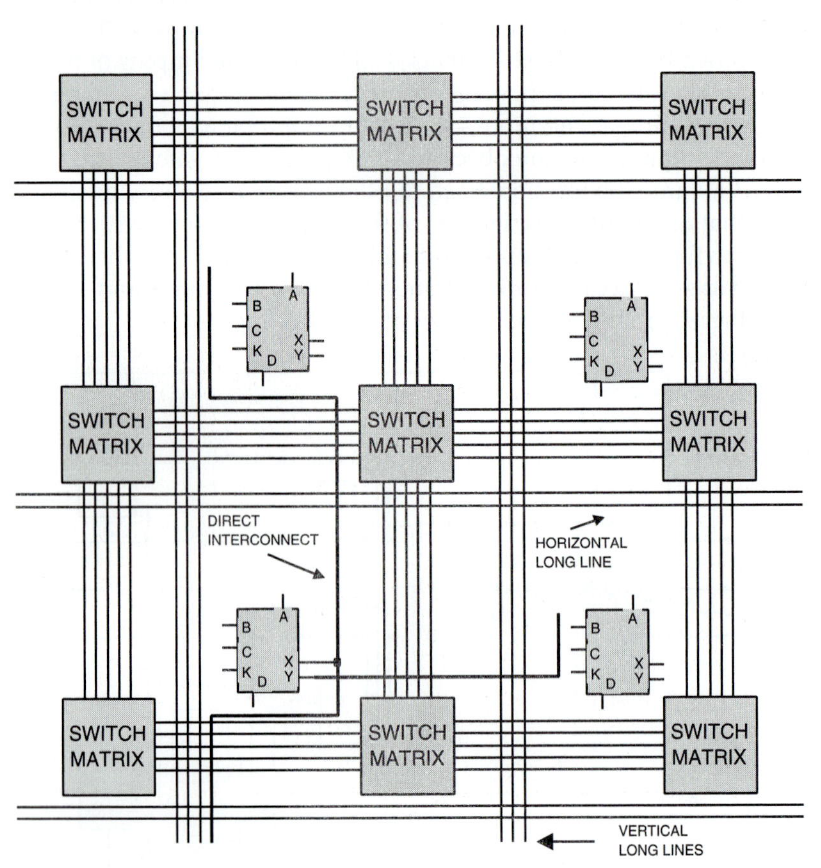

Figure 13.11 Interconnect structure of the Xilinx 3000 series.

Direct interconnect lines route between adjacent vertical and horizontal CLBs in the same column or row. These are relatively high-speed local connections through metal, but are not as fast as hard-wired metal connections because of the delay incurred by routing the signal paths through the transmission gates that configure the path. Direct interconnect lines do not use the switch matrices, which eliminates the delay incurred on paths going through a matrix. (See Xilinx documentation for the pinout conventions to establish local interconnect between CLBs.)

Long lines traverse the entire width and height of the interconnect area. These lines also bypass the switch matrices and support signals having a high fanout. They have relatively low skew and low delay. There are three long lines per column and two long lines per row of the device. The signals that drive long lines are buffered. Long lines can be driven by adjacent CLBs or IOBs and may interconnect to three-state buffers that are available to CLBs. Long lines provide three-state buses within the architecture, and implement wired-and logic. Each horizontal long line is driven by a three-state buffer and can be programmed to connect to a pull-up resistor, which pulls the line to a logical "1" if no driver is asserted on the line.

The interconnect resources include clock buffers having high fanout and low-skew dedicated clock lines over the area of the device. A global clock buffer (*GCLK*) connects to every CLB and IOB. An alternative buffer (*ACLK*) can be programmed to drive selected column lines. An on-chip oscillator can also drive a clock. Each IOB has a pair of inputs that route external clocks to the internal flip-flops and route through the IOB to the other IOBs and CLBs via the *.ok* and *.ik* ports.

A grid of switch matrices overlays the architecture to provide general purpose interconnect for branching and routing throughout the device. The elements of the grid are connected through these switch matrices to their nearest neighbors by five horizontal and five vertical lines. The switch matrices provide a limited interconnect between the paths they connect. Some of the paths have buffers to offset signal degradation due to the interconnect transmission gates.

13.4.3 XC3000 Switch Box Functional Configuration and Connectivity

The Xilinx 3000 switch box has a limited number of connections that can be made between a path into the switch box and a path out to a neighboring switch box. Figure 13.12 shows the options for connectivity through a switch box. The scheme enables the router to avoid blockage by switching to another track when bypassing several CLBs.

The programmable interconnect resources of the device connect CLBs and IOBs, either directly or through switch boxes. These resources consist of a grid of two layers of metal segments and programmable interconnect points (PIP) within switch boxes. A PIP is a CMOS transmission gate whose state (on or off) is determined by the content of a static RAM cell in the programmable memory, as shown in Figure 13.13. The connection is established when the transmission gate is on (i.e., a "1" is applied at the gate of the n-channel transistor, and a "0" is applied at the gate of the p-channel transistor). The interconnect path is established without altering the physical medium, as happens in a fuse-type interconnect technology. Thus, the device can be re-programmed by simply changing the content of the controlling memory cell.

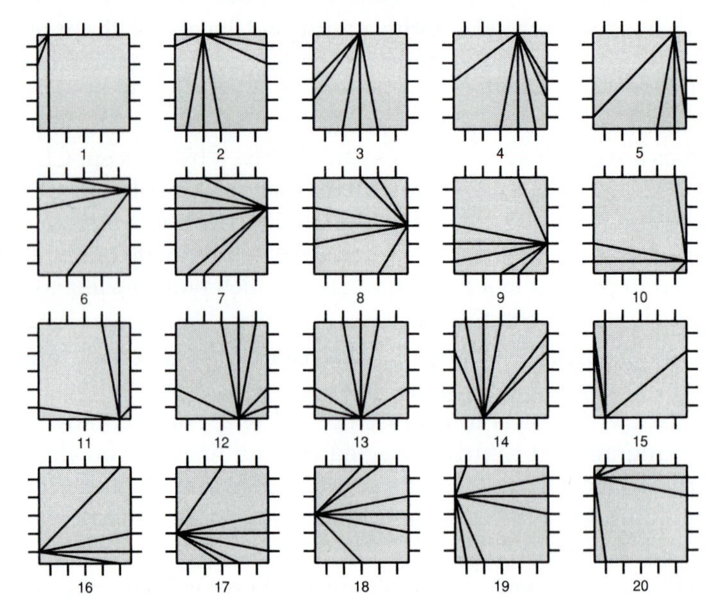

Figure 13.12 Paths through a Xilinx 3000 series switch box.

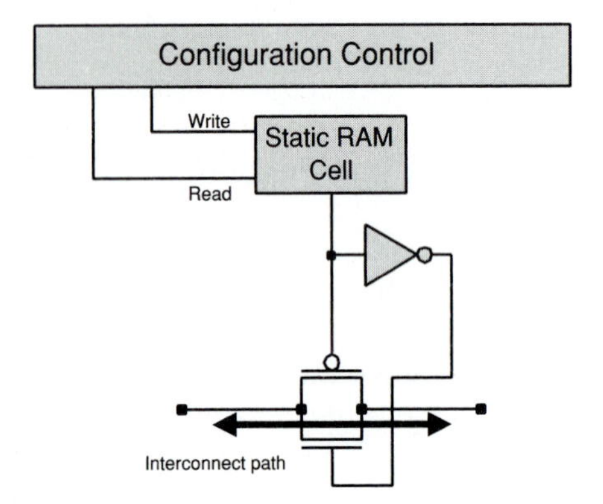

Figure 13.13 RAM Cell controlling a PIP transmission gate.

13.4.4 XC3000 I/O Block (IOB)

The XC3000 series parts have up to 168 physical I/O pins. Each of these pins (pads) is connected to a dedicated, programmable I/O block, as shown in Figure 13.14. The circuitry of the IOB connects the physical pin to a pair of diodes that provide protection against electrostatic discharge.

The I/O pad can be configured as an input, output, or bi-directional port. The input connection can be direct, latched, or registered. Input buffers provide compatibility with TTL and CMOS input signal levels. When configured as an input IOB, the signal at the pad is routed directly to the *.i* (*DIRECT IN*) port, or through a device configured as either a D-type flip-flop or latch. This path routes to the *.q* (*REGISTERED IN*) port of the IOB. When used as an output IOB the signal at the *.o* (OUT) input terminal of the IOB is routed, with programmable polarity, directly or through a D-type flip-flop, to a three-state output buffer driving the I/O pad. The registered output effectively eliminates interconnect delay in the clock-to-output time. The output has three-state control and a programmable slew rate for noise reduction.

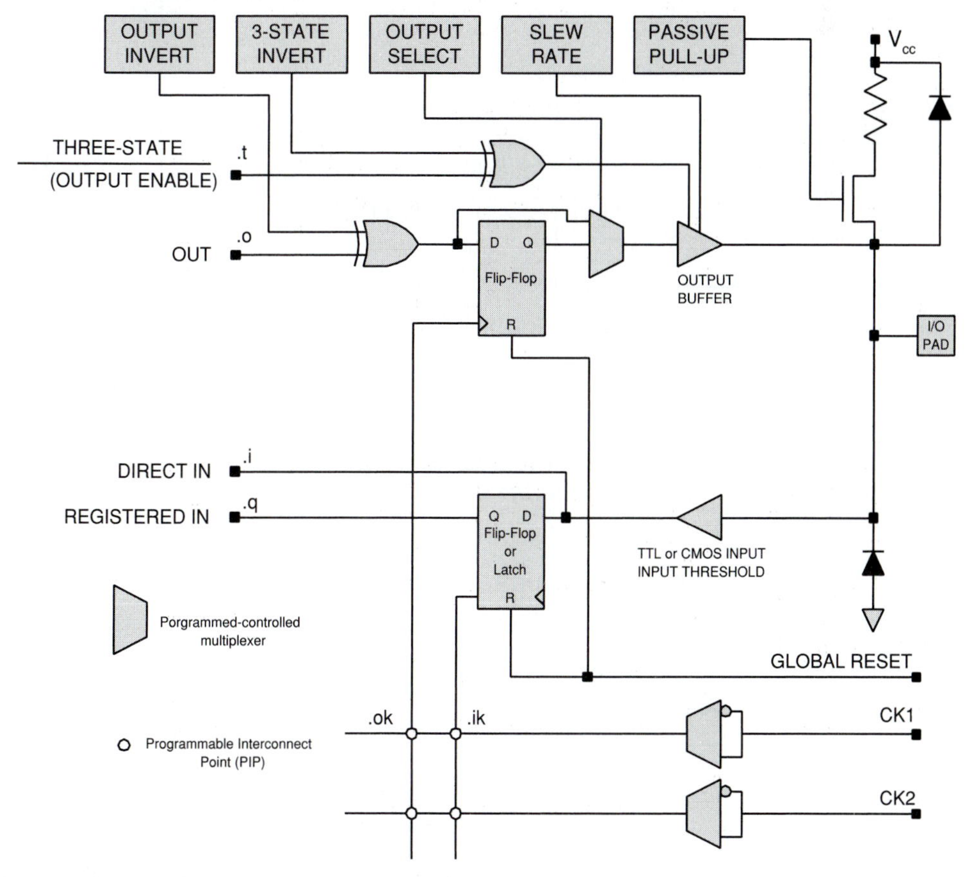

Figure 13.14 I/O block for the XC3000 family.

Devices in the Xilinx 3000 family can be programmed in a variety of modes. In a common configuration, shown in Figure 13.15, the FPGA controls a memory; alternatively, the FPGA can be under the control of another device (Figure 13.15b). In a third configuration an FPGA operates as a controlled unit in a Daisy-chain of devices (Figure 13.15c). For example, in the master mode, the FPGA can initiate either serial or byte-parallel transfer of configuration data from an external PROM; in slave mode the configuration program can be downloaded to the FPGA's configuration memory by an external controlling processor (e.g., a host workstation).

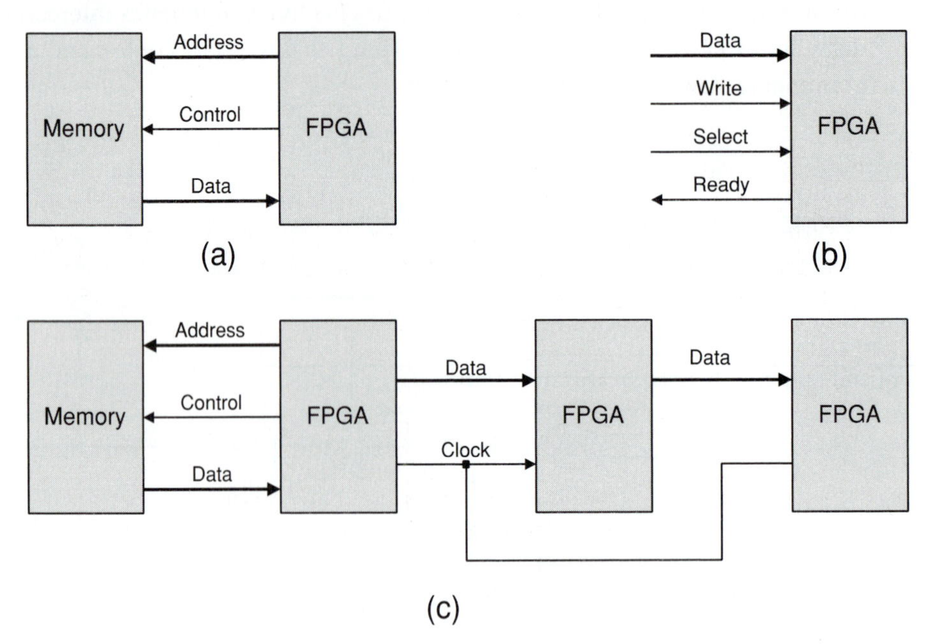

Figure 13.15 Connection modes for FPGAs: (a) Master mode (serial or parallel), (b) Peripheral mode, and (c) Slave mode for Daisy chains.

13.4.5 XC3000 Performance

The XC3000 family has evolved from an initial 1.2 microns process to a 0.8 micron process, with system speeds in excess of 100 MHz.[2]

13.5 THE XC4000 FPGA FAMILY

The Xilinx XC4000 family of devices has significantly increased density and improved performance over the XC3000 devices. This family continues to evolve with improvements in fundamental fabrication process. Its basic architecture will be discussed to point out key enhancements.

13.5.1 XC4000 Configurable Logic Block (CLB)

The configurable logic blocks of the XC4000 are more versatile than those of the XC3000. In the architecture shown in Figure 13.16, note that each CLB contains three function generators (*F*, *G*, and *H*). Each is based on a look-up table having 5ns delay independent of the function begin implemented. Two of the function generators (*F* and *G*) can generate any arbitrary function of four inputs, and the third (*H*) can generate any Boolean function of three inputs. The *H*-function block can get its inputs from the *F* and *G* LUTs, or from external inputs. The three function generators can be programmed to generate (1) three different functions of three independent sets of variables (two with four inputs and one with three inputs -- one function must be registered within the CLB), (2) an arbitrary function of five variables, (3) an arbitrary function of four variables together with some functions of six variables, and (4) some functions of nine variables.

Each CLB in the XC4000 series has two storage devices that can be configured as edge-triggered flip-flops with a common clock, or in the XC4000X they can be configured as flip-flops or as transparent latches with common clock (programmed for either edge and separately invertible) and enable. The storage elements can get their inputs from the function generators or the D_{in} input. The other element can get an external input from the *H1* input. The function generators can also drive two outputs directly (*X* and *Y*) and independently of the outputs of the storage elements. All of these outputs can be connected to the interconnect network. The storage elements are driven by a global set/reset during power-up; the global set/reset is programmed to match the programming of the local S/R control for a given storage element.

The *F* and *G* function generators of the XC4000 family have dedicated logic for fast carry and borrow logic, with dedicated routing to link the extra signal to the function generator in the adjacent CLB. The pre-built carry chain within a CLB can be used to add a pair of two-bit words in one CLB. One function generator (*F*) can be used to generate a0 + b0, and a second (*G*) can generate a1 + b1. The fast carry will forward the carry to the next CLB above or below. This feature is implemented with a hard macro and the graphical editor.

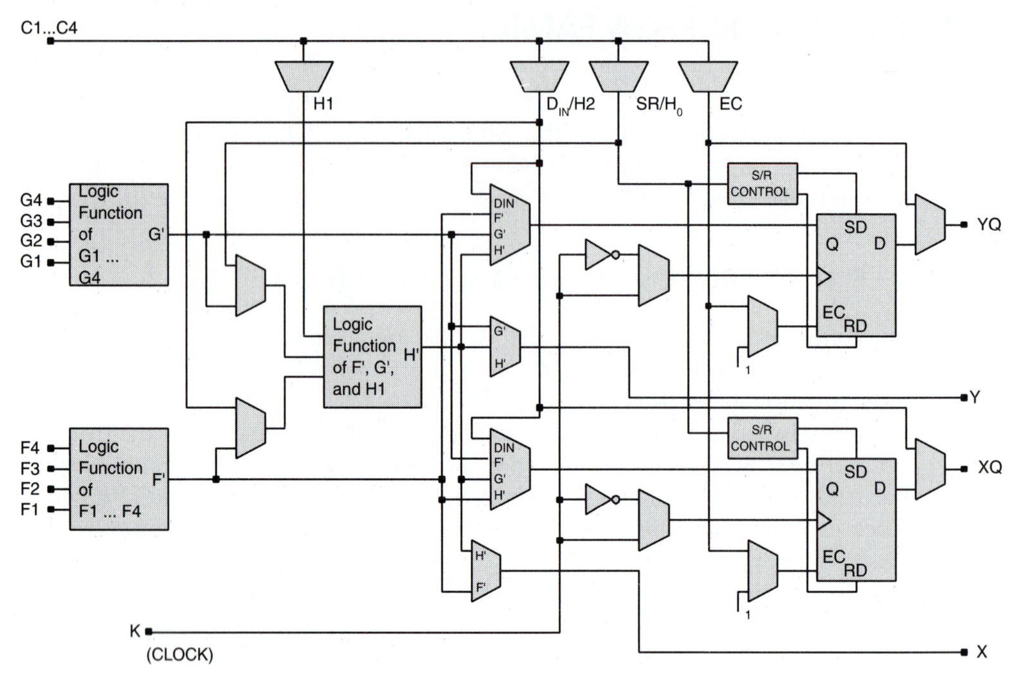

Figure 13.16 Architecture of the CLB in the XC4000 FPGA.

Each CLB has a pair of three-state buffers that can drive signals onto the nearest horizontal lines above or below the CLB. The three function generators within a CLB can be used as RAM (either a 16x2 dual port RAM or a 32x1 single-port RAM). A group of CLBs can form an array of memory.

13.5.2 XC4000 Interconnect Resources

The XC4000 series was designed with interconnect resources to minimize the resistance and capacitance of an average routed path. The device has three types of interconnect: single-length lines, double-length lines, and long lines. A grid of horizontal and vertical single-length lines connect an array of switch boxes. The boxes provide a reduced number of connections between signal paths within each box. In the XC4000, there is a richer set of connections between single-length lines and CLB inputs and outputs.

Double-length lines traverse the distance of two CLBs before entering a switch matrix, skipping every other CLB. These lines provide a more efficient implementation of intermediate-length connections by eliminating a switch matrix from the path, thereby reducing the delay of the path.

Long lines span the entire width of the array and they drive high-fanout control signals. Long vertical lines have programmable splitters that segment the lines and allow two independent routing channels spanning one-half of the array, but located in the same column. The routing resources are exploited automatically by the routing software. There are eight low-skew global buffers for clock distribution, and the skew on a global net is less than 2 ns. The XC4000 device also has an internally-generated clock.

Figure 13.17 illustrates the double-length interconnect lines of the XC4000 device.

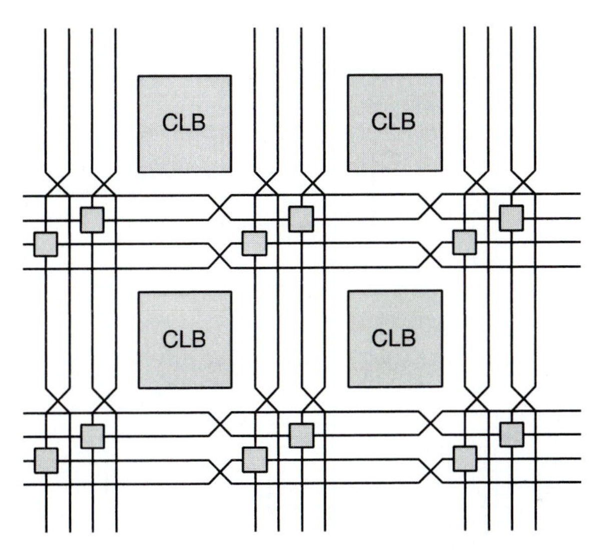

Figure 13.17 Double-length interconnects in the XC4000.

The XC4000 has more routing resources than the XC3000 and uses a simpler switch box with more PIPs. Also, its CLB can be used as an on-chip RAM or ROM. The device has boundary-scan compatibility. It accommodates wider decode logic and has more global clocks.

The connectivity of a CLB with neighboring switch matrices is shown in Figure 13.18.

The architecture of a PIP-based interconnection in a switch box is shown in Figure 13.19. The configuration of CMOS transmission gates determines the connection between a horizontal line and the opposite horizontal line, as well as the vertical lines at the connection. Each switch matrix PIP requires six pass transistors.

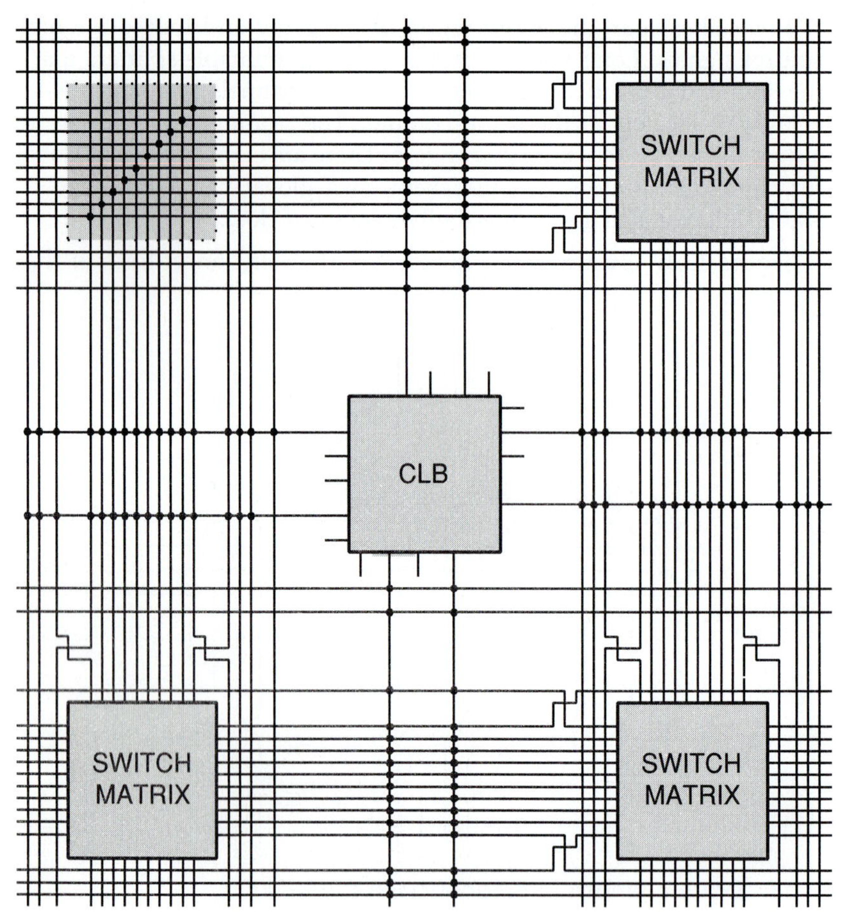

Figure 13.18 Connectivity of a CLB with its neighboring switch matrices.

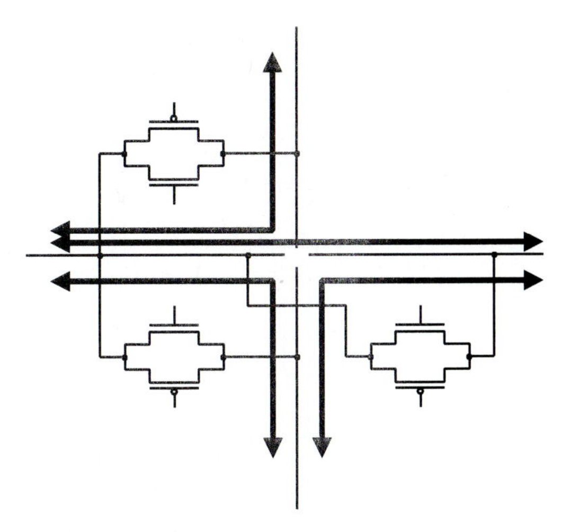

Figure 13.19 Circuit-level architecture of a programmable interconnect point (PIP) within a switch box.

13.5.3 XC4000 I/O Block (IOB)

Each programmable I/O pin of an XC4000 device has a programmable IOB with buffers for compatibility with TTL and CMOS signal levels. Figure 13.20 shows a simplified schematic for a XC4000 programmable IOB. It can be used as an input, output or bi-directional port. An IOB that is configured as an input can have direct, latched, or registered input. In an output configuration, the IOB has direct or registered output. The output buffer of an IOB has skew and slew rate control.

The XC4000 architecture has delay elements that compensate for the delay induced when a clock signal passes through a global buffer before reaching an IOB. This eliminates the hold condition on the data at an external pin. The three-state output of an IOB puts the output buffer in a high-impedance state. The output and enable for the output can be inverted. The slew rate of the output buffer can be controlled to minimize transients on the power bus when non-critical signals are switched. The IOB pin can be programmed for pull-up or pull-down to reduce power consumption and noise.

The XC4000X IOB includes features not found in the XC4000, notably a fast latch (not shown) on the input, which allows very fast capture of input data.

The registers available to the input and output path of an IOB are driven by separate, invertible clocks. There is a global set and reset.

The XC4000 has four edge decoders on each side of the chip. An edge decoder can accept up to 40 inputs from adjacent IOBs and 20 inputs from on-chip. These decoders provide fast decoding of wide address paths. Multiple

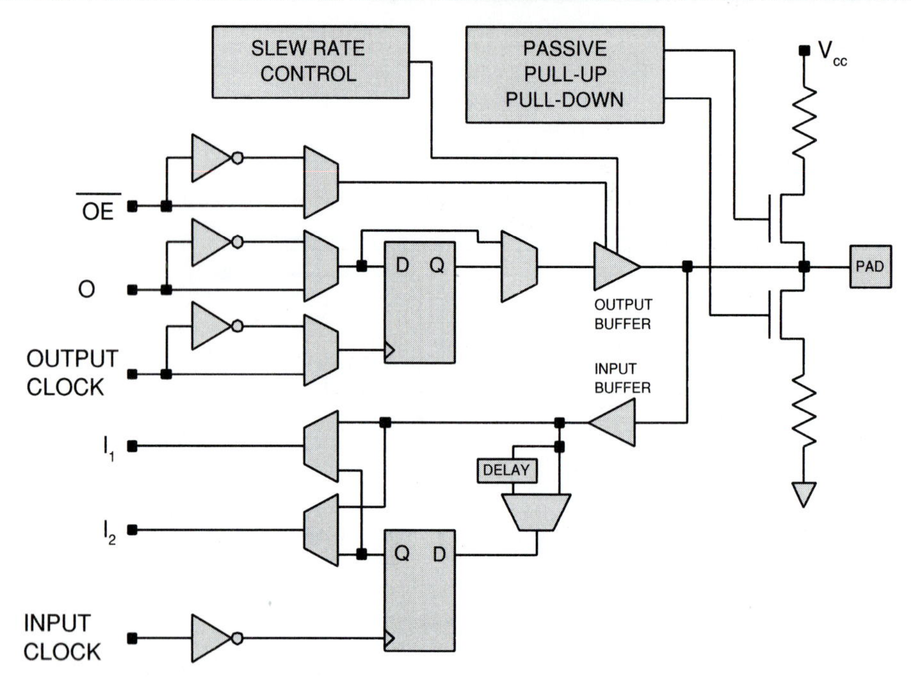

Figure 13.20 XC4000 series IOB.

CLBs might be needed to support decoding when the fan-in exceeds the width of a single CLB.

The XC4000 devices have embedded logic to support the IEEE 1149.1 (JTAG) boundary scan standard. There is an on-chip test access port (TAP controller), and the I/O cells can be configured as a shift register. Under test, the device can be checked to verify that all the pins on a PC board are properly connected by creating a serial chain of all the I/O pins of the chips on the board. A master three-state control signal puts all the IOBs in high impedance mode for board testing.

The XC4000EX series was introduced in February 1996. It provides a two-fold increase in routing resources over the earlier members of the XC4000 family. An additional 22 vertical lines are available in each column of CLBs. Another 12 quad lines have been added to each row and column to support fast global routing, illustrated in Figure 13.21. The IOBs have a dedicated early clock and fast-capture latch. The registers have a 4ns setup time and 6ns clock-to-output propagation delay.

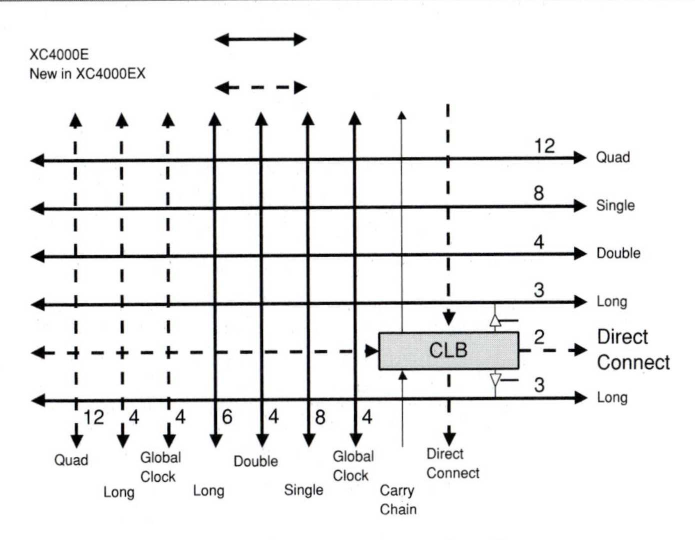

Figure 13.21 Routing resources of the XC4000E family.

13.5.4 Enhancements in the XC4000E and XC4000X Series

The XC4000E and XC4000X devices offer significant increases in speed and capacity, as well as architectural improvements over the basic XC4000 family. Devices in these families can run at system clock rates up to 80MHz, and internal frequencies can reach 150 MHz, as a result of sub-micron multi-layered metal processes. The XC4000XL (.35 micron feature size) operates at 3.3V with a system frequency of 80 MHz.

13.6 RAPID PROTOTYPING WITH VERILOG AND FPGAs

The technology of FPGAs and their supporting software tools make them an attractive target for rapid prototyping of a design. The objective of rapid prototyping is to create a working prototype as quickly as possible to meet market conditions and to support broader testing in the host environment.

Initially, the tools supporting FPGAs relied on schematic entry, but many vendors now support hardware description languages as well. As an example, the Xilinx FPGAs are supported by a robust suite of software tools covering design entry (by schematic or hardware description language), floorplanning, simulation, automatic block placement and routing of interconnects, timing verification, downloading of configuration data, and readback of the configuration bitstream. The tools ultimately produce the bitstream file that can be downloaded to the part to configure it on the host board. Tools that rely on HDL-based entry have a streamlined path of development that transforms, optimizes and maps an HDL behavioral description into an FPGA part. Extensive on-line and Web-based documentation accompanies the tools provided by Xilinx and partnering vendors.

Rapid prototyping bypasses the structural detail that is forced by schematic-based entry methods. If the place-and-route engine within the tool is allowed complete freedom it will generate an optimal assignment of pins. This freedom is curtailed when the part must fit into the socket of a previously-configured board. Ideally, the board is not configured until the FPGA has been fully designed. Constraining the pinout limits the optimization process, and may sacrifice performance. If feasible, careful pin assignment can lead to improved routing of the design. For example, the horizontal long lines in the XC3000 and XC4000 architectures have three-state buffers, which make them suitable for data buses. On the other hand, vertical long lines for clock-enable in the XC3000 and vertical carry chains in the XC4000 lead naturally to a vertical orientation of registers and counters. These architectural features suggest that data paths should be applied to the left and right sides of the part, and control lines to the top and bottom of the part when manual routing and pin assignment are necessary. The tool has maximum flexibility when no pins are pre-assigned, but other topological constraints may require that some pins be pre-assigned before routing. However, it is recommended that the unconstrained design be routed first to verify that it can meet timing specifications. If the unconstrained design does not, the constrained design will not meet them either.

Keeping a design synchronous with a single clock allows timing-driven routing tools to work more efficiently. The parts have clock enables, so there is no need to gate clocks in a design. (See Chapter 9 for a discussion of gated clocks.)

FPGAs are register-rich. Therefore, it is advantageous to employ one-hot encoding in state machines. This leads to simpler next-state and output logic. This form of encoding is sometimes referred to as "state-per-bit" encoding, because a unique single flip-flop is asserted for each state. Coding style affects the results of targeting a description into an FPGA. One notable example is in the description of a sequencer. If the count sequence does not have to be binary, a linear feedback shift register (LFSR) may be a more attractive alternatives because it requires less space and routes more efficiently than do binary counters. Designers should be aware that flip-flops in FPGAs tend to initialize to a cleared output during power-up. A state machine would have to anticipate this condition because it is not one of the explicit one-hot codes.

13.7 DESIGN EXERCISES

The software accompanying this text, the *Xilinx* Foundation Express™, Version 1.5, includes an environment for developing and debugging source code, embedded configurable logic blocks (Logic Blox™) for certain functionality, and a synthesis engine that optimizes and maps the design into a target Xilinx FPGA. The operation of this tool is well-documented in the accompanying users guide, including a tutorial that leads the reader through the design of a simple stopwatch. The use of the tool will not be covered here. The reader is urged to consult the users guide and to work the stopwatch tutorial to gain familiarity with the tool before attempting to use the tool on other designs.

The focus of the exercises here will be to couple the expressive power of the Verilog hardware description language with the synthesis and implementation capabilities of the student version of the *Xilinx* Foundation Express™ tool, which includes the widely-used synthesis technology of Synopsys™, Inc. The exercises can be implemented on the Xilinx demonstration board with the XC4003E part.

13.7.1 Microcontroller

Chapter 12 presented a description of a four-bit slice microcontroller. Use the Foundation Express™ tool to synthesize the design into a Xilinx FPGA (XC4003E). Be aware that a description that synthesizes under one tool might not synthesize under a different tool, even though the tool is supported by the same vendor. Make any necessary modifications to the description of the microcontroller. Using the reports generated by the tool, consider the FPGA resource requirements and performance of the synthesized microcontroller for a variety of word sizes formed by an aggregate of four-bit slice parts. Using the simulation capability within the Foundation Express™ tool, develop some stimulus files and verify the operation of the synthesized part. Create a schematic view of the layout of the design. Also save the design in a netlist format. Convert the netlist to a Verilog description, and verify the functionality of the design in the Silos-III environment, with the same stimulus testbench that was used to verify the behavioral model. Note: For post-layout simulation with Silos-III, any generic flip-flops in the netlist will have to be replaced by appropriate functional models.

13.7.2 Electronic Roulette Wheel

An electronic roulette wheel has a display consisting of two concentric rings of colored lights, as shown in Figure 13.22. One ring contains red lights; the other contains green lights. One red light is lit at a time, and the location of the lit light moves in sequence around the ring while the *play* button is depressed. The green lights are used to place bets. The Game Control Panel, also shown in Figure 13.22, has buttons to control the wheel and a digital readout of the value of the game. While the *play* button is depressed, the red lights are illuminated in sequence, one at a time, at a frequency determined by an external clock. When the play button is released, the movement halts. If the position of the lit red light at the end of play coincides with that of an illuminated green light the player wins an amount determined by the scoring rule described below. The frequency of the system clock should be high enough to ensure a random outcome.

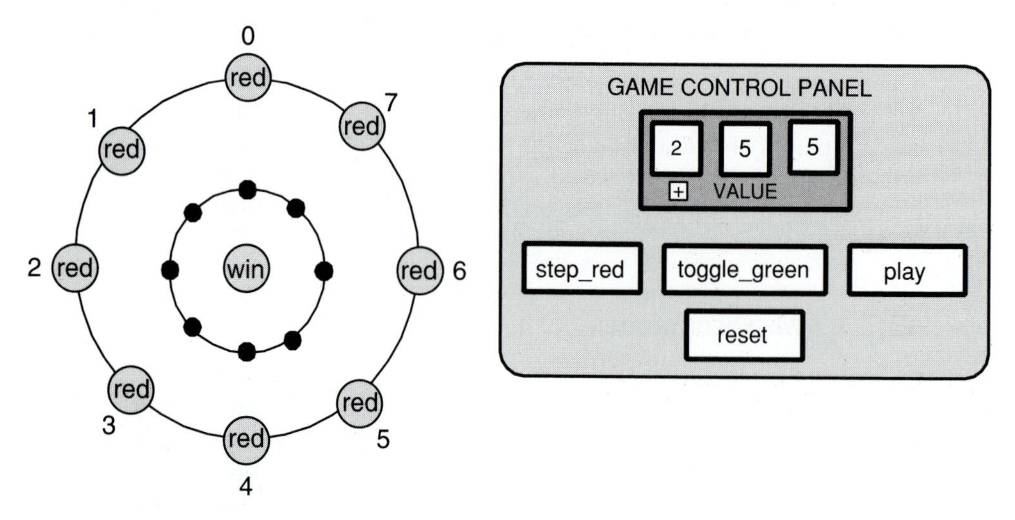

Figure 13.22 Display (a) and control panel (b) for an electronic roulette wheel.

A bet can be placed only when *reset* and *play* are not asserted. In this condition, each positive edge of *step_red* causes the location of the illuminated red light to advance by one position. If *play* and *reset* are not asserted, a positive edge of *toggle_grn* toggles the condition of the green light at the current position of the illuminated red light. A lit green light indicates the location of a placed bet. A player may place a bet at any or none of the positions. The payoff at the end of play depends upon the number of bets that the player has placed before *play* is asserted. Bets need not be removed at the end of play, but the player's value is increased or decreased depending upon the outcome at the end of play. The value register records the condition of the player's wins and/or losses. At the end of play the *win_yellow* light is illuminated if there is a win.

The Verilog fragment below declares the module information at the top level of the design.

```
module roulette_wheel (value, green_code, sign, win_yellow, red_index, init_value,
  clock, play, reset, toggle_grn, step_red);

  // Functionality: Top-level module of the roulette wheel chip. Responds to
  // commands from the Game Control Panel and generates signals that are used by
  // the Game Control Panel to control the various panel displays for the position
```

```
// and status lights, as well as the value of the game.

output [7:0]
   value,          // 8-bit representation of the the player's current value.
   green_code;     // 8-bit pattern indicating green LEDs that are lit.
                   // A '1' at a bit indicates a bet a that position.
output
   sign,           // A sign bit for the value of the game.
   win_yellow;     // A bit indicating win/lose at end of play.

output [2:0]
   red_index;      // A 3-bit pattern indicating the current position of the lit red light.

input [7:0]
   init_value;     // 8-bit representation of the initial
                   // value of the game (always positive).
input
   clock,          // System clock. Symmetric waveform, begins at zero.
                   // Generated at the control panel.
   play,           // Red wheel (lights) spins while
                   // play = 1 and reset not asserted.
                   // The lit red light advances by one position
                   // at each tick of the clock.
                   // Generated at the control panel.
   reset,          // Press reset to initialize the game to the
                   // initial value, position the lit red
                   // LED at position 0, turn off the green
                   // LEDs, and turn off the win_yellow LED.
                   // Generated at the control panel.

   toggle_grn,     // Press to toggle the green LED at
                   // the position of the lit red LED, while
                   // game is not in play and reset is not asserted.
                   // Generated at the control panel.

   step_red;       // Press to advance the lit red LED by one
                   // position when the game is not in play
                   // and reset is not asserted.
                   // Generated at the control panel.

// Your description goes here.

endmodule
```

The roulette wheel controller can be implemented as a single module or a hierarchy of modules. In the first case, the aim would be to develop a description of the behavior, without concern for structure. In the latter case, the roulette wheel's behavior could be decomposed into a group of modules which interact, each of which implements some aspect of the overall functionality.

The approach to be taken in this exercise is to decompose the overall behavior into the functional units shown in Figure 13.23. The *Spinning_Unit* is to control the position of the lit red light. While the synchronous *reset* is asserted, the *red_index* is reset to 3'b000. If *reset* is not asserted and *play* is asserted, the red light is to "move" synchronously with the clock (i.e., *red_index* is to increment at each clock cycle). Note that the end of play does not reset the wheel (i.e., the red light at the next "play" begins at the position determined by the last play).

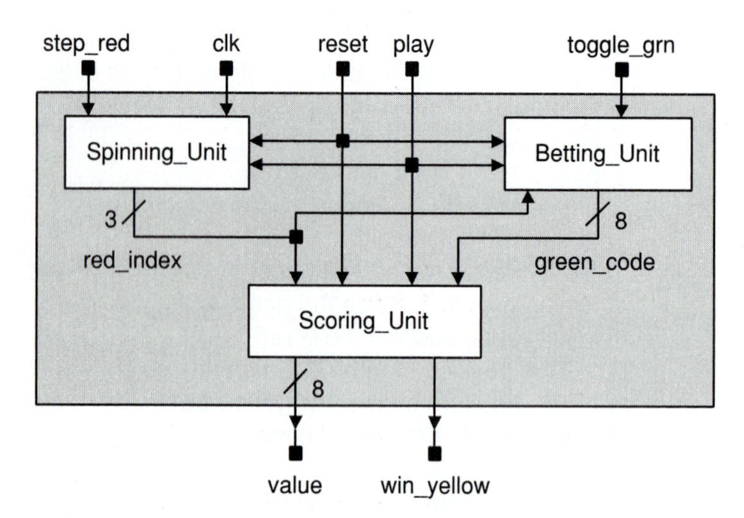

Figure 13.23 Functional decomposition of the roulette wheel.

The *Betting_Unit* is to create *green_code*, which is a pattern of eight bits. A one in a bit of *green_code* signifies a bet. The player places/removes a bet at the position corresponding to the lit red light by toggling on/off the *toggle_grn* signal if *step_red* and *play* and *reset* are not asserted. When *reset* and *play* are asserted, the position of the lit red light is advanced by toggling the *step_red* signal. Thus, a bet can be placed at any position. When *reset* is asserted, the *green_code* is reset to 8'b0.

The *Scoring_Unit* adjusts the value at the end of play according to Table 13.1.

win_yellow	Number of Bets	Payoff
0		-1
1	8,7,6,5	0
1	4	1
1	3	2
1	2	3
1	1	4

Table 13.1 *Payoff table for the roulette wheel.*

When designing this circuit, develop and verify each submodule separately. Then verify the integrated behavior. It might be convenient to develop a testbench in which the individual submodules are instantiated and tested progressively as they are developed. As the design evolves, the testbench can be expanded to accommodate additional tests and ultimately test the integrated description. Consider using the ***fork … join*** statement to create stimulus signal waveforms. Your test suite must document each test action and display output that partitions the data into segments corresponding to test actions. At a minimum, your testbench should consider the following details of operation:

a. Spinning Unit: Verify that *reset* overrides all other inputs and interrupts *play,* and that spinning resumes if *play* is asserted when *reset* is de-asserted. Verify that the wheel waits for the *play* button to be asserted and remains at its current position when *play* is de-asserted. Spinning should resume when *play* is asserted again. The position of the lit red light should be properly encoded by *red_index*.

b. Betting_Unit: Verify that *step_red* is ignored while *play* or *reset* is asserted. When *play* and *reset* are not asserted, a toggle of *step_red* should advance the location of the red light by one position. Verify that *toggle_grn* is ignored while *play* or *reset* is asserted, and that it toggles the green light at the position specified by *red_index* while *play* and *reset* are not asserted.

c. Scoring_Unit: Verify that the *Scoring_Unit* resets properly, indicates win/lose results, and adjusts the value and sign accordingly. Note: *init_value* is an arbitrary value passed to the roulette wheel from the environment. Be careful to treat the arithmetic of the scoring unit correctly. Although integers could be used to exploit built-in integer arithmetic, use register variables to produce smaller synthesized registers in hardware (integers are 32-bit values). This requires that you deal with the subtraction of value when the play results in a loss. Hint: Consider signed-magnitude arithmetic. The *sign* bit indicates whether the value is owed to the player (positive accumulated value) or indicates the amount the player is in debt to the system (negative accumulated value).

As additional exercises, consider the following options:

Option #1: Develop a single module implementing the roulette wheel and compare the results of synthesizing it with the result obtained with the decomposed design.

Option #2: Consider alternative scoring rules.

Option #3: Consider ways to randomize the light movement sequence.

Option #4: Parameterize the description.

Option #5: Design a multi-player version of the game.

Option #6: Consider ways to synchronize the activity of the asynchronous inputs. Develop an ASM chart describing the machine.

Option #7: Consider an alternative partition of the machine.

Option #8: Implement the machine in a single cyclic behavior.

After verifying your design, use the Foundation Express™ tool to synthesize the description into the Xilinx XC4003E FPGA. Choose pin assignments to map the red and green lights of the roulette wheel to the lights on the board (the board may have to be modified with additional parts; if necessary, adjust the sizes of variables to fit the resources on the prototype board.) Display the value of the game. Download the bitstream of the design to the board and verify that the synthesized circuit operates correctly.

13.7.3 Electronic Dice Game

Example 7.60 described an electronic dice game. Using the Foundation Express™ tool, synthesize this game into the Xilinx XC4003E FPGA. If necessary, modify your source description to fit the style required by the tool. Constrain the design by choosing a pin assignment to configure the displays and switches of the demonstration board to support the dice game. Download the bitstream to configure the demonstration board, and demonstrate that the game operates correctly.

13.8 SUMMARY

FPGAs provide an attractive technology for rapid prototyping of ASICs. Volatile FPGAs use CMOS transmission gates and memory to configure the FPGA to a particular application. The configuration is established by a bitstream that is downloaded from a host processor or onboard PROM. When power is removed from the board, the configuration is lost. The resources available in an FPGA vary from vendor to vendor, and between a vendor's part families. The Xilinx FPGAs discussed in this chapter offer a wide range of equivalent gate counts and capabilities. The tools that support FPGAs now include HDL editors, making possible rapid prototyping from a Verilog behavioral description.

REFERENCES

1. Oldfield, J.V. and Dorf, R. C., *Field-Programmable Gate Arrays,* Wiley Interscience, New York, 1995.
2. Trimberger, S.M., editor, *Field-Programmable Gate Array Technology,* Kluwer Academic Publishers, Boston, 1994.
3. Alfke, P., "Choosing a Xilinx Product Family", Xilinx Application Note, July 1998.
4. Donlin, M. "FPGAs Continue to Break Density Barriers", *Computer Design*, February 1996.

PROBLEMS

Note: Some of the exercises below may require that switches and other resources be added to the prototyping board.

1. Conduct a literature search to identify contemporary performance characteristics of antifuse, EPROM, and SRAM-based FPGAs.
2. Model, verify, and synthesize a circuit, *data_counter*, that counts and displays the number of ones in a parallel eight-bit word. The word is provided synchronously at the rising edge of a clock. Implement your design with the Foundation Express™ tool and create a working hardware prototype.
3. Synthesize both realizations of the FIFO discussed in Chapter 12, Problem 1, into an FPGA and compare their performance (speed) and use of resources.
4. Create a model for a 16-bit counter with synchronous load and reset. Synthesize the circuit into an FPGA to create a working prototype, and determine the maximum frequency at which the counter can operate. Also summarize the resources used in the implementation.
5. Create a 16-bit pipelined carry-look-ahead adder with four-bit slices. Synthesize the unit into an FPGA. Compare the performance of the synthesized circuit to that of a 16-bit rippled carry adder. Also summarize the resources used in the implementation.
6. Write, verify, and synthesize a behavioral model of a digital clock. The inputs to the model include a 60 Hz clock, controls to advance time by an increment of 10 minutes or 1 minute, and a control to select 12 hour or 24 hour time format. The outputs are to include a bit to indicate AM or PM, two seven-segment displays of the hour of the day in 12- or 24-hour formats, two seven-segment displays to indicate the minutes, and two seven-segment displays to indicate the seconds.
7. The figure below shows a simple CPU that is to be prototyped in an FPGA. The distribution of the *clk_dec*, *clk*, *reset*, and *fetch_mem* signals have been omitted to reduce the clutter in the schematic.

39

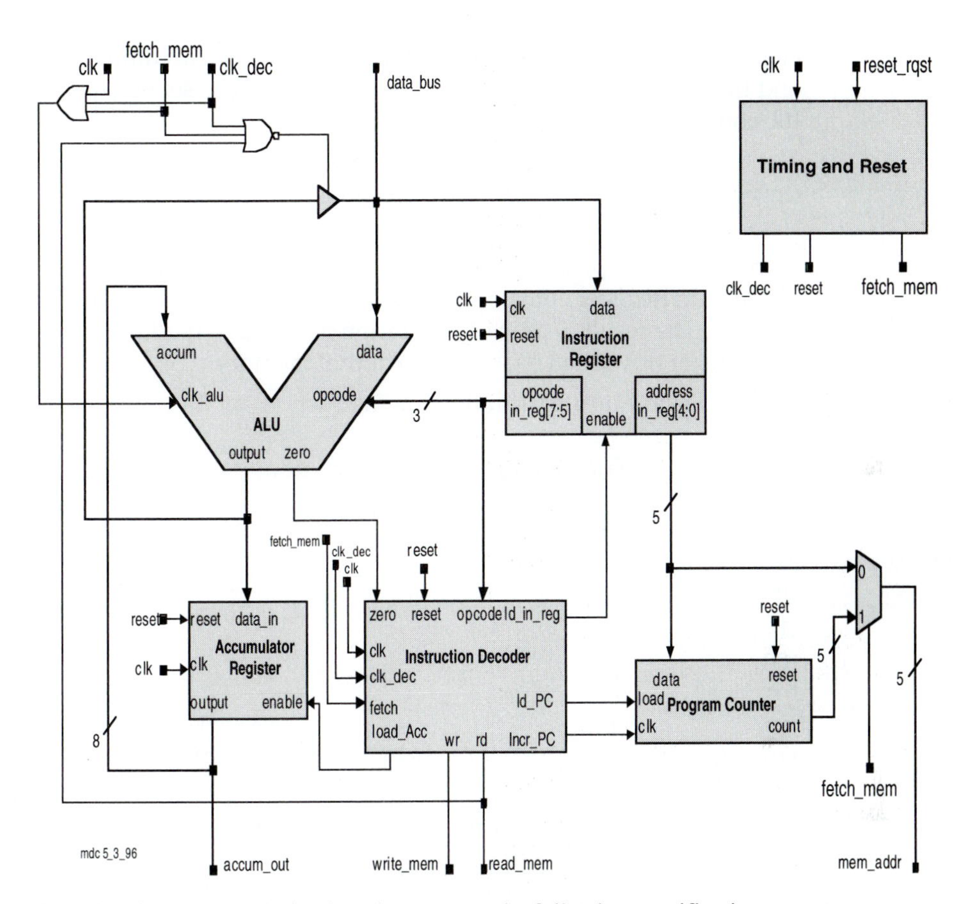

Specifications: Your design is to incorporate the following specifications:

a. The ALU is to implement the behaviors of the set of opcodes shown below:

Opcode	Functionality
00	Pass contents of accumulator
01	Pass contents of accumulator
10	Add data path and accumulator register
11	Bitwise and of data path and accumulator
100	Bitwise exclusive or of data path and accumulator
101	Pass contents of data path
110	Pass contents of accumulator
111	Pass contents of accumulator

The operations of the ALU are to be synchronized to the negative edge of its input clock (i.e., *clk_alu*), and the result of an operation is to be placed on the output of the ALU. At each cycle of its driving clock signal, the ALU operates on the contents of the accumulator and the data path. The zero bit of the ALU should be set whenever the contents of the accumulator are zero.

b. The CPU is to be driven by a master clock, denoted by *clk*. The *reset*, *fetch_mem* and *clk_dec* signals are to be derived from the master clock.

c. During the fetch phase of activity the data path to the ALU and Instruction Register contain the contents of the addressed memory location. The result of an ALU operation is stored in the Accumulator Register or written back to the external memory.

d. The CPU is to execute the following instruction set:

Opcode	Functionality
00	Halt
01	Skip if zero
10	add
11	and
100	xor
101	load
110	store
111	jump

e. The CPU is to interact with an external memory module.

f. The reset signal of the CPU is active low.

g. Implement the internal description of the CPU as a set of communicating behaviors (processes).

h. Use the signal labels given on the schematic.

Hints: The control signals generated by the Instruction Decoder must be carefully sequenced. When the signal *fetch_mem* is high, the address (from the program counter) must be set up, the instruction word must be fetched from memory, and the instruction must be loaded, followed by an idle period; when *fetch_mem* is low, the address (from the instruction register) must be set up, the operand must be fetched, the ALU operation must be performed, and the result of the operation must be stored. The signal *read_mem* is asserted on the *add, and, xor,* and load instructions. The signal *Inc_PC* will be asserted if the zero output of the ALU is set and the instruction is *skip*. The signal *load_ACC* will be asserted for the *add, and*, *xor*, and *load* operations. The signals *ld_PC* and *Incr_PC* will be asserted if the instruction is *jump*. The signal write_mem is asserted when the instruction is *store*. Model, verify, and synthesize the CPU into the Xilinx XC4003E FPGA using the Foundation Express™ tool. The timing relationships between *clk*, *clk_dec*, and *fetch_mem* are shown below, along with the instructions.

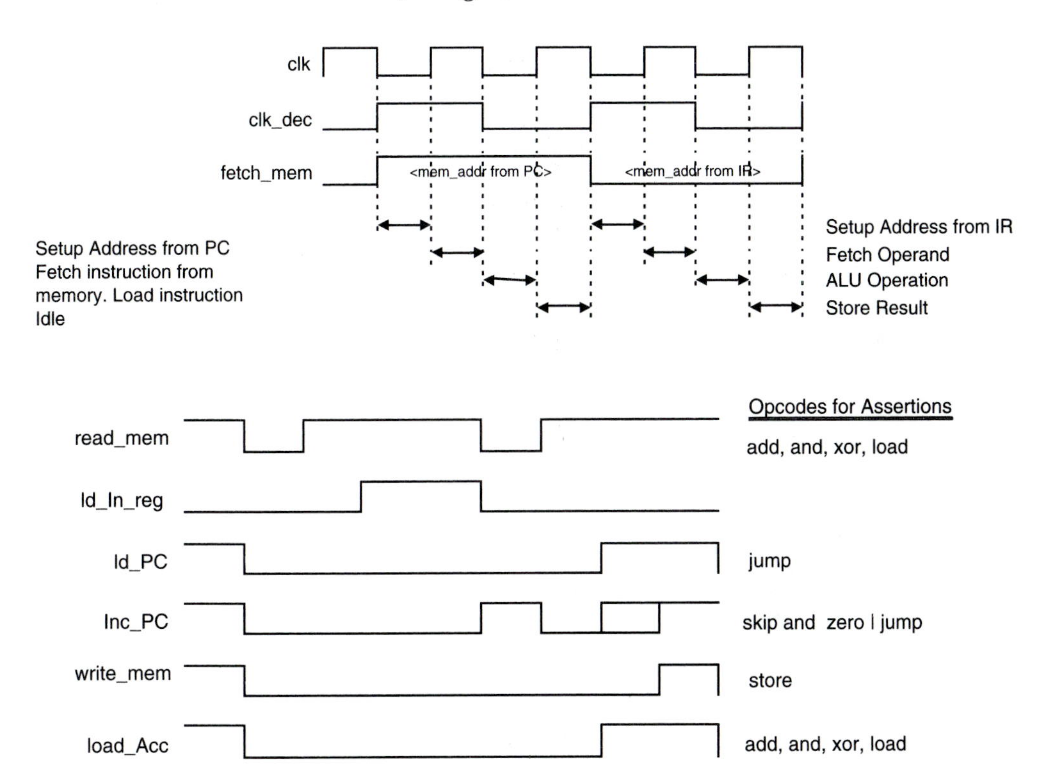

8. Write and verify a Verilog description of a digital ping-pong game. Synthesize the game into a Xilinx XC4003E FPGA using the Foundation Express™ tool, and create working prototype. When the game is in play, a string of LEDs are lit in a sequential manner from left to right or right to left. When a light is lit at the end of the string, the player is to depress a switch that returns the ball to the sender/opponent. If the switch is depressed within an acceptable interval of time, the ball returns; otherwise it is missed. When a player misses the ball, the opponent's score is incremented. The model is to include "crash-proof features" dealing with situations such as having multiple buttons depressed. There are a number of design issues that must be resolved. For example, what should happen if a player fails to respond? Should the ball be automatically served? Can play be suspended? Can speed be selected over the entire range of positions of the light? What does the machine do when the winning score is reached?

Design Specification. (Also consider crash-proof features).

a. The position of the ball is to be represented by a string of eight lights.
b. The speed of the ball is to be controlled by external buttons that are controlled by the players. The speed in each direction is independent of the speed in the opposite direction. A player must be able to toggle between speeds by pressing a button. If a player's selected speed is greater than the opponent's selected speed, the consequence of missing the opponent's return should be greater than it would be if the speeds were matched.
c. Each player's score is to be displayed in the range 0–21.
d. The speed of the ball in each direction is to be continuously displayed.
e. The system is to have a play/reset button.
f. The system is to be driven by an external clock.
g. A light is to be lit to identify the first player to reach the winning score of 15.
h. The output signal representing the score must be in a format suitable to drive a seven-segment LED.

9. Provide adequate documentation for any assumptions made to deal with contingencies or ambiguities in this specification. Provide a fully documented ASM chart for your design, testbench, results of simulation, and synthesis. Verify the design at frequencies that are appropriate for human interaction with the machine.

10. Create a working prototype of a bit-serial adder (see Problem 25 in Chapter 9).

APPENDIX A

PREDEFINED PRIMITIVES

Verilog has a set of 26 primitives for combinational and switch-level logic. The output terminals of an instantiated primitive are listed first in its primitive terminal list. The input terminals are listed last. The **buf**, **not**, **notif0**, and **notif1** primitives have a single input, but may have multiple scalar outputs. The other primitives may have multiple scalar inputs, but only one output. In the case of the three-state primitives (**bufif1**, **bufif0**, **tranif0**, **rtranif0**, **tranif1**, and **rtranif1**), the control input is the last input in the terminal list. If the inputs and outputs of a primitive are vectors, the output vector is formed on a bitwise basis from the input's vector. When a vector of primitives is instantiated, the ports may be vectors.

Primitives may be instantiated with propagation delay and may have strength assigned to their output net(s). Their input/output functionality in Verilog's four-valued logic system is defined by the truth tables shown below.

In the tables below the symbol "L" represents "0" or "z", and the symbol "H" represents "1" or "z". These additional symbols accommodate simulation results in which a signal can have a value of "0" or "z", or "1" or "z", respectively.

A.1 MULTI-INPUT COMBINATIONAL LOGIC GATES

The truth tables of Verilog's combinational logic gates are shown below for two inputs, but the gates may be instantiated with an arbitrary number of inputs.

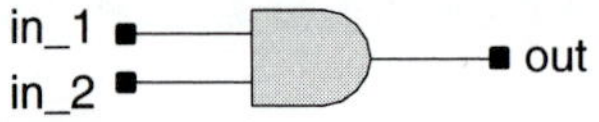

in_1
in_2
out

and	0	1	x	z
0	0	0	0	0
1	0	1	x	x
x	0	x	x	x
z	0	x	x	x

Figure A.1 Truth table for bitwise-and gate (**and**).
Terminal order: (out, in_1, in_2)

in_1
in_2
out

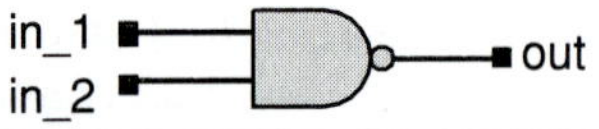

nand	0	1	x	z
0	1	1	1	1
1	1	0	x	x
x	1	x	x	x
z	1	x	x	x

Figure A.2 Truth table for bitwise-nand gate (**nand**).
Terminal order: (out, in_1, in_2)

in_1
in_2
out

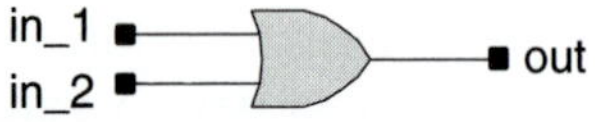

or	0	1	x	z
0	0	1	x	x
1	1	1	1	1
x	x	1	x	x
z	x	1	x	x

Figure A.3 Truth table for bitwise-or gate (**or**).
Terminal order: (out, in_1, in_2)

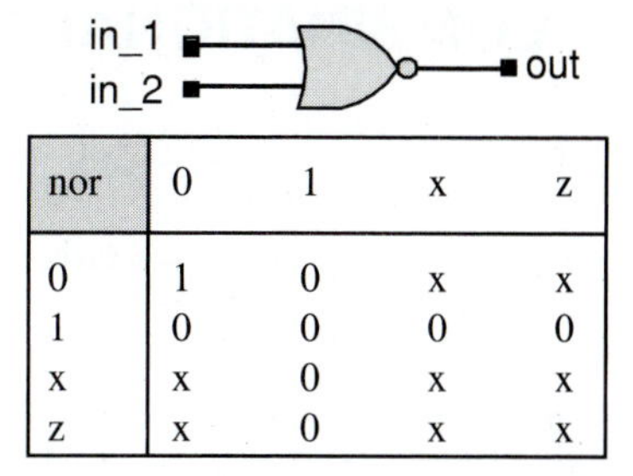

nor	0	1	x	z
0	1	0	x	x
1	0	0	0	0
x	x	0	x	x
z	x	0	x	x

Figure A.4 Truth table for bitwise-nor gate (**nor**).
Terminal order: (out, in_1, in_2)

xor	0	1	x	z
0	0	1	x	x
1	1	0	x	x
x	x	x	x	x
z	x	x	x	x

Figure A.5 Truth table for bitwise exclusive-or gate (**xor**).
Terminal order: (out, in_1, in_2)

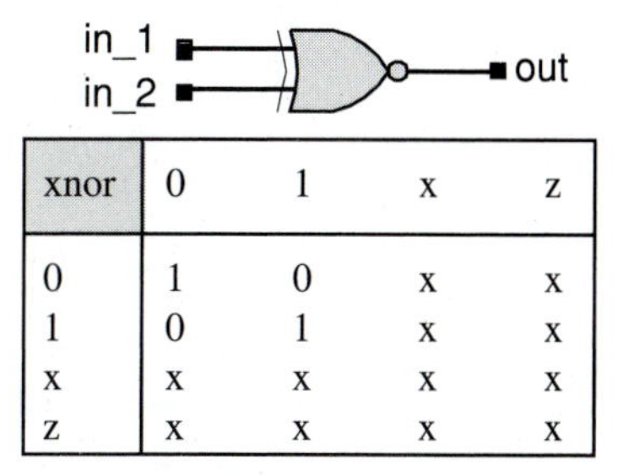

xnor	0	1	x	z
0	1	0	x	x
1	0	1	x	x
x	x	x	x	x
z	x	x	x	x

Figure A.6 Truth table for bitwise exclusive-nor gate (**xnor**).
Terminal order: (out, in_1, in_2)

A.2 MULTI-OUTPUT COMBINATIONAL GATES

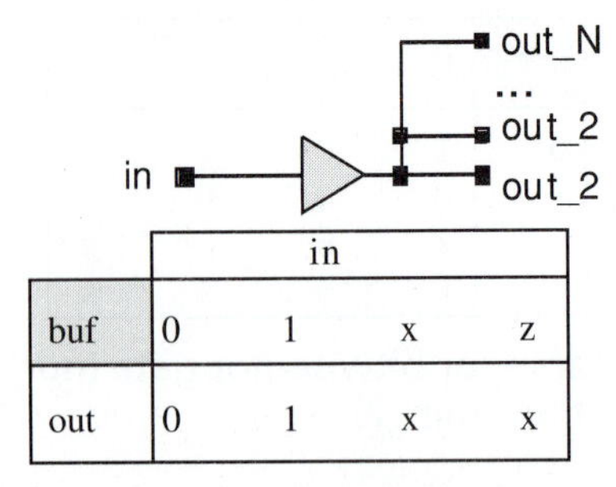

	in			
buf	0	1	x	z
out	0	1	x	x

Figure A.7 Truth table for bitwise buffer (**buf**).
Terminal order: (out_1, out_2, ..., out_N, in)

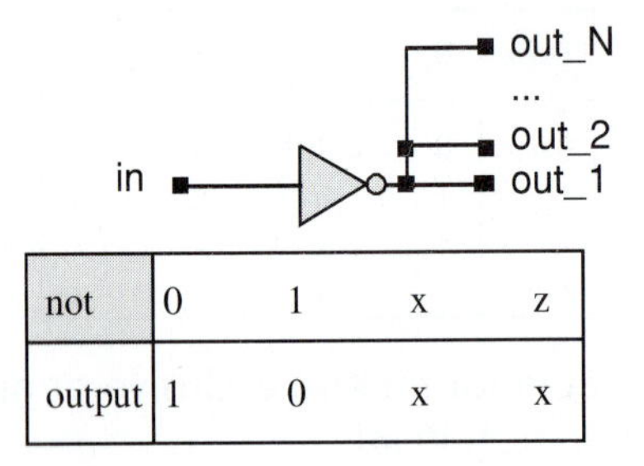

not	0	1	x	z
output	1	0	x	x

Figure A.8 Truth table for bitwise inverter (**not**).
Terminal order: (out_1, out_2, ..., out_N, in)

A.3 THREE-STATE GATES

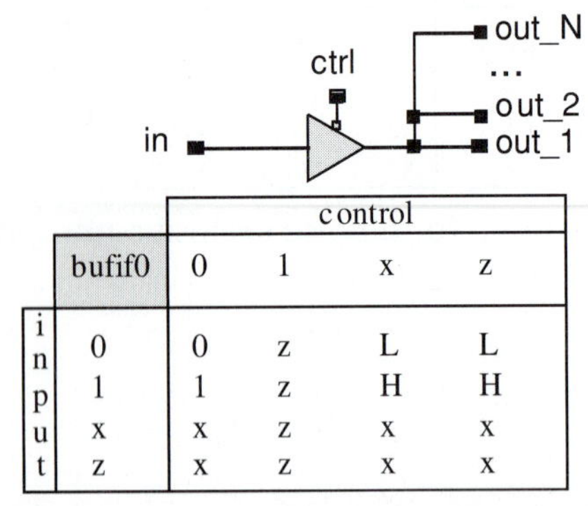

		control			
	bufif0	0	1	x	z
input	0	0	z	L	L
	1	1	z	H	H
	x	x	z	x	x
	z	x	z	x	x

Figure A.9 Truth table for bitwise three-state buffer (**bufif0**) gate with active-low enable.
Terminal order: (out_1, out_2, ..., out_N, in, ctrl)

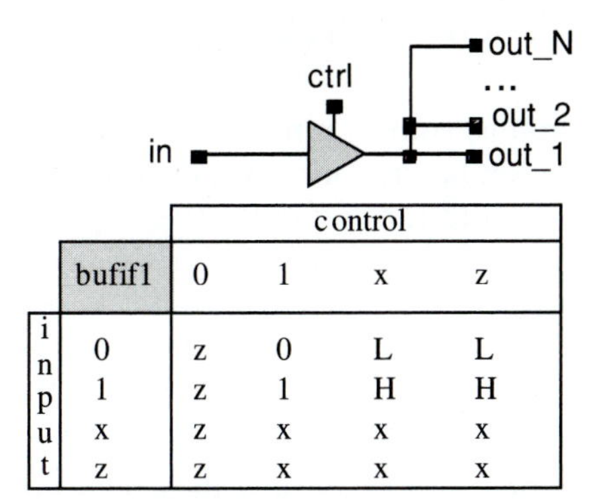

bufif1	control 0	1	x	z
input 0	z	0	L	L
1	z	1	H	H
x	z	x	x	x
z	z	x	x	x

Figure A.10 Truth table for bitwise three-state buffer (**bufif1**).
Terminal order: (out_1, out_2, ..., out_N, in, ctrl)

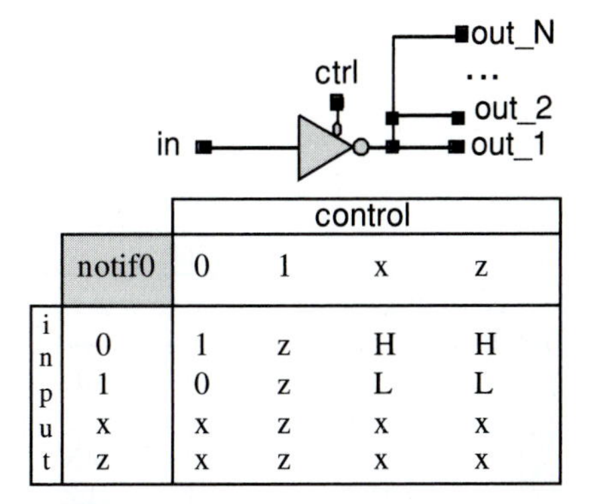

notif0	control 0	1	x	z
input 0	1	z	H	H
1	0	z	L	L
x	x	z	x	x
z	x	z	x	x

Figure A.11 Truth table for bitwise three-state inverter (**notif0**) with active-low enable.
Terminal order: (out_1, out_2, ..., out_N, in, ctrl)

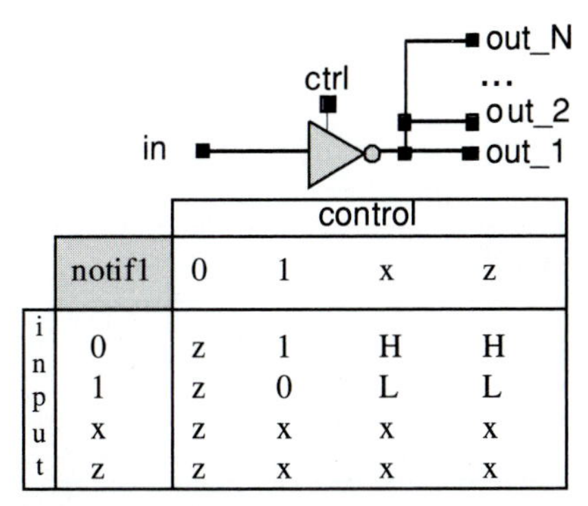

notif1	control 0	1	x	z
input 0	z	1	H	H
1	z	0	L	L
x	z	x	x	x
z	z	x	x	x

Figure A.12 Truth table for bitwise three-state inverter (**notif1**).
Terminal order: (out_1, out_2, ..., out_N, in, ctrl)

A.4 MOS TRANSISTOR SWITCHES

The **cmos**, **rcmos**, **nmos**, **rnmos**, **pmos**, and **rpmos** gates may be accompanied by a delay specification having one, two, or three values. A single value specifies the rising, falling, and turn-off delay (i.e., to the "z" state) of the output. A pair of values specifies the rising and falling delays, and the smaller of the two values determines the delay of transitions to "x" and "z". A triple of values specifies the rising, falling, and turn-off delay, and the smallest of the three values determines the transition to "x". Delays of transitions to "L" and "H" are the same as the delay of a transition to "x".

The rules in Chapter 11 specify how these gates reduce the strength of signals passing through them.

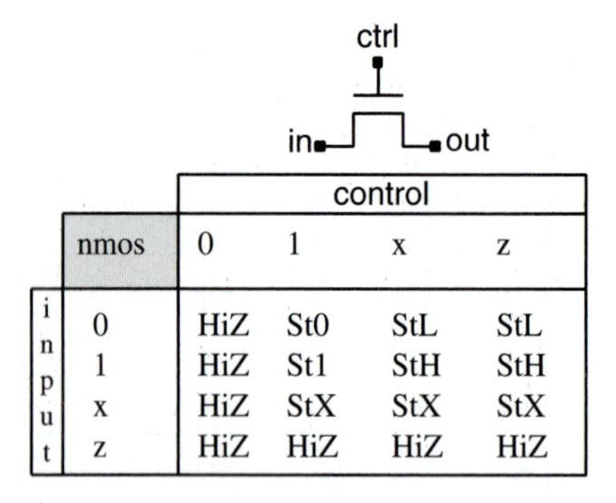

		control			
	nmos	0	1	x	z
input	0	HiZ	St0	StL	StL
	1	HiZ	St1	StH	StH
	x	HiZ	StX	StX	StX
	z	HiZ	HiZ	HiZ	HiZ

Figure A.13 **nmos** pass transistor switch.
Terminal order: (out, in, ctrl)

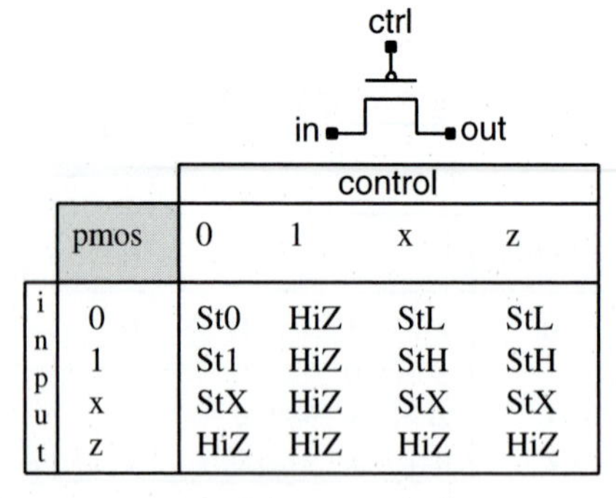

		control			
	pmos	0	1	x	z
input	0	St0	HiZ	StL	StL
	1	St1	HiZ	StH	StH
	x	StX	HiZ	StX	StX
	z	HiZ	HiZ	HiZ	HiZ

Figure A.14 **pmos** pass transistor switch.
Terminal order: (out, in, ctrl)

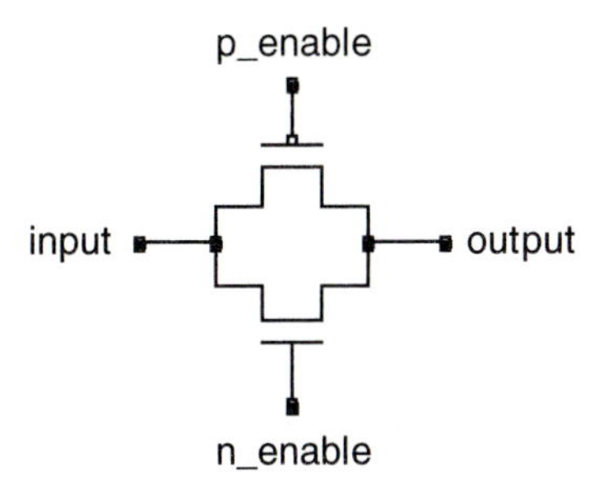

cmos	control	input			
n_enable	p_enable	0	1	x	z
0	0	St0	St1	StX	HiZ
0	1	HiZ	HiZ	HiZ	HiZ
0	x	StL	StH	StX	HiZ
0	z	StL	StH	StX	HiZ
1	0	St0	St1	StX	HiZ
1	1	St0	St1	StX	HiZ
1	x	St0	St1	StX	HiZ
1	z	St0	St1	StX	HiZ
x	0	St0	St1	StX	HiZ
x	1	StL	StH	StX	HiZ
x	x	StL	StH	StX	HiZ
x	z	StL	StH	StX	HiZ
z	0	St0	St1	StX	HiZ
z	1	StL	StH	StX	HiZ
z	x	StL	StH	StX	HiZ
z	z	StL	StH	StX	HiZ

Figure A.15 **cmos** transmission gate.
Terminal order: (output, input, n_enable, p_enable)

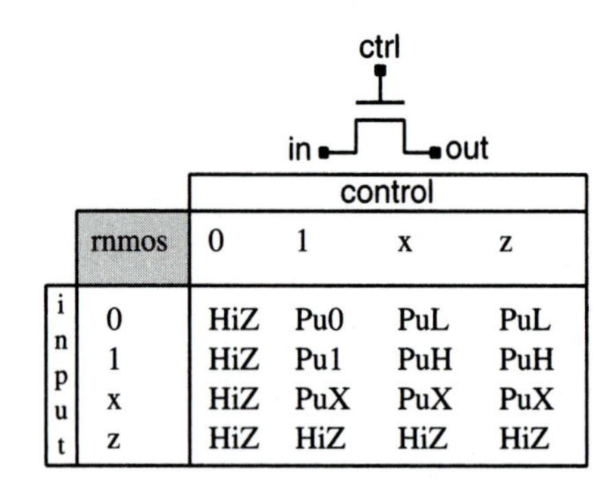

		control			
	rnmos	0	1	x	z
input	0	HiZ	Pu0	PuL	PuL
	1	HiZ	Pu1	PuH	PuH
	x	HiZ	PuX	PuX	PuX
	z	HiZ	HiZ	HiZ	HiZ

Figure A.16 High-resistance **rnmos** pass transistor switch. Terminal order: (out, in, ctrl)

ctrl

in out

		control			
	rpmos	0	1	x	z
input	0	Pu0	HiZ	PuL	PuL
	1	Pu1	HiZ	PuH	PuH
	x	PuX	HiZ	PuX	PuX
	z	HiZ	HiZ	HiZ	HiZ

Figure A.17 High-resistance **rpmos** transistor switch. Terminal order: (out, in, ctrl)

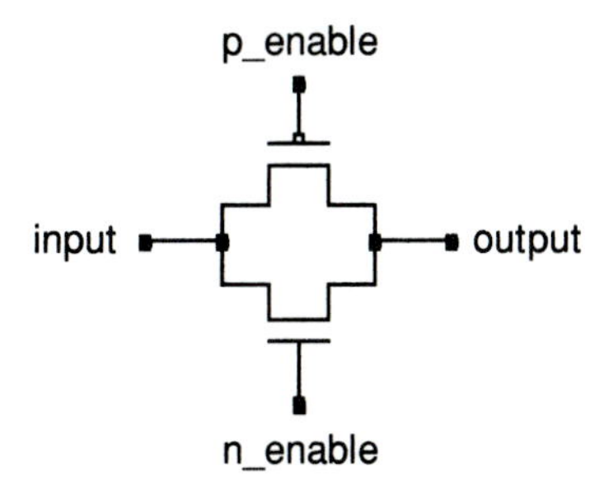

rcmos	control	input			
n_enable	p_enable	0	1	x	z
0	0	Pu0	Pu1	PuX	HiZ
0	1	HiZ	HiZ	HiZ	HiZ
0	x	PuL	PuH	PuX	HiZ
0	z	PuL	PuH	PuX	HiZ
1	0	Pu0	Pu1	PuX	HiZ
1	1	Pu0	Pu1	PuX	HiZ
1	x	Pu0	Pu1	Pux	HiZ
1	z	Pu0	Pu1	PuX	HiZ
x	0	Pu0	Pu1	PuX	HiZ
x	1	PuL	PuH	PuX	HiZ
x	x	PuL	PuH	PuX	HiZ
x	z	PuL	PuH	PuX	HiZ
z	0	Pu0	Pu1	PuX	HiZ
z	1	PuL	PuH	PuX	HiZ
z	x	PuL	PuH	PuX	HiZ
z	z	PuL	PuH	PuX	HiZ

Figure A.18 High-resistance **rcmos** transmission gate.
Terminal order: output, input, n_enable, p_enable

A.5 MOS PULL-UP/PULL-DOWN GATES

The pull-up (**pullup**) and pull-down (**pulldown**) gates place a constant value of "1" or "0" with strength "pull," respectively, on their output. This value is fixed for the duration of simulation, so no delay values may be specified for these gates. The default strength of these gates is "pull". Note: The **pulldown** and **pullup** gates are not to be confused with **tri0** and **tri1** nets. The latter are nets providing connectivity, and may have multiple drivers; the former are functional elements in the design. The **tri0** and **tri1** nets may have multiple drivers. The net driven by a **pullup** or **pulldown** gate may also have multiple drivers. Verilog's **pullup** and **pulldown** primitives can be used to model pull-up and pull-down devices in electrostatic-discharge circuitry tied to unused inputs on flip-flops. The devices are illustrated in Figures A.19–20.

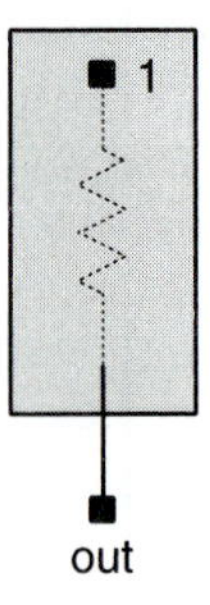

Figure A.19 Pull-up device.
Terminal order: (out)

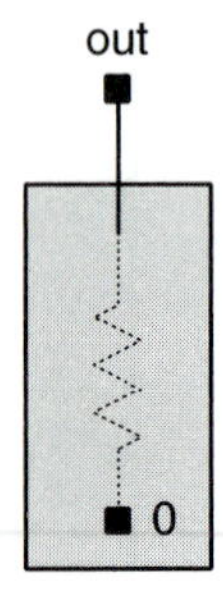

Figure A.20 Pull-down device.
Terminal order: (out)

A.6 MOS BI-DIRECTIONAL SWITCHES

Verilog includes six predefined bi-directional switch primitives: **tran**, **rtran**, **tranif0**, **rtranif0**, **tranif1**, and **rtranif1**. Bi-directional switches provide a layer of buffering on bi-directional signal paths between circuits. A signal passing through a bi-directional switch is not delayed (i.e., output transitions follow input transitions without delay). The rules in Chapter 11 govern the strength-reduction properties of these switches. (In general, the resistive versions of the switches reduce the strength of the signal passing through the switch.)

Note: The **tran** and **rtran** primitives model bi-directional pass gates, and may not have a delay specification. These bi-directional switches pass signals without delay. The **rtranif0**, **rtranif1**, **tranif1**, and **rtranif1** switches are accompanied by a delay specification, which specifies the turn-on and turn-off delays of the switch; the signal passing through the switch has no delay. A single value specifies both delays; a pair of values (turn-on, turn-off) specifies both delays, with the turn-on being the first item and turn-off being the second item. The default delay is zero. The terminal orders of the switches are shown in Figures A.21–26.

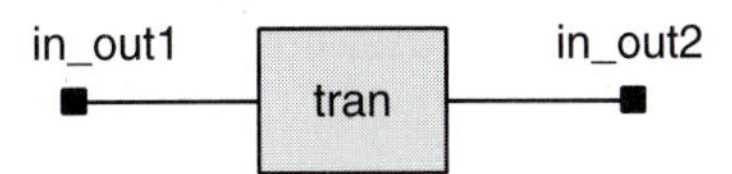

Figure A.21 Bi-directional switch: **tran**.
Terminal order: (in_out1, in_out2)

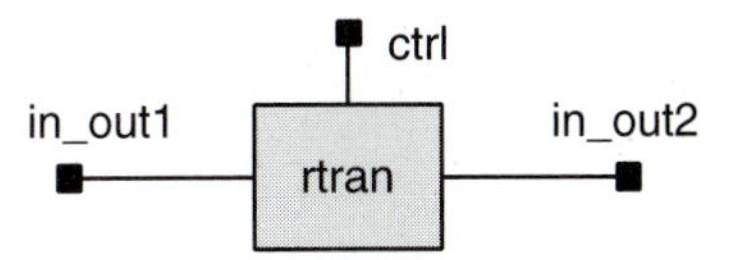

Figure A.22 Resistive bi-directional switch: **rtran**.
Terminal order: (in_out1, in_out2)

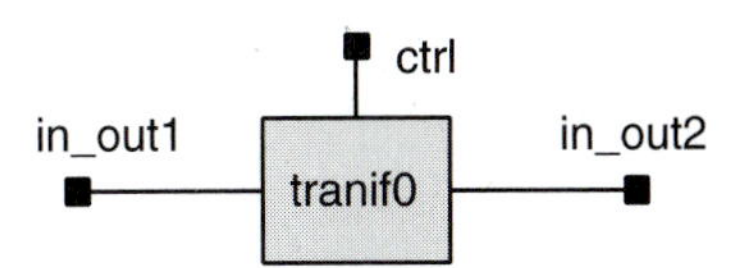

Figure A.23 Three-state bi-directional switch: **tranif0**.
Terminal order: (in_out1, in_out2, ctrl)

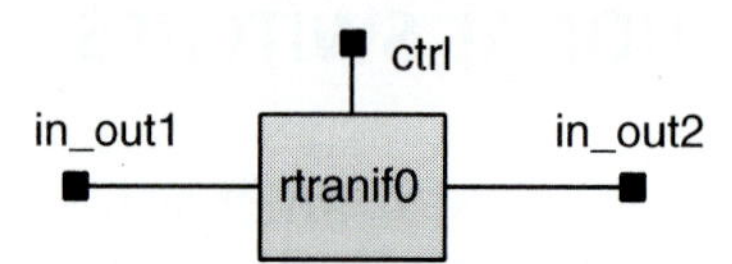

Figure A.24 Resistive three-state bi-directional switch: **rtranif0.**
Terminal order: (in_out1, in_out2, ctrl)

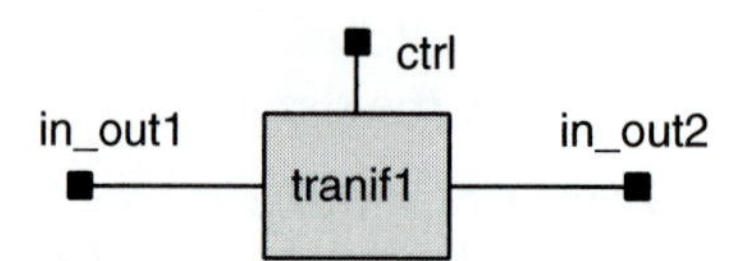

Figure A.25 Three-state bi-directional switch: **tranif1.**
Terminal order: (in_out1, in_out2, ctrl)

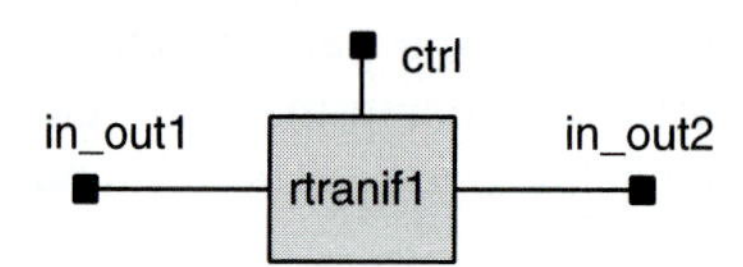

Figure A.26 Resistive three-state bi-directional switch: **rtranif1.**
Terminal order: (in_out1, in_out2, ctrl)

APPENDIX B

VERILOG KEYWORDS

Verilog keywords are predefined, lowercase, nonescaped identifiers that define the language constructs. An identifier may not be a keyword, and an escaped identifier is not treated as a keyword. In this text, Verilog keywords are printed in boldface.

always
and
assign
begin
buf
bufif0
bufif1
case
casex
casez
cmos
deassign
default
defparam
disable
edge
else
end
endcase
endfunction
endmodule
endprimitive
endspecify
endtable
endtask
event
for
force
forever
fork
function
highz0
highz1
if
initial
inout
input
integer
join
large
macromodule
medium
module
nand
negedge
nmos
nor
not
notif0
notif1or
output
parameter
pmos
posedge
primitive
pull0
pull1
pulldown
pullup
rcmos
real
realtime
reg
release
repeat
rnmos
rpmos
rtran
rtranif0
rtranif1
scalared
small
specify
specparam
strength
strong0
strong1
supply0
supply1
table
task
time
tran
tranif0
tranif1
tri
tri0
tri1
triand
trior
trireg
vectored
wait
wand
weak0
weak1
while
wire
wor
xnor
xor

APPENDIX C

VERILOG OPERATORS AND PRECENDENCE

Verilog operators create data objects from other data objects. Expressions combine operators with operands and functions in assignments to nets or registers, or in the argument of functions, ports, conditionals, or concatenations. The evaluation of an expression has a value that can always be expressed in bits in Verilog's four-valued logic system. An operand may be a net, register, number, or bit-select of a net or register, part-select of a net or register, memory element, or a concatenation of these. Figure C.1 lists related operators in groups. Note that operators are unary (one operand), binary (two operands), or ternary (three operands). For example, the "-" sign denotes subtraction when used with two operands (e.g., 5 –1) and denotes a negative value when used with a single operand (e.g., –3). The usage is clear from the context. All the reduction operators are unary. Note that the shift operator is a simple shift that fills in with zeros. The language does not include an arithmetic shift operator. The operands of the concatenation operator must be sized. The operators are grouped according to their precedence and are shown in Figure C.2. The operators within a group have the same functionality. Within an expression, the operators associate from left to right. When in doubt, use parentheses to specify an order of the evaluation of operands. Be aware that (1) arrays of real registers (memories) are not allowed, (2) real variables may not be the index of a bit-select or a part-select reference of a vector, (3) a variable declared as a real may not be referenced by an index or a part-select, and (4) an edge descriptor (e.g., **posedge**) may not be applied to a real register.

ARITHMETIC	
+	addition, unary sign
-	subtraction, unary sign
*	multiplication
/	division
%	modulus

REDUCTION	
&	reduction and
|	reduction or
~&	reduction nand
~|	reduction nor
^	reduction exclusive or
~^ or ^~	reduction exclusive nor

SPECIAL	
? :	conditional
{ }	concatenation

SHIFT	
<<	left shift
>>	right shift

RELATIONAL AND LOGICAL	
>	greater than
>=	greater than or equal to
<	less than
<=	less than or equal to
!	negation
&&	logical and
||	logical or
==	logical equality (identity)
!=	logical inequality
===	case equality (includes x, z match)
!==	case inequality (includes x, x match)

BIT-WISE	
~	bit-wise negation
&	bit-wise and
|	bit-wise inclusive or
^	bit-wise exclusive or
~&	bit-wise nand
~|	bit-wise nor
~^ or ^~	equivalence (bit-wise exclusive nor)

Figure C.1 Verilog operators.

Operator Symbol	Function	Precedence
+ - ! ~	Unary	Highest
+ - << >>	Binary Shift	
< <= > >= == != === !==	Relational Equality	
& ~& ^ ^~ | ~| && ||	Reduction Logical	
? :	Conditional	Lowest

Figure C.2 Precedence of the Verilog operators.

Appendix D

Backus–Naur (BNF) Formal Syntax Notation

The syntax of the Verilog language conforms to the following Backus–Naur Form (BNF) of formal syntax notation.

1. White space may be used to separate lexical tokens.

2. Name "::=" starts off the definition of a syntax construction item. Sometimes name contains embedded underscores "_". Also, the "::=:" may be found on the next line.

3. The vertical bar, " | ", introduces an alternative syntax definition, unless it appears in bold. (See next item.)

4. Name in bold text is used to denote reserved keywords, operators, and punctuation marks required in the syntax.

5. [item] is an optional item that may appear once or not at all.

6. {item} is an optional item that may appear once, more than once, or not at all. If the braces are in bold, they are part of the syntax.

7. *Name1_name2* is equivalent to the syntax construct item name2. The name1 (in italics) imparts some extra semantic information to name2. However, the item is defined by the definition of name2.

8. | .. is used in the non-appendix text to indicate that there are other alternatives, but that, due to space or expediency, they are not listed here.

APPENDIX E

SYSTEM TASKS AND FUNCTIONS

Verilog contains the predefined system tasks and functions shown in Figure E.1, including tasks for creating output from a simulation. The role of each task is summarized, but the more frequently used tasks and functions are described in more detail. For additional information, see the Language Reference Manual.

Abbreviations:	
LRM	Language Reference Manual
MCD	Multichannel descriptor.
VCD file	Value-change dump file. A file containing information about value changes on selected variables in the design stored by execution of value change dump system tasks during simulation.

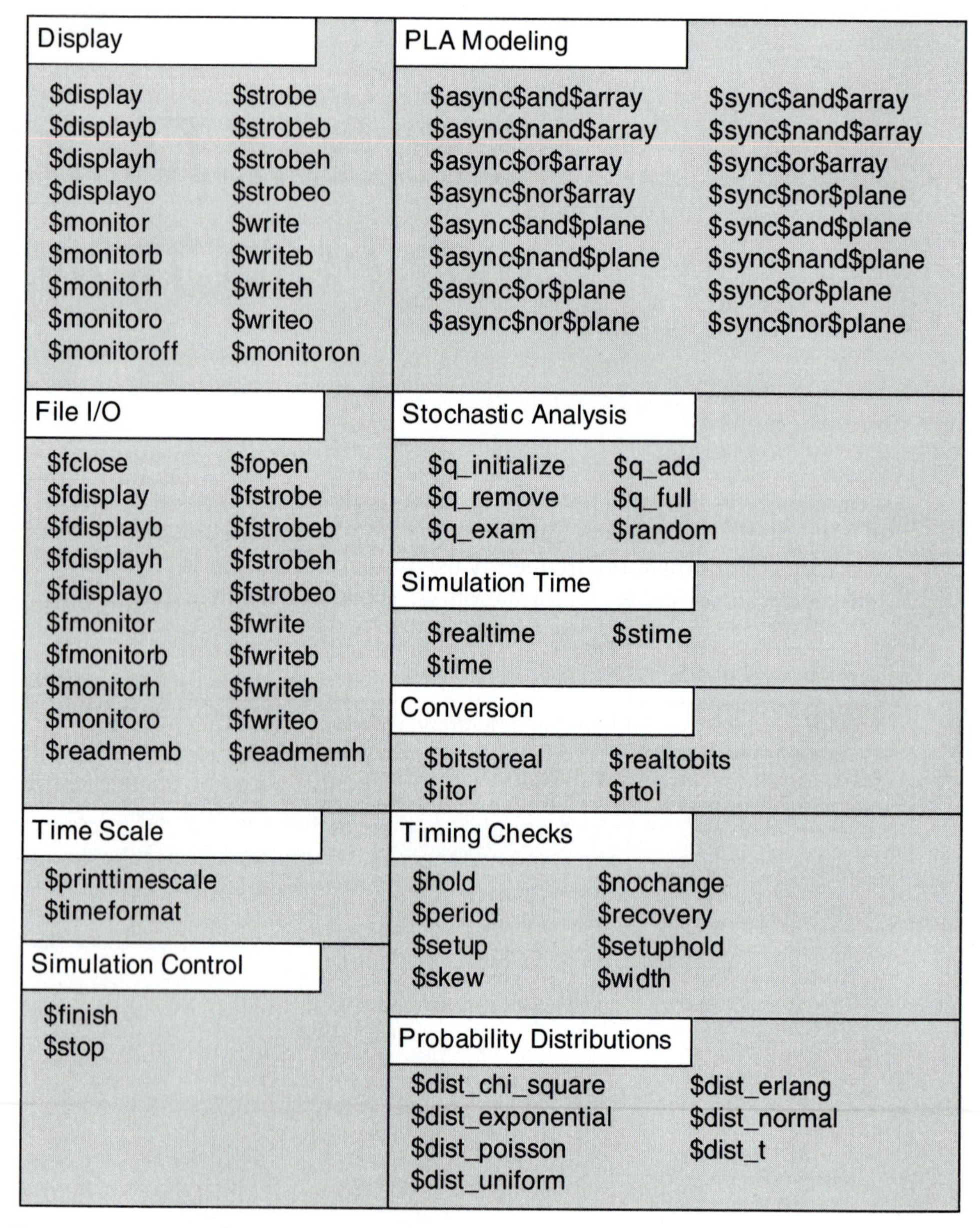

Figure E.1 Predefined system tasks.

E.1 DISPLAY TASKS

E.1.1 $display

$display displays information to standard output and adds a newline character to the end of its output. (See **$write**.)

Syntax:

display_tasks ::= display_task_name (list_of_arguments)

display_task_name ::= **$display** | **$displayb** | **$displayo** |
$displayh |**$write** | **$writeb** | **$writeo** | **$writeh**

The default format of an expression argument that has no format specification is decimal. The **$fdisplayb**, **$fdisplayo**, and **$fdisplayh** tasks specify binary, octal and hex default formats, respectively.

The only difference between the **$display** and **$write** tasks is that the **$display** tasks automatically add a newline character to the end of their output, and the **$write** tasks do not.

The arguments (parameters) are displayed in the order of their appearance in the parameter list and can be a quoted string, an expression that returns a value, or a null parameter. Strings are output literally, subject to any included escape sequences. Escape sequences may be inserted in a quoted string to display special characters or to specify the display format for a subsequent expression. Escape sequences are interpreted according to the following table:

\	The next character is a literal or nonprintable character.
%	The next character specifies the display format for a subsequent expression parameter. A expression parameter (following the string) must be supplied for each % character that appears in a string.
%%	Displays the % character.

A null parameter produces a single space character in the display.

Note: The **$display** task executes immediately when encountered in a behavior, and not necessarily after all simulation activity is complete. The **$strobe** tasks executes after all simulation activity is complete.

Note: If the host operating system buffers the text generated by **$write** instead of flushing it directly to the output, it is necessary to include an explicit new-line character (\n) in the **$write** task to immediately send the text to the output. Otherwise, use the **$display** system task.

Special Characters:

\n	New-line character
\t	Tab character
\\	The \ character
\"	The " character (Double quote)
\o	A character specified in one to three octal digits

Format Specifications:

The escape sequences shown in the next table specify the display format for a subsequent expression in a string parameter. Except for %m, a % character requires that a corresponding expression follow the string in the parameter list. The value of the expression replaces the format specification when the string is displayed. The decimal format is the default format.

%h or %H	Display in hexadecimal format
%d or %D	Display in decimal format
%o or %O	Display in octal format
%b or %B	Display in binary format
%c or %C	Display in ASCII character format
%v or %V	Display net signal value and strength
%m or %M	Display hierarchical name
%s or %S	Display as a string
%t or %T	Display in current time format
%e or %E	Display "real" value in exponential format
%f or %F	Display "real" value in decimal format
%g or %G	Display "real" value in exponential or decimal format, using the format that has the shorter printed output.

The %e, %f, and %g format specifications for real numbers have a minimum field width of 10 with three fractional digits (as in the C language). The %t format is used with the **$timeformat** system task, which specifies a time unit, time precision to fuse timing data from modules having different timing formats. See **$timeformat**.

Size of Displayed Data:

The values of expression arguments of the **$display** and **$write** tasks are sized automatically when written to a file or to a terminal output; i.e., the value stored as a bit pattern is converted to an appropriate number of characters in decimal, octal, or hex value. For example, an eight-bit word, having a maximum decimal value of 511, would require three decimal characters and two hex characters. Leading zeros are printed, except for decimal output. Insertion of a zero between the % character and the letter of the format specifier suppresses the automatic sizing of displayed data.

Display of High-Impedance and Unknown Values:

Decimal Format—When an expression containing unknown or high-impedance values is displayed in decimal format, a lowercase "x" ("z") indicates that all bits are unknown (high impedance), and an uppercase "X" ("Z") indicates that some bits are known. Decimal numbers are right-justified in a fixed-width field to conform to the output of the **$monitor** system task, which requires a fixed-columnar format.

Hexadecimal and Octal Formats—When an expression containing unknown or high-impedance values is displayed in hexadecimal (%h) (octal (%o)) format, each group of four (three) bits represents a single hexadecimal (octal) character. The convention for "x", "X", "z", and "Z" is the same as for the decimal format.

Binary Format—"x" and "z" characters.

Signal Strength Format —Character format. The first two characters are either a two-letter mnemonic or a pair of decimal digits indicating the signal's strength, as shown in the following table:

Mnemonic Level	Strength Name	Strength
Su	Supply drive	7
St	Strong drive	6
Pu	Pull Drive	5
La	Large capacitor	4
We	Weak drive	3

Mnemonic Level	Strength Name	Strength
Me	Medium capacitor	2
Sm	Small capacitor	1
Hi	High impedance	0

Driving strengths (Su, St, Pu, We) are associated with primitive gate outputs and with continuous assignment outputs. Charge storage strengths (La, Me, Sm) are associated with nets of type **trireg**. A mnemonic is used when a signal having a logic value of 0 or 1 has no range of strengths. If the signal has a range of strengths, the logic value is preceded by two decimal digits from the preceding table to indicate the maximum and minimum strength levels.

If a signal has an unknown logic value, a mnemonic is used when the 0 and 1 values of the signal have the same strength level. Otherwise, the signal value is displayed as "X" preceded by two decimal digits.

The high-impedance strength cannot have a known logic value; the only logic value allowed for this level is "Z".

A mnemonic is always used with the values "L" and "H" to indicate the strength level.

The third character indicates the signal's current logic value and may be any one of the following:

0	Denotes a logic 0 value
1	Denotes a logic 1 value
X	Denotes an unknown value
Z	Denotes a high-impedance value
L	Denotes a logic 0 or a high-impedance value
H	Denotes a logic 1 or a high-impedance value

Hierarchical Name Format—The %m format prints the hierarchical name of the module, task, function, or named block that invokes the system task containing the specifier. It can be used in conjunction with other system tasks to locate and report simulation activity, such as timing violations.

String Format—The %s format specifier prints ASCII codes as characters. The associated parameter is interpreted as a sequence of eight-bit hexadecimal ASCII code, with each eight bits representing a single character.

E.1.2 $monitor

$monitor continuously monitors and displays the values of any variables or expressions specified as parameters to the task. Parameters are specified in the same format as for **$display**. When a variable or expression changes, the simulator automatically displays the entire argument list at the end of the time step. If two or more variables change simultaneously, only one output is generated. Only one **$monitor** task display list may be active at a time, but new **$monitor** tasks can be invoked any number of times during a simulation. (See **$monitoron** and **$monitoroff**). For a discussion of interactive use of **$monitor**, see the LRM.

Syntax:

monitor_tasks ::= monitor_task_name [(list_of_arguments)]; | **$monitoron** | **$monitoroff**

monitor_task_name ::= **$monitor** | **$monitorb** | **$monitoro** | **$monitorh**

E.1.3 $monitoron

$monitoron controls a flag to re-enable a previously disabled **$monitor**. A re-enabled **$monitor** immediately prints a display, regardless of whether a value change has taken place.

Syntax: **$monitoron;**

E.1.4 $monitoroff

$monitoroff controls a flag to disable monitoring.

Syntax: **$monitoroff;**

E.1.5 $fmonitor

$fmonitor the counterpart of **$monitor** and is used to direct simulation results to a file.

Syntax:

$fmonitor ([multi_channel_descriptor], list_of_arguments);

The default format of an expression argument that has no format specification is decimal. The **$fmonitorb, $fmonitoro,** and **$fmonitorh** tasks specify binary, octal, and hex default formats, respectively.

E.1.6 $strobe

$strobe displays simulation data at a selected time, but after all simulation activity at that time is complete, immediately before time is advanced. Parameters are specified in the same format as for **$display**. (See **$fstrobe**.)

Syntax: **$strobe** (list_of_arguments);

E.1.7 $write

$write displays information to standard output without adding a new-line character to the end of its output. (See **$display** for additional details.)

Syntax: **$write** (list_of_arguments);

The default format of an expression argument that has no format specification is decimal. The companion **$writeb**, **$writeo**, and **$writeh** tasks specify binary, octal, and hex default formats, respectively.

E.2 FILE I/O TASKS

E.2.1 $fclose

$fclose closes the channels specified in the MCD and prevents further writing to the closed channels.

Syntax: file_closed_task ::= **$fclose** (MCD);

E.2.2 $fdisplay

$fdisplay is the counterpart of **$display** and is used to direct simulation data to a file.

Syntax:

$fdisplay ([multi_channel_descriptor], list_of_arguments);

E.2.3 $fopen

$fopen opens the file specified by a parameter and returns a 32-bit unsigned MCD (integer multi-channel-descriptor) uniquely associated with the file. Returns 0 if the file could not be opened. (See **$fclose**.)

Syntax:

```
file_open_function ::= integer multi_channel_descriptor = $fopen
  ("[name_of_file]");
```

Each bit of a MCD corresponds to a single output channel. The least significant bit (i.e., channel 0) refers to the standard output (i.e., log file and screen), unless redirected to a file. If a file cannot be opened, the MCD is returned as the value 0. The bits of the MCD are assigned in sequence, bits 1 through 31, as different files are referenced.

Note: The MCD feature of Verilog allows the same data to be written to multiple files by forming the bitwise-or of the MCD of the file. The file(s) receiving information can be altered dynamically and interactively during simulation.

E.2.4 $readmemb

$readmemb reads binary numbers from a text file and loads them into a Verilog memory, or sub-blocks of a memory, specified by an identifier.

Syntax:

```
$readmemb ("filename",
  memory_name [, start_addr [, finish_addr]]);
```

In this syntax, *memory_name* is the identifier of the memory that will be loaded with data from the file specified by *filename*, and *start_addr* is the address at which the first number is to be written. The default is the left-hand address given in the declaration of the memory. The numbers read from the file are assigned to successive locations in memory. Loading of numbers continues to *finish_addr* or until the memory is full. The file may contain addresses, which are denoted by the @ symbol followed by the number. When an address is encountered, the subsequent data is loaded into the memory, beginning at the address.

The text file may contain only white space, comments (either type), and binary numbers. (Also see **$readmemh**.) The length and base of the numbers may not be specified.

E.2.5 $readmemh

$readmemh reads hexadecimal numbers from a text file and loads then into a Verilog memory, or sub-blocks of a memory, specified by an identifier.

Syntax:

```
$readmemh ("filename",
  memname[. start_adddr [, finish_addr]]););
```

Example:

```
...
parameter ram_file_1 = "ram_data_file";
reg [15:0] RAM_1 [0:'hff];
initial $readmemh (ram_file_1, RAM_1);
```

In this syntax *memory_name* is the identifier of the memory that will be loaded with data from the file specified by *filename*, and *start_addr* is the address at which the first number is to be written. The default is the left-hand address given in the declaration of the memory. The numbers read from the file are assigned to successive locations in memory. Loading of numbers continues to *finish_addr* or until the memory is full. The file may contain addresses, which are denoted by the @ symbol followed by the number. When an address is encountered, the subsequent data is loaded into the memory, beginning at the address.

The text file may contain only white space, comments (either type), and hexadecimal numbers. (Also see **$readmemb**.) The length and base of the numbers may not be specified.

E.2.6 $fstrobe

$fstrobe writes simulation data to a file at a selected time, but after all simulation activity at that time is complete, immediately before time is advanced. Parameters are specified in the same format as for **$display**. (See **$strobe**.)

Syntax:

```
$fstrobe ([multi_channel_descriptor], list_of_arguments);
```

The default format of an expression argument that has no format specification is decimal. The **$fstrobeb**, **$fstrobeo**, and **$fstrobeh** tasks specify binary, octal, and hex default formats, respectively.

Note: The **$fdisplay** task executes immediately when encountered in a behavior, and not necessarily after all simulation activity is complete. The **$fstrobe** tasks executes after all simulation activity is complete.

E.2.7 $fwrite

$fwrite is the counterpart of **$write**, and it writes information to designated file output without adding a new-line character to the end of its output. (See **$write** for additional details.)

Syntax:

```
$write ([multi_channel_descriptor], list_of_arguments);
```

The default format of an expression argument that has no format specification is decimal. The companion **$fwriteb**, **$fwriteo**, and **$fwriteh** tasks specify binary, octal, and hex default formats, respectively.

E.3 TIME-SCALE TASKS

E.3.1 $printtimescale

The **$printtimescale** task displays the time unit and precision for a referenced module. If an optional argument is not given, the task displays the time unit and precision of the module that is the current scope.

Syntax:

$printtimescale [(hierarchical_name)]

E.3.2 $timeformat

$timeformat determines how the %t format specifier in system output tasks reports time information and specifies the time unit for delays entered interactively.

Syntax:

$timeformat [(units_number, precision_number,
suffix_string, minimum_field_width)];

The integer-valued *units_number* represents the time unit as follows:

Unit_number	Time Unit	Unit_number	Time Unit
0	1 s	-8	10 ns
-1	100 ms	-9	1 ns
-2	10 ms	-10	100 ps
-3	1 ms	-11	10 ps
-4	100 us	-12	1 ps
-5	10 us	-13	100 fs
-6	1 us	-14	10 s
-7	100 ns	-15	1 fs

This system task sets the time unit for subsequent delays entered interactively and sets the time unit, precision number (i.e., number of decimal digits to display), suffix string to be displayed after the time value, and a minimum field width for all %t formats specified in all modules that follow in the source description until another **$timeformat** system task is invoked. The default arguments of **$timeformat** are given in the following table:

Argument	Default
units_number	The smallest *time_precision* argument of all the time-scale compiler directives in the source description.
precision_number	0
suffix_string	A null character string.
minimum_field_width	20

Example: **$timeformat**(-10, 1, "x100 ps", 10); specifies that the output produced by a %t format specifier will be displayed in units of 100s of picoseconds.

E.4 SIMULATION CONTROL TASKS

E.4.1 $finish

$finish terminates simulation and returns control to the host operating system.

Syntax: **$finish**;

E.4.2 $finish(n)

$finish(*n*) terminates simulation and takes the following action, depending on the diagnostic control parameter, *n*:

n = 0 Prints nothing.
n = 1 Prints the simulation time and location
n = 2 Prints the simulation time and location and runs statistics.

Syntax: **$finish** (*n*);

E.4.3 $stop

$stop suspends simulation, issues an interactive prompt, and passes control to the user.

E.4.4 $stop(n)

Suspends simulation, issues and interactive prompt, and takes the following action, depending on the diagnostic control parameter, *n*:

n = 0 Prints nothing.
n = 1 Prints the simulation time and location.

n = 2 Prints simulation time and location, and CPU utilization and runs statistics.

E.5 PLA MODELING TASKS

Verilog includes a set of system tasks for modeling multi-input, multi-ouput programmable logic arrays (PLAs). PLAs implement two-level combinatonal logic by an array of **and**, **nand**, **or**, and **nor** logic planes. A "personality" file, or matrix, specifies the physical connections of transistors forming the product of inputs terms (cubes) and the sums of those products to form an output, whose value is the value of the Boolean expression.

Syntax:

pla_system_task ::= $array_type$logic$format (memory_name, input_terms, output_terms);
array_type ::= **sync** | **async**
logic ::= **and** | **or** | **nand** | **nor**
format ::= **array** | **plane**
memory_name ::= memory_identifier
input_terms ::= { scalar_variables }
output_terms ::= { scalar_variables }

See Figure E.1 for a list of the PLA system tasks.

Examples: The following statements illustrate calls to PLA system tasks describing synchronous and asynchronous arrays and planes:

```
$async$and$array (PLA_mem, {in0, in1, in2, in3, in4, in5, in6, in7}, {out0, out1,
  out2});
$sync$or$plane (PLA_mem, {in0, in1, in2, in3, in4, in5, in6, in7}, {out0, out1,
  out2});
$async$and$array (PLA_mem, {in0, in1, in2, in3, in4, in5, in6, in7}, {out0, out1,
  out2});
$async$and$array (PLA_mem, {in0, in1, in2, in3, in4, in5, in6, in7}, {out0, out1,
  out2});
```

The outputs of the asynchronous types are updated whenever an input signal changes value or whenever the personality matrix of the PLA changes during simulation. The synchronous types are updated when evaluated in a behavior. Both forms update their outputs with zero delay.

The personality matrix of a PLA specifies the cubes that are formed from the inputs to the PLA, and the expressions forming the outputs of the PLA. The data describing the personality are stored in a memory whose width accommodates the inputs and outputs of the PLA and whose depth accommodates the number of outputs.

There are two ways to load data into the personality matrix: (1) Using the **$readmemb** task, read the data from a file, and (2) load the data directly with procedural assignment statements. Both methods can be used at any time during a simulation to dynamically reconfigure the PLA.

There are two formats: **array** and **plane**. The **array** format stores either a 1 or 0 in memory to indicate whether a given input is in a cube and whether a given cube is in an output. For example, the **array** format shown next indicates that the cube *in1* & *in2* & *in3* is formed and used in *out1*, but not in *out2*. The cube *in1 & in3* is used in *out2*.

in1	in2	in3	out1	out2
1	1	1	1	0
1	0	1	0	1

Example: A combinational logic circuit forming three output functions from eight input variables is illustrated as follows:

```
module PLA_array (in0, in1, in2, in3, in4, in5, in6, in7, out0, out1, out2);
  input     in0, in1, in2, in3, in4, in5, in6, in7;
  output    out0, out1, out2;
  reg       out0, out1, out2;
  reg [0:10] PLA_mem [0:2];          // 3 functions of 8 variables
  initial begin
    $readmemb ("PLA_data.txt", PLA_mem);
    $async$and$array
    (PLA_mem, {in0, in1, in2, in3, in4, in5, in6, in7}, {out0, out1, out2});
  end
endmodule
```

The PLA is configured by the **initial** behavior at the beginning of the simulation. It reads the file "PLA_data.txt" and loads the data into the declared memory, *PLA_mem*. Note that the inputs and outputs are declared in ascending order. When an input to the module changes value the array is evaluated to form updated values of *out0*, *out1*, and *out2*. The personality file, "PLA_data.txt" would contain the following binary data:

```
11110000 100
10101010 011
00001111 111
01010101 101
```

The corresponding Boolean equations are as follows:

```
out0 = in0 & in1 & in2 & in3 + in 4 & in5 & in6 & in7 + in1 & in3 & in5 & in7
out1 = in1 & in3 & in5 & in7 + in4 & in5 & in6 & in7
out2 = in0 & in2 & in4 & in6 + in4 & in5 & in6 & in7 + in1 & in3 & in5 & in7
```

The array format requires that the complement of a literal be provided separately as an input if it is needed to form a cube. On the other hand, the **plane** format encodes the personality matrix, using the following table:

Table Entry	Interpretation
0	The complemented literal is used in the cube.
1	The literal is used in the cube.
x	The "worst case" of the input is used.
z	Don't care; the input has no significance.
?	Same as z.

Table E.1 *Personality matrix symbols for "plane" format.*

Example: Suppose the logic to be implemented in a PLA is described by the following statements:

```
out0 = in0 & ~ in2;
out1 = in0 & in1 & ~in3;
out2 = ~in0 & ~ in3;
```

In the **plane** format, the personality of the PLA is described by

```
3'b1?0?
3'b11?0
3'b0??0
```

Each row defines the conditions of the inputs that assert that output. For example, the inputs 1000 and 1101 will both assert the first output.

```
module PLA_plane (in0, in1, in2, in3, in4, in5, in6, in7, out0, out1, out2);
  input           in0, in1, in2, in3, in4, in5, in6, in7;
  output          out0, out1, out2;
  reg             out0, out1, out2;
  reg     [0:3]   PLA_mem [0:2];      // 3 functions of 4 variables
  reg     [0:4]   a;
  reg     [0:3]   b;

  initial begin

    $async$and$array
    (PLA_mem, {in0, in1, in2, in3, in4, in5, in6, in7}, {out0, out1, out2});
    // Load the personality matrix
    PLA_mem [0] = 4'b1?0?;
```

```
    PLA_mem [1] = 4'b11?0;
    PLA_mem [2] = 4'b0??0;
  end
endmodule
```

E.6 STOCHASTIC ANALYSIS TASKS

Verilog includes a set of system tasks for performing stochastic modeling and analysis.

E.6.1 $q_initialize

$q_initialize creates a first-in, first-out (FIFO) queue or a last-in, last-out (LIFO) queue.

Syntax: **$q_initialize** (q_id, q_type, max_length, status);

The parameters of the **$q_initialize** task are described in Table E.2.

Parameter	Function
q_id	An integer that uniquely identifies the queue.
q_type	An integer (1 or 2) that specifies a FIFO or LIFO queue.
max_length	An integer that specifies the maximum number of entries in the queue.
status	An integer value indicating a successful or failed attempt to create a queue.

*Table E.2 Parameters of **$q_initialize**.*

The integer parameter *status* indicates whether a queue has been formed successfully or not; its value can have the meanings shown in Table E.3.

Status value	Meaning
0	Queue has been formed successfully.
1	The queue is full; there is no room to add a value to the queue.
2	The *q_id* parameter is undefined.

*Table E.3 Status values for the **$q_initialize** task.*

Status value	Meaning
3	The queue is empty; there is no value to remove from the queue.
4	The specified queue type is unsupported and cannot be formed.
5	The specified queue length is <= 0; the queue cannot be created.
6	The parameter *q_id* is a duplicate; the queue cannot be formed.
7	There is insufficient memory to create the queue.

Table E.3 *Status values for the **$q_initialize** task. (Continued)*

E.6.2 $q_add

$q_add adds an entry to the queue.

Syntax: **$q_add** (q_id, job_id, inform_id, status);

q_id is an integer input specifying the queue that will receive the entry. The integer parameter *job_id* identifies the job, and the integer parameter *inform_id* has a user-defined meaning, such as the time required to complete a task in a system-level model of a processor. The *status* parameter is defined in Table E.3.

E.6.3 $q_remove

$q_remove removes an entry from the queue. *q_id* is an integer input specifying the queue from which an entry will be removed. The integer parameter *job_id* uniquely identifies the entry that is to be removed. The integer output parameter *inform_id* has a user-defined value; the value is assigned when the entry is originally made to the queue by the **$q_add** task. The *status* parameter is defined in Table E.3.

Syntax: **$q_remove** (q_id, job_id, inform_id, status);

E.6.4 $q_full

$q_full returns a value indicating whether the queue is not full (0) or full (1).

E.6.5 $q_exam

$q_exam provides statistical data about activity in the queue specified by *q_id*. A call to **$q_exam** returns *q_state_value* according to the specification of *q_stat_code*, as indicated in Table E.4.

Syntax: **$q_exam** (q_id, q_stat_code, q_stat_value, status);

q_stat_code	Information returned by *q_stat_value*
1	Length of the queue
2	Mean interarrival time
3	Maximum length of the queue
4	Shortest wait time ever
5	Longest wait time for remaining jobs
6	Average wait time in the queue

*Table E.4 Code for specifying statistic report by **$q_exam**.*

E.6.6 $random

$random is a system function that returns a new 32-bit signed random number (an integer) each time the function is called.

E.6.7 $random ([seed])

$random ([*seed*]) returns a new random number each time the function is called. The inout [*seed*] parameter controls the number that the function returns. It must be a register, integer, or time variable whose value is assigned before the function is called. It cannot be a real variable.

E.7 SIMULATION-TIME TASKS

The simulation time tasks return the value of time during simulation in a specified format.

E.7.1 $realtime

$realtime returns a time value as a real number that is scaled and rounded to the precision of the time unit of the module that invoked it. (See **$stime** and **$time**.)

E.7.2 $stime

$stime returns a 64-bit value, scaled and rounded to the time unit of the module that invoked the function. Returns zero if the time value is a fraction of an integer. (See **$time** and **$realtime**.)

E.7.3 $time

$time returns the current simulation time as a 64-bit real value scaled and rounded to the time unit of the module that invoked it. (See **$stime** and **$real-time**.)

E.8 CONVERSION TASKS

The conversion tasks included in Verilog convert the internal representation of a value stored in one format to a value stored in another format.

E.8.1 $bitstoreal

$bitstoreal converts a bit pattern to an equivalent real-number value.

E.8.2 $itor

$itor converts an integer to a real value (e.g., 43 becomes 43.0).

Syntax: **$itor** (integer_value);

E.8.3 $realtobits

$realtobits converts a real number to a 64-bit representation of its value. This task is used to pass real-valued information between ports.

E.8.4 $rtoi

$rtoi truncates the fraction from a real number to form an integer.

E.9 TIMING-CHECK TASKS

Tasks for timing checks support timing verification during simulation, i.e., dynamic timing analysis. See Chapter 7 (Section 7.15) for additional details.

E.9.1 $hold

$hold monitors and reports violations of a hold constraint between a reference event and a hold event.

E.9.2 $nochange

$nochange reports a timing violation if a data event occurs during the specified level of a control signal (reference event).

E.9.3 $period

$period monitors and reports violations of the period between successive occurrences of an edge-triggered event.

E.9.4 $recovery

$recovery reports a timing violation if the time interval between an edge-triggered reference event and a data event exceeds a limit.

E.9.5 $setup

$setup monitors and reports violations of setup constraint between a data event and a reference event.

E.9.6 $setuphold

$setuphold monitors and reports violations of setup and hold constraints between a data event and a reference event.

E.9.7 $skew

$skew monitors and reports violations of the skew between an edge-triggered reference event and an edge-triggered data event.

E.9.8 $width

$width monitors and reports violations of a constraint on the width of a signal pulse.

E.10 PROBABILTY DISTRIBUTIONS

The built-in functions for probability distributions of random numbers have integer parameters, except for the mean, degree of freedom, and k-stage parameters of the exponential, poisson, chi-square, and erlang functions. Those parameters must be positive integers. The seed parameter has mode **inout**. It will return the same value for the same seed. The seed is to be an integer that is initialized and then updated by the system.

E.10.1 $dist_chi_square

The task **$dist_chi_square** returns a pseudorandom number having a chi-square distribution.

Syntax: **$dist_chi_square** (seed, degree_of_freedom)

The distribution is over the interval specified by *start* and *end*, with *start* < *end*.

E.10.2 $dist_exponential

The task **$dist_exponential** returns a pseudorandom number having an exponential distribution.

Syntax: **$dist_exponential** (seed, mean)

E.10.3 $dist_poisson

The task **$dist_poisson** returns a pseudorandom number having a poisson distribution.

Syntax: **$poisson** (seed, mean)

E.10.4 $dist_uniform

The task $**dist_uniform** returns a pseudorandom number having a uniform distribution over the interval [*start, end*], with *start* < *end*.

Syntax: **$dist_uniform** (seed, start, end)

E.10.5 $dist_erlang

The task **$dist_erlang** returns a pseudorandom number having an erlang distribution.

Syntax: **$dist_erlang** (seed, k-stage, mean)

E.10.6 $dist_normal

The task **$dist_normal** returns a pseudorandom number having a normal distribution.

Syntax: **$dist_normal** (seed, mean, standard_deviation)

E.10.7 $dist_t

The task **$dist_t** returns a pseudorandom number having a "t" distribution.

Syntax: **$dist_t** (seed, degree_of_freedom)

E.11 VALUE-CHANGE DUMP (VCD) FILE TASKS

The value change dump tasks store information about variable changes in a value change dump (VCD) ASCII text file.

E.11.1 $dumpall

$dumpall creates a checkpoint to display the current value of the variables being recorded in the VCD file.

E.11.2 $dumpfile

$dumpfile specifies the name of the VCD file. Execution of **$dumpfile** creates the VCD. For example: **$dumpfile** ("dump_run1.dump").

E.11.3 $dumpflush

$dumpflush empties the buffer of the VCD file and ensures that all the data is transferred from the buffer to the VCD file.

E.11.4 $dumplimit

$dumplimit sets the size of the VCD file.

E.11.5 $dumpoff

$dumpoff stops a tool from saving value changes in the VCD file.

E.11.6 $dumpon

$dumpon enables recording to the VCD file.

E.11.7 $dumpvars

$dumpvars specifies the variables whose changes are to be recorded in the VCD file. The default is all variables.

E.12 ADDITIONAL (NONSTANDARD) TASKS

Many EDA tools use nonstandard tasks and functions that are not included in the IEEE 1364 standard. Some are documented in the LRM. A brief description of those tools is given below. Only the simplest options are shown.

E.12.1 $countdrivers

$countdrivers counts the drivers on a scalar net and identifies bus contention.

E.12.2 $getpattern

$getpattern provides fast processing of stimulus patterns that must be propagated to a large number of scalar inputs.

E.12.3 $incsave

$incsave is used in conjunction with **$save** to save only what has changed since the last invocation of **$save**. The subject file must be the one produced by the last **$save**. (See **$save** and **$restart**.)

Syntax: **$insave** ("[filename]");

E.12.4 $input

$input allows a command input text to come from a named file instead of from the terminal.

Syntax: **$input** ("[filename]");

This tool automatically switches control back to the terminal at the end of file.

E.12.5 $key

$key creates a file of all text that has been typed in from the standard input and information about asynchronous interrupts.

Syntax: **$key** [("file_name")];

E.12.6 $list

$list produces a list of the module, task, function, or named block that is defined as the current scope setting.

Syntax: **$list**;

E.12.7 $log

$log enables a log file to which a tool may copy all text that is printed to the standard output.

Syntax: **$log** [("filename")];

E.12.8 $nokey

$nokey disables output to the key file.

Syntax: **$nokey**;

E.12.9 $nolog

$nolog disables output to a log file.

Syntax: **$nolog**;

E.12.10 $reset

$reset enables a tool to be set at its "Time 0" state, i.e., at the beginning of simulation.

Syntax: **$reset**;

E.12.11 $reset_count

$reset_count counts the number of times a tool has been reset.

Syntax: **$reset_count**;

E.12.12 $reset_value

$reset_value communicates information from before a reset of a tool to the "Time 0" state to after the reset.

Syntax: **$reset_value**;

E.12.13 $restart

$restart restores a tool to a previously saved state, including interactive commands, from a specified file. (See **$save** and **$incsave**.) When **$restart** is used in conjunction with **$incsave,** the file(s) to which the file produced by **$incsave** is related must be present. Thus, a sequence of files created by **$save** and subsequent consecutive invocations of **$incsave** are required for a successful restart.

Syntax: **$restart** ("file_name");

E.12.14 $save

$save saves the state of a tool into a permanent host-operating-system file so that the tool state can be reloaded and resume processing where it left off. (See **$restart** and **$incsave**.)

Syntax: **$save** ("file_name");

E.12.15 $scale

$scale converts the time value from the time unit of one module (*hierarchical_name*) to the time unit of the module that invoked **$scale**.

Syntax: **$scale** (hierarchical_name);

E.12.16 $scope

$scope specifies a level of the hierarchy to be the interactive scope for identifying objects. The argument of the task is a single parameter specifying the complete hierarchical name of a module, task, function, or named block. The default setting of the interactive scope is the top-level module.

E.12.17 $showscopes

$showscopes lists modules, tasks, functions, and named blocks defined at the current scope level.

Syntax:

```
$showscopes  // Lists objects at the current scope level.
$showscopes (n)
                  // Lists objects at or below the current
                  // hierarchical level (integer n]0) or only
                  // at the current scope (n = 0).
```

E.12.18 $showvars

$showvars produces status information for scalar and vector variables.

Syntax:

```
$showvars   // Display status of all variables in the
            // current scope.
$showvars (list_of_vars)
            // Displays status of only the listed vars.
```

E.12.19 $sreadmemb

$sreadmemb takes memory data and addresses in binary format from a Verilog source character string and loads memory with the binary data.

Syntax:

$sreadmemb (mem_name, start_addr, finish_addr,
    string { , string2});
where:

mem_name	name of the memory structure
start_addr	memory start address
finish_add r	memory end address
stringN	string containing data to be placed in memory, starting at start_addr.

E.12.20 $sreadmemh

$sreadmemh takes memory data and addresses in hexadecimal format from a Verilog source character string and loads memory with the binary data.

Syntax:

$sreadmemh (mem_name, start_addr, finish_addr,
    string { , string});
where:

mem_name	name of the memory structure
start_addr	memory start address
finish_addr	memory end address
string	string containing data to be placed in memory, starting at start_addr.

See LRM for additional details.

APPENDIX F

VERILOG LANGUAGE FORMAL SYNTAX

This formal syntax specification is provided in BNF. It is reprinted from IEEE Std 1364-1995 IEEE *Standard Verilog Hardware Description Language Reference Manual* (LRM), Copyright 1995, by the Institute of Electrical and Electronics Engineers, Inc. The IEEE disclaims any responsibility or liability resulting from the placement and use in this publication. This information is reprinted with the permission of the IEEE.

F.1 Source Text

source_text ::= {description}
description ::=
    module_declaration
    | udp_declaration
module_declaration ::=
    module_keyword *module*_identifier [list_of_ports]; {module_item} **endmodule**
module_keyword ::= **module** | **macromodule**
list_of_ports ::= (port {, port})
port ::=
    [port_expression]
    | *port*_identifier ([port_expression])
port_expression ::=
    port_reference
    | {port_reference {, port_reference} }
port_reference ::=
    *port*_identifier

| *port*_identifier [constant_expression]
| *port*_identifier [*msb*_constant_expression : *lsb*_constant_expression]

module_item ::=
module_item_declaration
| parameter_override
| continuous_assign
| gate_instantiation
| udp_instantiation
| module_instantiation
| specify_block
| initial_construct
| always_construct

module_item_declaration ::=
parameter_declaration
| input_declaration
| output_declaration
| inout_declaration
| net_declaration
| reg_declaration
| integer_declaration
| real_declaration
| time_declaration
| realtime_declaration
| event_declaration
| task_declaration
| function_declaration

parameter_override ::= **defparam** list_of_param_assignments;

F.2 Declarations

parameter_declaration ::= **parameter** list_of_param_assignments;
list_of_param_assignments ::= param_assignment { , param_assignment}
param_assignment ::= *parameter*_identifier = constant_expression
input_declaration ::= **input** [range] list_of_port_identifiers;
output_declaration ::= **output** [range] list_of_port_identifiers ;
inout_declaration ::= **inout** [range] list_of_ port_identifiers ;
list_of_port_identifiers ::= *port*_identifier { , *port*_identifier}
reg declaration ::= **reg** [range] list_of_register_identifiers ;
time_declaration ::= **time** list_of_register_identifiers ;
integer_declaration ::= **integer** list_of_register_identifiers ;
real_declaration ::= **real** list_of_real_identifiers ;
realtime_declaration ::= **realtime** list_of_ real_identifiers ;
event_declaration ::= **event** *event*_identifier { , *event*_identifier} ;

list_of_real_identifiers ::= *real*_identifier { , *real*_identifier}
list_of_register_identifiers ::= register_name { , register_name}
    register_name ::=
        *register*_identifier
        | *memory*_identifier [*upper_limit*_constant_expression :
        *lower_limit*_constant_expression]
range ::= [*msb*_constant_expression : *lsb*_constant_expression]
net_declaration ::=
        net_type [**vectored** | **scalared**] [range] [delay3] list_of_net_identifiers ;
        | **trireg** [**vectored** | **scalared**] [charge_strength] [range] [delay3]
        list_of_net_identifiers ;
        | net_type [**vectored** | **scalared**] [drive_strength] [range] [delay3]
        list_of_net_decl_assignments ;
net_type ::= **wire** | **tri** | **tri1** | **supply0** | **wand** | **triand** | **tri0** | **supply1** | **wor** | **trior**
list_of_net_identifiers ::= *net*_identifier { , net_identifier }
drive_strength ::=
        (strength0, strength1)
        | (strength1, strength0)
        | (strength0, **highz1**)
        | (strength1, **highz0**)
        | (**highz1,** strength0)
        | (**highz0,** strength1)
strength0 ::= **supply0** | **strong0** | **pull0** | **weak0**
strength1 ::= **supply1** | **strong1** | **pull1** | **weak1**
charge_strength ::= **(small)** | **(medium)** | **(large)**
delay3 ::= # delay_value | # (delay_value [, delay_value [, delay_value]])
delay2 ::= # delay_value | # (delay_value [, delay_value])
delay_value ::= unsigned_number | parameter_identifier |
        constant_mintypmax_expression
list_of_net_decl_assignments ::= net_decl_assignment { , net_decl_assignment }
net_decl_assignment ::= *net*_identifier = expression
function_declaration ::=
        function [range_or_type] *function*_identifier ;
        function_item_declaration {function_item_declaration}
        statement
        endfunction
range_or_type ::= range | **integer** | **real** | **realtime** | **time**
function_item_declaration ::=
        block_item_declaration
        | input_declaration
task_declaration ::=
        task *task*_identifier ;
        { task_item_declaration }
        statement_or_ null

endtask
task_argument_declaration ::=
block_item_declaration
| output_declaration
| inout_declaration
block_item_declaration ::=
parameter_declaration
| reg_declaration
| integer_declaration
| real_declaration
| time_declaration
| realtime_declaration
| event_declaration

F.3 Primitive Instances

gate_instantiation ::=
n_input_gatetype [drive_strength] [delay2] n_input_gate_instance { , n_input_gate_instance } ;
| n_output_gatetype [drive_strength] [delay2] n_output_gate_instance { , n_output_gate_instance } ;
| enable_gatetype [drive_strength] [delay3] enable_gate_instance { , enable_gate_instance } ;
| mos_switchtype [delay3] mos_switch_instance { , mos_switch_instance } ;
| pass_switchtype pass_switch_instance { , pass_switch_instance };
| pass_en_switchtype [delay3] pass_en_switch_instance
{ , pass_en_switch_instance } ;
cmos_switchtype [delay3] cmos_switch_instance { , cmos_switch_instance) ;
| **pullup** [pullup_strength] pull_gate_instance { , pull_gate_instance};
| **pulldown** [pulldown_strength] pull_gate_instance { , pull_gate_instance } ;
n_input_gate_instance ::= [name_of_gate_instance] (output_terminal, input_terminal { , input_terminal })
n_output_gate_instance ::= [name_of_gate_instance] (output_terminal { , output_terminal } , input_terminal)
enable_gate_instance ::= [name_of_gate_instance] (output_terminal, input_terminal, enable_terminal)
mos_switch_instance ::= [name_of_gate_instance] (output_terminal, input_terminal , enable_terminal)
pass_switch_instance ::= [name_of_gate_instance] (inout_terminal, inout_terminal)
pass_enable_switch_instance ::= [name_of_gate_instance] (inout_terminal, inout_terminal, enable_terminal)
cmos_switch_instance ::= [name_of_gate_instance] (output_terminal, input_terminal, ncontrol_terminal, pcontrol_terminal)

pull_gate_instance ::= [name_of_gate_instance] (output_terminal)
name_of_gate_instance ::= *gate_instance*_identifier [range]
pullup_strength ::=
    (strength0, strength1)
    | (strength1, strength0)
    | (strength1)
pulldown_strength ::=
    (strength0, strength1)
    | (strength1, strength0)
    | (strength0)
input_terminal ::= *scalar*_expression
enable_terminal ::= *scalar*_expression
ncontrol_terminal ::= *scalar*_expression
pcontrol_terminal ::= *scalar*_expression
output_terminal ::= *terminal*_identifier | *terminal*_identifier [constant_expression]
inout_terminal ::= *terminal*_identifier | *terminal*_identifier [constant_expression]
n_input_gatetype ::= **and** | **nand** | **or** | **nor** | **xor** | **xnor**
n_output_gatetype ::= **buf** | **not**
enable_gatetype ::= **bufif0** | **bufif1** | **notif0** | **notif1**
mos_switchtype ::= **nmos** | **pmos** | **rnmos** | **rpmos** |
pass_switchtype ::= **tran** | **rtran**
pass_en_switchtype ::= **tranif0** | **tranif1** | **rtranif1** | **rtranif0**
cmos_switchtype ::= **cmos** | **rcmos**

F.4 Module Instantiation

module_instantiation ::=
*module*_identifier [parameter_value_assignment] module_instance { , module_instance} ;
parameter_value_assignment ::= # (expression { , expression})
module_instance ::= name_of_instance ([list_of_module_connections])
name_of_instance ::= *module_instance*_identifier [range]
list_of_module_connections ::=
    ordered_port_connection { , ordered_port_connection}
    | named_port_connection { , named_port_connection}
ordered_port_connection ::= [expression]
named_port_connection ::= . *port*_identifier ([expression])

F.5 UDP Declaration and Instantiation

udp_declaration ::=
    primitive *udp*_identifier (udp_port_list) ;
    udp_port_declaration {udp_port_declaration}
    udp_body

endprimitive
udp_port_list ::= *output_port*_identifier , *input_port*_identifier { , *input_port*_identifier}
udp_port_declaration ::=
output_declaration
| input_declaration
| reg_declaration
udp_body ::= combinational_body | sequential_body
combinational_body ::= **table** combinational_entry {combinational_entry} **endtable**
combinational_entry ::= level_input_list : output_symbol ;
sequential_body ::= [udp_initial_statement] **table** sequential_entry {sequential_entry} **endtable**
udp_initial_statement ::= **initial** *udp_output_port_* identifier = init_val ;
init_val ::= 1'b0 | 1'b1 | 1'bx | 1'bX | 1'B0 | 1'B1 | 1'Bx | 1'BX | 1 | 0
sequential_entry ::= seq_input_list : current_state : next_state ;
seq_input_list ::=level_input_list | edge_input_list
level_input_list ::= level_symbol { level_symbol}
edge_input_list ::= { level_symbol} edge_indicator {level_symbol}
edge_indicator ::= (level_symbol level_symbol) | edge_symbol
current_state ::= level_symbol |
next_state ::= output_symbol | -
output_symbol ::= **0** | **1** | **x** | **X**
level_symbol ::= **0** | **1** | **x** | **X** | **?** | **b** | **B**
edge_symbol ::= **r** | **R** | **f** | **F** | **p** | **P** | **n** | **N** | *
udp_instantiation ::= *udp*_identifier [drive_strength] [delay2] udp_instance { , udp_instance} ;
udp_instance ::= [name_of_udp_instance] (output_port_connection, input_port_connection { , input_port_connection})
name_udp_instance ::= *udp_instance*_identifier [range]

F.6 Behavioral Statements

continuous_assign ::= **assign** [drive_strength] [delay3] list_of_net_assignments ;
list_of_net_assignments ::= net_assignment { , net_assignment}
net_assignment ::= net_lvalue = expression

initial_construct ::= **initial** statement
always_construct ::= **always** statement

statement ::=
blocking_assignment ;
| non_blocking assignment ;
| procedural_continuous_assignments ;
| procedural_timing_control_statement

| conditional_statement
| case_statement
| loop_statement
| wait_statement
| disable_statement
| event_trigger
| seq_block
| par_block
| task_enable
| system_task_enable

statement_or_null ::= statement | ;
blocking_assignment ::= reg_1value<= [delay_or_event_control] expression
non_blocking assignment ::= reg_1value <= [delay_or_event_control] expression
procedural_continuous_assignment ::=
| **assign** reg_assignment ;
| **deassign** reg_lvalue ;
| **force** reg_assignment ;
| **force** net_assignment ;
| **release** reg_lvalue ;
| **release** net_lvalue ;
procedural_timing_control_statement ::=
delay_or_event_control statement_or_null
delay_or_event_control ::=
delay_control
| event_control
| **repeat** (expression) event_control
delay_control ::=
delay_value
| # (mintypmax_expression)
event_control ::=
@ *event*_identifier
| @ (event_expression)
event_expression ::=
expression
| *event*_identifier
| **posedge**_expression
| **negedge**_expression
| event_expression **or** event_expression
conditional_statement ::=
| **if (** expression) statement_or_null **[else** statement_or_null]
case_statement ::=
case (expression) case_item {case_item} **endcase**
| **casez (**expression) case_item {case_item} **endcase**

| **casex** (expression) case_item {cast_item) **endcase**
case_item ::=
expression { , expression} : statement_or_null
| **default** [:] statement_or_null
loop_statement ::=
| **forever** statement
| **repeat** (expression) statement
| **while (**expression) statement
| **for (** reg_assignment ; expression ; reg_assignment) statement
reg_assignment ::= reg_lvalue = expression
wait_statement ::= **wait** (expression) statement_or_null
event_trigger ::=
| **->** *event* identifier ;
disable_statement ::=
| **disable** *task*_identifier ;
| **disable** *block*_identifier ;
seq_block ::= **begin** [: block_identifier {block_item_declaration}] {statement} **end**
par_block ::= **fork** [: *block*_identifier { block_item_declaration}] {statement} **join**
task_enable ::= *task*_identifier [(expression { , expression })] ;
system_task_enable ::= system_task_name [(expression { , expression})] ;
system_task_name ::= $identifier [*Note:* The $ character may not be followed by a space.]

F.7 Specify Section

specify_block ::= **specify** [specify_item] **endspecify**
specify_item ::=
specparam_declaration
| path_declaration
| system_timing_check
specparam_declaration ::= **specparam** list_of_specparam_assignments ;
list_of_specparam_assignments ::= specparam_assignment { , specparam_assignment}
specparam_assignment
*specparam*_identifier = constant_expression
| pulse_control_specparam
pulse_control_specparam ::=
PATHPULSE$ = (*reject*_limit_value [, *error*_limit_value]);
|
**PATHPULSE$**specify_input_terminal_descriptor$specify_output_terminal
_descriptor = (*reject*_limit_value[, *error_limit*_value]);
limit_value ::= constant_mintypmax_expression
path_declaration ::= simple_path_declaration ;
| edge_sensitive_path_declaration ;

| state_dependent_path_declaration;
simple_path_declaration ::= parallel_path_description = path_delay_value
| full_path_description = path_delay_value
parallel_path_description ::= (specify_input_terminal_descriptor [polarity_operator] => specify_output_terminal_descriptor
full_path_description ::= (list_of_path_inputs [polarity_operator] *> list_of_path_outputs)
list_of_path_inputs ::= specify_input_terminal_descriptor { , specify_input_terminal}
list_of_path_outputs ::= specify_ output_terminal_descriptor { , specify_output_terminal_descriptor}
specify_input_terminal_descriptor ::=
input_identifier
| input_identifier [constant_expression]
| input_identifier [*msb*_constant_expression : *lsb*_constant_expression]
specify_output_terminal_descriptor ::=
output_identifier
| output_identifier [constant_expression]
| output_identifier [*msb*_constant_expression : *lsb*_constant_expression]

input_identifier ::= *input_port*_identifier | *inout_port*_identifier
output_identifier ::= *output_port*_identifier | *inout_port*_identifier
polarity_operator::= + | -
path_delay_value ::= list_of_path_delay_expressions | (list_of_path_delay_expressions)
list_of_path_delay_expressions ::=
*t*_path_delay_expression
| *trise*_path_delay_expression , *tfall*_path_delay_expression
| *trise*_path_delay_expression, *tfall*_path_delay_expression, *tz*_path_delay_expression
| *t01*_path_expression, *t10*_path_delay_expression, *t0z*_path_delay_expression, *tz1*_path_delay_expression, *t1z*_path_delay_expression, *tz0*_path_delay_expression
| *t01*_path_delay_expression, *t10*_path_delay_expression, *t0z*_path_delay_expression, *tz1*_path_delay_expression, *t1z*_path_delay_expression, *tz0*_path_delay_expression, *t0x*_path_delay_expression, *tx1*_path_delay_expression, *t1x*_path_delay_expression, *tx0*_path_delay_expression, *txz*_path_delay_expression, *tzx*_path_delay_expression
path_delay_expression ::= constant_mintypmax_expression
edge_sensitive_path_declaration ::= parallel_edge_sensitive_path_description = path_delay_value
| full_edge_sensitive_path_description = path_delay_value
parallel_edge_sensitive_path_description ::=

([edge_identifier] specify_input_terminal_descriptor =>
specify_output_terminal_descriptor [polarity_operator] :
data_source_expression))
full_edge_sensitive_path_description ::= ([edge_identifier] list_of_path inputs *>
list_of_path_outputs [polarity_operator] : data_source_expression))
data_source_expression ::= expression
edge_identifier ::= **posedge** | **negedge**
state_dependent_path_declaration ::=
if (conditional_expression) simple_path_declaration
| **if** (conditional_expression) edge_sensitive_path_declaration
| **ifnone** simple_path_declaration
system_timing_check::=
$setup (timing_check_event, timing_check_event, timing_check_limit
[, notify_register]) ;
| **$hold (**timing_check_event, timing_check_event, timing_check_ limit
[, notify_register]) ;
| **$period (**controlled_timing_check_event, timing_check_limit
[, notify_register]) ;
| **$width (**controlled_timing_check_event, timing_check_limit,
constant_expression [, notify_register]) ;
| **$skew** (timing_check_event, timing_check_event, timing_check_limit
[, notify_register]) ;
| **$recovery (**controlled_timing_check_event , timing_check_event ,
timing_check_limit [, notify_register]) ;
| **$setuphold** (timing_check_event, timing_check_event, timing_check_limit,
timing_check_limit [, notify_register]
timing_check_event ::=
[timing_check_event_control] specify_terminal_descriptor
[**&&&** timing_check_condition]
specify_terminal_descriptor ::=
specify_input_terminal_descriptor
| specify_output_terminal_descriptor
controlled_timing_check_event ::=
timing_check_event_control specify_terminal_descriptor
[**&&&** timing_check_condition]
timing_check_event_control ::=
posedge
| **negedge**
| edge_control_specifier
edge_control_specifier ::= **edge** [edge_descriptor [, edge_descriptor]]
edge_descriptor ::=
01
| **10**
| **0x**

| **xl**
| **1x**
| **x0**
timing_check_condition ::=
scalar_timing_check_condition
| (scalar_timing_check_condition)
scalar_timing_check_condition
expression
| ~ expression
| expression == scalar_constant
| expression === scalar_constant
| expression != scalar_constant
| expression !== scalar_constant
timing_check_limit ::= expression
scalar_constant ::=
1'b0 | **1'b1** | **1'B0** | **1'B1** | **'b0** | **'b1** | **'B0** | **'B1** | **1** | **0**
notify_register ::= *register*_identifier

F.8 Expressions

net_1value ::=
*net*_identifier
| *net*_identifier [expression]
| *net*_identifier [*msb*_constant_expression : lsb_constant_expression]
| *net*_concatenation
reg_1value ::=
*reg*_identifier
| *reg*_identifier [expression]
| *reg*_identifier [*msb*_constant_expression : *lsb*_constant_expression]
| *reg*_concatenation
constant_expression ::=
constant_primary
| unary_operator_constant_primary
| constant_expression binary_operator constant_expression
| constant_expression **?** constant_expression **:** constant_expression
| string
constant_primary ::=
number
| *parameter*_identifier
| *constant*_concatenation
| *constant*_multiple_concatenation
constant_mintypmax_expression ::=
constant_expression

| constant_expression : constant_expression : constant_expression
mintypmax_expression ::=
expression
| expression : expression : expression
expression ::=
primary
| unary_operator_primary
| expression binary_operator expression
| expression **?** expression **:** expression
| string
unary_operator ::=
+ | - | ! | ~ | & | ~& | | | ~| | ^ | ~^ | ^~
binary_operator ::=
+ | - | * | / | % | == | != | === | !== | && | ||
| < | <= | > | >= | & | | | ^ | ^~ | ~^ | >> | <<
primary ::=
number
| identifier
| identifier [expression]
| identifier [*msb*_constant_expression : *lsb*_constant_expression]
| concatenation
| multiple_concatenation
| function_call
| (mintypmax_expression)
number ::=
decimal_number
| octal_number
| binary_number
| hex_number
| real_number
real_number ::=
[sign] unsigned_number **.** unsigned_number
| [sign] unsigned_number [**.** unsigned_number] **e** [sign] unsigned_number
| [sign] unsigned_number [**.** unsigned_number] **E** [sign] unsigned_number
decimal_number ::=
[sign] unsigned_number
| [size] decimal_base unsigned_number
binary_number ::= [size] binary_base binary_digit { _ | binary_digit}
octal_number ::= [size] octal_base octal_digit { _ | octal_digit}
hex_number ::= [size] hex_base hex_digit { _ | hex_digit}
sign ::= + | -
size ::= unsigned_number
unsigned_number ::= decimal_digit { _ | decimal_digit}
decimal_base ::= 'd | 'D

binary_base ::= 'b | 'B
octal_base ::= 'o | 'O
hex_base ::= 'h | 'H
decimal_digit ::= 0 | 1 | 2 | 3 | 4 | 5 | 6 | 7 | 8 | 9
binary_digit ::= x | X | z | Z | 0 | 1
octal_digit ::= x | X | z | Z | 0 | 1 | 2 | 3 | 4 | 5 | 6 | 7 |
hex_digit ::= x | X | z | Z | 0 | 1 | 2 | 3 | 4 | 5 | 6 | 7 | 8 | 9 | a | b | c | d | e | f | A
  | B | C | D | E | F |

concatenation ::= {expression {, expression} }
Multiple_concatenation ::= {expression { expression { , expression }}}
function_call ::=
  *function*_identifier (expression { , expression})
  | name_of_system_function [(expression { , expression})]
name_of_system_function ::= $identifier
string ::= "{ Any_ASCII_Characters_except_new_line }"

Notes:

1) Embedded spaces are illegal.
2) The $ in *name_of_system_function* may not be followed by a space.

F.9 General

comment ::=
  short_comment
  | long_comment
short_comment ::= // comment_text \ n
long_comment ::= /* comment_text */
comment_text ::= {Any_ASCII_character }
identifier ::= IDENTIFIER [{ . IDENTIFIER }]
IDENTIFIER ::=
  simple_identifier
  | escaped_identifier
simple_identifier ::= [a-zA-Z][a-zA-Z_$]
escaped_identifier ::= \ {Any_ASCII_character_except_white_space} white_space
White_space ::= space | tab | newline

Note: The period in an identifier may not be preceded or followed by a space.

INDEX

Index of Verilog Modules and User-Defined Primitives

TABLES

SUBJECT INDEX

Note: Italicized names in the listing below are examples in the text.

D

E

L

M

XILINX STUDENT EDITION

Access the Xilinx Design Series Home Page for useful information and help

www.xilinx.com/programs/xds1.htm

This page will always contain the latest version of this information.

This CD-ROM set is designed for use with Prentice Hall's "Xilinx Design Series" educational products. This software is designed for first time installation of the student version of Xilinx Foundation Series Express F1.5 software, as well as to upgrade Xilinx Foundation Series F1.3 and F1.4 software to version F1.5 and to provide other electronic materials that may be useful. To properly execute the VHDL and Verilog synthesis software module-Synopsys FPGA Express ™ module of the Xilinx Foundation Series-software a "license.dat" file is required. If you are upgrading from 1.3 or 1.4 please access: http://www.xilinx.com/programs/xsefaq1.htm for information on making a simple modification to your existing license file. If this is a new installation then using the serial number from the coupon which is included in your Xilinx Design Series package from Prentice Hall (ISBN 0130205869), you can obtain a "license.dat" file from the Xilinx Design Series Home Page. Purchase of the Prentice Hall package is required to obtain a license file.

Basic installation instructions are printed here. Complete instructions are on the Design Environment CD-ROM disk 1 of 2 at D:\readme.wri. These instructions are complete at press time. Updated versions of these instructions may be found at

www.xilinx.com/programs/xds1.htm

Section I—Xilinx Foundation Series Express F1.5

Section II—Other tools

a. (Optional) Xilinx CORE Generator 1.5
b. (Optional) XESS Tools for XS40 or XS95 boards
c. (Optional) Aldec's Active-VHDL 3.1 simulator (Student Version)
d. (Optional) SimuCad's SILOSIII Verilog simulator (Demo Version)

Section III – License file setup

Section IV - (Optional) Adobe Acrobat and datasheets

To upgrade from F1.3 or F1.4 to 1.5, start at Section I A.

To install on a new computer, start at Section I B.

Section I—Xilinx Foundation Series Express F1.5

Section IA—Uninstall F1.3 or F1.4

1. Uninstall the Foundation Project Manager (Design Environment): Click on Start, Settings, Control Panel, Add/Remove Programs
2. Uninstall the Foundation Design Implementation Tools.
3. If uninstall does not work, then delete everything including and below C:\Xilinx directory (assuming that was the directory under which you installed your F1.3 or F1.4 Design Implementation Tools).

Section I.B—Install F1.5

1. Insert and install Disk 1 of 2, Design Environment.

   Begin installation: Click on Start, Run and select D:\ setup.exe When "Xilinx Foundation F1.5 Setup" window appears, click "Next" and then follow the prompts.

   Your CD KEY is: `FXAA2743390` Do not lose this number!

2. If you are upgrading go to http://www.xilinx.com/programs/xds1.htm for full instructions.

   Note: You may wish to deselect XC4000E and SpartanXL devices as this will save you over 50MB of disk space. You can run setup.exe again to add these devices later.

3. Insert and install Disk 2 of 2, Documentation.

   Begin installation: Click on Start, Run and select D:\ setup.exe

   When "Xilinx Foundation F1.5 Documentation Setup" window appears, click "Next" and then follow the prompts. The CD KEY is not required for this step.

4. To save disk space, you can deselect the "Multimedia Foundation Demo" in the Select Products to Install dialog box.

5. Under "Registry Settings Options" do not change any options.

6. When finished, reboot your computer to allow Registry settings to take place.

Section II—Other tools files for F1.5

There are other tools available on disk 2.

In general, the setup is similar to what you have just done with the foundation tools. Click on the Start button, then RUN and move to the directory for the tool that you wish to install. Double click on the "setup.exe" file and follow the prompts.

The labs in the Practical Xilinx Designer lab book may be found on disk 2 in D:\XESS\XLabs\... There is a subdirectory for each lab.

Information on system requirements, are on the Design Environment Disk CD-ROM disk 1 of 2 at D:\readme.wri as well as on the software packaging

The complete installation guide can be found on disk 2 at D:\Finstall\fndinst.pdf. Acrobat reader must be installed to read this file. Acrobat reader can be installed from the ACRO-READ directory on disk 2.

Section III—License file for F1.5

For version F1.5, the "license.dat" file is required to run the Synopsys FPGA Express VHDL and Verilog synthesis tools.

If you are upgrading an existing Xilinx Foundation Student Series F1.3 or F1.4 installation up to F1.5, you do not need to make any changes to your system. Your existing "license.dat" is sufficient. If using the Express tool see the next paragraph.

If you are installing the Xilinx Foundation Series Express F1.5 software on a new computer, or you just did not obtain a "license.dat" file previously, then go to the web to obtain your license file and read the instructions at: **http://www.xilinx.com/programs/xds1.htm**

Section IV—(Optional) Adobe Acrobat and datasheets

There are several documents included in Adobe Acrobat format on this CD-ROM set for your reference. For more comprehensive documentation on the Xilinx Foundation Series Express F1.5 software and installation, the commercial F1.5 Quick Start Guide is available at D:\installdoc\fndinst.pdf.

These are the key documents that are included on "Disk 2 of 2, Foundation Documentation":

D:\Applinx\Databook.pdf	Xilinx 1998 Databook
D:\Applinx\coregen.pdf	Xilinx Core Solutions Databook

Other key documents which can be found using the Dynatext Document viewer in the Xilinx Book directory:
Foundations Series Quickstart Guide
Foundation Series User Guide
Verilog Reference Guide
VHDL Reference Guide

If you need to install the Acrobat reader, run D:\ACRORED\WIN32\Ar32e30.exe and follow the instructions to install the Acrobat Reader software.

In addition, there are several lab projects available at the Xilinx University Program home page under "Presentation Materials and Lab Files".

See **www.xilinx.com/programs/univ.htm**

Minimum PC Requirements:

Pentium® Processor recommended   Windows '95 or NT 4.0

48MB RAM recommended 350MB (plus 100MB swap space) free hard drive space before installation.

Depending on options you select, the F1.5 Foundation Series Express "Base" installation will consume 250 to 500 Mbytes of hard drive space. Does not include Active-HDL or Silos III.

Please note that all support for the Xilinx Student Edition software is web based. Access: **http://www.xilinx.com/programs/univ.htm** for troubleshooting and other information.

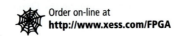

Foundation™ Series, Xilinx Student Edition Registration

KEEP THIS HALF OF THIS PAGE. This half of your Hardware Discount Coupon serves as your Xilinx Student Edition registration card. It has many purposes:

1) Your serial number, which is on the card inside the blisterpack, is required to obtain a license file from the Xilinx web site. The license file will enable your Xilinx Software to run on your computer.
2) Your serial number is required to authenticate your discount for the XS40-005XL and/or XS95-108 hardware development boards. The hardware board order form may be detached or copied from the preceding side of the page.
3) Your serial number may be used to authenticate future offers which may be announced.
4) You can use your serial number to obtain a Graduation Gift from Xilinx.

All of these benefits are explained on the Xilinx website. Register your software at this location now - http://www.xilinx.com/programs/xse1.htm

XILINIX END USER LICENSE

Attention: By opening this case, you are consenting to be bound by and are becoming a party to this agreement. If you do not agree to all the terms of this agreement, return the unopened case, along with the entire package withwithj proof of payment to authorized dealer where you took delivery within ten days; and get a full refund.

1. License. XILINX, Inc. ("XILINX") hereby grants you a nonexclumonstration, and system software included on this disk, diskette, tape or CD ROM, and reted documentation (the "Software") solely by the number of simultaneous users for which you have paid XILINX a license fee, and solely for your use in developing designs for XILINX Programmable Logic devices or internal business purposes. You own the media on which the Software is recorded, but XILINX and its licensors retain title to the Software and to any patents, copyrights, trade secrets and other intellectual property rights therein. This License allows you to use the Software on a single computer. In addition, you may make up to the number of the copies of the Software, if used on separate computers, or permit up to the number of simultaneous users to use the Software, if used in a network environment, as permitted in a separate written agreement between you and XILINX, and make one copy of the Software in machine-readable form for backup purposes only. You must reproduce on each copy of the Software the copyright and any other proprietary legends that were on the original copy of the Software. You may also transfer the Software, including any backup copy of the Software you may have made, the related documentation, and a copy of this License to another party provided the other party reads and agrees to accept the terms and conditions of this License prior to your transfer of the Software to the other party, and provided that you retain no copies of the Software yourself.

2. Restrictions. The Software contains copyrighted material, trade secrets, and other proprietary information. In order to protect them you may not decompile, reverse engineer, disassemble, or otherwise reduce the Software to a human-perceivable form. You may not modify or prepare derivative works of the Software in whole or in part. You may not publish any data or information that compares the performance of the Software with software created or distributed by others.

3. Termination. This License is effective until terminated. You may terminate this License at any time by destroying the Software and all copies thereof. This License will terminate immediately without notice from XILINX if you fail to comply with any provision of this License. Upon termination you must destroy the Software and all copies thereof.

4. Governmental Use. The Software is commercial computer software developed exclusively at Xilinx's expense. Accordingly, pursuant to the Federal Acquisition Regulations (FAR) Section 12.212 and Defense FAR Supplement Section 227.2702, use, duplication and disclosure of the Software by or for the Government is subject to the restrictions set forth in this License Agreement. Manufacturer is XILINX, INC., 2100 Logic Drive, San Jose, California 95124.

5. Limited Warranty and Disclaimer. XILINX warrants that, for a period of ninety (90) days from the date of delivery to you of the Software as evidenced by a copy of your receipt, the media on which the Software is furnished will, under normal use, be free from defects in material and workmanship. XILINX's and its Licensors' and Distributor's entire liability to you and your exclusive remedy under this warranty will be for XILINX, at its option, after return of the defective Software media, to either replace such media or to refund the purchase price paid therefor and terminate this Agreement. EXCEPT FOR THE ABOVE EXPRESS LIMITED WARRANTY, THE SOFTWARE IS PROVIDED TO YOU "AS IS". XILINX AND ITS LICENSORS AND DISTRIBUTORS MAKE AND YOU RECEIVE NO OTHER WARRANTIES OR CONDITIONS, EXPRESS, IMPLIED, STATUTORY OR OTHERWISE, AND XILINX SPECIFICALLY DISCLAIMS ANY IMPLIED WARRANTIES OF MERCHANTABILITY, NONINFRINGEMENT, OR FITNESS FOR A PARTICULAR PURPOSE. XILINX does not warrant that the functions contained in the Software will meet your requirements, or that the operation of the Software will be uninterrupted or error free, or that defects in the Software will be corrected. Furthermore, XILINX does not warrant or make any representations regarding use or the results of the use of the Software in terms of correctness, accuracy, reliability or otherwise.

6. Limitation of Liability. IN NO EVENT WILL XILINX OR ITS LICENSORS OR DISTRIBUTORS BE LIABLE FOR ANY LOSS OF DATA, LOST PROFITS, COST OF PROCUREMENT OF SUBSTITUTE GOODS OR SERVICES, OR FOR ANY SPECIAL, INCIDENTAL, CONSEQUENTIAL OR INDIRECT DAMAGES ARISING FROM THE USE OR OPERATION OF THE SOFTWARE OR ACCOMPANYING DOCUMENTATION, HOWEVER CAUSED AND ON ANY THEORY OF LIABILITY. THIS LIMITATION WILL APPLY EVEN IF XILINX HAS BEEN ADVISED OF THE POSSIBILITY OF SUCH DAMAGE. THIS LIMITATION SHALL APPLY NOTWITHSTANDING THE FAILURE OF THE ESSENTIAL PURPOSE OF ANY LIMITED REMEDIES HEREIN.

7. Export Restriction. You agree that you will not export or reexport the Software, reference images or accompanying documentation in any form without the appropriate United States and foreign government licenses. Your failure to comply with this provision is a material breach of this Agreement.

8. Third Party Beneficiary. You understand that portions of the Software and related documentation have been licensed to XILINX from third parties and that such third parties are intended third party beneficiaries of the provisions of this license Agreement.

9. General. This License shall be governed by the laws of the State of California, and without reference to conflict of laws principles. If for any reason a court of competent jurisdiction finds any provision of this License, or portion thereof, to be unenforceable, that provision of the License shall be enforced to the maximum extent permissible so as to effect the intent of the parties, and the remainder of this License shall continue in full force and effect. This License constitutes the entire agreement between the parties with respect to the use of this Software and related documentation, and supersedes all prior or contemporaneous understandings or agreements, written or oral, regarding such subject matter.